NATURAL HISTORY
UNIVERSAL LIBRARY

西方博物学大系

主编：江晓原

THE NATURALIST ON THE RIVER AMAZONS

亚马逊河上的博物学家

[英] 亨利·沃尔特·贝茨 著

华东师范大学出版社

图书在版编目(CIP)数据

亚马逊河上的博物学家 = The Naturalist on the River Amazons : 英文 /(英) 亨利 · 沃尔 · 贝茨(Henry Walter Bates)著. — 上海 : 华东师范大学出版社, 2019
(寰宇文献)
ISBN 978-7-5675-9541-5

Ⅰ. ①亚… Ⅱ. ①亨… Ⅲ. ①亚马逊河-博物学-英文
Ⅳ. ①P947.7 ②N91

中国版本图书馆CIP数据核字(2019)第156793号

亚马逊河上的博物学家
THE NATURALIST ON THE RIVER AMAZONS
[英] 贝茨 (Henry Walter Bates)

特约策划　黄曙辉　徐　辰
责任编辑　庞　坚
特约编辑　许　倩
装帧设计　刘怡霖

出版发行　华东师范大学出版社
社　　址　上海市中山北路3663号　邮编 200062
网　　址　www.ecnupress.com.cn
电　　话　021-60821666　行政传真　021-62572105
客服电话　021-62865537
门市（邮购）电话　021-62869887
地　　址　上海市中山北路3663号华东师范大学校内先锋路口
网　　店　http://hdsdcbs.tmall.com/

印 刷 者　北京虎彩文化传播有限公司
开　　本　787 × 1092 16开
印　　张　54
版　　次　2019年8月第1版
印　　次　2019年8月第1次
书　　号　ISBN 978-7-5675-9541-5
定　　价　980.00元 (精装全一册)

出 版 人　王　焰

(如发现本版图书有印订质量问题，请寄回本社客服中心调换或电话021-62865537联系)

《西方博物学大系》总序

江晓原

《西方博物学大系》收录博物学著作超过一百种，时间跨度为15世纪至1919年，作者分布于16个国家，写作语种有英语、法语、拉丁语、德语、弗莱芒语等，涉及对象包括植物、昆虫、软体动物、两栖动物、爬行动物、哺乳动物、鸟类和人类等，西方博物学史上的经典著作大备于此编。

中西方“博物”传统及观念之异同

今天中文里的“博物学“一词，学者们认为对应的英语词汇是Natural History，考其本义，在中国传统文化中并无现成对应词汇。在中国传统文化中原有“博物”一词，与“自然史”当然并不精确相同，甚至还有着相当大的区别，但是在“搜集自然界的物品”这种最原始的意义上，两者确实也大有相通之处，故以“博物学”对译Natural History一词，大体仍属可取，而且已被广泛接受。

已故科学史前辈刘祖慰教授尝言：古代中国人处理知识，如开中药铺，有数十上百小抽屉，将百药分门别类放入其中，即心安矣。刘教授言此，其辞若有憾焉——认为中国人不致力于寻求世界“所以然之理”，故不如西方之分析传统优越。然而古代中国人这种处理知识的风格，正与西方的博物学相通。

与此相对，西方的分析传统致力于探求各种现象和物体之间的相互关系，试图以此解释宇宙运行的原因。自古希腊开始，西方哲人即孜孜不倦建构各种几何模型，欲用以说明宇宙如何运行，其中最典型的代表，即为托勒密（Ptolemy）的宇宙体系。

比较两者，差别即在于：古代中国人主要关心外部世界“如何”运行，而以希腊为源头的西方知识传统（西方并非没有别的知识传统，只是未能光大而已）更关心世界“为何”如此运行。在线性发展无限进步的科学主义观念体系中，我们习惯于认为“为何”是在解决了“如何”之后的更高境界，故西方的分析传统比中国的传统更高明。

然而考之古代实际情形，如此简单的优劣结论未必能够成立。例如以天文学言之，古代东西方世界天文学的终极问题是共同的：给定任意地点和时刻，计算出太阳、月亮和五大行星（七政）的位置。古代中国人虽不致力于建立几何模型去解释七政“为何”如此运行，但他们用抽象的周期叠加（古代巴比伦也使用类似方法），同样能在足够高的精度上计算并预报任意给定地点和时刻的七政位置。而通过持续观察天象变化以统计、收集各种天象周期，同样可视之为富有博物学色彩的活动。

还有一点需要注意：虽然我们已经接受了用“博物学”来对译 Natural History，但中国的博物传统，确实和西方的博物学有一个重大差别——即中国的博物传统是可以容纳怪力乱神的，而西方的博物学基本上没有怪力乱神的位置。

古代中国人的博物传统不限于“多识于鸟兽草木之名”。体现此种传统的典型著作，首推晋代张华《博物志》一书。书名“博物”，其义尽显。此书从内容到分类，无不充分体现它作为中国博物传统的代表资格。

《博物志》中内容，大致可分为五类：一、山川地理知识；二、奇禽异兽描述；三、古代神话材料；四、历史人物传说；五、神仙方伎故事。这五大类，完全符合中国文化中的博物传统，深合中国古代博物传统之旨。第一类，其中涉及宇宙学说，甚至还有“地动”思想，故为科学史家所重视。第二类，其中甚至出现了中国古代长期流传的“守宫砂”传说的早期文献：相传守宫砂点在处女胳膊上，永不褪色，只有性交之后才会自动消失。第三类，古代神话传说，其中甚至包括可猜想为现代“连体人”的记载。第四类，各种著名历史人物，比如三位著名刺客的传说，此三名刺客及所刺对象，历史上皆实有其人。第五类，包括各种古代方术传说，比如中国古代房中养生学说，房中术史上的传说人物之一“青牛道士封君达”等等。前两类与西方的博物学较为接近，但每一类都会带怪力乱神色彩。

“所有的科学不是物理学就是集邮”

在许多人心目中，画画花草图案，做做昆虫标本，拍拍植物照片，这类博物学活动，和精密的数理科学，比如天文学、物理学等等，那是无法同日而语的。博物学显得那么的初级、简单，甚至幼稚。这种观念，实际上是将“数理程度”作为唯一的标尺，用来衡量一切知识。但凡能够使用数学工具来描述的，或能够进行物理实验的，那就是“硬”科学。使用的数学工具越高深越复杂，似乎就越“硬”；物理实验设备越庞大，花费的金钱越多，似乎就越“高端”、越“先进”……

这样的观念，当然带着浓厚的“物理学沙文主义”色彩，在很多情况下是不正确的。而实际上，即使我们暂且同意上述“物理学沙文主义”的观念，博物学的“科学地位”也仍然可以保住。作为一个学天体物理专业出身，因而经常徜徉在“物理学沙文主义”幻影之下的人，我很乐意指出这样一个事实：现代天文学家们的研究工作中，仍然有绘制星图，编制星表，以及为此进行的巡天观测等等活动，这些活动和博物学家“寻花问柳”，绘制植物或昆虫图谱，本质上是完全一致的。

这里我们不妨重温物理学家卢瑟福（Ernest Rutherford）的金句：“所有的科学不是物理学就是集邮（All science is either physics or stamp collecting）。”卢瑟福的这个金句堪称“物理学沙文主义”的极致，连天文学也没被他放在眼里。不过，按照中国传统的“博物”理念，集邮毫无疑问应该是博物学的一部分——尽管古代并没有邮票。卢瑟福的金句也可以从另一个角度来解读：既然在卢瑟福眼里天文学和博物学都只是“集邮”，那岂不就可以将博物学和天文学相提并论了？

如果我们摆脱了科学主义的语境，则西方模式的优越性将进一步被消解。例如，按照霍金（Stephen Hawking）在《大设计》（*The Grand Design*）中的意见，他所认同的是一种“依赖模型的实在论（model-dependent realism）”，即“不存在与图像或理论无关的实在性概念（There is no picture- or theory-independent concept of reality）”。在这样的认识中，我们以前所坚信的外部世界的客观性，已经不复存在。既然几何模型只不过是对外部世界图像的人为建构，则古代中国人干脆放弃这种建构直奔应用（毕竟在实际应用中我们只需要知道七政“如何”运行），又有何不可？

传说中的“神农尝百草”故事，也可以在类似意义下得到新的解读：“尝百草”当然是富有博物学色彩的活动，神农通过这一活动，得知哪些草能够治病，哪些不能，然而在这个传说中，神农显然没有致力于解释“为何”某些草能够治病而另一些则不能，更不会去建立“模型”以说明之。

“帝国科学”的原罪

今日学者有倡言“博物学复兴”者，用意可有多种，诸如缓解压力、亲近自然、保护环境、绿色生活、可持续发展、科学主义解毒剂等等，皆属美善。编印《西方博物学大系》也是意欲为“博物学复兴”添一助力。

然而，对于这些博物学著作，有一点似乎从未见学者指出过，而鄙意以为，当我们披阅把玩欣赏这些著作时，意识到这一点是必须的。

这百余种著作的时间跨度为15世纪至1919年，注意这个时间跨度，正是西方列强“帝国科学”大行其道的时代。遥想当年，帝国的科学家们乘上帝国的军舰——达尔文在皇家海军“小猎犬号”上就是这样的场景之一，前往那些已经成为帝国的殖民地或还未成为殖民地的“未开化”的遥远地方，通常都是踌躇满志、充满优越感的。

作为一个典型的例子，英国学者法拉在（Patricia Fara）《性、植物学与帝国：林奈与班克斯》（*Sex, Botany and Empire, The Story of Carl Linnaeus and Joseph Banks*）一书中讲述了英国植物学家班克斯（Joseph Banks）的故事。1768年8月15日，班克斯告别未婚妻，登上了澳大利亚军舰“奋进号”。此次“奋进号”的远航是受英国海军部和皇家学会资助，目的是前往南太平洋的塔希提岛（Tahiti，法属海外自治领，另一个常见的译名是“大溪地”）观测一次比较罕见的金星凌日。舰长库克（James Cook）是西方殖民史上最著名的舰长之一，多次远航探险，开拓海外殖民地。他还被认为是澳大利亚和夏威夷群岛的“发现”者，如今以他命名的群岛、海峡、山峰等不胜枚举。

当“奋进号”停靠塔希提岛时，班克斯一下就被当地美丽的土著女性迷昏了，他在她们的温柔乡里纵情狂欢，连库克舰长都看不下去了，“道德愤怒情绪偷偷溜进了他的日志当中，他发现自己根本不可能不去批评所见到的滥交行为”，而班克斯纵欲到了“连嫖妓都毫无激情”的地步——这是别人讽刺班克斯的说法，因为对于那时常年航行于茫茫大海上的男性来说，上岸嫖妓通常是一项能够唤起“激情”的活动。

而在“帝国科学”的宏大叙事中，科学家的私德是无关紧要的，人们关注的是科学家做出的科学发现。所以，尽管一面是班克斯在塔希提岛纵欲滥交，一面是他留在故乡的未婚妻正泪眼婆娑地“为远去的心上人绣织背心”，这样典型的“渣男”行径要是放在今天，非被互联网上的口水淹死不可，但是“班克斯很快从他们的分离之苦中走了出来，在外近三年，他活得倒十分滋润”。

法拉不无讽刺地指出了“帝国科学”的实质：“班克斯接管了当地的女性和植物，而库克则保护了大英帝国在太平洋上的殖民地。”甚至对班克斯的植物学本身也调侃了一番：“即使是植物学方面的科学术语也充满了性指涉。……这个体系主要依靠花朵之中雌雄生殖器官的数量来进行分类。”据说“要保护年轻妇女不受植物学教育的浸染，他们严令禁止各种各样的植物采集探险活动”。这简直就是将植物学看成一种“涉黄”的淫秽色情活动了。

在意识形态强烈影响着我们学术话语的时代，上面的故事通常是这样被描述的：库克舰长的“奋进号”军舰对殖民地和尚未成为殖民地的那些地方的所谓“访问”，其实是殖民者耀武扬威的侵略，搭载着达尔文的“小猎犬号”军舰也是同样行径；班克斯和当地女性的纵欲狂欢，当然是殖民者对土著妇女令人发指的蹂躏；即使是他采集当地植物标本的“科学考察”，也可以视为殖民者“窃取当地经济情报”的罪恶行为。

后来改革开放，上面那种意识形态话语被抛弃了，但似乎又走向了另一个极端，完全忘记或有意回避殖民者和帝国主义这个层面，只歌颂这些军舰上的科学家的伟大发现和成就，例如达尔文随着“小猎犬号”的航行，早已成为一曲祥和优美的科学颂歌。

其实达尔文也未能免俗，他在远航中也乐意与土著女性打打交道，当然他没有像班克斯那样滥情纵欲。在达尔文为“小猎犬号”远航写的《环球游记》中，我们读到：“回程途中我们遇到一群黑人姑娘在聚会，……我们笑着看了很久，还给了她们一些钱，这着实令她们欣喜一番，拿着钱尖声大笑起来，很远还能听到那愉悦的笑声。”

有趣的是，在班克斯在塔希提岛纵欲六十多年后，达尔文随着“小猎犬号”也来到了塔希提岛，岛上的土著女性同样引起了达尔文的注意，在《环球游记》中他写道：“我对这里妇女的外貌感到有些失望，然而她们却很爱美，把一朵白花或者红花戴在脑后的发髻上……”接着他以居高临下的笔调描述了当地女性的几种发饰。

用今天的眼光来看，这些在别的民族土地上采集植物动物标本、测量地质水文数据等等的“科学考察”行为，有没有合法性问题？有没有侵犯主权的问题？这些行为得到当地人的同意了吗？当地人知道这些行为的性质和意义吗？他们有知情权吗？……这些问题，在今天的国际交往中，确实都是存在的。

也许有人会为这些帝国科学家辩解说：那时当地土著尚在未开化或半开化状态中，他们哪有“国家主权”的意识啊？他们也没有制止帝国科学家的考察活动啊。但是，这样的辩解是无法成立的。

姑不论当地土著当时究竟有没有试图制止帝国科学家的“科学考察”行为，现在早已不得而知，只要殖民者没有记录下来，我们通常就无法知道。况且殖民者有军舰有枪炮，土著就是想制止也无能为力。正如法拉所描述的：“在几个塔希提人被杀之后，一套行之有效的易货贸易体制建立了起来。”

即使土著因为无知而没有制止帝国科学家的“科学考察”行为，这事也很像一个成年人闯进别人的家，难道因为那家只有不懂事的小孩子，闯入者就可以随便打探那家的隐私、拿走那家的东西、甚至将那家的房屋土地据为己有吗？事实上，很多情况下殖民者就是这样干的。所以，所谓的“帝国科学”，其实是有着原罪的。

如果沿用上述比喻，现在的局面是，家家户户都不会只有不懂事的孩子了，所以任何外来者要想进行“科学探索”，他也得和这家主人达成共识，得到这家主人的允许才能够进行。即使这种共识的达成依赖于利益的交换，至少也不能单方面强加于人。

博物学在今日中国

博物学在今日中国之复兴，北京大学刘华杰教授提倡之功殊不可没。自刘教授大力提倡之后，各界人士纷纷跟进，仿佛昔日蔡锷在云南起兵反袁之“滇黔首义，薄海同钦，一檄遥传，景从恐后”光景，这当然是和博物学本身特点密切相关的。

无论在西方还是在中国，无论在过去还是在当下，为何博物学在它繁荣时尚的阶段，就会应者云集？深究起来，恐怕和博物学本身的特点有关。博物学没有复杂的理论结构，它的专业训练也相对容易，至少没有天文学、物理学那样的数理“门槛”，所以和一些数理学科相比，博物学可以有更多的自学成才者。这次编印的《西方博物学大系》，卷帙浩繁，蔚为大观，同样说明了这一点。

最后，还有一点明显的差别必须在此处强调指出：用刘华杰教授喜欢的术语来说，《西方博物学大系》所收入的百余种著作，绝大部分属于“一阶”性质的工作，即直接对博物学作出了贡献的著作。事实上，这也是它们被收入《西方博物学大系》的主要理由之一。而在中国国内目前已经相当热的博物学时尚潮流中，绝大部分已经出版的书籍，不是属于“二阶”性质（比如介绍西方的博物学成就），就是文学性的吟风咏月野草闲花。

要寻找中国当代学者在博物学方面的“一阶”著作，如果有之，以笔者之孤陋寡闻，唯有刘华杰教授的《檀岛花事——夏威夷植物日记》三卷，可以当之。这是刘教授在夏威夷群岛实地考察当地植物的成果，不仅属于直接对博物学作出贡献之作，而且至少在形式上将昔日“帝国科学”的逻辑反其道而用之，岂不快哉！

2018 年 6 月 5 日
于上海交通大学
科学史与科学文化研究院

亨利·沃尔特·贝茨
（Henry Walter Bates）

亨利·沃尔特·贝茨（Henry Walter Bates，1825—1892），英国博物学家和探险家。出生于莱斯特一个纺织业主家庭。自幼热爱收集，喜欢研究昆虫。没有受过高等教育，小学毕业后一边当学徒一边自学。1848 年，他和动物学家阿尔弗雷德·罗素·华莱士一起前往南美洲的亚马逊河流域进行科学考察，为大英博物馆收集哺乳动物、鸟类、昆虫和植物标本，一去就是十年。期间，他不但收集了大量生物标本，还发现一些昆虫为保护自身，演化出具有防御功能的拟态性状，对以达尔文为代表的演化论学说有一定的推动作用。为纪念他的发现，动物的这一性状被称为“贝茨拟态”。

《亚马逊河上的博物学家》出版于 1862 年，是贝茨毕生唯一的著作。此书是在达尔文的帮助和推动下才得以付梓的。贝茨在书中除详细记述自己在南美洲的十载科考经历与发现外，还满怀深情地记录了与当地土人的交流与友谊。在那个博物学具有明显帝国主义倾向的年代，贝茨是少有的异类，他没有种族与地域偏见，视南美原住民为友人，也因此得到后者的全力协助，得以完成长年累月的野外考察工作。可以说，这不仅是一部博物学著作，更是一部了解 19 世纪前半叶亚马逊河流域风土人情的人文佳作。今据原版影印行世。

ADVENTURE WITH CURL-CRESTED TOUCANS.

Frontispiece to Vol. I.

THE

NATURALIST ON THE RIVER AMAZONS,

A RECORD OF ADVENTURES, HABITS OF ANIMALS, SKETCHES OF BRAZILIAN AND INDIAN LIFE, AND ASPECTS OF NATURE UNDER THE EQUATOR, DURING ELEVEN YEARS OF TRAVEL.

BY HENRY WALTER BATES.

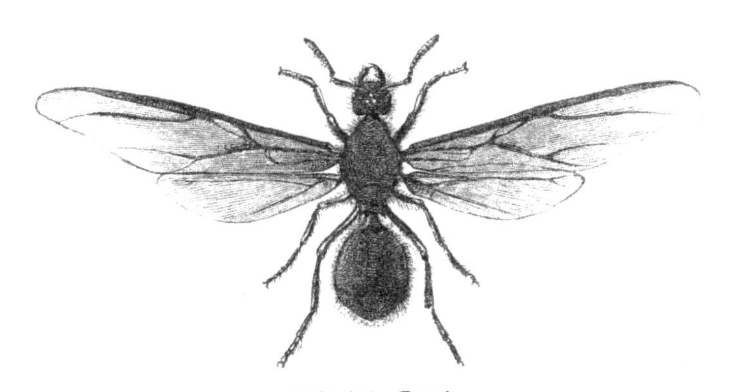

Saüba Ant.—Female.

IN TWO VOLUMES.—VOL. I.

LONDON:
JOHN MURRAY, ALBEMARLE STREET.
1863.

[*The Right of Translation is Reserved.*]

PREFACE.

In the autumn of 1847 Mr. A. R. Wallace, who has since acquired wide fame in connection with the Darwinian theory of Natural Selection, proposed to me a joint expedition to the river Amazons, for the purpose of exploring the Natural History of its banks; the plan being to make for ourselves a collection of objects, dispose of the duplicates in London to pay expenses, and gather facts, as Mr. Wallace expressed it in one of his letters, "towards solving the problem of the origin of species," a subject on which we had conversed and corresponded much together. We met in London, early in the following year, to study South American animals and plants at the principal collections; and in the month of April, as related in the following narrative, commenced our journey.

My companion left the country at the end of four years; and, on arriving in England, published a narrative of his voyage, under the title of "Travels on the Amazons and Rio Negro." I remained seven years longer, returning home in July, 1859; and having taken, after the first two years, a different route from that of my friend, an account of my separate travels and experiences seems not an inappropriate offering to the public.

When I first arrived in England, being much depressed in health and spirits after eleven years' residence within four degrees of the equator, the last three of which were spent in the wild country 1400 miles from the sea-coast, I saw little prospect of ever giving my narrative to the world; and indeed, after two years had elapsed, had almost abandoned the intention of doing so. At that date I became acquainted with Mr. Darwin, who, having formed a flattering opinion of my ability for the task, strongly urged me to write a book, and reminded me of it months afterwards, when, after having made a commencement, my half-formed resolution began to give way. Under this encouragement the arduous task is at length accomplished. It seems necessary to make this statement, as it explains why so long a time has intervened between my arrival in England and the publication of my book.

The collections that I made during the whole eleven years were sent, at intervals of a few months, to London for distribution, except a set of species reserved for my own study, which remained with me, and always accompanied me in my longer excursions. With the exception of a few living plants and specimens in illustration of

Economical and Medicinal Botany, these collections embraced only the Zoological productions of the region. The following is an approximative enumeration of the total number of species of the various classes which I obtained :—

Mammals	52
Birds	360
Reptiles	140
Fishes	120
Insects	14,000
Mollusks	35
Zoophytes	5
	14,712

The part of the Amazons region where I resided longest being unexplored country to the Naturalist, no less than 8000 of the species here enumerated were *new to science,* and these are now occupying the busy pens of a number of learned men in different parts of Europe to describe them. The few new mammals have been named by Dr. Gray; the birds by Dr. Sclater; the zoophytes by Dr. Bowerbank; and the more numerous novelties in reptiles and fishes are now in course of publication by Dr. Günther.

A word will perhaps be here in place with reference to what has become of these large collections. It will be an occasion for regret to many Naturalists to learn that a complete set of the species has nowhere been preserved, seeing that this would have formed a fair illustration of the Fauna of a region not likely to be explored again for the same purpose in our time. The limited means of a private traveller do not admit of his keeping, for a purely scientific end, a large collection. A considerable number, from many of the consignments which arrived in London from time to time, were chosen for the British Museum, so that the largest set next to my own is contained in our National Collection; but this probably comprises less than half the total number of species obtained. My very complete private collection of insects of nearly all the orders, which was especially valuable as containing the various connecting varieties, ticketed with their exact localities for the purpose of illustrating the formation of races, does not now exist in its entirety, a few large groups having passed into private hands in different parts of Europe.

With regard to the illustrations with which my book is adorned, it requires to be mentioned that the Natural History subjects have been drawn chiefly from specimens obtained by me, and the others by able artists partly from my own slight sketches. Messrs. Wolf and Zwecker have furnished most of the larger ones, which give an accurate idea of the objects and scenes they represent: for the smaller ones, many of which, for example the fishes, reptiles, and insects, are drawn with extreme care, I am indebted to Mr. E. W. Robinson.

Leicester, January, 1863.

CONTENTS OF VOL. I.

CHAPTER I.

PARÁ.

CHAPTER II.

PARÁ—*continued.*

CHAPTER III.

PARÁ—*concluded.*

CHAPTER IV.

THE TOCANTINS AND CAMETÁ.

CHAPTER V.

CARIPÍ AND THE BAY OF MARAJÓ.

CHAPTER VI.

THE LOWER AMAZONS—PARÁ TO OBYDOS.

CHAPTER VII.

THE LOWER AMAZONS—OBYDOS TO MANAOS, OR THE BARRA OF THE RIO NEGRO.

LIST OF ILLUSTRATIONS.

VOL. I.

THE

NATURALIST ON THE AMAZONS.

CHAPTER I.

PARÁ.

Arrival—Aspect of the Country—The Pará River—First walk in the Suburbs of Pará—Free Negroes—Birds, Lizards, and Insects of the Suburbs—Leaf-carrying Ant—Sketch of the climate, history, and present condition of Pará.

I EMBARKED at Liverpool, with Mr. Wallace, in a small trading vessel, on the 26th of April, 1848; and, after a swift passage from the Irish Channel to the equator, arrived, on the 26th of May, off Salinas. This is the pilot-station for vessels bound to Pará, the only port of entry to the vast region watered by the Amazons. It is a small village, formerly a missionary settlement of the Jesuits, situated a few miles to the eastward of the Pará river. Here the ship anchored in the open sea, at a distance of six miles from the shore, the shallowness of the water far out around the mouth of the great river not permitting in safety a nearer approach; and the signal was hoisted for a pilot. It was with deep interest that my companion and myself, both now about to see and examine the beauties of a tropical country for the first time, gazed on the land, where I, at least, eventually spent eleven of the best years of my life. To the eastward the country was not remarkable in appearance, being slightly undulating, with bare sand-hills and scattered trees; but to the westward, stretching towards the mouth of the river, we could see through the captain's glass a long line of forest, rising apparently out of the water; a densely-packed mass of tall trees, broken into groups, and finally into single trees, as it dwindled away in the distance. This was the frontier, in this direction, of the great primæval forest characteristic of this region, which contains so many wonders in its recesses, and clothes the whole surface of the country for two thousand miles from this point to the foot of the Andes.

On the following day and night we sailed, with a light wind, partly aided by the tide, up the Pará river. Towards evening we passed Vigia and Colares, two fishing villages, and saw many native canoes, which seemed like toys beneath the lofty walls of dark forest. The air was excessively close, the sky overcast, and sheet lightning played almost incessantly around the horizon, an appropriate greeting on the threshold of a country lying close under the equator! The evening was calm, this being the season when the winds are not strong, so we glided along in a noiseless manner, which contrasted pleasantly with the unceasing turmoil to which we had been lately accustomed on the Atlantic. The immensity of the river struck us greatly, for although sailing sometimes at a distance of eight or nine miles from the eastern bank, the opposite shore was at no time visible. Indeed, the Pará river is 36 miles in breadth at its mouth ; and at the city of Pará, nearly 70 miles from the sea, it is 20 miles wide ; but at that point a series of islands commences which contracts the river view in front of the city.

It will be well to explain here that the Pará river is not, strictly speaking, one of the mouths of the Amazons. It is made to appear so on many of the maps in common use, because the channels which connect it with the main river are there given much broader than they are in reality, conveying the impression that a large body of water finds an outlet from the main river into the Pará. It is doubtful, however, if there be any considerable stream of water flowing constantly downward through these channels. The whole of the district traversed by them consists of a complex group of low islands formed of river deposit, between which is an intricate net-work of deep and narrow channels. The land probably lies somewhat lower here than it does on the sea coast, and the tides meet about the middle of the channels ; but the ebb and flow are so complicated that it is difficult to ascertain whether there is a constant line of current in one direction. A flow down one of the channels is in some cases diverted into an ebb through other ramifications. In travelling from the Pará to the main Amazons, I have always followed the most easterly channel, and there the flow of the tide always causes a strong upward current ; it is said that this is not so perceptible in other channels, and that the flow never overpowers the stream of water coming from the main river; this would seem to favour the opinion of those geographers who believe the Pará to be one of the mouths of the King of Rivers.

The channels of which we are speaking, at least those straighter ones which trading vessels follow in the voyage from Pará to the Amazons, are about 80 miles in length; but for many miles of their course they are not more than 100 yards in breadth. They are of great depth, and in many places are so straight and regular that they appear like artificial canals. The great river steamers which now run regularly to the interior, in some places brush the overhanging trees with their paddle-boxes on each side as they pass. The whole of the region is one vast wilderness of the most luxuriant tropical vegetation, the strangest forms of palm trees of some score of different species forming a great proportion of the mass. I shall, however, have to allude again to the wonderful beauty of these romantic channels, when I arrive at that part of my narrative.

The Pará river, on this view, may be looked upon as the common fresh-water estuary of the numerous rivers which flow into it from the south; the chief of which is the Tocantins, a stream 1600 miles in length, and about 10 miles in breadth at its mouth. The estuary forms, then, a magnificent body of water 160 miles in length, and eight miles in breadth at its abrupt commencement, where it receives the channels just described. There is a great contrast in general appearance between the Pará and the main Amazons. In the former the flow of the tide always creates a strong current upwards, whilst in the Amazons the turbid flow of the mighty stream overpowers all tides, and produces a constant downward current. The colour of the water is different, that of the Pará being of a dingy orange-brown, whilst the Amazons has an ochreous or yellowish clay tint. The forests on their banks have a different aspect. On the Pará the infinitely diversified trees seem to rise directly out of the water; the forest frontage is covered with greenery, and wears a placid aspect, whilst the shores of the main Amazons are encumbered with fallen trunks, and are fringed with a belt of broad-leaved grasses. The difference is partly owing to the currents, which on the main river tear away the banks, and float out to sea an almost continuous line of dead trees and other debris of its shores.

We may, however, regard the combined mouths of the Pará and the Amazons with their archipelago of islands as forming one immense river delta, each side of which measures 180 miles—an area about equal to the southern half of England and Wales. In the middle of it lies the island of Marajo, which is as large as Sicily. The land is low and flat, but it does not consist entirely of alluvium or river deposit; in many parts the surface is rocky; rocks also form reefs in the middle of the Pará river. The immense volumes of fresh water which are poured through these broad embouchures, the united contributions of innumerable streams, fed by drenching tropical rains, prevent them from becoming salt-water estuaries. The water is only occasionally a little brackish near Pará, at high spring tides. Indeed, the fresh water tinges the sea along the shores of Guiana to a distance of nearly 200 miles from the mouth of the river.

On the morning of the 28th of May we arrived at Pará. The appearance of the city at sunrise was pleasing in the highest degree. It is built on a low tract of land, having only one small rocky elevation at its southern extremity; it therefore affords no amphitheatral view from the river; but the white buildings roofed with red tiles, the numerous towers and cupolas of churches and convents, the crowns of palm trees reared above the buildings, all sharply defined against the clear blue sky, give an appearance of lightness and cheerfulness which is most exhilarating. The perpetual forest hems the city in on all sides landwards; and towards the suburbs, picturesque country houses are seen scattered about, half buried in luxuriant foliage. The port was full of native canoes and other vessels, large and small; and the ringing of bells and firing of rockets, announcing the dawn of some Roman Catholic festival day, showed that the population was astir at that early hour.

We went ashore in due time, and were kindly received by Mr. Miller, the consignee of the vessel, who invited us to make his house our home until we could obtain a suitable residence. On landing, the hot moist mouldy air, which seemed to strike from the ground and walls, reminded me of the atmosphere of tropical stoves at Kew. In the course of the afternoon a heavy shower fell, and in the evening, the atmosphere having been cooled by the rain, we walked about a mile out of town to the residence of an American gentleman to whom our host wished to introduce us.

The impressions received during this first walk can never wholly fade from my mind. After traversing the few streets of tall, gloomy, convent-looking buildings near the port, inhabited chiefly by merchants and shopkeepers, along which idle soldiers, dressed in shabby uniforms, carrying their muskets carelessly over their arms, priests, negresses with red water-jars on their heads, sad-looking Indian women carrying their naked children astride on their hips, and other samples of the motley life of the place, were seen, we passed down a long narrow street leading to the suburbs. Beyond this, our road lay across a grassy common into a picturesque lane leading to the virgin forest. The long street was inhabited by the poorer class of the population. The houses were of one story only, and had an irregular and mean appearance. The windows were without glass, having, instead, projecting lattice casements. The street was unpaved and inches deep in loose sand. Groups of people were cooling themselves outside their doors: people of all shades in colour of skin, European, Negro and Indian, but chiefly an uncertain mixture of the three. Amongst them were several handsome women, dressed in a slovenly manner, barefoot or shod in loose slippers; but wearing richly-decorated ear-rings, and around their necks strings of very large gold beads. They had dark expressive eyes, and remarkably rich heads of hair. It was a mere fancy, but I thought the mingled squalor, luxuriance and beauty of these women were pointedly in harmony with the rest of the scene; so striking, in the view, was the mixture of natural riches and human poverty. The houses were mostly in a dilapidated condition, and signs of indolence and neglect were everywhere visible. The wooden palings which surrounded the weed-grown gardens were strewn about, broken; and hogs, goats and ill-fed poultry, wandered in and out through the gaps. But amidst all, and compensating every defect, rose the overpowering beauty of the vegetation. The massive dark crowns of shady mangos were seen everywhere amongst the dwellings, amidst fragrant blossoming orange, lemon, and many other tropical fruit trees; some in flower, others in fruit, at varying stages of ripeness. Here and there, shooting above the more dome-like and sombre trees, were the smooth columnar stems of palms, bearing aloft their magnificent crowns of finely-cut fronds. Amongst the latter the slim assai-palm was especially noticeable; growing in groups of four or five; its smooth, gently-curving stem, twenty to thirty feet high, terminating in a head of feathery foliage, inexpressibly light and elegant in outline. On the boughs of the taller and more ordinary-looking trees sat tufts of curiously-leaved parasites. Slender woody lianas hung in festoons from the branches, or were suspended in the form of cords and ribbons; whilst luxuriant creeping plants overran alike tree-trunks, roofs and walls, or toppled over palings in copious profusion of foliage. The superb banana (Musa paradisiaca), of which I had always read as forming one of the charms of tropical vegetation, here grew with great luxuriance: its glossy velvety-green leaves, twelve feet in length, curving over the roofs of verandahs in the rear of every house. The shape of the leaves, the varying shades of green which they present when lightly moved by the wind, and especially the contrast they afford in colour and form to the more sombre hues and more rounded outline of the other trees, are quite sufficient to account for the charm of this glorious tree. Strange forms of vegetation drew our attention at almost every step. Amongst them were the different kinds of Bromelia, or pine-apple plants, with their long, rigid, sword-shaped leaves, in some species jagged or toothed along their edges. Then there was the bread-fruit tree—an importation, it is true ; but remarkable from its large, glossy, dark green, strongly digitated foliage, and its interesting history. Many other trees and plants, curious in leaf, stem, or manner of growth, grew on the borders of the thickets along which lay our road ; they were all attractive to new comers, whose last country ramble of quite recent date was over the bleak moors of Derbyshire on a sleety morning in April.

As we continued our walk the brief twilight commenced, and the sounds of multifarious life came from the vegetation around. The whirring of cicadas ; the shrill stridulation of a vast number and variety of field crickets and grasshoppers,—each species sounding its peculiar note ; the plaintive hooting of tree frogs—all blended together in one continuous ringing sound,—the audible expression of the teeming profusion of Nature. As night came on, many species of frogs and toads in the marshy places joined in the chorus : their croaking and drumming, far louder than anything I had before heard in the same line, being added to the other noises, created an almost deafening din. This uproar of life, I afterwards found, never wholly ceased, night or day: in course of time I became, like other residents, accustomed to it. It is, however, one of the peculiarities of a tropical—at least, a Brazilian—climate which is most likely to surprise a stranger. After my return to England the death-like stillness of summer days in the country appeared to me as strange as the ringing uproar did on my first arrival at Pará. The object of our visit being accomplished, we returned to the city. The fire flies were then out in great numbers, flitting about the sombre woods, and even the frequented streets. We turned into our hammocks, well pleased with what we had seen, and full of anticipation with regard to the wealth of natural objects we had come to explore.

During the first few days, we were employed in landing our baggage and arranging our extensive apparatus. We then accepted the invitation of Mr. Miller to make use of his rocinha, or country-house in the suburbs, until we finally decided on a residence. Upon this we made our first essay in housekeeping. We bought cotton hammocks, the universal substitute for beds in this country, cooking utensils and crockery, and then engaged a free negro, named Isidoro, as cook and servant-of-all-work. Isidoro had served Englishmen in this capacity before, and, although he had not picked up two words of English, he thought he had a great talent for understanding and making himself understood; in his efforts to do which he was very amusing. Having no other medium through which we could make known our wants, we progressed rapidly in learning Portuguese. I was quite surprised to find little or no trace in Isidoro of that baseness of character which I had read of as being the rule amongst negroes in a slave country. Isidoro was an old man, with an anxious, lugubrious expression of countenance, and exhibited signs of having been overworked in his younger days, which I understood had been passed in slavery. The first traits I perceived in him were a certain degree of self-respect and a spirit of independence: these I found afterwards to be by no means rare qualities among the free negroes. Some time after he had entered our service, I scolded him one morning about some delay in getting breakfast. It happened that it was not his fault, for he had been detained, much against his will, at the shambles. He resented the scolding, not in an insolent way, but in a quiet, respectful manner, and told me how the thing had occurred; that I must not expect the same regularity in Brazil which is found in England, and that "paciencia" was a necessary accomplishment to a Brazilian traveller. There was nothing ridiculous about Isidoro; there was a gravity of demeanour and sense of propriety about him which would have been considered becoming in a serving-man in any country. This spirit of self-respect is, I think, attributable partly to the lenient treatment which slaves have generally received from their white masters in this part of Brazil, and partly to the almost total absence of prejudice against coloured people amongst the inhabitants. This latter is a very hopeful state of things. It seems to be encouraged by the governing class in Brazil ; and, by drawing together the races and classes of the heterogeneous population, will doubtless lead to the most happy results. I had afterwards, as I shall have to relate in the course of my narrative, to number free negroes amongst my most esteemed friends : men of temperate, quiet habits, desirous of mental and moral improvement, observant of the minor courtesies of life, and quite as trustworthy, in more important matters, as the whites and half-castes of the province. Isidoro was not, perhaps, scrupulously honest in small matters : scrupulous honesty is a rare quality in casual servants anywhere. He took pains to show that he knew he had made a contract to perform certain duties, and he tried, evidently, to perform them to the best of his ability.

Our first walks were in the immediate suburbs of Pará. The city lies on a corner of land formed by the junction of the river Guamá with the Pará. As I have said before, the forest, which covers the whole country, extends close up to the city streets ; indeed, the town is built on a tract of cleared land, and is kept free from the jungle only by the constant care of the Government. The surface, though everywhere low, is slightly undulating, so that areas of dry land alternate throughout with areas of swampy ground, the vegetation and animal tenants of the two being widely different. Our residence lay on the side of the city nearest the Guamá, on the borders of one of the low and swampy areas which here extend over a portion of the suburbs. The tract of land is intersected by well-macadamized suburban roads, the chief of which, the Estrada das Mongubeiras (the Monguba road), about a mile long, is a magnificent avenue of silk-cotton trees (Bombax monguba and B. ceiba), huge trees whose trunks taper rapidly from the ground upwards, and whose flowers before opening look like red balls studding the branches. This fine road was constructed under the governorship of the Count dos Arcos, about the year 1812. At right angles to it run a number of narrow green lanes, and the whole district is drained by a system of small canals or trenches through which the tide ebbs and flows, showing the lowness of the site. Before I left the country, other enterprising presidents had formed a number of avenues lined with cocoa-nut palms, almond and other trees, in continuation of the Monguba road, over the more elevated and drier ground to the northeast of the city. On the high ground the vegetation has an aspect quite different from that which it presents in the swampy parts. Indeed, with the exception of the palm trees, the suburbs here have an aspect like that of a village green at home. The soil is sandy, and the open commons are covered with a short grassy and shrubby vegetation. Beyond this, the land again descends to a marshy tract, where, at the bottom of the moist hollows, the public wells are situated. Here all the linen of the city is washed by hosts of noisy negresses, and here also the water-carts are filled—painted hogsheads on wheels, drawn by bullocks. In early morning, when the sun sometimes shines through a light mist, and everything is dripping with moisture, this part of the city is full of life: vociferous negroes and wrangling

Gallegos,* the proprietors of the water-carts, are gathered about, jabbering continually, and taking their morning drams in dirty wine-shops at the street corners.

Along these beautiful roads we found much to interest us during the first few days. Suburbs of towns, and open, sunny, cultivated places in Brazil, are tenanted by species of animals and plants which are mostly different from those of the dense primeval forests. I will, therefore, give an account of what we observed of the animal world during our explorations in the immediate neighbourhood of Pará.

The number and beauty of the birds and insects did not at first equal our expectations. The majority of the birds we saw were small and obscurely coloured; they were indeed similar, in general appearance, to such as are met with in country places in England. Occasionally a flock of small parroquets, green, with a patch of yellow on the forehead, would come at early morning to the trees near the Estrada. They would feed quietly, sometimes chattering in subdued tones, but setting up a harsh scream, and flying off, on being disturbed. Humming-birds we did not see at this time, although I afterwards found them by hundreds when certain trees were in flower. Vultures we only saw at a distance, sweeping round at a great height, over the public slaughter-houses. Several flycatchers, finches, ant-thrushes, a tribe of plainly-coloured birds, intermediate in structure between flycatchers and thrushes, some of which startle the new-comer by their extra-

* Natives of Galicia, in Spain, who follow this occupation in Lisbon and Oporto, as well as at Pará.

ordinary notes emitted from their places of concealment in the dense thickets ; and also tanagers, and other small birds, inhabited the neighbourhood. None of these had a pleasing song, except a little brown wren (Troglodytes furvus), whose voice and melody resemble those of our English robin. It is often seen hopping and climbing about the walls and roofs of houses and on trees in their vicinity. Its song is more frequently heard in the rainy season, when the Monguba trees shed their leaves. At those times the Estrada das Mongubeiras has an appearance quite unusual in a tropical country. The tree is one of the few in the Amazons region which sheds all its foliage before any of the new leaf-buds expand. The naked branches, the soddened ground matted with dead leaves, the grey mist veiling the surrounding vegetation, and the cool atmosphere soon after sunrise, all combine to remind one of autumnal mornings in England. Whilst loitering about at such times, in a half-oblivious mood, thinking of home, the song of this bird would create for the moment a perfect illusion. Numbers of tanagers frequented the fruit and other trees in our garden. The two principal kinds which attracted our attention were the Rhamphocœlus Jacapa and the Tanagra Episcopus. The females of both are dull in colour. The male of Jacapa has a beautiful velvety purple and black plumage, the beak being partly white. The same sex in Episcopus is of a pale blue colour, with white spots on the wings. In their habits they both resemble the common house-sparrow of Europe, which does not exist in South America, its place being in some measure filled by these familiar tanagers. They are just as lively, restless, bold, and wary; their notes are very similar; chirping and inharmonious, and they seem to be almost as fond of the neighbourhood of man. They do not, however, build their nests on houses.

Another interesting and common bird was the Japím, a species of Cassicus (C. icteronotus). It belongs to the same family of birds as our starling, magpie and rook. It has a rich yellow and black plumage, remarkably compact and velvety in texture. The shape of its head and its physiognomy are very similar to those of the magpie; it has light gray eyes, which give it the same knowing expression. It is social in its habits; and builds its nest, like the English rook, on trees in the neighbourhood of habitations. But the nests are quite differently constructed, being shaped like purses, two feet in length, and suspended from the slender branches all round the tree, some of them very near the ground. The entrance is on the side near the bottom of the nest. This bird is a great favourite with the Brazilians of Pará: it is a noisy, stirring, babbling creature, passing constantly to and fro, chattering to its comrades, and is very ready at imitating other birds, especially the domestic poultry of the vicinity. There was at one time a weekly newspaper published at Pará, called "The Japim;" the name being chosen, I suppose, on account of the babbling propensities of the bird. Its eggs are nearly round, and of a bluish-white colour, speckled with brown.

Of other vertebrate animals we saw very little except of the lizards. These are sure to attract the attention of the new comer from Northern Europe, by reason of their strange appearance, great numbers, and variety. The species which are seen crawling over the walls of buildings in the city, are different from those found in the forest or in the interior of houses. They are unpleasant-looking animals, with colours assimilated to those of the dilapidated stone and mud walls on which they are seen. The house lizards belong to a peculiar family, the Geckos. They are found even in the best-kept houses, most frequently on the walls and ceilings : they are generally motionless by day, being active only at night. They are of speckled gray, or ashy colours. The structure of their feet is beautifully adapted for clinging to and running over smooth surfaces ; the underside of their toes being expanded into cushions, beneath which folds of skin form a series of flexible plates. By means of this apparatus they can walk or run across a smooth ceiling with their backs downwards, the plated soles, by quick muscular action, exhausting and admitting air alternately. These Geckos are very repulsive in appearance. The Brazilians give them the name of Osgas, and firmly believe them to be poisonous ; they are, however, harmless creatures. The species found in houses are small ; I have seen others of great size, in crevices of tree trunks in the forest. Sometimes Geckos are found with forked tails ; this results from the budding of a rudimentary tail at the side, from an injury done to the member. A slight rap will cause their tails to snap off ; the loss being afterwards partially repaired by a new growth. The tails of lizards seem to be almost useless appendages to the animals. I used often to amuse myself in the suburbs, whilst resting in the verandah of our house during the heat of midday, by watching the variegated green, brown, and yellow ground-lizards. They would come nimbly forward, and commence grubbing with their fore feet and snouts around the roots of herbage, searching for insect larvæ. On the slightest alarm they would scamper off; their tails cocked up in the air as they waddled awkwardly away, evidently an incumbrance to them in their flight.

Next to the birds and lizards, the insects of the suburbs of Pará deserve a few remarks. The species observed in the weedy and open places, as already remarked, were generally different from those which dwell in the shades of the forest. It is worthy of notice that those species which have the widest distribution in America, and which have the closest affinity to those of the tropics of the old world, are such as occur in open sunny places near towns. The general appearance of the insects and birds belonging to such situations is very similar to that of European species. This resemblance, however, is, in many cases, one of analogy only; that is, the species are similar in size, form, and colours, but belong to widely different genera. Thus, all the small carnivorous beetles seen running along sandy pathways, look precisely like the Amaræ, those oval coppery beetles which are seen in similar situations in England. But they belong to quite another genus—namely, Selenophorus, the genus Amara being unknown in Tropical America. In butterflies, again, we saw a small species of Erycinidæ flying about low shrubs in grassy places, which was extremely similar in colours to the European Nemeobius Lucina. The Pará insect, however, belongs to a genus far removed in all essential points of structure from Nemeobius; namely, to Lemonias, being the L. epulus. It is worthy of note that all the old-world representatives, both tropical and temperate, of this beautiful family of butterflies belong to the same group as the English Nemeobius Lucina; whilst the few species inhabiting North America belong wholly to South American types.

Facts of this kind, and there are many of them, would seem to show that it is not wholly the external conditions of light, heat, moisture, and so forth, which determine the general aspect of the animals of a country. It is a notion generally entertained that the superior size and beauty of tropical insects and birds are immediately due to the physical conditions of a tropical climate, or are in some way directly connected with them. I think this notion is an incorrect one, and that there are other causes more powerful than climatal conditions which affect the dress of species. To test this we ought to compare the members of those genera which are common to two regions; say, to Northern Europe and equinoctial America, and ascertain which climate produces the largest and most beautifully-coloured species. We should thus see the supposed effects of climate on nearly-allied congeners, that is, creatures very similarly organised. In the first family of the order Coleoptera, for instance, the tiger-beetles (Cicindelidæ), there is one genus, Cicindela, common to the two regions. The species found in the Amazons Valley have precisely the same habits as their English brethren, running and flying over sandy soils in the bright sunshine. About the same number is found in each of the two countries : but all the Amazonian species are far smaller in size and more obscure in colour than those inhabiting Northern Europe; none being at all equal in these respects to the common English Cicindela campestris, the handsome light-green tiger-beetle, spotted with white, which is familiar to country residents of Natural History tastes in most parts of England. In butterflies I find there are eight genera common to the two regions we are thus pitting against each other. Of these, three only (Papilio, Pieris and Thecla) are represented by handsomer species in Amazonia than in Northern Europe. Three others (Lycæna, Melitæa and Apatura) yield far more beautiful and larger forms in England than in the Amazonian plains; as to the remaining two (Pamphila and Pyrgus) there is scarcely any difference. There is another and hitherto neglected fact which I would strongly press upon those who are interested in these subjects. This is, that it is almost always the *males only* which are beautiful in colours. The brilliant dress is rarely worn by both sexes of the same species: if climate has any direct influence in this matter, why have not both sexes felt its effects, and why are the males of genera living under our gloomy English skies adorned with bright colours?

The tropics, it is true, have a vastly greater total number of handsome butterflies than the temperate zones; but it must be borne in mind that they contain a far greater number of genera and species altogether.

It holds good in all families that the two sexes of the more brilliantly-coloured kinds are seldom equally beautiful; the females being often quite obscure in dress. There is a very large number of dull-coloured species in tropical countries. The tropics have also species in which the contrast between the sexes is greater than in any species of temperate zones; in some cases the males have been put in one genus and the females in another, so great is the difference between them. There are species of larger size, but at the same time there are others of smaller size in the same families in tropical than in temperate latitudes. If we reflect on all these facts, we must come to the conclusion, that climate, to which we are naturally at first sight inclined to attribute much, has little or no direct influence in the matter. Mr. Darwin was led to the same conclusion many years ago, when comparing the birds, plants, and insects of the Galapagos islands, situated under the equator, with those of Patagonia and Tropical America. The abundance of food, the high temperature, absence of seasons of extreme cold and dearth, and the variety of stations, all probably operate in favouring the existence of a greater number and variety of species in tropical than in temperate latitudes. This, perhaps, is all we can say with regard to the influence of climatal conditions. The causes which have produced the great beauty that astonishes us, if we really wish to investigate them, must be sought in other directions. I think that the facts above mentioned are calculated to guide us in the search. They show, for instance, that beauty of form and colour is not peculiar to one zone, but is producible under any climate where a number of species of a given genus lead a flourishing existence. The ornamental dress is generally the property of one sex to the exclusion of the other, and the cases of widest contrast between the two are exhibited in those regions where life is generally more active and prolific. All this points to the mutual relations of the species, and especially to those between the sexes, as having far more to do in the matter than climate.

In the gardens, numbers of fine showy butterflies were seen. There were two swallow-tailed species, similar in colours to the English Papilio Machaon; a white Pieris (P. Monuste), and two or three species of brimstone and orange coloured butterflies, which do not belong, however, to the same genus as our English species. In weedy places a beautiful butterfly, with eye-like spots on its wings, was common, the Junonia Lavinia, the only Amazonian species which is at all nearly related to our Vanessas, the Admiral and Peacock butterflies. One day we made our first acquaintance with two of the most beautiful productions of nature in this department; namely the Helicopis Cupido and Endymion. A little beyond our house, one of the narrow green lanes which I have already mentioned diverged from the Monguba avenue, and led, between enclosures overrun with a profusion of creeping plants and glorious flowers, down to a moist hollow, where there was a public well in a picturesque nook, buried in a grove of Mucajá palm-trees. On the tree-trunks, walls, and palings, grew a great quantity of climbing

Pothos plants, with large glossy heart-shaped leaves. These plants were the resort of these two exquisite species, and we captured a great number of specimens. They are of extremely delicate texture. The wings are cream-coloured; the hind pair have several tail-like appendages, and are spangled beneath as if with silver. Their flight is very slow and feeble; they seek the protected under-surface of the leaves, and in repose close their wings over the back, so as to expose the brilliantly spotted under-surface.

I will pass over the many other orders and families of insects, and proceed at once to the ants. These were in great numbers everywhere, but I will mention here only two kinds. We were amazed at seeing ants an inch and a quarter in length, and stout in proportion, marching in single file through the thickets. These belonged to the species called Dinoponera grandis. Its colonies consist of a small number of individuals, and are established about the roots of slender trees. It is a stinging species, but the sting is not so severe as in many of the smaller kinds. There was nothing peculiar or attractive in the habits of this giant among the ants. Another far more interesting species was the Saüba (Œcodoma cephalotes). This ant is seen everywhere about the suburbs, marching to and fro in broad columns. From its habit of despoiling the most valuable cultivated trees of their foliage, it is a great scourge to the Brazilians. In some districts it is so abundant that agriculture is almost impossible, and everywhere complaints are heard of the terrible pest.

The workers of this species are of three orders, and

vary in size from two to seven lines; some idea of them may be obtained from the accompanying wood-cut. The true working-class of a colony is formed by the small-

Saüba or Leaf-carrying ant.—1, Worker-minor; 2, Worker-major; 3, Subterranean worker.

sized order of workers, the worker-minors as they are called (Fig. 1). The two other kinds, whose functions, as we shall see, are not yet properly understood, have enormously swollen and massive heads; in one (Fig. 2), the head is highly polished; in the other (Fig. 3), it is opaque and hairy. The worker-minors vary greatly in size, some being double the bulk of others. The entire body is of very solid consistence, and of a pale reddish-brown colour. The thorax or middle segment is armed with three pairs of sharp spines; the head, also, has a pair of similar spines proceeding from the cheeks behind.

In our first walks we were puzzled to account for large mounds of earth, of a different colour from the surrounding soil, which were thrown up in the plantations and woods. Some of them were very extensive, being forty yards in circumference, but not more than two feet in height. We soon ascertained that these were the work of the Saübas, being the outworks, or domes, which overlie and protect the entrances to their vast subterranean galleries. On close examination, I found the earth of which they are composed to consist of very minute granules, agglomerated without cement, and forming many rows of little ridges and turrets. The difference in colour from the superficial soil of the vicinity is owing to their being formed of the undersoil, brought up from a considerable depth. It is very rarely that the ants are seen at work on these mounds; the entrances seem to be generally closed; only now and then, when some particular work is going on, are the galleries opened. The entrances are small and numerous; in the larger hillocks it would require a great amount of excavation to get at the main galleries; but I succeeded in removing portions of the dome in smaller hillocks, and then I found that the minor entrances converged, at the depth of about two feet, to one broad elaborately-worked gallery or mine, which was four or five inches in diameter.

This habit in the Saüba ant of clipping and carrying away immense quantities of leaves has long been recorded in books on natural history. When employed on this work, their processions look like a multitude of animated leaves on the march. In some places I found an accumulation of such leaves, all circular pieces, about the size of a sixpence, lying on the pathway, unattended by ants, and at some distance from any colony. Such heaps are always found to be removed when the place is revisited the next day. In course of time I had plenty of opportunities of seeing them at work. They mount the tree in multitudes, the individuals being all worker-minors. Each one places itself on the surface of a leaf, and cuts with its sharp scissor-like jaws a nearly semicircular incision on the upper side; it then takes the edge between its jaws, and by a sharp jerk detaches the piece. Sometimes they let the leaf drop to the ground, where a little heap accumulates, until carried off by another relay of workers; but, generally, each marches off with the piece it has operated upon, and as all take the same road to their colony, the path they follow becomes in a short time smooth and bare, looking like the impression of a cart-wheel through the herbage.

It is a most interesting sight to see the vast host of busy diminutive labourers occupied on this work. Unfortunately they choose cultivated trees for their purpose. This ant is quite peculiar to Tropical America, as is the entire genus to which it belongs; it sometimes despoils the young trees of species growing wild in its native forests; but it seems to prefer, when within reach, plants imported from other countries, such as the coffee and orange trees. It has not hitherto been shown satisfactorily to what use it applies the leaves. I discovered it only after much time spent in investigation. The leaves are used to thatch the domes which cover the entrances to their subterranean dwellings, thereby protecting from the deluging rains the young broods in the nests beneath. The larger mounds, already described, are so extensive that few persons would attempt to remove them for the purpose of examining their interior; but smaller hillocks, covering other entrances to the same system of tunnels and chambers may be found in sheltered places, and these are always thatched with leaves, mingled with granules of earth. The heavily-laden workers, each carrying its segment of leaf vertically, the lower edge secured in its mandibles, troop up and cast their burthens on the hillock; another relay of labourers place the leaves in position, covering them with a layer of earthy granules, which are brought one by one from the soil beneath.

The underground abodes of this wonderful ant are known to be very extensive. The Rev. Hamlet Clark has related that the Saüba of Rio de Janeiro, a species closely allied to ours, has excavated a tunnel under the bed of the river Parahyba, at a place where it is as broad as the Thames at London Bridge. At the Magoary rice mills, near Pará, these ants once pierced the embankment of a large reservoir: the great body of water which it contained escaped before the damage could be repaired. In the Botanic Gardens, at Pará, an enterprising French gardener tried all he could think of to extirpate the Saüba. With this object he made fires over some of the main entrances to their colonies, and blew the fumes of sulphur down the galleries by means of bellows. I saw the smoke issue from a great number of outlets, one of which was 70 yards distant from the place where the bellows were used. This shows how extensively the underground galleries are ramified.

Besides injuring and destroying young trees by despoiling them of their foliage, the Saüba ant is troublesome to the inhabitants from its habit of plundering the stores of provisions in houses at night, for it is even more active by night than in the day-time. At first I was inclined to discredit the stories of their entering habitations and carrying off grain by grain the farinha or mandioca meal, the bread of the poorer classes of Brazil. At length, whilst residing at an Indian village on the Tapajos, I had ample proof of the fact. One night my servant woke me three or four hours before sunrise by calling out that the rats were robbing the farinha baskets. The article at that time was scarce and dear. I got up, listened, and found the noise was very unlike that made by rats. So I took the light and went into the store-room, which was close to my sleeping-place. I there found a broad column of Saüba ants, consisting of thousands of individuals, as busy as possible, passing to and fro between the door and my precious baskets. Most of those passing outwards were laden each with a grain of farinha, which was, in some cases, larger and many times heavier than the bodies of the carriers. Farinha consists of grains of similar size and appearance to the tapioca of our shops; both are products of the same root, tapioca being the pure starch, and farinha the starch mixed with woody fibre, the latter ingredient giving it a yellowish colour. It was amusing to see some of the dwarfs, the smallest members of their family, staggering along, completely hidden under their load. The baskets, which were on a high table, were entirely covered with ants, many hundreds of whom were employed in snipping the dry leaves which served as lining. This produced the rustling sound which had at first disturbed us. My servant told me that they would carry off the whole contents of the two baskets (about two bushels) in the course of the night, if they were not driven off; so we tried to exterminate them by killing them with our wooden clogs. It was impossible, however, to prevent fresh hosts coming in as fast as we killed their companions. They returned the next night; and I was then obliged to lay trains of gunpowder along their line, and blow them up. This, repeated many times, at last seemed to intimidate them, for we were free from their visits during the remainder of my residence at the place. What they did with the hard dry grains of mandioca I was never able to ascertain, and cannot even conjecture. The meal contains no gluten, and therefore would be useless as cement. It contains only a small relative portion of starch, and, when mixed with water, it separates and falls away like so much earthy matter. It may serve as food for the subterranean workers. But the young or larvæ of ants are usually fed by juices secreted by the worker nurses.

Ants, it is scarcely necessary to observe, consist, in each species, of three sets of individuals, or, as some express it, of three sexes—namely, males, females, and workers; the last-mentioned being undeveloped females. The perfect sexes are winged on their first attaining the adult state; they alone propagate their kind, flying away, previous to the act of reproduction, from the nest in which they have been reared. This winged state of the perfect males and females, and the habit of flying abroad before pairing, are very important points in the economy of ants; for they are thus enabled to intercross with members of distant colonies which swarm at the same time, and thereby increase the vigour of the race, a proceeding essential to the prosperity of any species. In many ants, especially those of tropical climates, the workers, again, are of two classes, whose structure and functions are widely different. In some species they are wonderfully unlike each other, and constitute two well-defined forms of workers. In others, there is a gradation of individuals between the two extremes. The curious differences in structure and habits between these two classes form an interesting, but very difficult, study. It is one of the great peculiarities of the Saüba ant to possess *three* classes of workers. My investigations regarding them were far from complete; I will relate, however, what I have observed on the subject.

When engaged in leaf-cutting, plundering farinha, and other operations, two classes of workers are always seen (Figs. 1 and 2, page 24). They are not, it is true, very sharply defined in structure, for individuals of intermediate grades occur. All the work, however, is done by the individuals which have small heads (Fig. 1), whilst those which have enormously large heads, the worker-majors (Fig. 2), are observed to be simply walking about. I could never satisfy myself as to the function of these worker-majors. They are not the soldiers or defenders of the working portion of the community, like the armed class in the Termites, or white ants; for they never fight. The species has no sting, and does not display active resistance when interfered with. I once imagined they exercised a sort of superintendence over the others; but this function is entirely unnecessary in a community where all work with a precision and regularity resembling the subordinate parts of a piece of machinery. I came to the conclusion, at last, that they have no very precisely defined function. They cannot, however, be entirely useless to the community, for the sustenance of an idle class of such bulky individuals would be too heavy a charge for the species to sustain. I think they serve, in some sort, as passive instruments of protection to the real workers. Their enormously large, hard, and indestructible heads may be of use in protecting them against the attacks of insectivorous animals. They would be, on this view, a kind of "pièces de resistance," serving as a foil against onslaughts made on the main body of workers.

The third order of workers is the most curious of all. If the top of a small, fresh hillock, one in which the thatching process is going on, be taken off, a broad cylindrical shaft is disclosed, at a depth of about two feet from the surface. If this be probed with a stick, which may be done to the extent of three or four feet without touching bottom, a small number of colossal fellows (Fig. 3) will slowly begin to make their way up the smooth sides of the mine. Their heads are of the same size as those of the class fig. 2; but the front is clothed with hairs, instead of being polished, and they have in the middle of the forehead a twin ocellus, or simple eye, of quite different structure from the ordinary compound eyes, on the sides of the head. This frontal eye is totally wanting in the other workers, and is not known in any other kind of ant. The apparition oi these strange creatures from the cavernous depths of the mine reminded me, when I first observed them, of the Cyclopes of Homeric fable. They were not very pugnacious, as I feared they would be, and I had no difficulty in securing a few with my fingers. I never saw them under any other circumstances than those here related, and what their special functions may be I cannot divine.

The whole arrangement of a Formicarium, or ant-colony, and all the varied activity of ant-life, are directed to one main purpose:—the perpetuation and dissemination of the species. Most of the labour which we see performed by the workers has for its end the sustenance and welfare of the young brood, which are helpless grubs. The true females are incapable of attending to the wants of their offspring; and it is on the poor sterile workers, who are denied all the other pleasures of maternity, that the entire care devolves. What a wonderfully-organised community is that of the ant! The workers are also the chief agents in carrying out the different migrations of the colonies, which are of vast importance to the dispersal and consequent prosperity of the species. The successful *début* of the winged males and females depends likewise on the workers. It is amusing to see the activity and excitement which reign in an ant's nest when the exodus of the winged individuals is taking place. The workers clear the roads of exit, and show the most lively interest in their departure, although it is highly improbable that any of them will return to the same colony. The swarming or exodus of the winged males and females of the Saüba ant takes place in January and February, that is, at the commencement of the rainy season. They come out in the evening in vast numbers, causing quite a commotion in the streets and lanes. They are of very large size, the female measuring no less than

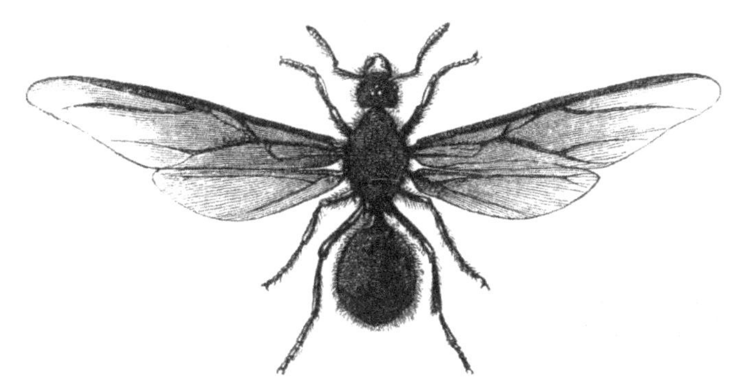

Saüba Ant.—Female.

two-and-a-quarter inches in expanse of wing; the male is not much more than half this size. They are so eagerly preyed upon by insectivorous animals that on the morning after their flight not an individual is to be seen, a few impregnated females alone escaping the slaughter to found new colonies.

At the time of our arrival, Pará had not quite recovered from the effects of a series of revolutions, brought about by the hatred which existed between the native Brazilians and the Portuguese; the former, in the end, calling to their aid the Indian and mixed coloured population. The number of inhabitants of the city had decreased, in consequence of these disorders, from 24,500 in 1819, to 15,000 in 1848. Although the public peace had not been broken for twelve years before the date of our visit, confidence was not yet completely restored, and the Portuguese merchants and tradesmen would not trust themselves to live at their beautiful country-houses or rocinhas which lie embosomed in the luxuriant shady gardens around the city. No progress had been made in clearing the second-growth forest which had grown over the once cultivated grounds and now reached the end of all the suburban streets. The place had the aspect of one which had seen better days ; the public buildings, including the palaces of the President and Bishop, the cathedral, the principal churches and convents, all seemed constructed on a scale of grandeur far beyond the present requirements of the city. Streets full of extensive private residences built in the Italian style of architecture, were in a neglected condition, weeds and flourishing young trees growing from large cracks in the masonry. The large public squares were over-grown with weeds and impassable on account of the swampy places which occupied portions of their areas. Commerce, however, was now beginning to revive, and before I left the country I saw great improvements, as I shall have to relate towards the conclusion of this narrative.

The province of which Pará is the capital, was, at the time I allude to, the most extensive in the Brazilian empire, being about 1560 miles in length from east to west, and about 600 in breadth. Since that date—namely in 1853—it has been divided into two by the separation of the Upper Amazons as a distinct province. It formerly constituted a section, capitania, or governorship of the Portuguese colony. Originally it was well peopled by Indians, varying much in social condition according to their tribe, but all exhibiting the same general physical characters, which are those of the American red man, somewhat modified by long residence in an equatorial forest country. Most of the tribes are now extinct or forgotten, at least those which originally peopled the banks of the main river, their descendants having amalgamated with the white and negro immigrants :* many still exist, however, in their original state on the Upper Amazons and most of the branch rivers. On this account Indians in this province are far more numerous than elsewhere in Brazil, and the Indian element may be said to prevail in the mongrel population, the negro proportion being much smaller than in South Brazil.

The city is built on the best available site for a port of entry to the Amazons region, and must in time become a vast emporium ; for the northern shore of the main river, where alone a rival capital could be founded, is much more difficult of access to vessels,

* The mixed breeds which now form, probably, the greater part of the population, have each a distinguishing name. Mameluco denotes the offspring of White with Indian ; Mulatto, that of White with Negro ; Cafuzo, the mixture of the Indian and Negro ; Curiboco, the cross between the Cafuzo and the Indian ; Xibaro, that between the Cafuzo and Négro. These are seldom, however, well-demarcated, and all shades of colour exist ; the names are generally applied only approximatively. The term Creolo is confined to negroes born in the country. The civilised Indian is called Tapuyo or Caboclo.

and is besides extremely unhealthy. Although lying so near the equator (1° 28′ S. lat.) the climate is not excessively hot. The temperature during three years only once reached 95° of Fahrenheit. The greatest heat of the day, about 2 p.m., ranges generally between 89° and 94°; but on the other hand, the air is never cooler than 73°, so that a uniformly high temperature exists, and the mean of the year is 81°. North American residents say that the heat is not so oppressive as it is in summer in New York and Philadelphia. The humidity is, of course, excessive, but the rains are not so heavy and continuous in the wet season as in many other tropical climates. The country had for a long time a reputation for extreme salubrity. Since the small-pox in 1819, which attacked chiefly the Indians, no serious epidemic had visited the province. We were agreeably surprised to find no danger from exposure to the night air or residence in the low swampy lands. A few English residents, who had been established here for twenty or thirty years, looked almost as fresh in colour as if they had never left their native country. The native women, too, seemed to preserve their good looks and plump condition until late in life. I nowhere observed that early decay of appearance in Brazilian ladies, which is said to be so general in the women of North America. Up to 1848 the salubrity of Pará was quite remarkable for a city lying in the delta of a great river in the middle of the tropics and half surrounded by swamps. It did not much longer enjoy its immunity from epidemics. In 1850 the yellow fever visited the province for the first time, and carried off in a few weeks more than four per cent. of the population.* One disease after another succeeded, until in 1855 the cholera swept through the country and caused fearful havoc. Since then, the healthfulness of the climate has been gradually restored, and it is now fast recovering its former good reputation. Pará is free from serious endemic disorders, and was once a resort of invalids from New York and Massachusetts. The equable temperature, the perpetual verdure, the coolness of the dry season when the sun's heat is tempered by the strong sea-breezes and the moderation of the periodical rains, make the climate one of the most enjoyable on the face of the earth.

The province is governed, like all others in the empire, by a President, as chief civil authority. At the time of our arrival he held also, exceptionally, the chief military command. This functionary, together with the head of the police administration and the judges, is nominated by the central Government at Rio Janeiro. The municipal and internal affairs are managed by a provincial assembly elected by the people. Every villa or borough throughout the province also possesses its municipal council, and in thinly-populated districts, the inhabitants choose every four years a justice of the peace who adjudicates in small disputes between neighbours. A system of popular education exists, and every village has its school of first letters, the master being paid by the

* Relatorio of the President, Jeronymo Francisco Coelho, 1850. From January 1 to July 31, 1850, 12,000 persons, in the city of Pará alone, fell ill out of a population of 16,000, but only 506 died.

government, the salary amounting to about 70*l.*, or the same sum as the priests receive. Besides common schools a well-endowed classical seminary is maintained at Pará, to which the sons of most of the planters and traders in the interior are sent to complete their education. The province returns its quota of members every four years to the lower and upper houses of the imperial parliament. Every householder has a vote. Trial by jury has been established, the jurymen being selected from householders, no matter what their race or colour; and I have seen the white merchant, the negro husbandman, the mameluco, the mulatto and the Indian, all sitting side by side on the same bench. Altogether the constitution of government in Brazil seems to combine happily the principles of local self-government and centralisation, and only requires a proper degree of virtue and intelligence in the people to lead the nation to great prosperity.

The province of Pará, or, as we may now say, the two provinces of Pará and the Amazons contain an area of 800,000 square miles; the population of which is only about 230,000, or in the ratio of one person to four square miles. The country is covered with forests, and the soil fertile in the extreme even for a tropical country. It is intersected throughout by broad and deep navigable rivers. It is the pride of the Paraenses to call the Amazons the Mediterranean of South America. It perhaps deserves the name, for not only have the main river and its principal tributaries an immense expanse of water bathing the shores of extensive and varied regions, but there is also throughout a system of backchannels, connected with the main rivers by narrow outlets and linking together a series of lakes, some of which are fifteen, twenty, and thirty miles in length. The whole Amazons valley is thus covered by a network of navigable waters, forming a vast inland freshwater sea with endless ramifications rather than a river.

The city of Pará was founded in 1615, and was a place of considerable importance towards the latter half of the eighteenth century, under the government of the brother of Pombal, the famous Portuguese statesman. The province was the last in Brazil to declare its independence of the mother country and acknowledge the authority of the first emperor, Don Pedro. This was owing to the great numbers and influence of the Portuguese, and the rage of the native party was so great in consequence, that immediately after independence was proclaimed in 1823, a counter revolution broke out, during which many hundred lives were lost and much hatred engendered. The antagonism continued for many years, partial insurrections taking place when the populace thought that the immigrants from Portugal were favoured by the governors sent from the capital of the empire. At length, in 1835, a serious revolt took place which in a short time involved the entire province. It began by the assassination of the President and the leading members of the government; the struggle was severe, and the native party in an evil hour called to their aid the ignorant and fanatic mongrel and Indian population. The cry of death to the Portuguese was soon changed to death to the freemasons, then a powerfully-organised society embracing the greater part of the male white inhabitants. The victorious native party endeavoured to establish a government of their own. After this state of things had endured six months, they accepted a new President sent from Rio Janeiro, who, however, again irritated them by imprisoning their favourite leader, Vinagre. The revenge which followed was frightful. A vast host of half-savage coloured people assembled in the retired creeks behind Pará, and on a day fixed, after Vinagre's brother had sent a message three times to the President demanding, in vain, the release of their leader, the whole body poured into the city through the gloomy pathways of the forest which encircles it. A cruel battle, lasting nine days, was fought in the streets; an English, French, and Portuguese man-of-war, from the side of the river, assisting the legal authorities. All the latter, however, together with every friend of peace and order, were finally obliged to retire to an island a few miles distant. The city and province were given up to anarchy; the coloured people, elated with victory, proclaimed the slaughter of all whites, except the English, French, and American residents. The mistaken principals, who had first aroused all this hatred of races, were obliged now to make their escape. In the interior the supporters of lawful authority, including, it must be stated, whole tribes of friendly Indians and numbers of the better disposed negroes and mulattos, concentrated themselves in certain strong positions and defended themselves, until the reconquest of the capital and large towns of the interior, in 1836, by a force sent from Rio Janeiro, after ten months of anarchy.

Years of conciliatory government, the lesson learnt by the native party and the moderation of the Portuguese, aided by the natural indolence and passive goodness of the Paraenses of all classes and colours, were only beginning to produce their good effects about the time I am speaking of. Life, however, was now and had been for some time quite safe throughout the country. Some few of the worst characters had been transported or imprisoned, and the remainder after being pardoned were converted once more into quiet and peaceable citizens.

I resided at Pará nearly a year and a half altogether, returning thither and making a stay of a few months after each of my shorter excursions into the interior,* until the 6th of November, 1851, when I started on my long voyage to the Tapajos and the Upper Amazons, which occupied me seven years and a half. I became during this time tolerably familiar with the capital of the Amazons region, and its inhabitants. Compared with other Brazilian seaport towns, I was always told, Pará shone to great advantage. It was cleaner, the suburbs were fresher, more rural and much pleasanter on account of their verdure, shade, and magnificent vegetation. The people were simpler, more peaceable and friendly in their manners and dispositions, and assassinations, which give the southern provinces so ill a reputation, were almost unknown. At the same time

* The following were the excursions alluded to:—Aug. 26 to Sept. 30, 1848, I went to the Arroyos cataracts on the Tocantins. Dec. 8, 1848, to Feb. 11, 1849, I visited Caripí on the Bahia of Marajo. June 8 to July 21, 1849, I visited Cametá and the lower part of the Tocantins. Lastly, from Sept. 22, 1849, to April 19, 1851, I made a preliminary voyage to Obydos, the Rio Negro, and Ega.

the Pará people were much inferior to Southern Brazilians in energy and industry. Provisions and house rents being cheap and the wants of the people few—for they were content with food and lodging of a quality which would be spurned by paupers in England—they spent the greater part of their time in sensual indulgences and in amusements which the government and wealthier citizens provided for them gratis. The trade, wholesale and retail, was in the hands of the Portuguese, of whom there were about 2500 in the place. Many handicrafts were exercised by coloured people, mulattos, mamelucos, free negroes and Indians. The better sort of Brazilians dislike the petty details of shopkeeping, and if they cannot be wholesale merchants prefer the life of planters in the country however small may be the estate and the gains. The negroes constituted the class of field-labourers and porters; Indians were universally the watermen, and formed the crews of the numberless canoes of all sizes and shapes which traded between Pará and the interior. The educated Brazilians, not many of whom are of pure Caucasian descent—for the immigration of Portuguese, for many years, has been almost exclusively of the male sex—are courteous, lively, and intelligent people. They were gradually weaning themselves of the ignorant, bigoted notions which they inherited from their Portuguese ancestors, especially those entertained with regard to the treatment of women. Formerly the Portuguese would not allow their wives to go into society, or their daughters to learn reading and writing. In 1848, Brazilian ladies were only just beginning to emerge from this inferior position, and Brazilian fathers were opening their eyes to the advantages of education for their daughters. Reforms of this kind are slow. It is, perhaps, in part owing to the degrading position always held by women, that the relations between the sexes were and are still on so unsatisfactory a footing, and private morality at so low an ebb in Brazil. In Pará I believe that an improvement is now taking place, but formerly promiscuous intercourse seemed to be the general rule amongst all classes, and intrigues and love-making the serious business of the greater part of the population. That this state of things is a necessity depending on the climate and institutions I do not believe, as I have resided at small towns in the interior, where the habits, and the general standard of morality of the inhabitants, were as pure as they are in similar places in England.

CHAPTER II.

PARÁ—*continued.*

The swampy forests of Pará—A Portuguese landed proprietor—Country house at Nazareth—Life of a Naturalist under the equator—The drier virgin forests—Magoary—Retired creeks—Aborigines.

AFTER having resided about a fortnight at Mr. Miller's rocinha we heard of another similar country-house to be let, much better situated for our purpose, in the village of Nazareth, a mile and a half from the city and close to the forest. The owner was an old Portuguese gentleman named Danin, who lived at his tile manufactory at the mouth of the Una, a small river lying two miles below Pará. We resolved to walk to his place through the forest, a distance of three miles, although the road was said to be scarcely passable at this season of the year, and the Una much more easily accessible by boat. We were glad, however, of this early opportunity of traversing the rich swampy forest which we had admired so much from the deck of the ship; so, about eleven o'clock one sunny morning, after procuring the necessary information about the road, we set off in that direction. This part of the forest afterwards became one of my best hunting-grounds. I will narrate the incidents of the walk, giving my first impressions and some remarks on the wonderful vegetation. The forest is very similar on most of the low lands, and therefore one description will do for all.

On leaving the town we walked along a straight, suburban road constructed above the level of the surrounding land. It had low swampy ground on each side, built upon, however, and containing several spacious rocinhas which were embowered in magnificent foliage. Leaving the last of these, we arrived at a part where the lofty forest towered up like a wall five or six yards from the edge of the path to the height of, probably, 100 feet. The tree trunks were only seen partially here and there, nearly the whole frontage from ground to summit being covered with a diversified drapery of creeping plants, all of the most vivid shades of green; scarcely a flower to be seen, except in some places a solitary scarlet passion-flower set in the green mantle like a star. The low ground on the borders between the forest wall and the road, was encumbered with a tangled mass of bushy and shrubby vegetation, amongst which prickly mimosas were very numerous covering the other bushes in the same way as brambles do in England. Other dwarf mimosas trailed along the ground close to the edge of the road, shrinking at the slightest touch of the feet as we passed by. Cassia trees, with their elegant pinnate foliage and conspicuous yellow flowers, formed a great proportion of the lower trees, and arborescent arums grew in groups around the swampy hollows. Over the whole fluttered a larger number of brilliantly-coloured butterflies than we had yet seen; some wholly orange or yellow (Callidryas), others with excessively elongated wings, sailing horizontally through the air, coloured black, and varied with blue, red, and yellow (Heliconii). One magnificent grassy-green species (Colænis Dido) especially attracted our attention. Near the ground hovered many other smaller species very similar in appearance to those found at home, attracted by the flowers of numerous leguminous and other shrubs. Besides butterflies, there were few other insects except dragonflies, which were in great numbers, similar in shape to English species, but some of them looking conspicuously different on account of their fiery red colours.

After stopping repeatedly to examine and admire we at length walked onward. The road then ascended slightly, and the soil and vegetation became suddenly altered in character. The shrubs here were grasses, Cyperaceæ and other plants, smaller in foliage than those growing in moist grounds. The forest was second growth, low, consisting of trees which had the general aspect of laurels and other evergreens in our gardens at home: the leaves glossy and dark green. Some of them were elegantly veined and hairy (Melastomæ), whilst many, scattered amongst the rest, had smaller foliage (Myrtles), but these were not sufficient to subtract much from the general character of the whole.

The sun, now, for we had loitered long on the road, was exceedingly powerful. The day was most brilliant; the sky without a cloud. In fact, it was one of those glorious days which announce the commencement of the dry season. The radiation of heat from the sandy ground was visible by the quivering motion of the air above it. We saw or heard no mammals or birds; a few cattle belonging to an estate down a shady lane were congregated, panting, under a cluster of wide-spreading trees. The very soil was hot to our feet, and we hastened onward to the shade of the forest which we could see not far ahead. At length, on entering it, what a relief! We found ourselves in a moderately broad pathway or alley, where the branches of the trees crossed overhead and produced a delightful shade. The woods were at first of second growth, dense, and utterly impenetrable; the ground, instead of being clothed with grass and shrubs as in the woods of Europe, was everywhere carpeted with Lycopodiums (Selaginellæ). Gradually the scene became changed. We descended slightly from an elevated, dry, and sandy area to a low and swampy one; a cool air breathed on our faces, and a mouldy smell of rotting vegetation greeted us. The trees were now taller, the underwood less dense, and we could obtain glimpses into the wilderness on all sides. The leafy crowns of the trees, scarcely two of which could be seen together of the same kind, were now far away above us, in another world as it were. We could only see at times, where there was a break above, the tracery of the foliage against the clear blue sky. Sometimes the leaves were palmate, or of the shape of large outstretched hands; at others, finely cut or feathery like the leaves of Mimosæ. Below, the tree trunks were everywhere linked together by sipós; the woody, flexible stems of climbing and creeping trees, whose foliage is far away above, mingled with that of the taller independent trees. Some were twisted in strands like cables, others had thick stems contorted in every variety of shape, entwining snake-like round the tree trunks or forming gigantic loops and coils among the larger branches ; others, again, were of zigzag shape, or indented like the steps of a staircase, sweeping from the ground to a giddy height.

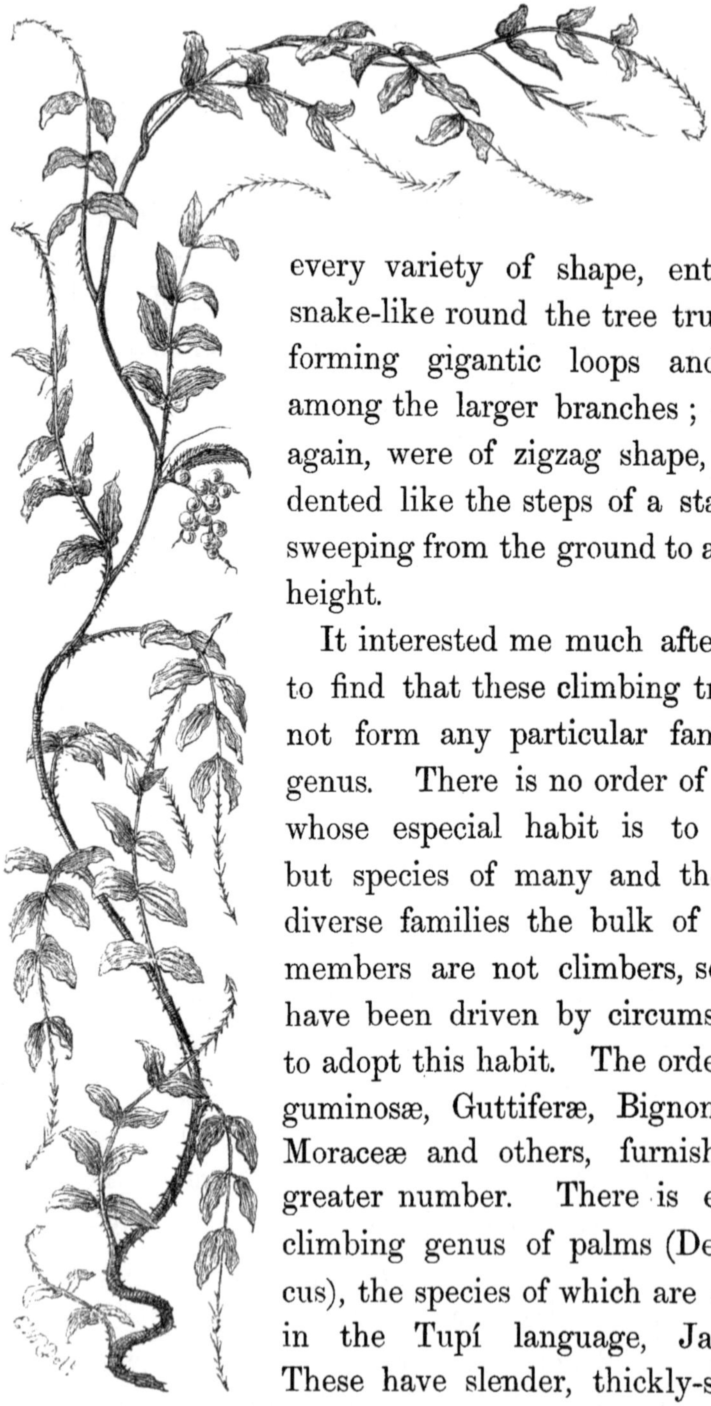

It interested me much afterwards to find that these climbing trees do not form any particular family or genus. There is no order of plants whose especial habit is to climb, but species of many and the most diverse families the bulk of whose members are not climbers, seem to have been driven by circumstances to adopt this habit. The orders Leguminosæ, Guttiferæ, Bignoniaceæ, Moraceæ and others, furnish the greater number. There is even a climbing genus of palms (Desmoncus), the species of which are called, in the Tupí language, Jacitára. These have slender, thickly-spined, and flexuous stems, which twine

Climbing Palm (Desmoncus).

about the taller trees from one to the other, and grow to an incredible length. The leaves, which have the ordinary pinnate shape characteristic of the family, are emitted from the stems at long intervals, instead of being collected into a dense crown, and have at their tips a number of long recurved spines. These structures are excellent contrivances to enable the trees to secure themselves by in climbing, but they are a great nuisance to the traveller, for they sometimes hang over the pathway and catch the hat or clothes, dragging off the one or tearing the other as he passes. The number and variety of climbing trees in the Amazons forests are interesting, taken in connection with the fact of the very general tendency of the animals, also, to become climbers.

All the Amazonian, and in fact all South American, monkeys are climbers. There is no group answering to the baboons of the Old World, which live on the ground. The Gallinaceous birds of the country, the representatives of the fowls and pheasants of Asia and Africa, are all adapted by the position of the toes to perch on trees, and it is only on trees, at a great height, that they are to be seen. A genus of Plantigrade Carnivora, allied to the bears (Cercoleptes), found only in the Amazonian forests, is entirely arboreal, and has a long flexible tail like that of certain monkeys. Many other similar instances could be enumerated, but I will mention only the Geodephaga, or carnivorous ground beetles, a great proportion of whose genera and species in these forest regions are, by the structure of their feet, fitted to live exclusively on the branches and leaves of trees.

Many of the woody lianas suspended from trees are not climbers but the air-roots of epiphytous plants (Aroideæ), which sit on the stronger boughs of the trees above, and hang down straight as plumb-lines. Some are suspended singly, others in leashes; some reach halfway to the ground and others touch it, striking their rootlets into the earth. The underwood in this part of the forest was composed partly of younger trees of the same species as their taller neighbours, and partly of palms of many species, some of them twenty to thirty feet in height, others small and delicate, with stems no thicker than a finger. These latter (different kinds of Bactris) bore small bunches of fruit, red or black, often containing a sweet grape-like juice.

Further on the ground became more swampy, and we had some difficulty in picking our way. The wild banana (Urania Amazonica) here began to appear, and, as it grew in masses, imparted a new aspect to the scene. The leaves of this beautiful plant are like broad sword-blades, eight feet in length and a foot broad; they rise straight upwards, alternately, from the top of a stem five or six feet high. Numerous kinds of plants with leaves similar in shape to these but smaller, clothed the ground. Amongst them were species of Marantaceæ, some of which had broad glossy leaves, with long leaf-stalks radiating from joints in a reed-like stem. The trunks of the trees were clothed with climbing ferns, and Pothos plants with large, fleshy, heart-shaped leaves. Bamboos and other tall grass and reed-like plants arched over the pathway. The appearance of this part of the forest was strange in the extreme; description can convey no adequate idea of it. The reader who has visited Kew may form some notion by conceiving a vegetation like that in the great palm-house spread over a large tract of swampy ground, but he must fancy it mingled with large exogenous trees similar to our oaks and elms covered with creepers and parasites, and figure to himself the ground encumbered with fallen and rotting trunks, branches, and leaves; the whole illuminated by a glowing vertical sun, and reeking with moisture.

In these swampy shades we were afraid at each step of treading on some venomous reptile. On this first visit, however, we saw none, although I afterwards found serpents common here. We perceived no signs of the larger animals and saw very few birds. Insects were more numerous, especially butterflies. The most conspicuous species was a large, glossy, blue and black Morpho (M. Achilles, of Linnæus), which measures six inches or more in expanse of wings. It came along the alley at a rapid rate and with an undulating flight, but diverged into the thicket before reaching the spot where we stood. Another was the very handsome Papilio Sesostris, velvety black in colour, with a large silky green patch on its wings. It is the male only which is so coloured; the female being plainer, and so utterly unlike its partner, that it was always held to be a different species until proved to be the same. Several other kinds allied to this inhabit almost exclusively these moist shades. In all of them the males are brilliantly coloured and widely different from the females. Such are P. Æneas, P. Vertumnus, and P.

E 2

Lysander, all velvety black, with patches of green and crimson on their wings. The females of these species do not court the company of the males, but are found slowly flying in places where the shade is less dense. In the moist parts great numbers of males are seen, often four species together, threading the mazes of the forest, and occasionally rising to settle on the scarlet flowers of climbers near the tops of the trees. Occasionally a stray one is seen in the localities which the females frequent. In the swampiest parts, we saw numbers of the Epicalia ancea, one of the most richly-coloured of the whole tribe of butterflies, being black, decorated with broad stripes of pale blue and orange. It delighted to settle on the broad leaves of the Uraniæ and similar plants where a ray of sunlight shone, but it was excessively wary, darting off with lightning speed when approached.

To obtain a fair notion of the number and variety of the animal tenants of these forests, it is necessary to follow up the research month after month and explore them in different directions and at all seasons. During several months I used to visit this district two or three days every week, and never failed to obtain some species new to me, of bird, reptile, or insect. It seemed to be an epitome of all that the humid portions of the Pará forests could produce. This endless diversity, the coolness of the air, the varied and strange forms of vegetation, the entire freedom from mosquitos and other pests, and even the solemn gloom and silence, combined to make my rambles through it always pleasant as well as profitable. Such places are paradises to a naturalist, and if he be of a contemplative turn there is no situation more favourable for his indulging the tendency. There is something in a tropical forest akin to the ocean in its effects on the mind. Man feels so completely his insignificance there, and the vastness of nature. A naturalist cannot help reflecting on the vegetable forces manifested on so grand a scale around him. A German traveller, Burmeister, has said that the contemplation of a Brazilian forest produced on him a painful impression, on account of the vegetation displaying a spirit of restless selfishness, eager emulation, and craftiness. He thought the softness, earnestness, and repose of European woodland scenery were far more pleasing, and that these formed one of the causes of the superior moral character of European nations.

In these tropical forests each plant and tree seems to be striving to outvie its fellow, struggling upwards towards light and air—branch, and leaf, and stem—regardless of its neighbours. Parasitic plants are seen fastening with firm grip on others, making use of them with reckless indifference as instruments for their own advancement. Live and let live is clearly not the maxim taught in these wildernesses. There is one kind of parasitic tree, very common near Pará, which exhibits this feature in a very prominent manner. It is called the Sipó Matador, or the Murderer Liana. It belongs to the fig order, and has been described and figured by Von Martius in the Atlas to Spix and Martius's Travels. I observed many specimens. The base of its stem would be unable to bear the weight of the upper growth; it is obliged, therefore, to support itself on a tree of another species. In this it is not essentially different from other climbing trees and plants, but the way the matador sets about it is peculiar, and produces certainly a disagreeable impression. It springs up close to the tree on which it intends to fix itself, and the wood of its stem grows by spreading itself like a plastic mould over one side of the trunk of its supporter. It then puts forth, from each side, an arm-like branch, which grows rapidly, and looks as though a stream of sap were flowing and hardening as it went. This adheres closely to the trunk of the victim and the two arms meet on the opposite side and blend together. These arms are put forth at somewhat regular intervals in mounting upwards, and the victim, when its strangler is full-grown, becomes tightly clapsed by a number of inflexible rings. These rings gradually grow larger as the Murderer flourishes, rearing its crown of foliage to the sky mingled with that of its neighbour, and in course of time they kill it by stopping the flow of its sap. The strange spectacle then remains of the selfish parasite clasping in its arms the lifeless and decaying body of its victim, which had been a help to its own growth. Its ends have been served—it has flowered and fruited, reproduced and disseminated its kind; and now, when the dead trunk moulders away, its own end approaches; its support is gone, and itself also falls.

The Murderer Sipó merely exhibits, in a more conspicuous manner than usual, the struggle which necessarily exists amongst vegetable forms in these crowded forests, where individual is competing with individual and species with species, all striving to reach light and air in order to unfold their leaves and perfect their organs of fructification. All species entail in their successful struggles the injury or destruction of many of their neighbours or supporters, but the process is not in others so speaking to the eye as it is in the case of the Matador. The efforts to spread their roots are as strenuous in some plants and trees, as the struggle to mount upwards is in others. From these apparent strivings result the buttressed stems, the dangling air roots, and other similar phenomena. The competition amongst organised beings has been prominently brought forth in Darwin's "Origin of Species;" it is a fact which must be always kept in view in studying these subjects. It exists everywhere, in every zone, in both the animal and vegetable kingdoms. It is doubtless most severe, on the whole, in tropical countries, but its display in vegetable forms in the forest is no exceptional phenomenon. It is only more conspicuously exhibited, owing perhaps to its affecting principally the vegetative organs—root, stem, and leaf—whose growth is also stimulated by the intense light, the warmth, and the humidity. The competition exists also in temperate countries, but it is there concealed under the external appearance of repose which vegetation wears. It affects, in this case, perhaps more the reproductive than the vegetative organs, especially the flowers, which it is probable are far more general decorations in the woodlands of high latitudes than in tropical forests. This, however, is a difficult subject, and one which requires much further investigation.

I think there is plenty, in tropical nature, to counteract any unpleasant impression which the reckless energy of the vegetation might produce. There is the incomparable beauty and variety of the foliage, the vivid colours, the richness and exuberance everywhere displayed, which make, in my opinion, the richest woodland scenery in Northern Europe a sterile desert in comparison. But it is especially the enjoyment of life manifested by individual existences which compensates for the destruction and pain caused by the inevitable competition. Although this competition is nowhere more active, and the dangers to which each individual is exposed nowhere more numerous, yet nowhere is this enjoyment more vividly displayed. If vegetation had feeling, its vigorous and rapid growth, uninterrupted by the cold sleep of winter, would, one would think, be productive of pleasure to its individuals. In animals, the mutual competition may be greater, the predacious species more constantly on the alert, than in temperate climates; but there is at the same time no severe periodical struggle with inclement seasons. In sunny nooks, and at certain seasons, the trees and the air are gay with birds and insects, all in the full enjoyment of existence; the warmth, the sunlight, and the abundance of food producing their results in the animation and sportiveness of the beings congregated together. We ought not to leave out of sight, too, the sexual decorations—the brilliant colours and ornamentation of the males, which, although existing in the fauna of all climates, reach a higher degree of perfection in the tropics than elsewhere. This seems to point to the pleasures of the pairing seasons. I think it is a childish notion that the beauty of birds, insects, and other creatures is given to please the human eye. A little observation and reflection show that this cannot be the case, else why should one sex only be richly ornamented, the other clad in plain drab and gray? Surely, rich plumage and song, like all the other endowments of species, are given them for their own pleasure and advantage. This, if true, ought to enlarge our ideas of the inner life and mutual relations of our humbler fellow creatures!

We at length emerged from the forest, on the banks of the Una, near its mouth. It was here about one hundred yards wide. The residence of Senhor Danin stood on the opposite shore; a large building, whitewashed and red-tiled as usual, raised on wooden piles above the humid ground. The second story was the part occupied by the family, and along it was an open verandah where people, male and female, were at work. Below were several negroes employed carrying clay on their heads. We called out for a boat, and one of them crossed over to fetch us. Senhor Danin received us with the usual formal politeness of the Portuguese; he spoke English very well, and after we had arranged our business we remained conversing with him on various topics connected with the country. Like all employers in this province he was full of one topic—the scarcity of hands. It appeared that he had made great exertions to introduce white labour but had failed, after having brought numbers of men from Portugal and other countries under engagement to work for him. They all left him one by one soon after their arrival. The abundance of unoccupied land, the liberty that exists, a state of things produced by the half-wild canoe-life of the people, and the ease with which a mere subsistence can be obtained with moderate work, tempt even the best-disposed to quit regular labour as soon as they can. He complained also of the dearness of slaves, owing to the prohibition of the African traffic, telling us that formerly a slave could be bought for 120 dollars, whereas they are now difficult to procure at 400 dollars.

Mr. Danin told us that he had travelled in England and the United States, and that he had now two sons completing their education in those countries. I afterwards met with many enterprising persons of Mr. Danin's order, both Brazilians and Portuguese; their great ambition is to make a voyage to Europe or North America, and to send their sons to be educated there. The land on which his establishment is built, he told us, was an artificial embankment on the swamp; the end of the house was built on a projecting point overlooking the river, so that a good view was obtained, from the sitting rooms, of the city and the shipping. We learnt there was formerly a large and flourishing cattle estate on this spot, with an open grassy space like a park. On Sundays gay parties of 40 or 50 persons used to come by land and water, in carriages and gay galliotas, to spend the day with the hospitable owner. Since the political disorders which I have already mentioned, decay had come upon this as on most other large establishments in the country. The cultivated grounds, and the roads leading to them, were now entirely overgrown with dense forest. When we were ready to depart, Senhor Danin lent a canoe and two negroes to take us to the city, where we arrived in the evening after a day rich in new experiences.

Shortly afterwards we took possession of our new residence. The house was a square building, consisting of four equal-sized rooms; the tiled roof projected all round, so as to form a broad verandah, cool and pleasant to sit and work in. The cultivated ground, which appeared as if newly cleared from the forest, was planted with fruit trees and small plots of coffee and mandioca. The entrance to the grounds was by an iron-grille gateway from a grassy square, around which were built the few houses and palm-thatched huts which then constituted the village. The most important building was the chapel of our Lady of Nazareth, which stood opposite our place. The saint here enshrined was a great favourite with all orthodox Paraenses, who attributed to her the performance of many miracles. The image was to be seen on the altar, a handsome doll about four feet high, wearing a silver crown and a garment of blue silk, studded with golden stars. In and about the chapel were the offerings that had been made to her, proofs of the miracles which she had performed. There were models of legs, arms, breasts, and so forth, which she had cured. But most curious of all was a ship's boat, deposited here by the crew of a Portuguese vessel which had foundered, a year or two before our arrival, in a squall off Cayenne; part of them having been saved in the boat, after invoking the protection of the saint here enshrined. The annual festival in honour of our Lady of Nazareth is the greatest of the Pará holidays; many persons come to it from the neighbouring city of Maranham, 300 miles distant. Once the president ordered the mail steamer to be delayed two days at Pará for the convenience of these visitors. The popularity of the festa is partly owing to the beautiful weather that prevails when it takes place, namely, in the middle of the fine season, on the ten days preceding the full moon in October or November. Pará is then seen at its best. The weather is not too dry, for three weeks never follow in succession without a shower; so that all the glory of verdure and flowers can be enjoyed with clear skies. The moonlit nights are then especially beautiful; the atmosphere is transparently clear, and the light sea-breeze produces an agreeable coolness.

We now settled ourselves for a few months' regular work. We had the forest on three sides of us; it was the end of the wet season; most species of birds had finished moulting, and every day the insects increased in number and variety. Behind the rocinha, after several days' exploration, I found a series of pathways through the woods, which led to the Una road; about half way was the house in which the celebrated travellers Spix and Martius resided during their stay at Pará, in 1819. It was now in a neglected condition, and the plantations were overgrown with bushes. The paths hereabout were very productive of insects, and being entirely under shade were very pleasant for strolling. Close to our doors began the main forest road. It was broad enough for two horsemen abreast, and branched off in three directions; the main line going to the village of Ourem, a distance of 50 miles. This road formerly extended to Maranham, but it had been long in disuse and was now grown up, being scarcely passable between Pará and Ourem.

Our researches were made in various directions along these paths, and every day produced us a number of new and interesting species. Collecting, preparing our specimens, and making notes, kept us well occupied. One day was so much like another, that a general description of the diurnal round of incidents, including the sequence of natural phenomena, will be sufficient to give an idea of how days pass to naturalists under the equator.

We used to rise soon after dawn, when Isidoro would go down to the city, after supplying us with a cup of coffee, to purchase the fresh provisions for the day. The two hours before breakfast were devoted to ornithology. At that early period of the day the sky was invariably cloudless (the thermometer marking 72° or 73° Fahr.); the heavy dew or the previous night's rain, which lay on the moist foliage, becoming quickly dissipated by the glowing sun, which rising straight out of the east, mounted rapidly towards the zenith. All nature was fresh, new leaf and flower-buds expanding rapidly. Some mornings a single tree would appear in flower amidst what was the preceding evening a uniform green mass of forest—a dome of blossom suddenly created as if by magic. The birds were all active; from the wild-fruit trees, not far off, we often heard the shrill yelping of the Toucans (Rhamphastos vitellinus). Small flocks of parrots flew over on most mornings, at a great height, appearing in distinct relief against the blue sky, always two by two chattering to each other, the pairs being separated by regular intervals; their bright colours, however, were not apparent at that height. After breakfast we devoted the hours from 10 a.m. to 2 or 3 p.m. to entomology; the best time for insects in the forest being a little before the greatest heat of the day. We did not find them at all numerous, although of great variety as to species. The only kinds that appeared in great numbers of individuals were ants, termites, and certain species of social wasps; in the open grounds dragon-flies were also amongst the most abundant kinds of insects. Beetles were certainly much lower in the proportion of individuals to species than they are in England, and this led us to the conclusion that the ants and termites here must perform many of the functions in nature which in temperate climates are the office of Coleoptera. As to butterflies, I extract the following note from many similar ones in my journal. "On Tuesday, collected 46 specimens, of 39 species. On Wednesday, 37 specimens, of 33 species, 27 of which are different from those taken on the preceding day." The number of specimens would be increased if I had reckoned all the commonest species seen, but still the fact is well established, that there is a great paucity of individuals compared with species in both Lepidoptera and Coleoptera. We rarely saw caterpillars. After several years' observation, I came to the conclusion that the increase of these creatures was checked by the close persecution of insectivorous animals, which are excessively numerous in this country. The check operates at all periods of life—on the eggs, the larvæ, and the perfect insects.

The heat increased rapidly towards two o'clock (92° and 93° Fahr.), by which time every voice of bird or mammal was hushed; only in the trees was heard at intervals the harsh whirr of a cicada. The leaves, which were so moist and fresh in early morning, now became lax and drooping; the flowers shed their petals. Our neighbours the Indian and Mulatto inhabitants of the open palm-thatched huts, as we returned home fatigued with our ramble, were either asleep in their hammocks or seated on mats in the shade, too languid even to talk. On most days in June and July a heavy shower would fall some time in the afternoon, producing a most welcome coolness. The approach of the rain-clouds was after a uniform fashion very interesting to observe. First, the cool sea-breeze, which commenced to blow about 10 o'clock, and which had increased in force with the increasing power of the sun, would flag and finally die away. The heat and electric tension of the atmosphere would then become almost insupportable. Languor and uneasiness would seize on every one; even the denizens of the forest betraying it by their motions. White clouds would appear in the east and gather into cumuli, with an increasing blackness along their lower portions. The whole eastern horizon would become almost suddenly black, and this would spread upwards, the sun at length becoming obscured. Then the rush of a mighty wind is heard through the forest, swaying the tree-tops; a vivid flash of lightning bursts forth, then a crash of thunder, and down streams the deluging rain. Such storms soon cease, leaving bluish-black motionless clouds in the sky until night. Meantime all nature is refreshed; but heaps of flower-petals and fallen leaves are seen under the trees. Towards evening life revives again, and the ringing uproar is resumed from bush and tree. The following morning the sun again rises in a cloudless sky, and so the cycle is completed; spring, summer, and autumn, as it were, in one tropical day. The days are more or less like this throughout the year in this country. A little difference exists between the dry and wet seasons; but generally, the dry season, which lasts from July to December, is varied with showers, and the wet, from January to June, with sunny days. It results from this, that the periodical phenomena of plants and animals do not take place at about the same time in all species, or in the individuals of any given species, as they do in temperate countries. Of course there is no hybernation; nor, as the dry season is not excessive, is there any æstivation as in some tropical countries. Plants do not flower or shed their leaves, nor do birds moult, pair, or breed simultaneously. In Europe, a woodland scene has its spring, its summer, its autumnal, and its winter aspects. In the equatorial forests the aspect is the same or nearly so every day in the year: budding, flowering, fruiting, and leaf shedding are always going on in one species or other. The activity of birds and insects proceeds without interruption, each species having its own separate times; the colonies of wasps, for instance, do not die off annually, leaving only the queens, as in cold climates; but the succession of generations and colonies goes on incessantly. It is never either spring, summer, or autumn, but each day is a combination of all three. With the day and night always of equal length, the atmospheric disturbances of each day neutralising themselves before each succeeding morn; with the sun in its course proceeding mid-way across the sky and the daily temperature the same within two or three degrees throughout the year—how grand in its perfect equilibrium and simplicity is the march of Nature under the equator!

Our evenings were generally fully employed preserving our collections, and making notes. We dined at four, and took tea about seven o'clock. Sometimes we walked to the city to see Brazilian life or enjoy the pleasures of European and American society. And so the time passed away from June 15th to August 26th. During this period we made two excursions of greater length to the rice and saw-mills of Magoary, an establishment owned by an American gentleman, Mr. Upton, situated on the banks of a creek in the heart of the forest, about 12 miles from Pará. I will narrate some of the incidents of these excursions, and give an account of the more interesting observations made on the Natural History and inhabitants of these interior creeks and forests.

Our first trip to the mills was by land. The creek on whose banks they stand, the Iritirí, communicates with the river Pará, through another larger creek, the Magoary; so that there is a passage by water; but this is about 20 miles round. We started at sunrise, taking Isidoro with us. The road plunged at once into the forest after leaving Nazareth, so that in a few minutes we were enveloped in shade. For some distance the woods were of second growth, the original forest near the town having been formerly cleared or thinned. They were dense and impenetrable on account of the close growth of the young trees and the mass of thorny shrubs and creepers. These thickets swarmed with ants and ant-thrushes ; they were also frequented by a species of puff-throated manikin, a little bird which flies occasionally across the road, emitting a strange noise, made, I believe, with its wings, and resembling the clatter of a small wooden rattle.

A mile or a mile and a half further on, the character of the woods began to change, and we then found ourselves in the primæval forest. The appearance was greatly different from that of the swampy tract I have already described. The land was rather more elevated and undulating; the many swamp plants with their long and broad leaves were wanting, and there was less underwood, although the trees were wider apart. Through this wilderness the road continued for seven or eight miles. The same unbroken forest extends all the way to Maranham and in other directions, as we were told, a distance of about 300 miles southward and eastward of Pará. In almost every hollow part the road was crossed by a brook, whose cold, dark, leaf-stained waters were bridged over by tree trunks. The ground was carpeted, as usual, by Lycopodiums, but it was also encumbered with masses of vegetable *débris* and a thick coating of dead leaves. Fruits of many kinds were scattered about, amongst which were many sorts of beans, some of the pods a foot long, flat and leathery in texture, others hard as stone. In one place there was a quantity of large empty wooden vessels, which Isidoro told us fell from the Sapucaya tree. They are called Monkey's drinking-cups (Cuyas de Macaco), and are the capsules which contain the nuts sold under the name just mentioned, in Covent Garden Market. At the top of the vessel is a circular hole, in which a natural lid fits neatly. When the nuts are ripe this lid becomes loosened, and the heavy cup falls with a crash, scattering the nuts over the ground. The tree which yields the nut (Lecythis ollaria), is of immense height. It is closely allied to the Brazil-nut tree (Bertholletia excelsa), whose seeds are also enclosed in large woody vessels; but these have no lid, and fall entire to the ground. This is the reason why the one kind of nut is so much dearer than the other. The Sapucaya is not less abundant, probably, than the Bertholletia, but its nuts in falling are scattered about and eaten by wild animals; whilst the full capsules of Brazil-nuts are collected entire by the natives.

What attracted us chiefly were the colossal trees. The general run of trees had not remarkably thick stems; the great and uniform height to which they grow without emitting a branch, was a much more noticeable feature than their thickness; but at intervals of a furlong or so a veritable giant towered up. Only one of these monstrous trees can grow within a given space; it monopolises the domain, and none but individuals of much inferior size can find a footing near it. The cylindrical trunks of these larger trees were generally about 20 to 25 feet in circumference. Von Martius mentions having measured trees in the Pará district belonging to various species (Symphonia coccinea, Lecythis sp. and Cratæva Tapia), which were 50 to 60 feet in girth at the point where they become cylindrical. The height of the vast column-like stems could not be less than 100 feet from the ground to their lowest branch. Mr. Leavens, at the saw-mills, told me they frequently squared logs for sawing 100 feet long, of the Pao d'Arco and the Massaranduba. The total height of these trees, stem and crown together, may be estimated at from 180 to 200 feet: where one of them stands, the vast dome of foliage rises above the other forest trees as a domed cathedral does above the other buildings in a city.

A very remarkable feature in these trees is the growth of buttress-shaped projections around the lower part of their stems. The spaces between these buttresses, which are generally thin walls of wood, form spacious chambers, and may be compared to stalls in a stable: some of them are large enough to hold half-a-dozen persons. The purpose of these structures is as obvious, at the first glance, as that of the similar props of brickwork which support a high wall. They are not peculiar to one species, but are common to most of the larger forest trees. Their nature and manner of growth are explained when a series of young trees of different ages is examined. It is then seen that they are the roots which have raised themselves ridge-like out of the earth; growing gradually upwards as the increasing height of the tree required augmented support. Thus they are plainly intended to sustain the massive crown and trunk in these crowded forests, where lateral growth of the roots in the earth is rendered difficult by the multitude of competitors.

The other grand forest trees whose native names we learnt, were the Moira-tinga (the White or King-tree), probably the same as, or allied to, the Mora excelsa, which Sir Robert Schomburgk discovered in British Guiana; the Samaüma (Eriodendron Samauma) and the Massaranduba, or Cow-tree. The last-mentioned is the most remarkable. We had already heard a good deal about this tree, and about its producing from its bark a copious supply of milk as pleasant to drink as that of the cow. We had also eaten its fruit in Pará, where it is sold in the streets by negro market women; and had heard a good deal of the durableness in water of its timber. We were glad, therefore, to see this wonderful tree growing in its native wilds. It is one of the largest of the forest monarchs, and is peculiar in appearance on account of its deeply-scored reddish and ragged bark. A decoction of the bark, I was told, is used as a red dye for cloth. A few days afterwards we tasted its milk, which was drawn from dry logs that had been standing many days in the hot sun, at the saw-mills. It was pleasant with coffee, but had a slight rankness when drunk pure; it soon thickens to a glue, which is excessively tenacious, and is often used to cement broken crockery. I was told that it was not safe to drink much of it, for a slave had recently nearly lost his life through taking it too freely.

In some parts of the road ferns were conspicuous objects. But I afterwards found them much more numerous on the Maranham-road, especially in one place where the whole forest glade formed a vast fernery; the ground was covered with terrestrial species, and the tree trunks clothed with climbing and epiphytous kinds. I saw no tree ferns in the Pará district; they belong to hilly regions; some occur, however, on the Upper Amazons.

Such were the principal features in the vegetation of the wilderness; but where were the flowers? To our great disappointment we saw none, or only such as were insignificant in appearance. Orchids are very rare in the dense forests of the low lands. I believe it is now tolerably well ascertained that the majority of forest trees in equatorial Brazil have small and inconspicuous flowers. Flower-frequenting insects are also rare in the forest. Of course they would not be found where their favourite food was wanting, but I always noticed that even where flowers occurred in the forest, few or no insects were seen upon them. In the open country or campos of Santarem on the Lower Amazons, flowering trees and bushes are more abundant, and there a large number of floral insects are attracted. The forest bees of South America belonging to the genera Melipona and Euglossa are more frequently seen feeding on the sweet sap which exudes from the trees, or on the excrement of birds on leaves, than on flowers.

We were disappointed also in not meeting with any of the larger animals in the forest. There was no tumultuous movement, or sound of life. We did not see or hear monkeys, and no tapir or jaguar crossed our path. Birds, also, appeared to be exceedingly scarce. We heard, however, occasionally, the long-drawn, wailing note of the Inambú, a kind of partridge (Crypturus cinereus ?) ; and, also, in the hollows on the banks of the rivulets, the noisy notes of another bird, which seemed to go in pairs, amongst the tree-tops, calling to each other as they went. These notes resounded through the solitude. Another solitary bird had a most sweet and melancholy song ; it consisted simply of a few notes, uttered in a plaintive key, commencing high, and descending by harmonic intervals. It was probably a species of warbler of the genus Trichas. All these notes of birds are very striking and characteristic of the forest.

I afterwards saw reason to modify my opinion, founded on these first impressions, with regard to the amount and variety of animal life in this and other parts of the Amazonian forests. There is, in fact, a great variety of mammals, birds, and reptiles, but they are widely scattered, and all excessively shy of man. The region is so extensive, and uniform in the forest clothing of its surface, that it is only at long intervals that animals are seen in abundance when some particular spot is found which is more attractive than others. Brazil, moreover, is throughout poor in terrestrial mammals, and the species are of small size ; they do not, therefore, form a conspicuous feature in its forests. The huntsman would be disappointed who expected to find here flocks of animals similar to the buffalo herds of North America, or the swarms of antelopes and herds of ponderous pachyderms of Southern Africa. The largest and most interesting portion of the Brazilian mammal fauna is arboreal in its habits ; this feature of the animal denizens of these forests I have already alluded to. The most *intensely* arboreal animals in the world are the South American monkeys of the family Cebidæ, many of which have a fifth hand for climbing in their prehensile tails, adapted for this function by their strong muscular development, and the naked palms under their tips. This seems to teach us that the South American fauna has been slowly adapted to a forest life, and, therefore, that extensive forests must have always existed since the region was first peopled by mammalia. But to this subject, and to the natural history of the monkeys, of which thirty-eight species inhabit the Amazon region, I shall have to return.

We often read, in books of travels, of the silence and gloom of the Brazilian forests. They are realities, and the impression deepens on a longer acquaintance. The few sounds of birds are of that pensive or mysterious character which intensifies the feeling of solitude rather than imparts a sense of life and cheerfulness. Sometimes, in the midst of the stillness, a sudden yell or scream will startle one ; this comes from some defenceless fruit-eating animal, which is pounced upon by a tiger-cat or stealthy boa-constrictor. Morning and evening the howling monkeys make a most fearful and harrowing noise, under which it is difficult to keep up one's buoyancy of spirit. The feeling of inhospitable wildness which the forest is calculated to inspire, is increased tenfold under this fearful uproar. Often, even in the still hours of midday, a sudden crash will be heard resounding afar through the wilderness, as some great bough or entire tree falls to the ground. There

INTERIOR OF PRIMÆVAL FOREST ON THE AMAZONS.

Vol. I., page 72.

are, besides, many sounds which it is impossible to account for. I found the natives generally as much at a loss in this respect as myself. Sometimes a sound is heard like the clang of an iron bar against a hard, hollow tree, or a piercing cry rends the air ; these are not repeated, and the succeeding silence tends to heighten the unpleasant impression which they make on the mind. With the natives it is always the Curupíra, the wild man or spirit of the forest, which produces all noises they are unable to explain. Myths are the rude theories which mankind, in the infancy of knowledge, invent to explain natural phenomena. The Curupíra is a mysterious being, whose attributes are uncertain, for they vary according to locality. Sometimes he is described as a kind of orang-otang, being covered with long, shaggy hair, and living in trees. At others he is said to have cloven feet, and a bright red face. He has a wife and children, and sometimes comes down to the roças to steal the mandioca. At one time I had a Mameluco youth in my service, whose head was full of the legends and superstitions of the country. He always went with me into the forest ; in fact, I could not get him to go alone, and whenever we heard any of the strange noises mentioned above, he used to tremble with fear. He would crouch down behind me, and beg of me to turn back. He became easy only after he had made a charm to protect us from the Curupíra. For this purpose he took a young palm leaf, plaited it, and formed it into a ring, which he hung to a branch on our track.

At length, after a six hours' walk, we arrived at our

destination, the last mile or two having been again through second-growth forest. The mills formed a large pile of buildings, pleasantly situated in a cleared tract of land, many acres in extent, and everywhere surrounded by the perpetual forest. We were received in the kindest manner by the overseer, Mr. Leavens, who showed us all that was interesting about the place, and took us to the best spots in the neighbourhood for birds and insects. The mills were built a long time ago by a wealthy Brazilian. They had belonged to Mr. Upton for many years. I was told that when the dark-skinned revolutionists were preparing for their attack on Pará, they occupied the place, but not the slightest injury was done to the machinery or building, for the leaders said it was against the Portuguese and their party that they were at war, not against the other foreigners.

The creek Iritirí at the mills is only a few yards wide ; it winds about between two lofty walls of forest for some distance, then becomes much broader, and finally joins the Magoary. There are many other ramifications, creeks or channels, which lead to retired hamlets and scattered houses, inhabited by people of mixed white, Indian, and negro descent. Many of them did business with Mr. Leavens, bringing for sale their little harvests of rice, or a few logs of timber. It was interesting to see them in their little, heavily-laden montarias. Sometimes the boats were managed by handsome, healthy young lads, loosely clad in straw hat, white shirt, and dark blue trousers, turned up to the knee. They steered, paddled, and managed the varejaō (the boating pole), with much grace and dexterity.

We made many excursions down the Iritirí, and saw much of these creeks; besides, our second visit to the mills was by water. The Magoary is a magnificent channel; the different branches form quite a labyrinth, and the land is everywhere of little elevation. All these smaller rivers, throughout the Pará estuary, are of the nature of creeks. The land is so level, that the short local rivers have no sources and downward currents like rivers as we generally understand them. They serve the purpose of draining the land, but instead of having a constant current one way, they have a regular ebb and flow with the tide. The natives call them, in the Tupí language, Igarapés, or canoe-paths. The igarapés and furos or channels, which are infinite in number in this great river delta, are characteristic of the country. The land is everywhere covered with impenetrable forests; the houses and villages are all on the waterside, and nearly all communication is by water. This semi-aquatic life of the people is one of the most interesting features of the country. For short excursions, and for fishing in still waters, a small boat, called montaria, is universally used. It is made of five planks; a broad one for the bottom, bent into the proper shape by the action of heat, two narrow ones for the sides, and two small triangular pieces for stem and stern. It has no rudder; the paddle serves for both steering and propelling. The montaria takes here the place of the horse, mule, or camel of other regions. Besides one or more montarias, almost every family has a larger canoe, called Igaritè. This is fitted with two masts, a rudder, and keel, and has an arched awning or cabin near the stern, made of a framework of tough lianas, thatched with palm leaves. In the igarité they will cross stormy rivers fifteen or twenty miles broad. The natives are all boat-builders. It is often remarked, by white residents, that an Indian is a carpenter and shipwright by intuition. It is astonishing to see in what crazy vessels these people will risk themselves. I have seen Indians cross rivers in a leaky montaria, when it required the nicest equilibrium to keep the leak just above water; a movement of a hair's breadth would send all to the bottom, but they managed to cross in safety. They are especially careful when they have strangers under their charge, and it is the custom of Brazilian and Portuguese travellers to leave the whole management to them. When they are alone they are more reckless, and often have to swim for their lives. When a squall overtakes them as they are crossing in a heavily-laden canoe, they all jump overboard and swim about until the heavy sea subsides, when they re-embark.

A few words on the aboriginal population of the Pará estuary will here not be out of place. The banks of the Pará were originally inhabited by a number of distinct tribes, who, in their habits, resembled very much the natives of the sea-coast from Maranham to Bahia. It is related that one large tribe, the Tupinambas, migrated from Pernambuco to the Amazons. One fact seems to be well-established, namely, that all the coast tribes were far more advanced in civilisation, and milder in their manners, than the savages who inhabited the interior lands of Brazil. They were settled in villages, and addicted to agriculture. They navigated the rivers in large canoes, called ubás, made of immense hollowed-out tree trunks; in these they used to go on war expeditions, carrying in the prows their trophies and calabash rattles, whose clatter was meant to intimidate their enemies. They were gentle in disposition, and received the early Portuguese settlers with great friendliness. The inland savages, on the other hand, led a wandering life, as they do at the present time, only coming down occasionally to rob the plantations of the coast tribes, who always entertained the greatest enmity towards them.

The original Indian tribes of the district are now either civilised, or have amalgamated with the white and negro immigrants. Their distinguishing tribal names have long been forgotten, and the race bears now the general appellation of Tapuyo, which seems to have been one of the names of the ancient Tupinambas. The Indians of the interior, still remaining in the savage state, are called by the Brazilians Indios, or Gentios (Heathens). All the semi-civilised Tapuyos of the villages, and in fact the inhabitants of retired places generally, speak the Lingoa geral, a language adapted by the Jesuit missionaries from the original idiom of the Tupinambas. The language of the Guaranis, a nation living on the banks of the Paraguay, is a dialect of it, and hence it is called by philologists the Tupi-Guarani language; printed grammars of it are always on sale at the shops of the Pará booksellers. The fact of one language having been spoken over so wide an extent of country as that from the Amazons to Paraguay, is quite an isolated one in this country, and points to considerable migrations of the Indian tribes in former times. At present the languages spoken by neighbouring tribes on the banks of the interior rivers are totally distinct; on the Juruá, even scattered hordes belonging to the same tribe are not able to understand each other.

The civilised Tapuyo of Pará, differs in no essential point, in physical or moral qualities, from the Indian of the interior. He is more stoutly built, being better fed than some of them; but in this respect there are great differences amongst the tribes themselves. He presents all the chief characteristics of the American red man. The skin of a coppery brown colour, the features of the face broad, and the hair black, thick, and straight. He is generally about the middle height, thick-set, has a broad muscular chest, well-shaped but somewhat thick legs and arms, and small hands and feet. The cheek bones are not generally prominent; the eyes are black, and seldom oblique like those of the Tartar races of Eastern Asia, which are supposed to have sprung from the same original stock as the American red man. The features exhibit scarcely any mobility of expression; this is connected with the excessively apathetic and undemonstrative character of the race. They never betray, in fact they do not feel keenly, the emotions of joy, grief, wonder, fear, and so forth. They can never be excited to enthusiasm; but they have strong affections, especially those connected with family. It is commonly stated by the whites and negroes that the Tapuyo is ungrateful. Brazilian mistresses of households, who have much experience of Indians, have always a long list of instances to relate to the stranger, showing their base ingratitude. They certainly do not appear to remember or think of repaying benefits, but this is probably because they did not require, and do not value such benefits as their would-be masters confer upon them. I have known instances of attachment and fidelity on the part of Indians towards their masters, but these are exceptional cases. All the actions of the Indian show that his ruling desire is to be let alone ; he is attached to his home, his quiet monotonous forest and river life ; he likes to go to towns occasionally, to see the wonders introduced by the white man, but he has a great repugnance to living in the midst of the crowd ; he prefers handicraft to field labour, and especially dislikes binding himself to regular labour for hire. He is shy and uneasy before strangers, but if they visit his abode, he treats them well, for he has a rooted appreciation of the duty of hospitality ; there is a pride about him, and being naturally formal and polite, he acts the host with great dignity. He withdraws from towns as soon as the stir of civilisation begins to make itself felt. When we first arrived at Pará many Indian families resided there, for the mode of living at that time was more like that of a large village than a city ; as soon as river steamers and more business activity were introduced, they all gradually took themselves away.

These characteristics of the Pará Indians are applicable, of course, to some extent, to the Mamelucos, which now constitute a great proportion of the population. The inflexibility of character of the Indian, and his total inability to accommodate himself to new arrangements, will infallibly lead to his extinction, as immigrants, endowed with more supple organisations, increase, and civilisation advances in the Amazon region. But, as the different races amalgamate readily, and the offspring of white and Indian often become distinguished Brazilian citizens, there is little reason to regret the fate of the race. Formerly the Indian was harshly treated, and even now he is so in many parts of the interior. But, according to the laws of Brazil, he is a free citizen, having equal privileges with the whites; and there are very strong enactments providing against the enslaving and ill-treatment of the Indians. The residents of the interior, who have no higher principles to counteract instinctive selfishness or antipathy of race, cannot comprehend why they are not allowed to compel Indians to work for them, seeing that they will not do it of their own accord. The inevitable result of the conflict of interests between a European and a weaker indigenous race, when the two come in contact, is the sacrifice of the latter. In the Pará district, the Indians are no longer enslaved, but they are deprived of their lands, and this they feel bitterly, as one of them, an industrious and worthy man, related to me. Is not a similar state of things now exhibited in New Zealand, between the Maoris and the English colonists? It is interesting to read of the bitter contests that were carried on from the year 1570 to 1759, between the Portuguese immigrants in Brazil, and the Jesuit and other missionaries. They were similar to those which have recently taken place in South Africa, between the Boers and the English missionaries, but they were on a much larger scale. The Jesuits, as far as I could glean from tradition and history, were actuated by the same motives as our missionaries; and they seemed like them to have been, in great measure, successful in teaching the pure and elevated Christian morality to the simple natives. But the attempt was vain to protect the weaker race from the inevitable ruin which awaited it in the natural struggle with the stronger one; which, although calling itself Christian, seemed to have stood in need of missionary instruction quite as much as the natives themselves. In 1759, the white colonists finally prevailed, the Jesuits were forced to leave the country, and the 51 happy mission villages went to ruin. Since then, the aboriginal race has gone on decreasing in numbers under the treatment which it has received; it is now, as I have already stated, protected by the laws of the central government.

On our second visit to the mills, we stayed ten days. There is a large reservoir and also a natural lake near the place both containing aquatic plants, whose leaves rest on the surface like our water lilies, but they are not so elegant as our nymphæa, either in leaf or flower. On the banks of these pools grow quantities of a species of fan-leaved palm-tree, the Caraná, whose stems are surrounded by whorls of strong spines. I sometimes took a montaria, and paddled myself alone down the creek. One day I got upset, and had to land on a grassy slope leading to an old plantation, where I ran about naked whilst my clothes were being dried on a bush. The creek Iritirí is not so picturesque as many others which I subsequently explored. Towards the Magoary the banks at the edge of the water are clothed with mangrove bushes, and beneath them the muddy banks into which the long roots that hang down from the fruit before it leaves the branches strike their fibres, swarm with crabs. On the lower branches the beautiful bird, Ardea helias is found. This is a small heron of exquisitely graceful shape and mien; its plumage is minutely variegated with bars and spots of many colours, like the wings of certain kinds of moths. It is difficult to see the bird in the woods, on account of its sombre colours, and the shadiness of its dwelling-places; but its note, a soft long-drawn whistle, often betrays its hiding-place. I was told by the Indians that it builds in trees, and that the nest, which is made of clay, is beautifully constructed. It is a favourite pet-bird of the Brazilians, who call it Pavaõ (pronounced Pavaong), or peacock. I often had opportunities of observing its habits. It soon becomes tame, and walks about the floors of houses picking-up scraps of food, or catching insects, which it secures by walking gently to the place where they settle, and spearing them with its long, slender beak. It allows itself to be handled by children, and will answer to its name "Pavaõ! Pavaõ!" walking up with a dainty, circumspect gait, and taking a fly or beetle from the hand.

We made several shorter excursions in the neighbourhood. There was a favourite young negro slave named Hilario (anglicised to Larry), who took an interest in our pursuit. He paddled me one day over the lake, where we shot a small alligator and several Piosocas

(Parra Jacana), a waterfowl having very long legs and toes, which give it the appearance of walking on stilts, as it stalks about, striding from one water-lily leaf to another. I was surprised to find no coleopterous insects on the aquatic plants. The situation appeared to be as favourable for them as possibly could be. In England such a richly-mantled pool would have yielded an abundance of Donaciæ, Chrysomelæ, Cassidæ, and other beetles ; here I could not find a single specimen. Neither could I find any water-beetles ; the only exception was a species of Gyrinus, about the same size as G. natator, the little shining whirligig-beetle of Europe, which was seen in small groups in shady corners, spinning round on the surface of the water precisely as its congener does in England. The absence of leaf-eating beetles on the water plants, I afterwards found was general throughout the country. A few are found on large grasses, and Marantaceous plants in some places, but these are generally concealed in the sharp folds of the leaves, and are almost all very flat in shape.* I, therefore, conclude that the aquatic plants in open places in this country are too much exposed to the sun's heat to admit of the existence of leaf-eating beetles.

Larry told me the Indian names, and enumerated the properties of a number of the forest trees. One of these was very interesting—viz., the Jutahí, which yields the gum copal, called by the natives Jutahí-síca. There are several species of it, as appears at once from

* The species belong to the families Hispidæ and Cassidiadæ, and to the genera Cephaloleia, Arescus, Himatidium, Homalispa. Carnivorous beetles, also flat in shape, sometimes accompany them.

the nature of the fruit. They belong to the order Leguminosæ: the pods are woody and excessively hard; inside they contain a number of beans, enveloped in a sweet yellowish floury substance, which is eaten by the inhabitants. The shell burns with a clear flame. Some of the species have large pods, others small oval ones, containing only one bean. The trees are amongst the largest in the forest, growing from 150 to 180 feet in height: the bark is similar to that of our oak. The leaves are in pairs, whence arises the botanical name of the genus, Hymenæa. The resin which the various species produce exudes from wounds or gashes made in the bark: but I was told that the trees secrete it also spontaneously from the base of the trunk within, and that large lumps are found in the earth amongst the roots when a tree is uprooted by storms. In the resin, ants and other insects are sometimes embedded, precisely as they are in amber, which substance the Jutahí-síca often resembles, at least in colour and transparency.

During these rambles by land and water we increased our collections considerably. Before we left the mills we arranged a joint excursion to the Tocantins. Mr. Leavens wished to ascend that river to ascertain if the reports were true, that cedar grew abundantly between the lowermost cataract and the mouth of the Araguaya, and we agreed to accompany him. Whilst we were at the mills, a Portuguese trader arrived with a quantity of worm-eaten logs of this cedar, which he had gathered from the floating timber in the current of the main Amazons. The tree producing this wood, which is named cedar on account of the similarity of its aroma to that of the true cedars, is not, of course, a coniferous tree, as no member of that class is found in equatorial America, at least in the Amazons region. It is, according to Von Martius, the Cedrela odorata, an exogen belonging to the same order as the mahogany tree. The wood is light, and the tree is therefore, on falling into the water, floated down with the river currents. It must grow in great quantities somewhere in the interior, to judge from the number of uprooted trees annually carried to the sea, and as the wood is much esteemed for cabinet work and canoe building, it is of some importance to learn where a regular supply can be obtained. We were glad, of course, to arrange with Mr. Leavens, who was familiar with the language, and an adept in river-navigation; so we returned to Pará to ship our collections for England, and prepare for the journey to a new region.

CHAPTER III.

PARÁ—*concluded.*

Religious holidays—Marmoset monkeys—Serpents—Insects of the forest—Relations of the Fauna of the Pará district.

BEFORE leaving the subject of Pará, where I resided, as already stated, in all eighteen months, it will be necessary to give a more detailed account of several matters connected with the customs of the people and the Natural History of the neighbourhood, which have hitherto been only briefly mentioned. I reserve an account of the trade and improved condition of Pará in 1859 for the end of this narrative.

During the first few weeks of our stay many of those religious festivals took place, which occupied so large a share of the time and thoughts of the people. These were splendid affairs, wherein artistically-planned processions through the streets, accompanied by thousands of people; military displays; the clatter of fireworks, and the clang of military music, were superadded to pompous religious services in the churches. To those who had witnessed similar ceremonies in the Southern countries of Europe, there would be nothing remarkable perhaps in these doings, except their taking place amidst the splendours of tropical nature; but to me they were full of novelty, and were besides interesting as exhibiting much that was peculiar in the manners of the people. The festivals celebrate either the anniversaries of events concerning saints, or those of the more important transactions in the life of Christ. To them have been added, since the Independence, many gala days connected with events in the Brazilian national history; but these have all a semi-religious character. The holidays had become so numerous, and interfered so much with trade and industry towards the year 1852, that the Brazilian Government were obliged to reduce them; obtaining the necessary permission from Rome to abolish several which were of minor importance. Many of those which have been retained are declining in importance since the introduction of railways and steam boats, and the increased devotion of the people to commerce; at the time of our arrival, however, they were in full glory. The way they were managed was in this fashion. A general manager or "Juiz" for each festa was elected by lot every year in the vestry of the church, and to him were handed over all the paraphernalia pertaining to the particular festival which he was chosen to manage; the image of the saint, the banners, silver crowns and so forth. He then employed a number of people to go the round of the parish and collect alms, towards defraying the expenses. It was considered that the greater the amount of money spent in wax candles, fireworks, music and feasting, the greater the honour done to the saint. If the Juiz was a rich man, he seldom sent out alms-gatherers, but celebrated the whole affair at his own expense, which was sometimes to the extent of several hundred pounds. Each festival lasted nine days (a *novena*), and in many cases refreshments for the public were provided every evening. In the smaller towns a ball took place two or three evenings during the novena, and on the last day there was a grand dinner. The priest, of course, had to be paid very liberally, especially for the sermon delivered on the Saint's-day or termination of the festival, sermons being extra duty in Brazil.

There was much difference as to the accessories of these festivals between the interior towns and villages and the capital; but little or no work was done anywhere whilst they lasted, and they tended much to demoralise the people. It is soon perceived that religion is rather the amusement of the Paraenses than their serious exercise. The ideas of the majority evidently do not reach beyond the belief that all the proceedings are, in each case, in honour of the particular wooden image enshrined at the church. The uneducated Portuguese immigrants seemed to me to have very degrading notions of religion. I have often travelled in the company of these shining examples of European enlightenment. They generally carry with them, wherever they go, a small image of some favourite saint in their trunks, and when a squall or any other danger arises, their first impulse is to rush to the cabin, take out the image and clasp it to their lips, whilst uttering a prayer for protection. The negroes and mulattos are similar in this respect to the low Portuguese, but I think they show a purer devotional feeling ; and in conversation I have always found them to be more rational in religious views than the lower orders of Portuguese. As to the Indians ; with the exception of the more civilized families residing near the large towns, they exhibit no religious sentiment at all. They have their own patron saint, St. Thomé, and celebrate his anniversary in the orthodox way, for they are fond of observing all the formalities ; but they think the feasting to be of equal importance with the church ceremonies. At some of the festivals, masquerading forms a large part of the proceedings, and then the Indians really shine. They get up capital imitations of wild animals, dress themselves to represent the Caypór and other fabulous creatures of the forest, and act their parts throughout with great cleverness. When St. Thomé's festival takes place, every employer of Indians knows that all his men will get drunk. The Indian, generally too shy to ask directly for cashaça (rum), is then very bold ; he asks for a frasco at once (two-and-a-half bottles), and says, if interrogated, that he is going to fuddle in honour of St. Thomé.

In the city of Pará, the provincial government assists to augment the splendour of the religious holidays. The processions which traverse the principal streets consist, in the first place, of the image of the saint, and those of several other subordinate ones belonging to the same church ; these are borne on the shoulders of respectable householders, who volunteer for the purpose : sometimes you will see your neighbour the grocer or the carpenter groaning under the load. The priest and his crowd of attendants precede the images, arrayed in embroidered robes, and protected by magnificent sunshades—no useless ornament here, for the heat is very great when the sun is not obscured. On each side of the long line the citizens walk, clad in crimson silk cloaks, and holding each a large lighted wax candle. Behind follows a regiment or two of foot soldiers with their bands of music, and last of all the crowd: the coloured people being cleanly dressed and preserving a grave demeanour. The women are always in great force, their luxuriant black hair decorated with jasmines, white orchids and other tropical flowers. They are dressed in their usual holiday attire, gauze chemises and black silk petticoats; their necks are adorned with links of gold beads, which when they are slaves are generally the property of their mistresses, who love thus to display their wealth.

At night, when festivals are going on in the grassy squares around the suburban churches, there is really much to admire. A great deal that is peculiar in the land and the life of its inhabitants can be seen best at those times. The cheerful white church is brilliantly lighted up, and the music, not of a very solemn description, peals forth fiom the open windows and doors. Numbers of young gaudily-dressed negresses line the path to the church doors with stands of liqueurs, sweetmeats, and cigarettes, which they sell to the outsiders. A short distance off is heard the rattle of dice-boxes and roulette at the open-air gambling-stalls. When the festival happens on moonlit nights, the whole scene is very striking to a new-comer. Around the square are groups of tall palm trees, and beyond it, over the illuminated houses, appear the thick groves of mangoes near the suburban avenues, from which comes the perpetual ringing din of insect life. The soft tropical moonlight lends a wonderful charm to the whole. The inhabitants are all out, dressed in their best. The upper classes, who come to enjoy the fine evening and the general cheerfulness, are seated on chairs around the doors of friendly houses. There is no boisterous conviviality, but a quiet enjoyment seems to be felt everywhere, and a gentle courtesy rules amongst all classes and colours. I have seen a splendidly-dressed colonel, from the President's palace, walk up to a mulatto, and politely ask his permission to take a light from his cigar. When the service is over, the church bells are set ringing, a shower of rockets mounts upwards, the bands strike up, and parties of coloured people in the booths begin their dances. About ten o'clock the Brazilian national air is played, and all disperse quietly and soberly to their homes.

At the festival of Corpus Christi there was a very pretty arrangement. The large green square of the Trinidade was lighted up all round with bonfires. On one side a fine pavilion was erected, the upright posts consisting of real fan-leaved palm trees—the Mauritia flexuosa, which had been brought from the forest, stems and heads entire, and fixed in the ground. The booth was illuminated with coloured lamps, and lined with red and white cloth. In it were seated the ladies, not all of pure Caucasian blood, but presenting a fine sample of Pará beauty and fashion.

The grandest of all these festivals is that held in

honour of Our Lady of Nazareth: it is, I believe, peculiar to Pará. As I have said before, it falls in the second quarter of the moon, about the middle of the dry season—that is, in October or November—and lasts, like the others, nine days. On the first day a very extensive procession takes place, starting from the Cathedral, whither the image of the saint had been conveyed some days previous, and terminating at the chapel or hermitage, as it is called, of the saint at Nazareth, a distance of more than two miles. The whole population turns out on this occasion. All the soldiers, both of the line and the National Guard, take part in it, each battalion accompanied by its band of music. The civil authorities, also, with the President at their head, and the principal citizens, including many of the foreign residents, join in the line. The boat of the shipwrecked Portuguese vessel is carried after the saint on the shoulders of officers or men of the Brazilian navy, and along with it are borne the other symbols of the miracles which Our Lady is supposed to have performed. The procession starts soon after the sun's heat begins to moderate—that is, about half-past four o'clock in the afternoon. When the image is deposited in the chapel the festival is considered to be inaugurated, and the village every evening becomes the resort of the pleasure-loving population, the holiday portion of the programme being preceded, of course, by a religious service in the chapel. The aspect of the place is then that of a fair, without the humour and fun, but, at the same time, without the noise and coarseness of similar holidays in England. Large rooms are set apart for panoramic and other exhibitions, to which the public is admitted gratis. In the course of each evening, large displays of fireworks take place, all arranged according to a published programme of the festival.

The various ceremonies which take place during Lent seemed to me the most impressive, and some of them were exceedingly well-arranged. The people, both performers and spectators, conduct themselves with more gravity on these occasions, and there is no holiday-making. Performances, representing the last events in the life of Christ, are enacted in the churches or streets, in such a way as to remind one of the old miracle plays or mysteries. A few days before Good Friday, a torchlight procession takes place by night from one church to another, in which is carried a large wooden image of Christ bent under the weight of the cross. The chief members of the Government assist, and the whole slowly moves to the sounds of muffled drums. A double procession is managed a few days afterwards. The image of St. Mary is carried in one direction, and that of the Saviour in another. Both meet in the middle of one of the most beautiful of the churches, which is previously filled to excess with the multitudes anxious to witness the affecting meeting of mother and son a few days before the crucifixion. The two images are brought face to face in the middle of the church, the crowd falls prostrate, and a lachrymose sermon is delivered from the pulpit. The whole thing, as well as many other spectacles arranged during the few succeeding days, is highly theatrical, and well calculated to excite the religious emotions of the people, although, perhaps, only temporarily. On Good Friday the bells do not ring, all musical sounds are interdicted, and the hours, night and day, are announced by the dismal noise of wooden clappers, wielded by negroes stationed near the different churches. A sermon is delivered in each church. In the middle of it, a scroll is suddenly unfolded from the pulpit, on which is an exaggerated picture of the bleeding Christ. This act is accompanied by loud groans, which come from stout-lunged individuals concealed in the vestry and engaged for the purpose. The priest becomes greatly excited, and actually sheds tears. On one of these occasions I squeezed myself into the crowd, and watched the effect of the spectacle on the audience. Old Portuguese men and Brazilian women seemed very much affected—sobbing, beating their breasts, and telling their beads. The negroes behaved themselves with great propriety, but seemed moved more particularly by the pomp, the gilding, the dresses, and the general display. Young Brazilians laughed. Several aborigines were there, coolly looking on. One old Indian, who was standing near me, said, in a derisive manner, when the sermon was over, "It's all very good; better it could not be" (Está todo bom; melhor naõ pude ser).

The negroes of Pará are very devout. They have built, by slow degrees, a fine church, as I was told, by their own unaided exertions. It is called Nossa Senhora do Rosario, or Our Lady of the Rosary. During the first weeks of our residence at Pará, I frequently observed a line of negroes and negresses, late at night, marching along the streets, singing a chorus. Each carried on his or her head a quantity of building materials—stones, bricks, mortar, or planks. I found they were chiefly slaves, who, after their hard day's work, were contributing a little towards the construction of their church. The materials had all been purchased by their own savings. The interior was finished about a year afterwards, and is decorated, I thought, quite as superbly as the other churches which were constructed, with far larger means, by the old religious orders more than a century ago. Annually, the negroes celebrate the festival of Nossa Senhora do Rosario, and generally make it a complete success.

I will now add a few more notes which I have accumulated on the subject of the natural history, and then we shall have done, for the present, with Pará and its neighbourhood.

I have already mentioned that monkeys were rare in the immediate vicinity of Pará. I met with three species only in the forest near the city; they are shy animals, and avoid the neighbourhood of towns, where they are subject to much persecution by the inhabitants, who kill them for food. The only kind which I saw frequently was the little Midas ursulus, one of the Marmosets, a family peculiar to tropical America, and differing in many essential points of structure and habits from all other apes. They are small in size, and more like squirrels than true monkeys in their manner of climbing. The nails, except those of the hind thumbs, are long and claw-shaped like those of squirrels, and the thumbs of the fore extremities, or hands, are not opposable to the other fingers. I do not mean to convey that they have a near relationship to squirrels, which belong to the Rodents, an inferior order of mammals; their resemblance to those animals is merely a superficial one. They have two molar teeth less in each jaw than the Cebidæ, the other family of American monkeys; they agree with them, however, in the sideway position of the nostrils, a character which distinguishes both from all the monkeys of the old world. The body is long and slender, clothed with soft hairs, and the tail, which is nearly twice the length of the trunk, is not prehensile. The hind limbs are much larger in volume than the anterior pair. The Midas ursulus is never seen in large flocks; three or four is the greatest number observed together. It seems to be less afraid of the neighbourhood of man than any other monkey. I sometimes saw it in the woods which border the suburban streets, and once I espied two individuals in a thicket behind the English consul's house at Nazareth. Its mode of progression along the main boughs of the lofty trees is like that of squirrels; it does not ascend to the slender branches, or take those wonderful flying leaps which the Cebidæ do, whose prehensile tails and flexible hands fit them for such headlong travelling. It confines itself to the larger boughs and trunks of trees, the long nails being of great assistance to the creature, enabling it to cling securely to the bark; and it is often seen passing rapidly round the perpendicular cylindrical trunks. It is a quick, restless, timid little creature, and has a great share of curiosity, for when a person passes by under the trees along which a flock is running, they always stop for a few moments to have a stare at the intruder. In Pará, Midas ursulus is often seen in a tame state in the houses of the inhabitants. When full grown it is about nine inches long, independently of the tail, which measures fifteen inches. The fur is thick, and black in colour, with the exception of a reddish-brown streak down the middle of the back. When first taken, or when kept tied up, it is very timid and irritable. It will not allow itself to be approached, but keeps retreating backwards when any one attempts to coax it. It is always in a querulous humour, uttering a twittering, complaining noise ; its dark, watchful eyes, expressive of distrust, observant of every movement which takes place near it. When treated kindly, however, as it generally is in the houses of the natives, it becomes very tame and familiar. I once saw one as playful as a kitten, running about the house after the negro children, who fondled it to their heart's content. It acted somewhat differently towards strangers, and seemed not to like them to sit in the hammock which was slung in the room, leaping up, trying to bite, and otherwise annoying them. It is generally fed on sweet fruits, such as the banana ; but it is also fond of insects, especially soft-bodied spiders and grasshoppers, which it will snap up with eagerness when within reach. The expression of countenance in these small monkeys is intelligent and pleasing. This is partly owing to the open facial angle, which is given as one of 60° ; but the quick movements of the head, and the way they have of inclining it to one side when their curiosity is excited, contribute very much to give them a knowing expression. Anatomists who have dissected species of Midas tell us that the brain is of a very low type, as far as the absence of convolutions goes, the surface being as smooth as that of a squirrel's. I should conclude, at once, that this character is an unsafe guide in judging on the mental qualities of these animals; in mobility of expression of countenance, intelligence, and general manners, these small monkeys resemble the higher apes far more than they do any Rodent animal with which I am acquainted.

On the Upper Amazons I once saw a tame individual of the Midas leoninus, a species first described by Humboldt, which was still more playful and intelligent than the one just described. This rare and beautiful little monkey is only seven inches in length, exclusive of the tail. It is named leoninus on account of the long brown mane which depends from the neck, and which gives it very much the appearance of a diminutive lion. In the house where it was kept, it was familiar with every one; its greatest pleasure seemed to be to climb about the bodies of different persons who entered. The first time I went in, it ran across the room straightway to the chair on which I had sat down, and climbed up to my shoulder; arrived there, it turned round and looked into my face, showing its little teeth, and chattering, as though it would say, "Well, and how do *you* do?" It showed more affection towards its master than towards strangers, and would climb up to his head a dozen times in the course of an hour, making a great show every time of searching there for certain animalcula. Isidore Geoffroy St. Hilaire relates of a species of this genus, that it distinguished between different objects depicted on an engraving. M. Audouin showed it the portraits of a cat and a wasp; at these it became much terrified: whereas, at the sight of a figure of a grasshopper or beetle, it precipitated itself on the picture, as if to seize the objects there represented.

Although monkeys are now rare in a wild state near Pará, a great number may be seen semi-domesticated in the city. The Brazilians are fond of pet animals. Monkeys, however, have not been known to breed in captivity in this country. I counted, in a short time, thirteen different species, whilst walking about the Pará streets, either at the doors or windows of houses, or in the native canoes. Two of them I did not meet with afterwards in any other part of the country. One of these was the well-known Hapale Jacchus, a little creature resembling a kitten, banded with black and gray all over the body and tail, and having a fringe of long white hairs surrounding the ears. It was seated on the shoulder of a young mulatto girl, as she was walking along the street, and I was told had been captured in the island of Marajo. The other was a species of Cebus, with a remarkably large head. It had ruddy-brown fur, paler on the face, but presenting a blackish tuft on the top of the forehead.

In the wet season serpents are common in the neighbourhood of Pará. One morning, in April, 1849, after a night of deluging rain, the lamplighter, on his rounds to extinguish the lamps, knocked me up to show me a boa-constrictor he had just killed in the Rua St. Antonio,

H 2

not far from my door. He had cut it nearly in two with a large knife, as it was making its way down the sandy street. Sometimes the native hunters capture boa-constrictors alive in the forest near the city. We bought one which had been taken in this way, and kept it for some time in a large box under our verandah. This is not, however, the largest or most formidable serpent found in the Amazons region. It is far inferior, in these respects, to the hideous Sucurujú, or Water Boa (Eunectes murinus), which sometimes attacks man; but of this I shall have to give an account in a subsequent chapter.

It frequently happened, in passing through the thickets, that a snake would fall from the boughs close to me. Once I got for a few moments completely entangled in the folds of one, a wonderfully slender kind, being nearly six feet in length, and not more than half an inch in diameter at its broadest part. It was a species of Dryophis. The majority of the snakes seen were innocuous. One day, however, I trod on the tail of a young serpent belonging to a very poisonous kind, the Jararaca (Craspedocephalus atrox). It turned round and bit my trousers; and a young Indian lad, who was behind me, dexterously cut it through with his knife before it had time to free itself. In some seasons snakes are very abundant, and it often struck me as strange that accidents did not occur more frequently than was the case.

Amongst the most curious snakes found here were the Amphisbænæ, a genus allied to the slow-worm of Europe. Several species occur at Pará. Those brought to me were generally not much more than a foot in length. They are of cylindrical shape, having, properly speaking, no neck, and the blunt tail which is only about an inch in length, is of the same shape as the head.

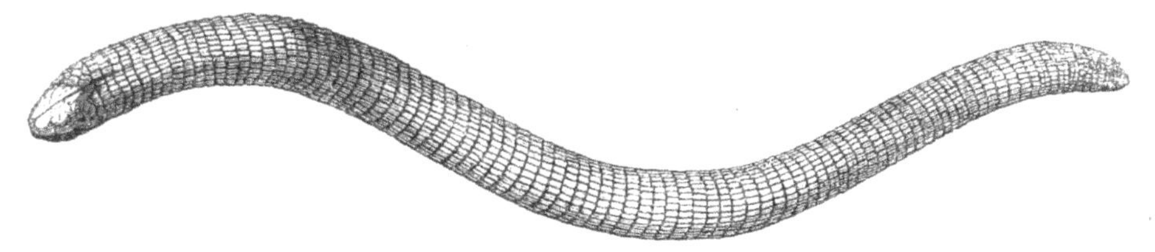

Amphisbæna.

This peculiar form added to their habit of wriggling backwards as well as forwards, has given rise to the fable that they have two heads, one at each extremity. They are extremely sluggish in their motions, and are clothed with scales that have the form of small imbedded plates arranged in rings round the body. The eye is so small as to be scarcely perceptible. They live habitually in the subterranean chambers of the Saüba ant; only coming out of their abodes occasionally in the night time. The natives call the Amphisbæna the "Mai das Saübas," or Mother of the Saübas, and believe it to be poisonous, although it is perfectly harmless. It is one of the many curious animals which have become the subject of mythical stories with the natives They say the ants treat it with great affection, and that, if the snake be taken away from a nest, the Saübas will forsake the spot. I once took one quite whole out of the body of a young Jararaca, the poisonous species already alluded to, whose body was so distended with its contents that the skin was stretched out to a film over the contained Amphisbæna. I was, unfortunately, not able to ascertain the exact relation which subsists between these curious snakes and the Saüba ants. I believe, however, they feed upon the Saübas, for I once found remains of ants in the stomach of one of them. Their motions are quite peculiar; the undilatable jaws, small eyes and curious plated integument also distinguish them from other snakes. These properties have evidently some relation to their residence in the subterranean abodes of ants. It is now well ascertained by naturalists, that some of the most anomalous forms amongst Coleopterous insects are those which live solely in the nests of ants, and it is curious that an abnormal form of snakes should also be found in the society of these insects.

The neighbourhood of Pará is rich in insects. I do not speak of the quantity of individuals, which is probably less than one meets with, excepting ants and Termites, in summer days in temperate latitudes; but the variety, or in other words, the number of species is very great. It will convey some idea of the diversity of butterflies when I mention that about 700 species of that tribe are found within an hour's walk of the town; whilst the total number found in the British Islands does not exceed 66, and the whole of Europe supports only 390. Some of the most showy species, such as the swallow-tailed kinds, Papilio Polycaon, Thoas, Torquatus, and others, are seen flying about the streets and gardens; sometimes they come through the open windows, attracted by flowers in the apartments. Those species of Papilio which are most characteristic of the country, so conspicuous in their velvety-black, green, and rose-coloured hues, which Linnæus, in pursuance of his elegant system of nomenclature,—naming the different kinds after the heroes of Greek mythology,—called Trojans, never leave the shades of the forest. The splendid metallic blue Morphos, some of which measure seven inches in expanse, are generally confined to the shady alleys of the forest. They sometimes come forth into the broad sunlight. When we first went to look at our new residence in Nazareth, a Morpho Menelaus, one of the most beautiful kinds, was seen flapping its huge wings like a bird along the verandah. This species, however, although much admired, looks dull in colour by the side of its congener, the Morpho Rhetenor, whose wings, on the upper face, are of quite a dazzling lustre. Rhetenor usually prefers the broad sunny roads in the forest, and is an almost unattainable prize, on account of its lofty flight; for it very rarely descends nearer the ground than about twenty feet. When it comes sailing along, it occasionally flaps its wings, and then the blue surface flashes in the sunlight, so that it is visible a quarter of a mile off. There is another species of this genus, of a satiny-white hue, the Morpho Eugenia; this is equally difficult to obtain; the male only has the satiny lustre, the female being of a pale-lavender colour. It is in the height of the dry season that the greatest number and variety of butterflies are found in the woods; especially when a shower falls at intervals of a few days. An infinite number of curious and rare species may then be taken, most diversified in habits, mode of flight, colours, and markings: some yellow, others bright red, green, purple, and blue, and many bordered or spangled with metallic lines and spots of a silvery or golden lustre. Some have wings transparent as glass; one of these clear wings is especially beautiful, namely, the Hetaira Esmeralda; it has one spot only of opaque colouring on its wings, which is of a violet and rose hue; this is the only part visible when the insect is flying low over dead leaves in the gloomy shades where alone it is found, and it then looks like a wandering petal of a flower.

Moths also are of great variety at Pará; but most of them are diurnal in their time of flight and keep company with the butterflies. I never succeeded in finding many moths at night. In situations such as gardens and wood sides, where so many are to be seen in England, scarcely a single individual is to be found. I attribute this scarcity of nocturnal moths to the multitude of night-flying insectivorous animals, chiefly bats and goat-suckers, which perpetually haunt the places where they would be found. On the open commons a moth is seen flying about in broad daylight which is scarcely distinguishable from the common English Plusia Gamma. Several times I found the Erebus strix expanded over the trunks of trees, to the bark of which it is assimilated in colour. This is one of the largest moths known, some specimens measuring nearly a foot in expanse. Along the narrow paths in the forests, an immense number of clear-winged moths are found in the day-time; mostly coloured like wasps, bees, ichneumon flies, and other Hymenopterous insects. Some species of the same family have opaque wings, and wear the livery of different species of beetles; these hold their wings in repose, in a closed position over their bodies, so that they look like the wing-cases of the beetles they deceptively imitate.

The Libellulidæ, or Dragonflies, are almost equally conspicuous with the butterflies in open, sunny places. More than a hundred different kinds are found near Pará; the numerous ditches and pools being, doubtless, favourable to their increase, for the adolescent states of the dragonfly are passed in an element different from that in which the adult exists. The species are not all confined to open, sunny places. Some are adapted to live only in the darkest shades of the forest, and these are, perhaps, the most beautiful, being brightly coloured and more delicate in structure than the others. One of them, the Chalcopteryx rutilans, is seen only near the shady rivulets which cross the solitary Magoary road. Its fore-wings are quite transparent, whilst the hind-wings have a dark ground-colour, which glitters with a violet and golden refulgence. All the kinds of dragonflies wage an unceasing war with day-flying winged insects, and I am inclined to think that they commit as much destruction in this way as birds do. I have often observed them chasing butterflies. They are not always successful in capturing them, for some of their intended victims, by a dodging manner of flight, contrive to escape their clutches. When a dragonfly seizes its prey, he retires to a tree, and there, seated on a branch, devours the body at his leisure. The different species consume great quantities of small flies, especially during the brief twilight, when large flocks of the hawk-like creatures congregate to chase them over the swamps and about the tree-tops.

Bees and wasps are not especially numerous near Pará, and I will reserve an account of their habits for a future chapter. Many species of Mygale, those monstrous hairy spiders, half a foot in expanse, which attract the attention so much in museums, are found in sandy places at Nazareth. The different kinds have the most diversified habits. Some construct, amongst the tiles or thatch of houses, dens of closely-woven web, which, in texture, very much resembles fine muslin ; these are often seen crawling over the walls of apartments. Others build similar nests in trees, and are known to attack birds. One very robust fellow, the Mygale Blondii, burrows into the earth, forming a broad, slanting gallery, about two feet long, the sides of which he lines beautifully with silk. He is nocturnal in his habits. Just before sunset he may be seen keeping watch within the mouth of his tunnel, disappearing suddenly when he hears a heavy foot-tread near his hiding-place. The number of spiders ornamented with showy colours was somewhat remarkable. Some double themselves up at the base of leaf-stalks, so as to resemble flower-buds, and thus deceive the insects on which they prey. The most extraordinary-looking spider was a species of Acrosoma, which had two

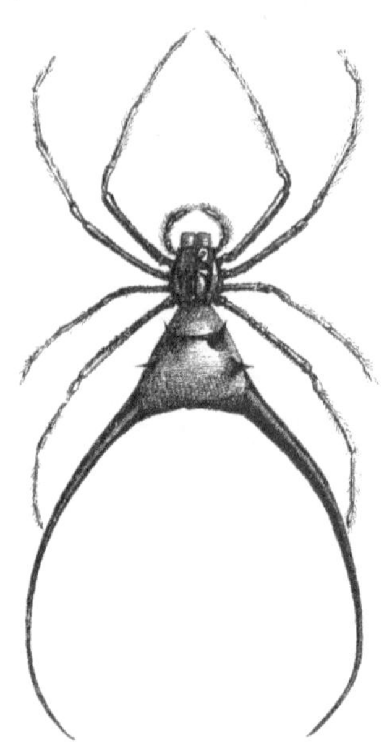

Acrosoma arcuatum.

curved bronze-coloured spines, an inch and a half in length, proceeding from the tip of its abdomen. It spins a large web, the monstrous appendages being apparently no impediment to it in its work; but what their use can be I am unable to divine.

Coleoptera, or beetles, at first seemed to be very scarce. This apparent scarcity has been noticed in other equatorial countries and arises, probably, from the great heat of the sun not permitting them to exist in exposed situations, where they form such conspicuous objects in Europe. Many hundred species of the different families can be found, when they are patiently searched for in the shady places to which they are confined. It is vain to look for the Geodephaga, or carnivorous beetles, under stones, or anywhere, indeed, in open, sunny places. The terrestrial forms of this interesting family, which abound in England and temperate countries generally, are scarce in the neighbourhood of Pará, in fact, I met with only four or five species; on the other hand the purely arboreal kinds were rather numerous. The contrary of this happens in northern latitudes, where the great majority of the species and genera are exclusively terrestrial. The arboreal forms are distinguished by the structure of the feet, which have broad spongy soles and toothed claws enabling them to climb over and cling to branches and leaves. The remarkable scarcity of ground beetles is, doubtless, attributable to the number of ants and Termites which people every inch of surface in all shady places, and which would most likely destroy the larvæ of Coleoptera. These active creatures have the same functions as Coleoptera, and thus render their existence unnecessary. The large proportion of climbing forms of carnivorous beetles is an interesting fact, because it affords another instance of the arboreal character which animal forms tend to assume in equinoctial America, a circumstance which points to the slow adaptation of the Fauna to a forest-clad country throughout an immense lapse of geological time.

The large collections which I made of the animal productions of Pará, especially of insects, enabled me to arrive at some conclusions regarding the relations of the Fauna of the south side of the Amazons Delta to those of neighbouring regions. It is generally allowed that Guiana and Brazil, to the north and south of the Pará district, form two distinct provinces, as regards their animal and vegetable inhabitants. By this it is meant that the two regions have a very large number of forms peculiar to themselves, and which are supposed not to have been derived from other quarters during modern geological times. Each may be considered as a centre of distribution in the latest process of dissemination of species over the surface of tropical America. Pará lies midway between the two centres, each of which has a nucleus of elevated table-land, whilst the intermediate river-valley forms a wide extent of low-lying country. It is, therefore, interesting to ascertain from which the latter received its population, or whether it contains so large a number of endemic species as would warrant the conclusion that it is itself an independent province. To assist in deciding such questions as these, we must compare closely the species found in the district with those of the other contiguous regions, and endeavour to ascertain whether they are identical, or only slightly modified, or whether they are highly peculiar.

Von Martius, when he visited this part of Brazil forty years ago, coming from the south, was much struck with the dissimilarity of the animal and vegetable productions to those of other parts of Brazil. In fact, the Fauna of Pará, and the lower part of the Amazons, has no close relationship with that of Brazil proper ; but it has a very great affinity with that of the coast region of Guiana, from Cayenne to Demerara. If we may judge from the results afforded by the study of certain families of insects, no peculiar Brazilian forms are found in the Pará district ; whilst more than one-half the total number are essentially Guiana species, being found nowhere else but in Guiana and Amazonia. Many of them, however, are modified from the Guiana type, and about one-seventh seem to be restricted to Pará. These endemic species are not highly peculiar, and they may be yet found over a great part of Northern Brazil when the country is better explored. They do not warrant us in concluding that the district forms an independent province, although they show that its Fauna is not wholly derivative, and that the land is probably not entirely a new formation. From all these facts, I think we must conclude that the Pará district belongs to the Guiana province, and that, if it is newer land than Guiana, it must have received the great bulk of its animal population from that region. I am informed by Dr. Sclater that similar results are derivable from the comparison of the birds of these countries.

The interesting problem, how has the Amazons Delta been formed? receives light through this comparison of Faunas. Although the portion of Guiana in question is considerably nearer Pará than are the middle and southern parts of Brazil, yet it is separated from it by two wide expanses of water, which must serve as a barrier to migration in many cases. On the contrary, the land of Brazil proper is quite continuous from Rio Janeiro and Bahia up to Pará; and there are no signs of a barrier ever having existed between these places within recent geological epochs. Some of the species common to Pará and Guiana are not found higher up the river where it is narrower, so they could not have passed round in that direction. The question here arises, has the mouth of the Amazons always existed as a barrier to migration since the present species of the contiguous regions came into existence? It is difficult to decide the question; but the existing evidence goes far to show that it has not. If the mouth of the great river, which, for a long distance, is 170 miles broad, had been originally a wide gulf, and had become gradually filled up by islands formed of sediment brought down by the stream, we should have to decide that an effectual barrier had indeed existed. But the delta of the Amazons is not an alluvial formation like those of the Mississippi and the Nile. The islands in its midst and the margins of both shores have a foundation of rocks, which lie either bare or very near the surface of the soil. This is especially the case towards the sea-coast. In ascending the river southward and south-westward, a great extent of country is traversed which seems to have been made up wholly of river deposit, and here the land lies somewhat lower than it does on the sea-coast. The rocky and sandy country of Marajo and other islands of the delta towards the sea, is so similar in its physical configuration to the opposite mainland of Guiana that Von Martius concluded the whole might have been formerly connected, and that the Amazons had forced a way to the Atlantic through what was, perhaps, a close series of islands, or a continuous line of low country.

CHAPTER IV.

THE TOCANTINS AND CAMETÁ.

Preparations for the journey—The bay of Goajará—Grove of fan-leaved palms—The lower Tocantins—Sketch of the river—Vista alegre—Baiaõ—Rapids—Boat journey to the Guariba falls—Native life on the Tocantins—Second journey to Cametá.

August 26*th*, 1848.—Mr. Wallace and I started to-day on the excursion which I have already mentioned as having been planned with Mr. Leavens, up the river Tocantins, whose mouth lies about forty-five miles in a straight line, but eighty miles following the bends of the river channels, to the south-west of Pará. This river, as before stated, has a course of 1600 miles, and stands third in rank amongst the streams which form the Amazons system. The preparations for the journey took a great deal of time and trouble. We had first to hire a proper vessel, a two-masted *vigilinga* twenty-seven feet long, with a flat prow and great breadth of beam and fitted to live in heavy seas; for, although our voyage was only a river trip, there were vast sea-like expanses of water to traverse. It was not decked over, but had two arched awnings formed of strong wicker-work, and thatched with palm leaves. We had then to store it with provisions for three months, the time we at first intended to be away; procure the necessary passports; and, lastly, engage a crew. Mr. Leavens, having had much experience in the country, managed all these matters. He brought two Indians from the rice-mills, and these induced another to enrol himself. We, on our parts, took our cook Isidoro, and a young Indian lad, named Antonio, who had attached himself to us in the course of our residence at Nazareth. Our principal man was Alexandro, one of Mr. Leavens's Indians. He was an intelligent and well-disposed young Tapuyo, an expert sailor, and an indefatigable hunter. To his fidelity we were indebted for being enabled to carry out any of the objects of our voyage. Being a native of a district near the capital, Alexandro was a civilized Tapuyo, a citizen as free as his white neighbours. He spoke only Portuguese. He was a spare-built man, rather under the middle height, with fine regular features, and, what was unusual in Indians, the upper lip decorated with a moustache. Three years afterwards I saw him at Pará in the uniform of the National Guard, and he called on me often to talk about old times. I esteemed him as a quiet, sensible, manly young fellow.

We set sail in the evening, after waiting several hours in vain for one of our crew. It was soon dark, the wind blew stiffly, and the tide rushed along with great rapidity, carrying us swiftly past the crowd of vessels which were anchored in the port. The canoe rolled a good deal. After we had made five or six miles of way the tide turned, and we were obliged to cast anchor. Not long after, we lay ourselves down all three together on the mat, which was spread over the floor of our cabin, and soon fell asleep.

On awaking at sunrise the next morning, we found ourselves gliding upwards with the tide, along the Bahia or Bay, as it is called, of Goajará. This is a broad channel lying between the mainland and a line of islands which extends some distance beyond the city. Into it three large rivers discharge their waters, namely, the Guamá, the Acará, and the Mojú; so that it forms a kind of sub-estuary within the grand estuary of Pará. It is nearly four miles broad. The left bank, along which we were now sailing, was beautiful in the extreme; not an inch of soil was to be seen; the water frontage presented a compact wall of rich and varied forest, resting on the surface of the stream. It seemed to form a finished border to the water scene, where the dome-like, rounded shapes of exogenous trees which constituted the mass formed the ground-work, and the endless diversity of broad-leaved Heliconiæ and Palms—each kind differing in stem, crown, and fronds—the rich embroidery. The morning was calm and cloudless; and the slanting beams of the early sun, striking full on the front of the forest, lighted up the whole most gloriously. The only sound of life which reached us was the call of the Serracúra (Gallinula Cayennensis), a kind of wild-fowl; all else was so still that the voices of boatmen could be plainly heard from canoes passing a mile or two distant from us. The sun soon gains great power on the water, but with it the sea-breeze increases in strength, moderating the heat, which would otherwise be almost insupportable. We reached the end of the Goajará about midday, and then entered the narrower channel of the Mojú. Up this we travelled, partly rowing and partly sailing between the same unbroken walls of forest, until the morning of the 28th.

August 29th.—The Mojú, a stream little inferior to the Thames in size, is connected about 20 miles from its mouth by means of a short artificial canal with a small stream, the Igarapé-mirim, which flows the opposite way into the water-system of the Tocantins. Small vessels like ours take this route in preference to the stormy passage by way of the main river, although the distance is considerably greater. We passed through the canal yesterday, and to-day have been threading our way through a labyrinth of narrow channels; their banks all clothed with the same magnificent forest; but agreeably varied by houses of planters and settlers. We passed many quite large establishments, besides one pretty little village, called Santa Anna. All these channels are washed through by the tides,—the ebb, contrary to what takes place in the short canal, setting towards the Tocantins. The water is almost tepid (77° Fahr.), and the rank vegetation all around seems reeking with moisture. The country however, as we were told, is perfectly healthy. Some of the houses are built on wooden piles driven into the mud of the swamp.

In the afternoon we reached the end of the last channel, called the Anapú, which runs for several miles between two unbroken lines of fan-leaved palms, forming with their straight stems colossal palisades.

On rounding a point of land we came in full view of the Tocantins. The event was announced by one of our Indians, who was on the look-out at the prow, shouting, "La está o Paraná-uassú!" "Behold, the great river!" It was a grand sight—a broad expanse of dark waters dancing merrily to the breeze; the opposite shore, a narrow blue line, miles away.

We went ashore on an island covered with palm-trees, to make a fire and boil our kettle for tea. I wandered a short way inland, and was astounded at the prospect. The land lay below the upper level of the daily tides, so that there was no underwood, and the ground was bare. The trees were almost all of one species of Palm, the gigantic fan-leaved Mauritia flexuosa; on the borders only was there a small number of a second kind, the equally remarkable Ubussú palm, Manicaria saccifera. The Ubussú has erect, uncut leaves, twenty-five feet long, and six feet wide, all arranged round the top of a four-feet high stem, so as to form a figure like that of a colossal shuttlecock. The fan-leaved palms, which clothed nearly the entire islet, had huge cylindrical smooth stems, three feet in diameter, and about a hundred feet high. The crowns were formed of enormous clusters of fan-shaped leaves, the stalks alone of which measured seven to ten feet in length. Nothing in the vegetable world could be more imposing than this grove of palms. There was no underwood to obstruct the view of the long perspective of towering columns. The crowns, which were densely packed together at an immense height overhead, shut out the rays of the sun; and the gloomy solitude beneath, through which the sound of our voices seemed to reverberate, could be compared to nothing so well as a solemn temple. The fruits of the two palms were scattered over the ground; those of the Ubussú adhere together by twos and threes, and have a rough, brown-coloured shell; the fruit of the Mauritia, on the contrary, is of a bright red hue, and the skin is impressed with deep crossing lines, which give it a resemblance to a quilted cricket-ball.

About midnight, the tide being favourable and the breeze strong, we crossed the river, taking it in a slanting direction, a distance of sixteen miles, and arrived at eight o'clock the following morning at Cametá. This is a town of some importance, pleasantly situated on the somewhat high terra firma of the left bank of the Tocantins. I will defer giving an account of the place till the end of this narrative of our Tocantins voyage. We lost here another of our men, who got drinking with some old companions ashore, and were obliged to start on the difficult journey up the river with two hands only, and they in a very dissatisfied humour with the prospect.

The river view from Cametá is magnificent. The town is situated, as already mentioned, on a high bank, which forms quite a considerable elevation for this flat country, and the broad expanse of dark-green waters is studded with low, palm-clad islands, the prospect down river, however, being clear, or bounded only by a sea-like horizon of water and sky. The shores are washed by the breeze-tossed waters into little bays and creeks, fringed with sandy beaches. The Tocantins has been likened, by Prince Adalbert of Prussia, who crossed its mouth in 1846, to the Ganges. It is upwards of ten miles in breadth at its mouth ; opposite Cametá it is five miles broad. Mr. Burchell, the well-known English traveller, descended the river from the mining provinces of interior Brazil some years before our visit. Unfortunately, the utility of this fine stream is impaired by the numerous obstructions to its navigation in the shape of cataracts and rapids, which commence, in ascending, at about 120 miles above Cametá, as will be seen in the sequel.

Aug. 30th.—Arrived, in company with Senhor Laroque, an intelligent Portuguese merchant, at Vista Alegre, fifteen miles above Cametá. This was the residence of Senhor Antonio Ferreira Gomez, and was a fair sample of a Brazilian planter's establishment in this part of the country. The buildings covered a wide space, the dwelling-house being separated from the place of business, and as both were built on low, flooded ground, the communication between the two was by means of a long wooden bridge. From the office and visitors' apartments a wooden pier extended into the river. The whole was raised on piles above high-water mark. There was a rude mill for grinding sugar-cane, worked by bullocks, but cashaça, or rum, was the only article manufactured from the juice. Behind the buildings was a small piece of ground cleared from the forest, and planted with fruit-trees, orange, lemon, genipapa, goyava, and others ; and beyond this, a broad path through a neglected plantation of coffee and cacao, led to several large sheds, where the farinha, or mandiocca meal, was manufactured.

The plantations of mandiocca are always scattered about in the forest, some of them being on islands in the middle of the river. Land being plentiful, and the plough, as well as, indeed, nearly all other agricultural implements, unknown, the same ground is not planted three years together; but a new piece of forest is cleared every alternate year, and the old clearing suffered to relapse into jungle.

We stayed here two days, sleeping ashore in the apartment devoted to strangers. As usual in Brazilian houses of the middle class, we were not introduced to the female members of the family, and, indeed, saw nothing of them except at a distance. In the forest and thickets about the place we were tolerably successful in collecting, finding a number of birds and insects which do not occur at Pará. I saw here, for the first time, the sky-blue Chatterer (Ampelis cotinga). It was on the topmost bough of a very lofty tree, and completely out of the reach of an ordinary fowling-piece. The beautiful light-blue colour of its plumage was plainly discernible at that distance. It is a dull, quiet bird. A much commoner species was the Cigana or Gipsy (Opisthocomus cristatus), a bird belonging to the same order, Gallinacea, as our domestic fowl. It is about the size of a pheasant; the plumage is dark brown, varied with reddish, and the head is adorned with a crest of long feathers. It is a remarkable bird in many respects. The hind toe is not placed high above the level of the other toes, as it is in the fowl-order generally, but lies on the same plane with them; the shape of the foot becomes thus suited to the purely arboreal habits of the bird, enabling it to grasp firmly the branches of trees. This is a distinguishing character of all the birds in equinoctial America which represent the fowl and pheasant tribes of the old world, and affords another proof of the adaptation of the Fauna to a forest region. The Cigana lives in considerable flocks on the lower trees and bushes bordering the streams and lagoons, and feeds on various wild fruits, especially the sour Goyava (Psidium sp.). The natives say it devours the fruit of arborescent Arums (Caladium arborescens), which grow in crowded masses around the swampy banks of lagoons. Its voice is a harsh, grating hiss ; it makes the noise when alarmed, all the individuals sibilating as they fly heavily away from tree to tree, when disturbed by passing canoes. It is polygamous, like other members of the same order. It is never, however, by any chance, seen on the ground, and is nowhere domesticated. The flesh has an unpleasant odour of musk combined with wet hides—a smell called by the Brazilians catinga ; it is, therefore, uneatable. If it be as unpalateable to carnivorous animals as it is to man, the immunity from persecution which it would thereby enjoy would account for its existing in such great numbers throughout the country.

A great number of the insects which we found here were different from those of Pará. Species characteristic of the one locality were replaced by allied species in the other, a fact which would tend to the conclusion that the Tocantins serves, to some extent, as a barrier to migration. This was especially the case with the Papilios of the group which wear a livery of black, green, and red. P. Echelus of this group, which is so common at Pará, was here absent, and its place supplied by the closely related P. Æneides. Both have the same habits, and seem to fill similar spheres in the natural economy of the two districts. Another handsome butterfly taken here was a member of the Erycinidæ family, the Alesa Prema, which is of a dazzling emerald-green colour chequered with black. I caught here a young Iguana; Iguanas, however, are extremely common everywhere throughout the country. They are especially numerous in the neighbourhood of villages, where they climb about fruit-trees overrun with creepers. The eggs, which are oblong, and about an inch and a half in length, are laid in hollow trees, and are very pleasant eating taken raw and mixed with farinha. The colour of the skin in the Iguana changes like that of the chameleon; in fact, it is called chameleon by the Portuguese. It grows to a length of five feet, and becomes enormously fat. This lizard is interesting to English readers on account of its relationship to the colossal fossil reptile of the Wealden, the Iguanodon. The Iguana is one of the stupidest animals I ever met with. The one I caught dropped helplessly from a tree just ahead of me; it turned round for a moment to have an idiotic stare at the intruder, and then set off running along the pathway. I ran after it, and it then stopped as a timid dog would do, crouching down, and permitting me to seize it by the neck and carry it off.

We lost here another of our crew; and thus, at the commencement of our voyage, had before us the prospect of being forced to return, from sheer want of hands to manage the canoe. Senhor Gomez, to whom we had brought letters of introduction from Senhor Joaõ Augusto Correia, a Brazilian gentleman of high standing at Pará, tried what he could do to induce the canoe-men of his neighbourhood to engage with us, but it was a vain endeavour. The people of these parts seemed to be above working for wages. They are naturally indolent, and besides, have all some little business or plantation of their own, which gives them a livelihood with independence. It is difficult to obtain hands under any circumstances, but it was particularly so in our case, from being foreigners, and suspected, as was natural amongst ignorant people, of being strange in our habits. At length, our host lent us two of his slaves to help us on another stage, namely, to the village of Baiaõ, where we had great hopes of having this, our urgent want, supplied by the military commandant of the district.

Sept. 2nd.—The distance from Vista Alegre to Baiaõ is about twenty-five miles. We had but little wind, and our men were therefore obliged to row the greater part of the way. The oars used in such canoes as ours are made by tying a stout paddle to the end of a long pole by means of woody lianas. The men take their stand on a raised deck, formed by a few rough planks placed over the arched covering in the fore part of the vessel, and pull with their back to the stern. We started at 6 a.m., and about sunset reached a point where the west channel of the river, along which we had been travelling since we left Cametá, joined a broader middle one, and formed with it a great expanse of water. The islands here seem to form two pretty regular lines, dividing the great river into three channels. As we progressed slowly, we took the montaria, and went ashore, from time to time, to the houses, which were numerous on the river banks as well as on the larger islands. In low situations they had a very unfinished appearance, being mere frameworks raised high on wooden piles, and thatched with the leaves of the Ubussú palm. In their construction another palm-tree is made much use of, viz., the Assai (Euterpe oleracea). The outer part of the stem of this species is hard and tough as horn; it is split into narrow planks, and these form a great portion of the walls and flooring. The residents told us that the western channel becomes nearly dry in the middle of the fine season, but that at high water, in April

Assai Palm (Euterpe oleracea).

and May, the river rises to the level of the house-floors. The river bottom is everywhere sandy, and the country perfectly healthy. The people seemed to be all contented and happy, but idleness and poverty were exhibited by many unmistakeable signs. As to the flooding of their island abodes, they did not seem to care about that at all. They seem to be almost amphibious, or as much at home on the water as on land. It was really alarming to see men and women and children, in little leaky canoes laden to the water-level with bag and baggage, crossing broad reaches of river. Most of them have houses also on the terra firma, and reside in the cool palm-swamps of the Ygapó islands, as they are called, only in the hot and dry season. They live chiefly on fish, shellfish (amongst which were large Ampullariæ, whose flesh I found, on trial, to be a very tough morsel), the never-failing farinha, and the fruits of the forest. Amongst the latter the fruits of palm-trees occupied the chief place. The Assai is the most in use, but this forms a universal article of diet in all parts of the country. The fruit, which is perfectly round, and about the size of a cherry, contains but a small portion of pulp lying between the skin and the hard kernel. This is made, with the addition of water, into a thick, violet-coloured beverage, which stains the lips like blackberries. The fruit of the Mirití is also a common article of food, although the pulp is sour and unpalatable, at least to European tastes. It is boiled, and then eaten with farinha. The Tucumá (Astrocaryum tucuma), and the Mucujá (Acrocomia lasiospatha), grow only on the main land. Their fruits yield a yellowish, fibrous pulp, which the natives eat in the same way as the Miritî. They contain so much fatty matter, that vultures and dogs devour them greedily.

Early on the morning of September 3rd we reached the right or eastern bank, which is here from forty to sixty feet high. The houses were more substantially built than those we had hitherto seen. We succeeded in buying a small turtle ; most of the inhabitants had a few of these animals, which they kept in little inclosures made with stakes. The people were of the same class everywhere, Mamelucos. They were very civil ; we were not able, however, to purchase much fresh food from them. I think this was owing to their really not having more than was absolutely required to satisfy their own needs. In these districts, where the people depend for animal food solely on fishing, there is a period of the year when they suffer hunger, so that they are disposed to prize highly a small stock when they have it. They generally answered in the negative when we asked, money in hand, whether they had fowls, turtles, or eggs to sell. "Naõ ha, sinto que naõ posso lhe ser bom ;" or, "Naõ ha, meu coracaõ." "We have none ; I am sorry I cannot oblige you ;" or, "There is none, my heart."

Sept. 3rd to *7th.*—At half-past eight a.m. we arrived at Baiaõ, which is built on a very high bank, and contains about 400 inhabitants. We had to climb to the village up a ladder, which is fixed against the bank, and, on arriving at the top, took possession of a room, which Senhor Seixas had given orders to be prepared for us. He himself was away at his sitio, and would not be here until the next day. We were now quite dependent on him for men to enable us to continue our voyage, and so had no remedy but to wait his leisure. The situation of the place, and the nature of the woods around it, promised well for novelties in birds and insects; so we had no reason to be vexed at the delay, but brought our apparatus and store-boxes up from the canoe, and set to work.

The easy, lounging life of the people amused us very much. I afterwards had plenty of time to become used to tropical village life. There is a free, familiar, pro bono publico style of living in these small places, which requires some time for a European to fall into. No sooner were we established in our rooms, than a number of lazy young fellows came to look on and make remarks, and we had to answer all sorts of questions. The houses have their doors and windows open to the street, and people walk in and out as they please; there is always, however, a more secluded apartment, where the female members of the families reside. In their familiarity there is nothing intentionally offensive, and it is practised simply in the desire to be civil and sociable. A young Mameluco, named Soares, an Escrivaõ, or public clerk, took me into his house to show me his library. I was rather surprised to see a number of well-thumbed Latin classics, Virgil, Terence, Cicero's Epistles, and Livy. I was not familiar enough, at this early period of my residence in the country, with Portuguese to converse freely with Senhor Soares, or ascertain what use he made of these books; it was an unexpected sight, a classical library in a mud-plastered and palm-thatched hut on the banks of the Tocantins.

The prospect from the village was magnificent, over the green wooded islands, far away to the grey line of forest on the opposite shore of the Tocantins. We were now well out of the low alluvial country of the Amazons proper, and the climate was evidently much drier than it is near Pará. They had had no rain here for many weeks, and the atmosphere was hazy around the horizon; so much so that the sun, before setting, glared like a blood-red globe. At Pará this never happens; the stars and sun are as clear and sharply defined when they peep above the distant tree-tops as they are at the zenith. This beautiful transparency of the air arises, doubtless, from the equal distribution through it of invisible vapour. I shall ever remember, in one of my voyages along the Pará river, the grand spectacle that was once presented at sunrise. Our vessel was a large schooner, and we were bounding along before a spanking breeze which tossed the waters into foam, when the day dawned. So clear was the air, that the lower rim of the full moon remained sharply defined until it touched the western horizon, whilst, at the same time, the sun rose in the east. The two great orbs were visible at the same time, and the passage from the moonlit night to day was so gentle, that it seemed to be only the brightening of dull weather. The woods around Baiaõ were of second growth, the ground having been formerly cultivated. A great number of coffee and cotton trees grew amongst the thickets. A fine woodland pathway extends for miles over the high, undulating bank, leading from one house to another along the edge of the cliff. I went into several of them, and talked to their inmates. They were all poor people. The men were out fishing, some far away, a distance of many days' journey; the women plant mandiocca, make the farinha, spin and weave cotton, manufacture soap of burnt cacao shells and andiroba oil, and follow various other domestic employments. I asked why they allowed their plantations to run to waste. They said that it was useless trying to plant anything hereabout; the Saüba ant devoured the young coffee-trees, and every one who attempted to contend against this universal ravager was sure to be defeated. The country, for many miles along the banks of the river, seemed to be well peopled. The inhabitants were nearly all of the tawny-white Mameluco class. I saw a good many mulattos, but very few negroes and Indians, and none that could be called pure whites.

When Senhor Seixas arrived, he acted very kindly. He provided us at once with two men, killed an ox in our honour, and treated us altogether with great consideration. We were not, however, introduced to his family. I caught a glimpse once of his wife, a pretty little Mameluco woman, as she was tripping with a young girl, whom I supposed to be her daughter, across the back yard. Both wore long dressing-gowns, made of bright-coloured calico print, and had long wooden tobacco-pipes in their mouths. The room in which we slept and worked had formerly served as a storeroom for cacao, and at night I was kept awake for hours by rats and cockroaches, which swarm in all such places. The latter were running about all over the walls;

now and then one would come suddenly with a whirr full at my face, and get under my shirt if I attempted to jerk it off. As to the rats, they were chasing one another by dozens all night long, over the floor, up and down the edges of the doors, and along the rafters of the open roof.

September 7th.—We started from Baiaõ at an early hour. One of our new men was a good-humoured, willing young mulatto, named José; the other was a sulky Indian called Manoel, who seemed to have been pressed into our service against his will. Senhor Seixas, on parting, sent a quantity of fresh provisions on board. A few miles above Baiaõ the channel became very shallow; we got aground several times, and the men had to disembark and shove the vessel off. Alexandro here shot several fine fish, with bow and arrow. It was the first time I had seen fish captured in this way. The arrow is a reed, with a steel barbed point, which is fixed in a hole at the end, and secured by fine twine made from the fibres of pine-apple leaves. It is only in the clearest water that fish can be thus shot; and the only skill required is to make, in taking aim, the proper allowance for refraction.

The next day before sunrise a fine breeze sprung up, and the men awoke and set the sails. We glided all day through channels between islands with long, white, sandy beaches, over which, now and then, aquatic and wading birds were seen running. The forest was low, and had a harsh, dry aspect. Several palm trees grew here which we had not before seen. On low bushes, near the water, pretty, red-headed tanagers (Tanagra gularis) were numerous, flitting about and chirping like sparrows. About half-past four p.m., we brought to at the mouth of a creek or channel, where there was a great extent of sandy beach. The sand had been blown by the wind into ridges and undulations, and over the moister parts large flocks of sandpipers were running about. Alexandro and I had a long ramble over the rolling plain, which came as an agreeable change after the monotonous forest scenery amid which we had been so long travelling. He pointed out to me the tracks of a huge jaguar on the sand. We found here, also, our first turtle's nest, and obtained 120 eggs from it, which were laid at a depth of nearly two feet from the surface, the mother first excavating a hole, and afterwards covering it up with sand. The place is discoverable only by following the tracks of the turtle from the water. I saw here an alligator for the first time, which reared its head and shoulders above the water just after I had taken a bath near the spot. The night was calm and cloudless, and we employed the hours before bed-time in angling by moonlight.

On the 10th we reached a small settlement called Patos, consisting of about a dozen houses, and built on a high, rocky bank, on the eastern shore. The rock is the same nodular conglomerate which is found at so many places, from the sea-coast to a distance of 600 miles up the Amazons. Mr. Leavens made a last attempt here to engage men to accompany us to the Araguaya; but it was in vain; not a soul could be induced by any amount of wages to go on such an expedition. The reports as to the existence of cedar were very vague. All said that the tree was plentiful somewhere, but no one could fix on the precise locality. I believe that the cedar grows, like all other forest trees, in a scattered way, and not in masses anywhere. The fact of its being the principal tree observed floating down with the current of the Amazons is to be explained by its wood being much lighter than that of the majority of trees. When the banks are washed away by currents, trees of all species fall into the river; but the heavier ones, which are the most numerous, sink, and the lighter, such as the cedar, alone float down to the sea.

Mr. Leavens was told that there were cedar trees at Trocará, on the opposite side of the river, near some fine rounded hills covered with forest, visible from Patos; so there we went. We found here several families encamped in a delightful spot. The shore sloped gradually down to the water, and was shaded by a few wide-spreading trees. There was no underwood. A great number of hammocks were seen slung between the tree-trunks, and the litter of a numerous household lay scattered about. Women, old and young, some of the latter very good-looking, and a large number of children, besides pet animals, enlivened the encampment. They were all half-breeds, simple, well-disposed people, and explained to us that they were inhabitants of Cametá, who had come thus far, eighty miles, to spend the summer months. The only motive they could give for coming was, that "it was so hot in the town in the veraō (summer), and they were all so fond of fresh fish." Thus these simple folks think nothing of leaving home

K 2

and business to come on a three months' pic-nic. It is the annual custom of this class of people throughout the province to spend a few months of the fine season in the wilder parts of the country. They carry with them all the farinha they can scrape together, this being the only article of food necessary to provide. The men hunt and fish for the day's wants, and sometimes collect a little India-rubber, sarsaparilla, or copaiba oil, to sell to traders on their return; the women assist in paddling the canoes, do the cooking, and sometimes fish with rod and line. The weather is enjoyable the whole time, and so days and weeks pass happily away.

One of the men volunteered to walk with us into the forest, and show us a few cedar-trees. We passed through a mile or two of spiny thickets, and at length came upon the banks of the rivulet Trocará, which flows over a stony bed, and, about a mile above its mouth, falls over a ledge of rocks, thus forming a very pretty cascade. In the neighbourhood, we found a number of specimens of a curious land-shell, a large flat Helix, with a labyrinthine mouth (Anastoma). We learnt afterwards that it was a species which had been discovered a few years previously by Dr. Gardner, the botanist, on the upper part of the Tocantins.

At Patos we stayed three days. In the woods, we found a number of conspicuous insects new to us. Three species of Pieris were the most remarkable. We afterwards learnt that they occurred also in Venezuela and in the south of Brazil; but they are quite unknown in the alluvial plains of the Amazons. We saw, for the first time, the splendid Hyacinthine macaw (Macrocercus hyacinthinus, Lath., the Araruna of the natives), one of the finest and rarest species of the Parrot family. It only occurs in the interior of Brazil, from 16° S. lat. to the southern border of the Amazons valley. It is three feet long from the beak to the tip of the tail, and is entirely of a soft hyacinthine blue colour, except round the eyes, where the skin is naked and white. It flies in pairs, and feeds on the hard nuts of several palms, but especially of the Mucujá (Acrocomia lasiospatha). These nuts, which are so hard as to be difficult to break with a heavy hammer, are crushed to a pulp by the powerful beak of this macaw.

Mr. Leavens was thoroughly disgusted with the people of Patos. Two men had come from below with the intention, I believe, of engaging with us, but they now declined. The inspector, constable, or governor of the place appeared to be a very slippery customer, and I fancy discouraged the men from going, whilst making a great show of forwarding our views. These outlying settlements are the resort of a number of idle worthless characters. There was a kind of festival going on, and the people fuddled themselves with caxirí, an intoxicating drink invented by the Indians. It is made by soaking mandioca cakes in water until fermentation takes place, and tastes like new beer.

Being unable to obtain men, Mr. Leavens now gave up his project of ascending the river as far as the Araguaya. He assented to our request, however, to ascend to the cataracts near Arroyos. We started therefore from Patos with a more definite aim before us than we had hitherto had. The river became more picturesque as we advanced. The water was very low, it being now the height of the dry season; the islands were smaller than those further down, and some of them were high and rocky. Bold wooded bluffs projected into the stream, and all the shores were fringed with beaches of glistening white sand. On one side of the river there was an extensive grassy plain or campo with isolated patches of trees scattered over it. On the 14th and following day we stopped several times to ramble ashore. Our longest excursion was to a large shallow lagoon, choked up with aquatic plants, which lay about two miles across the campo. At a place called Juquerapuá we engaged a pilot to conduct us to Arroyos, and a few miles above the pilot's house, arrived at a point where it was not possible to advance further in our large canoe on account of the rapids.

September 16th. Embarked at six a.m. in a large montaria which had been lent to us for this part of our voyage by Senhor Seixas, leaving the vigilinga anchored close to a rocky islet, named Santa Anna, to await our return. Isidoro was left in charge, and we were sorry to be obliged to leave behind also our mulatto José, who had fallen ill since leaving Baiaō. We had then remaining only Alexandro, Manoel, and the pilot, a sturdy Tapuyo named Joaquim; scarcely a sufficient crew to paddle against the strong currents.

At ten a.m. we arrived at the first rapids, which are called Tapaiunaquára. The river, which was here about a mile wide, was choked up with rocks, a broken ridge passing completely across it. Between these confused piles of stone the currents were fearfully strong and formed numerous eddies and whirlpools. We were obliged to get out occasionally and walk from rock to rock, whilst the men dragged the canoe over the obstacles. Beyond Tapaiunaquára, the stream became again broad and deep, and the river scenery was beautiful in the extreme. The water was clear and of a bluish-green colour. On both sides of the stream stretched ranges of wooded hills, and in the middle picturesque islets rested on the smooth water, whose brilliant green woods fringed with palms formed charming bits of foreground to the perspective of sombre hills fading into grey in the distance. Joaquim pointed out to us grove after grove of Brazil nut trees (Bertholletia excelsa) on the mainland. This is one of the chief collecting grounds for this nut. The tree is one of the loftiest in the forest, towering far above its fellows; we could see the woody fruits, large and round as cannon-balls, dotted over the branches. The currents were very strong in some places, so that during the greater part of the way the men preferred to travel near the shore, and propel the boat by means of long poles.

We arrived at Arroyos about four o'clock in the afternoon, after ten hours' hard pull. The place consists simply of a few houses built on a high bank, and forms a station where canoe-men from the mining countries of the interior of Brazil stop to rest themselves before or after surmounting the dreaded falls and rapids of Guaribas, situated a couple of miles further up. We dined ashore, and in the evening again embarked to visit the falls. The vigorous and successful way in which our men battled with the terrific currents excited our astonishment. The bed of the river, here about a mile wide, is strewn with blocks of various sizes, which lie in the most irregular manner, and between them rush currents of more or less rapidity. With an accurate knowledge of the place and skilful management, the falls can be approached in small canoes by threading the less dangerous channels. The main fall is about a quarter of a mile wide; we climbed to an elevation overlooking it, and had a good view of the cataract. A body of water rushes with terrific force down a steep slope, and boils up with deafening roar around the boulders which obstruct its course. The wildness of the whole scene was very impressive. As far as the eye could reach, stretched range after range of wooded hills, scores of miles of beautiful wilderness, inhabited only by scanty tribes of wild Indians. In the midst of such a solitude the roar of the cataract seemed fitting music.

September 17*th*. We commenced early in the morning our downward voyage. Arroyos is situated in about 4° 10′ S. lat; and lies, therefore, about 130 miles from the mouth of the Tocantins. Fifteen miles above Guaribas another similar cataract called Tabocas lies across the river. We were told that there were in all fifteen of these obstructions to navigation between Arroyos and the mouth of the Araguaya. The worst was the Inferno, the Guaribas standing second to it in evil reputation. Many canoes and lives have been lost here, most of the accidents arising through the vessels being hurled against an enormous cubical mass of rock called the Guaribinha, which we, on our trip to the falls in the small canoe, passed round with the greatest ease about a quarter of a mile below the main falls. This, however, was the dry season; in the time of full waters a tremendous current sets against it. We descended the river rapidly, and found it excellent fun shooting the rapids. The men seemed to delight in choosing the swiftest parts of the current; they sang and yelled in the greatest excitement, working the paddles with great force, and throwing clouds of spray above us as we bounded downwards. We stopped to rest at the mouth of a rivulet named Caganxa. The pilot told us that gold has been found in the bed of this brook; so we had the curiosity to wade several hundred yards through the icy cold waters in search of it. Mr. Leavens seemed very much interested in the matter; he picked up all the shining stones he could espy in the pebbly bottom, in hopes of finding diamonds also. There is, in fact, no reason why both gold and diamonds should not be found here, the hills being a continuation of those of the mining countries of interior Brazil, and the brooks flowing through the narrow valleys between them.

On arriving at the place where we had left our canoe, we found poor José the mulatto much worse, so we hastened on to Juquerapuá to procure aid. An old half-caste woman took charge of him; she made poultices of the pulp of a wild fruit, administered cooling draughts made from herbs which grew near the house, and in fact acted the part of nurse admirably.

We stayed at this place all night and part of the following day, and I had a stroll along a delightful pathway, which led over hill and dale, two or three miles through the forest. I was surprised at the number and variety of brilliantly-coloured butterflies; they were all of small size, and started forth at every step I took, from the low bushes which bordered the road. I first heard here the notes of a trogon; it was seated alone on a branch, at no great elevation; a beautiful bird, with glossy-green back and rose-coloured breast (probably Trogon melanurus). At intervals it uttered, in a complaining tone, a sound resembling the words "quá, quá." It is a dull inactive bird, and not very ready to take flight when approached. In this respect, however, the trogons are not equal to the jacamars, whose stupidity in remaining at their posts, seated on low branches in the gloomiest shades of the forest, is somewhat remarkable in a country where all other birds are exceedingly wary. One species of jacamar was not uncommon here (Galbula viridis); I sometimes saw two or three together seated on a slender branch silent and motionless with the exception of a slight movement of the head; when an insect flew past within a short distance, one of the birds would dart off, seize it, and return again to its sitting place. The trogons are found in the tropics of both hemispheres; the jacamars, which are clothed in plumage of the most beautiful golden-bronze and steel colours, are peculiar to tropical America.

September 18th. We stayed only twenty-four hours at Juquerapuá, and then resumed our downward journey.

I was sorry to be obliged to leave this beautiful, though almost uninhabited, country so soon, our journey through it having been a mere tourist's gallop. Its vegetable and animal productions, of which we had obtained merely a glimpse, so to speak, were evidently different from those of the alluvial plains of the Amazons. The time we had spent, however, was too short for making a sufficient collection of specimens and facts to illustrate the amount and nature of the difference between the two faunas: a subject of no small importance as being calculated to throw light on the migrations of species across the equator in South America. In the rocky pools near Juquerapuá we found many species of fresh-water shells, and each of us, Mr. Leavens included, made a large collection of them. One was a turret-shaped univalve, a species of Melania, every specimen of which was worn at the apex; we tried in vain to get a perfect specimen. In the crystal waters the fishes could be seen as plainly as in an aquarium. One kind especially attracted our attention, a species of Diodon, which was not more than three inches long and of a pretty green colour banded with black; the natives call it Mamayacú. It is easily caught, and when in the hand distends itself, becoming as round as a ball. This fish amuses the people very much; when a person gets corpulent, they tell him he is as fat as a Mamayacú.

At night I slept ashore as a change from the confinement of the canoe, having obtained permission from Senhor Joaquim to sling my hammock under his roof. The house, like all others in these out-of-the-way parts of the country, was a large, open, palm-thatched shed, having one end inclosed by means of partitions also made of palm-leaves, so as to form a private apartment. Under the shed were placed all the household utensils; earthenware jars, pots, and kettles, hunting and fishing implements, paddles, bows and arrows, harpoons, and so forth. One or two common wooden chests serve to contain the holiday clothing of the females; there is no other furniture except a few stools and the hammock which answers the purposes of chair and sofa. When a visitor enters he is asked to sit down in a hammock; persons who are on intimate terms with each other recline together in the same hammock, one at each end; this is a very convenient arrangement for friendly conversation. There are neither tables nor chairs; the cloth for meals is spread on a mat, and the guests squat round in any position they choose. There is no cordiality of manners, but the treatment of the guests shows a keen sense of the duties of hospitality on the part of the host. There is a good deal of formality in the intercourse of these half-wild mamelucos which, I believe, has been chiefly derived from their Indian forefathers, although a little of it may have been copied from the Portuguese.

A little distance from the house were the open sheds under which the farinha for the use of the establishment was manufactured. In the centre of each shed stood the shallow pans, made of clay and built over ovens, where the meal is roasted. A long flexible cylinder made of the peel of a marantaceous plant, plaited into the proper form, hung suspended from a beam; it is in this that the pulp of the mandioca is pressed, and from it the juice, which is of a highly poisonous nature, although the pulp is wholesome food, runs into pans placed beneath to receive it. A wooden trough, such as is used in all these places for receiving the pulp before the poisonous matter is extracted, stood on the ground, and from the posts hung the long wicker-work baskets, or aturás, in which the women carry the roots from the roça or clearing; a broad ribbon made from the inner bark of the monguba tree is attached to the rims of the baskets, and is passed round the forehead of the carriers, to relieve their backs in supporting the heavy load. Around the shed were planted a number of banana and other fruit trees; amongst them were the never-failing capsicum-pepper bushes brilliant as holly-trees at Christmas time with their fiery red fruit, and lemon trees; the one supplying the pungent the other the acid for sauce to the perpetual meal of fish. There is never in such places any appearance of careful cultivation, no garden or orchard; the useful trees are surrounded by weeds and bushes, and close behind rises the everlasting forest.

There were other strangers under Senhor Joaquim's roof besides myself; mulattos, mamelucos, and Indians, so we formed altogether a large party. Houses occur at rare intervals in this wild country, and hospitality is freely given to the few passing travellers. After a frugal supper, a large wood fire was lighted in the middle of the shed, and all turned into their hammocks and began to converse. A few of the party soon dropped asleep; others, however, kept awake until a very late hour telling stories. Some related adventures which had happened to them whilst hunting or fishing; others recounted myths about the Curupíra, and other demons or spirits of the forest. They were all very appropriate to the time and place, for now and then a yell or a shriek resounded through the gloomy wilderness around the shed. One old parchment-faced fellow, with a skin the colour of mahogany, seemed to be a capital story-teller; but I was sorry I did not know enough of the language to follow him in all the details which he gave. Amongst other things he related an adventure he had once had with a jaguar. He got up from his hammock in the course of the narrative to give it the greater effect by means of gestures; he seized a bow and a large taquará arrow to show how he slew the beast, imitated its hoarse growl, and danced about the fire like a demon.

In descending the river we landed frequently, and Mr. Wallace and I lost no chance of adding to our collections; so that before the end of our journey we had got together a very considerable number of birds, insects, and shells chiefly taken, however, in the low country. Leaving Baiaõ we took our last farewell of the limpid waters and varied scenery of the upper river, and found ourselves again in the humid flat region of the Amazons valley. We sailed down this lower part of the river by a different channel from the one we travelled along in ascending, and frequently went ashore on the low islands in mid-river. As already stated, these are covered with water in the wet season;

but at this time, there having been three months of fine weather, they were dry throughout, and by the subsidence of the waters placed four or five feet above the level of the river. They are covered with a most luxuriant forest, comprising a large number of india-rubber trees. We found several people encamped here, who were engaged in collecting and preparing the rubber, and thus had an opportunity of observing the process.

The tree which yields this valuable sap is the Siphonia elastica, a member of the Euphorbiaceous order; it belongs, therefore, to a group of plants quite different from that which furnishes the caoutchouc of the East Indies and Africa. This latter is the product of different species of Ficus, and is considered, I believe, in commerce an inferior article to the india-rubber of Pará. The Siphonia elastica grows only on the lowlands in the Amazons region; hitherto the rubber has been collected chiefly in the islands and swampy parts of the mainland within a distance of fifty to a hundred miles to the west of Pará; but there are plenty of untapped trees still growing in the wilds of the Tapajos, Madeira, Juruá, and Jauarí, as far as 1800 miles from the Atlantic coast. The tree is not remarkable in appearance; in bark and foliage it is not unlike the European ash; but the trunk, like that of all forest trees, shoots up to an immense height before throwing off branches. The trees seem to be no man's property hereabout. The people we met with told us they came every year to collect rubber on these islands, as soon as the waters had subsided, namely, in August, and remained till January or February. The process is very simple. Every morning each person, man or woman, to whom is allotted a certain number of trees, goes the round of the whole and collects in a large vessel the milky sap which trickles from gashes made in the bark on the preceding evening, and which is received in little clay cups, or in ampullaria shells stuck beneath the wounds. The sap, which at first is of the consistence of cream, soon thickens; the collectors are provided with a great number of wooden moulds of the shape in which the rubber is wanted, and when they return to the camp they dip them into the liquid, laying on, in the course of several days, one coat after another. When this is done the substance is white and hard; the proper colour and consistency are given by passing it repeatedly through a thick black smoke obtained by burning the nuts of certain palm trees,* after which process the article is ready for sale. India-rubber is known throughout the province only by the name of seringa, the Portuguese word for syringe; it owes this appellation to the circumstance that it was in this form only that the first Portuguese settlers noticed it to be employed by the aborigines. It is said that the Indians were first taught to make syringes of rubber by seeing natural tubes formed by it when the spontaneously-flowing sap gathered round projecting twigs. Brazilians of all classes still use it extensively in the form of syringes, for injections form a great feature in the popular system of cures; the rubber for this

* The species I have seen used for this purpose are Maximiliana regia; Attalea excelsa; and Astrocaryum murumurum.

purpose is made into a pear-shaped bottle, and a quill fixed in the long neck.*

September 24th.—Opposite Cametá the islands are all planted with cacao, the tree which yields the chocolate nut. The forest is not cleared for the purpose, but the cacao plants are stuck in here and there almost at random amongst the trees. There are many houses on the banks of the river, all elevated above the swampy soil on wooden piles, and furnished with broad ladders by which to mount to the ground floor. As we passed by in our canoe we could see the people at their occupations in the open verandahs, and in one place saw a ball going on in broad daylight; there were fiddles and guitars hard at work, and a number of lads in white shirts and trousers dancing with brown damsels clad in showy print dresses. The cacao tree produces a curious impression on account of the flowers and fruit growing directly out of the trunk and branches. There is a whole group of wild fruit trees which have the same habit in this country. In the wildernesses where the cacao is planted, the collecting of the fruit is dangerous from the number of poisonous snakes which inhabit the places. One day, when we were running our montaria to a landing-place, we saw a large serpent on the trees overhead, as we were about to brush past; the boat was stopped just in the nick of time, and

* India-rubber is now one of the chief articles of export from Pará, and the government derives a considerable revenue from it. In value it amounts to one-third the total sum of exports. Thus in 1857 the amount was £139,000, the total exports being £450,720. In 1858, the rubber exported amounted to £123,000 and the total exports to £356,000.

Mr. Leavens brought the reptile down with a charge of shot.

September 26th.—At length we got clear of the islands, and saw once more before us the sea-like expanse of waters which forms the mouth of the Tocantins. The river had now sunk to its lowest point, and numbers of fresh-water dolphins were rolling about in shoaly places. There are here two species, one of which was new to science when I sent specimens to England; it is called the Tucuxí (Steno tucuxi of Gray). When it comes to the surface to breathe, it rises horizontally, showing first its back fin; draws an inspiration, and then dives gently down, head foremost. This mode of proceeding distinguishes the Tucuxí at once from the other species, which is called Bouto or porpoise by the natives (Inia Geoffroyi of Desmarest). When this rises the top of the head is the part first seen; it then blows, and immediately afterwards dips head downwards, its back curving over, exposing successively the whole dorsal ridge with its fin. It seems thus to pitch heels over head, but does not show the tail fin. Besides this peculiar motion, it is distinguished from the Tucuxí by its habit of generally going in pairs. Both species are exceedingly numerous throughout the Amazons and its larger tributaries, but they are nowhere more plentiful than in the shoaly water at the mouth of the Tocantins, especially in the dry season. In the Upper Amazons a third pale flesh-coloured species is also abundant (the Delphinus pallidus of Gervais). With the exception of a species found in the Ganges, all other varieties of dolphin inhabit exclusively the sea. In the broader parts of the Amazons, from its mouth to a distance of fifteen hundred miles in the interior, one or other of the three kinds here mentioned are always heard rolling, blowing, and snorting, especially at night, and these noises contribute much to the impression of sea-wide vastness and desolation which haunts the traveller. Besides dolphins in the water, frigate birds in the air are characteristic of this lower part of the Tocantins. Flocks of them were seen the last two or three days of our journey, hovering above at an immense height. Towards night we were obliged to cast anchor over a shoal in the middle of the river to await the ebb tide. The wind blew very strongly, and this, together with the incoming flow, caused such a heavy sea that it was impossible to sleep. The vessel rolled and pitched until every bone in our bodies ached with the bumps we received, and we were all more or less sea-sick. On the following day we entered the Anapu, and on the 30th September, after threading again the labyrinth of channels communicating between the Tocantins and the Moju, arrived at Pará.

I will now give a short account of Cametá, the principal town on the banks of the Tocantins, which I visited for the second time, in June, 1849; Mr. Wallace, in the same month, departing from Pará to explore the rivers Guamá and Capim. I embarked as passenger in a Cametá trading vessel, the St. John, a small schooner of thirty tons burthen. I had learnt by this time that the only way to attain the objects for which I had come to this country was to accustom myself to the ways of life of the humbler classes of the inhabitants. A traveller on the Amazons gains little by being furnished with letters of recommendation to persons of note, for in the great interior wildernesses of forest and river the canoe-men have pretty much their own way; the authorities cannot force them to grant passages or to hire themselves to travellers, and therefore a stranger is obliged to ingratiate himself with them in order to get conveyed from place to place. I thoroughly enjoyed the journey to Cametá; the weather was again beautiful in the extreme. We started from Pará at sunrise on the 8th of June, and on the 10th emerged from the narrow channels of the Anapú into the broad Tocantins. The vessel was so full of cargo, that there was no room to sleep in the cabin; so we passed the nights on deck. The captain or supercargo, called in Portuguese *cabo*, was a mameluco, named Manoel, a quiet, good-humoured person, who treated me with the most unaffected civility during the three days' journey. The pilot was also a mameluco, named John Mendez, a handsome young fellow, full of life and spirit. He had on board a wire guitar or viola, as it is here called; and in the bright moonlight nights, as we lay at anchor hour after hour waiting for the tide, he enlivened us all with songs and music. He was on the best of terms with the cabo, both sleeping in the same hammock slung between the masts. I passed the nights wrapped in an old sail outside the roof of the cabin. The crew, five in number, were Indians and half-breeds, all of whom treated their two superiors with the most amusing familiarity, yet I never sailed in a better managed vessel than the St. John.

In crossing to Cametá we had to await the flood-tide in a channel called Entre-as-Ilhas, which lies between two islands in mid-river, and John Mendez, being in good tune, gave us an extempore song, consisting of a great number of verses. The crew lay about the deck listening, and all joined in the chorus. Some stanzas related to me, telling how I had come all the way from "Ingalaterra" to skin monkeys and birds and catch insects; the last-mentioned employment of course giving ample scope for fun. He passed from this to the subject of political parties in Cametá; and then, as all the hearers were Cametaenses and understood the hits, there were roars of laughter, some of them rolling over and over on the deck, so much were they tickled. Party spirit runs high at Cametá, not merely in connection with local politics, but in relation to affairs of general concern, such as the election of members to the Imperial Parliament, and so forth. This political strife is partly attributable to the circumstance that a native of Cametá, Dr. Angelo Custodio Correia, had been in almost every election one of the candidates for the representation of the province. I fancied these shrewd but unsophisticated canoe-men saw through the absurdities attending these local contests, and hence their inclination to satirise them; they were, however, evidently partisans of Dr. Angelo. The brother of Dr. Angelo, Joaō Augusto Correia, a distinguished merchant, was an active canvasser. The party of the Correias was the

Liberal, or, as it is called throughout Brazil, the Santa Luzia faction; the opposite side, at the head of which was one Pedro Moraes, was the Conservative, or Saquarema party. I preserved one of the stanzas of the song, which, however, does not contain much point; it ran thus:—

> Ora paná, tana paná, paná taná,
> Joaõ Augusto hé bonito e homem pimpaõ,
> Mas Pedro hé feio e hum grande ladraõ,
> (Chorus) Ora paná, &c.

> John Augustus is handsome and as a man ought to be,
> But Peter is ugly and a great thief.
> (Chorus) Ora paná, &c.

The canoe-men of the Amazons have many songs and choruses, with which they are in the habit of relieving the monotony of their slow voyages, and which are known all over the interior. The choruses consist of a simple strain, repeated almost to weariness, and sung generally in unison, but sometimes with an attempt at harmony. There is a wildness and sadness about the tunes which harmonise well with, and in fact are born of, the circumstances of the canoe-man's life; the echoing channels, the endless gloomy forests, the solemn nights, and the desolate scenes of broad and stormy waters and falling banks. Whether they were invented by the Indians or introduced by the Portuguese it is hard to decide, as many of the customs of the lower classes of Portugese are so similar to those of the Indians that they have become blended with them. One of the commonest songs is very wild and pretty. It has for refrain the words "Mai, Mai,"

"Mother, Mother," with a long drawl on the second word. The stanzas are very variable; the best wit on board starts the verse, improvising as he goes on, and the others join in the chorus. They all relate to the lonely river life and the events of the voyage; the shoals, the wind; how far they shall go before they stop to sleep, and so forth. The sonorous native names of places, Goajará, Tucumandúba, &c., add greatly to the charm of the wild music. Sometimes they bring in the stars thus:—

A lua está sahindo,
Mai, Mai!
A lua está sahindo,
Mai, Mai!
As sete estrellas estaõ chorando,
Mai, Mai!
Por s'acharem desamparados,
Mai, Mai!

The moon is rising,
Mother, Mother!
The moon is rising,
Mother, mother!
The seven stars (Pleiades) are weeping,
Mother, Mother!
To find themselves forsaken,
Mother, mother!

I fell asleep about ten o'clock, but at four in the morning John Mendez woke me, to enjoy the sight of the little schooner tearing through the waves before a spanking breeze. The night was transparently clear and almost cold, the moon appeared sharply defined against the dark blue sky, and a ridge of foam marked where the prow of the vessel was cleaving its way through the water. The men had made a fire in the galley to make tea of an acid herb, called *erva cidreira,* a quantity of which they had gathered at the last landing-place, and the flames sparkled cheerily upwards. It is at such times as these that Amazon travelling is enjoyable, and one no longer wonders at the love which many, both natives and strangers, have for this wandering life. The little schooner sped rapidly on with booms bent and sails stretched to the utmost. Just as day dawned, we ran with scarcely slackened speed into the port of Cametá, and cast anchor.

I stayed at Cametá until the 16th of July, and made a considerable collection of the natural productions of the neighbourhood. The town in 1849 was estimated to contain about 5000 inhabitants, but the municipal district of which Cametá is the capital numbered 20,000 ; this, however, comprised the whole of the lower part of the Tocantins, which is the most thickly populated part of the province of Pará. The productions of the district are cacao, india-rubber, and Brazil nuts. The most remarkable feature in the social aspect of the place is the hybrid nature of the whole population, the amalgamation of the white and Indian races being here complete. The aborigines were originally very numerous on the western bank of the Tocantins, the principal tribe having been the Camútas, from which the city takes its name. They were a superior nation, settled, and attached to agriculture, and received with open arms the white immigrants who were attracted to the district by its fertility, natural beauty, and the healthfulness of the climate. The Portuguese settlers were nearly all males, the Indian women were good-looking, and made excellent wives; so the natural result has been, in the course of two centuries, a complete blending of the two races. There is now, however, a considerable infusion of negro blood in the mixture, several hundred African slaves having been introduced during the last seventy years. The few whites are chiefly Portuguese, but there are also two or three Brazilian families of pure European descent. The town consists of three long streets, running parallel to the river, with a few shorter ones crossing them at right angles. The houses are very plain, being built, as usual in this country, simply of a strong framework, filled up with mud, and coated with white plaster. A few of them are of two or three storeys. There are three churches, and also a small theatre, where a company of native actors at the time of my visit were representing light Portuguese plays with considerable taste and ability. The people have a reputation all over the province for energy and perseverance; and it is often said, that they are as keen in trade as the Portuguese. The lower classes are as indolent and sensual here as in other parts of the province, a moral condition not to be wondered at in a country where perpetual summer reigns, and where the necessaries of life are so easily obtained. But they are light-hearted, quick-witted, communicative, and hospitable. I found here a native poet, who had written some pretty verses, showing an appreciation of the natural beauties of the country, and was told that the Archbishop of Bahia, the primate of Brazil, was a native of Cametá. It is interesting to find the mamelucos displaying talent and enterprise, for it shows that degeneracy does not necessarily result from the mixture of white and Indian blood. The Cametaenses boast, as they have a right to do, of theirs being the only large town which resisted successfully the anarchists in the great rebellion of 1835-6. Whilst the whites of Pará were submitting to the rule of half-savage revolutionists, the mamelucos of Cametá placed themselves under the leadership of a courageous priest, named Prudencio; armed themselves, fortified the place, and repulsed the large forces which the insurgents of Pará sent to attack the place. The town not only became the refuge for all loyal subjects, but was a centre whence large parties of volunteers sallied forth repeatedly to attack the anarchists in their various strongholds.

The forest behind Cametá is traversed by several broad roads, which lead over undulating ground many miles into the interior. They pass generally under shade, and part of the way through groves of coffee and orange trees, fragrant plantations of cacao, and tracts of second-growth woods. The narrow brook-watered valleys, with which the land is intersected, alone have remained clothed with primæval forest, at least near the town. The houses along these beautiful roads belong chiefly to mameluco, mulatto, and Indian families, each of which has its own small plantation. There are only a few planters with larger establishments and these have seldom more than a dozen slaves. Besides the main roads, there are endless bye-paths which thread the forest, and communicate with isolated houses. Along these the traveller may wander day after day without leaving the shade, and everywhere meet with cheerful, simple, and hospitable people.

Soon after landing I was introduced to the most distinguished citizen of the place, Dr. Angelo Custodio Correia, whom I have already mentioned. This excellent man was a favourable specimen of the highest class of native Brazilians. He had been educated in Europe, was now a member of the Brazilian Parliament, and had been twice President of his native province. His manners were less formal, and his goodness more thoroughly genuine, perhaps, than is the rule generally with Brazilians. He was admired and loved, as I had ample opportunity of observing, throughout all Amazonia. He sacrificed his life in 1855, for the good of his fellow-townsmen, when Cametá was devastated by the cholera; having stayed behind with a few heroic spirits to succour invalids and direct the burying of the dead, when nearly all the chief citizens had fled from the place. After he had done what he could, he embarked for Pará, but was himself then attacked with cholera and died on board the steamer before he reached the capital. Dr. Angelo received me with the usual kindness which he showed to all strangers. He procured me, unsolicited, a charming country house, free of rent, hired a mulatto servant for me, and thus relieved me of the many annoyances and delays attendant on a first arrival in a country town where even the name of an inn is unknown. The rocinha thus given up for my residence belonged to a friend of his, Senhor José Raimundo Furtado, a stout florid-complexioned gentleman, such a one as might be met with any day in a country town in England. To him also I was indebted for many acts of kindness.

The rocinha was situated near a broad grassy road bordered by lofty woods, which leads from Cametá to the Aldeia, a village two miles distant. My first walks were along this road. From it branches another similar, but still more picturesque road, which runs to Curimá and Pacajá, two small settlements, several miles distant, in the heart of the forest. The Curimá road is beautiful in the extreme. About half a mile from the house where I lived it crosses a brook flowing through a deep dell, by means of a long rustic wooden bridge. The virgin forest is here left untouched; numerous groups of slender palms, mingled with lofty trees overrun with creepers and parasites, fill the shady glen and arch over the bridge, forming one of the most picturesque scenes imaginable. On the sunny slopes near this place, I found a great number of new and curious insects. A little beyond the bridge there was an extensive grove of orange and other trees, which also yielded me a rich harvest. The Aldeia road runs parallel to the river, the land from the border of the road to the indented shore of the Tocantins forming a long slope, which was also richly wooded; this slope was threaded by numerous shady paths and abounded in beautiful insects and birds. At the opposite or southern end of the town there was a broad road called the Estrada da Vacaria; this ran along the banks of the Tocantins at some distance from the river, and continued over hill and dale, through bamboo thickets and palm swamps, for about fifteen miles.

I found at Cametá an American, named Bean, who had been so long in the country that he had almost forgotten his mother tongue. He knew the neighbourhood well, and willingly accompanied me as guide in many long excursions. I was astonished in my walks with him at the universal friendliness of the people. We were obliged, when rambling along the intricate pathways through the woods, occasionally to pass the houses of settlers. The good people, most of whom knew Bean, always invited us to stop. The master of the house would step out first and insist on our walking in to take some refreshment; at the same moment I generally espied the female members of the family hurrying to the fireplace to prepare the inevitable cup of coffee. After conversing a little with the good folks we would take our leave, and then came the parting present—a bunch of bananas, a few eggs, or fruits of one kind or other. It would have been cruel to refuse these presents, but they were sometimes so inconvenient to us that we used to pitch them into the thickets as soon as we were out of sight of the donors.

One day we embarked in a montaria to visit a widow lady, named Dona Paulina, to whom Bean was going to be married, and who lived on one of the islands in mid-river, about ten miles above Cametá. The little boat had a mast and sail, the latter of which was of very curious construction. It was of the shape which sailors call shoulder-of-mutton sail, and was formed of laths of pith split from the leaf stalks of the Jupatí palm (Raphia tædigera). The laths were strung together so as to form a mat, and the sail was hoisted or lowered by means of a rope attached to the top. The same material serves for many purposes; partitions and even the external walls of houses of the poorer classes are often made of it. It fell to my charge to manage the sail during our voyage, whilst Bean steered, but when in the middle of the broad river the halyard broke, and in endeavouring to mend it we nearly upset the boat, for the wind blew strongly and the waves ran high. We fortunately met, soon afterwards, a negro who was descending in a similar boat to ours, and who, seeing our distress, steered towards us and kindly supplied us with a new rope. We stayed a day and night on the island. The house was of a similar description to those I have already described as common on the low islands of the Tocantins. The cacaoal which surrounded it consisted of about 10,000 trees, which I was astonished to hear produced altogether only 100 arrobas or 3200 pounds of the chocolate nut per annum. I had seen trees on the main land, which having been properly attended to, produced yearly thirty-two pounds each, or 100 times as much as those of Dona Paulina's cacaoal; the average yield in plantations on the Amazons near Santarem is 700 arrobas to 10,000 trees. Agriculture was evidently in a very low condition hereabout; the value of a cacao estate was very trifling, each tree being worth only forty reis or one penny, this including the land on which the plantation stands. A square league of country planted with cacao could thus be bought for 40*l.* or 50*l.* sterling. The selling price of cacao is very fluctuating; 3,500 reis, or about eight shillings the arroba of 32lbs., may be taken as the average. The management of a plantation requires very few hands; the tree yields three crops a-year, namely, one each in March, June, and September; but the June crop often fails, and those of the other months are very precarious. In the intervals between harvest-times the plantations require weeding; the principal difficulty is to keep the trees free from woody creepers and epiphytes, but especially from parasitic plants of the Loranthaceæ group, the same family to which our miseltoe belongs, and which are called "pés de passarinho," or "little birds' feet," from their pretty orange and red flowers resembling in shape and arrangement the three toes of birds. When the fruit is ready for gathering, neighbours help each other, and so each family is able to manage its own little plantation without requiring slaves. It appeared to me that cacao-growing would be an employment well suited to the habits and constitutions of European immigrants. All the work is done under shade; but it would yield a poor livelihood unless a better style of cultivation and preparation were introduced than that now prevailing here. The fruit is of oblong shape, and six to eight inches in length; the seeds are enveloped in a mass of white pulp which makes a delicious lemonade when mixed with water, and when boiled down produces an excellent jelly.

I found many interesting insects in the cacaoal; the most handsome was the Salamis jucunda, a magnificent butterfly with sickle-shaped wings, which flies with great rapidity, but is readily taken when quietly feeding on decaying cacao fruits. The island was three or four miles long and about a mile broad, and was situated in the central part of the river. The view from Dona Paulina's house was limited by the western row of islets, this middle channel being about a mile broad; not a glimpse was obtainable of the main land on either side, and each island was a mass of greenery, towering to a great height, and seeming to repose on the surface of the water. The house was in a very dilapidated condition; but Dona Paulina, who was a simple, good-natured little woman, with her slaves, tried to make us as comfortable as the circumstances permitted. At night it rained heavily, and the water poured through the broken tiles on to my hammock, so I was obliged to get up and shift my quarters; but this is a common incident in Brazilian houses.

The next day we crossed the river to the main land, to the house of Dona Paulina's father, where we slept, and on the following morning started to walk to Cametá through the forest, a distance of nine miles. The road was sometimes tolerably good, at others it was a mere track, and twice we had to wade through swamps which crossed the path. We started at six a.m., but did not reach Cametá until nine at night.

In the course of our walk I chanced to verify a fact relating to the habits of a large hairy spider of the genus Mygale, in a manner worth recording. The species was M. avicularia, or one very closely allied to it; the individual was nearly two inches in length of body, but the legs expanded seven inches, and the entire body and legs were covered with coarse grey and reddish hairs.

BIRD-KILLING SPIDER (MYGALE AVICULARIA) ATTACKING FINCHES.

Vol. I., page 161.

I was attracted by a movement of the monster on a tree-trunk; it was close beneath a deep crevice in the tree, across which was stretched a dense white web. The lower part of the web was broken, and two small birds, finches, were entangled in the pieces; they were about the size of the English siskin, and I judged the two to be male and female. One of them was quite dead, the other lay under the body of the spider not quite dead, and was smeared with the filthy liquor or saliva exuded by the monster. I drove away the spider and took the birds, but the second one soon died. The fact of species of Mygale sallying forth at night, mounting trees, and sucking the eggs and young of humming-birds, has been recorded long ago by Madame Merian and Palisot de Beauvois; but, in the absence of any confirmation, it has come to be discredited. From the way the fact has been related it would appear that it had been merely derived from the report of natives, and had not been witnessed by the narrators. Count Langsdorff, in his "Expedition into the Interior of Brazil," states that he totally disbelieved the story. I found the circumstance to be quite a novelty to the residents hereabout. The Mygales are quite common insects: some species make their cells under stones, others form artistical tunnels in the earth, and some build their dens in the thatch of houses. The natives call them Aranhas caranguejeiras, or crab-spiders. The hairs with which they are clothed come off when touched, and cause a peculiar and almost maddening irritation. The first specimen that I killed and prepared was handled incautiously, and I suffered terribly for three days afterwards. I think this is not owing to any poisonous quality residing in the hairs, but to their being short and hard, and thus getting into the fine creases of the skin. Some Mygales are of immense size. One day I saw the children belonging to an Indian family who collected for me with one of these monsters secured by a cord round its waist, by which they were leading it about the house as they would a dog.

The only monkeys I observed at Cametá were the Couxio (Pithecia Satanas), a large species, clothed with long brownish-black hair, and the tiny Midas argentatus. The Couxio has a thick bushy tail; the hair of the head sits on it like a cap, and looks as if it had been carefully combed. It inhabits only the most retired parts of the forest, on the terra firma, and I observed nothing of its habits. The little Midas argentatus is one of the rarest of the American monkeys. I have not heard of its being found anywhere except near Cametá. I once saw three individuals together running along a branch in a cacao grove near Cametá; they looked like white kittens: in their motions they resembled precisely the Midas ursulus already described. I saw afterwards a pet animal of this species, and heard that there were many so kept, and that they were esteemed as choice treasures. The one I saw was full-grown, but it measured only seven inches in length of body. It was covered with long, white, silky hairs, the tail was blackish, and the face nearly naked and flesh-coloured. It was a most timid and sensitive little thing. The woman who owned it carried it constantly in her bosom, and no money would induce her to part with her pet. She called it Mico. It fed from her mouth and allowed her to fondle it freely, but the nervous little creature would not permit strangers to touch it. If any one attempted to do so it shrank back, the whole body trembling with fear, and its teeth chattered, whilst it uttered its tremulous frightened tones. The expression of its features was like that of its more robust brother Midas ursulus; the eyes, which were black, were full of curiosity and mistrust, and it always kept them fixed on the person who attempted to advance towards it.

In the orange groves and other parts humming-birds were plentiful, but I did not notice more than three species. I saw a little pigmy belonging to the genus Phaethornis one day in the act of washing itself in a brook. It was perched on a thin branch, whose end was under water. It dipped itself, then fluttered its wings and pruned its feathers, and seemed thoroughly to enjoy itself alone in the shady nook which it had chosen—a place overshadowed by broad leaves of ferns and Heliconiæ. I thought as I watched it that there was no need for poets to invent elves and gnomes whilst Nature furnishes us with such marvellous little sprites ready to hand.

My return journey to Pará afforded many incidents characteristic of Amazonian travelling. I left Cametá on the 16th of July. My luggage was embarked in the morning in the Santa Rosa, a vessel of the kind called cuberta, or covered canoe. The cuberta is very much used on these rivers. It is not decked, but the sides forward are raised and arched over so as to admit of cargo being piled high above the water-line. At the stern is a neat square cabin, also raised, and between the cabin and covered fore part is a narrow piece decked over, on which are placed the cooking arrangements. This is called the tombadilha or quarter-deck, and when the canoe is heavily laden it goes under water as the vessel heels over to the wind. There are two masts, rigged with fore and aft sails. The foremast has often besides a main and top sail. The fore part is planked over at the top, and on this raised deck the crew work the vessel, pulling it along when there is no wind, by means of the long oars already described.

As I have just said, my luggage was embarked in the morning. I was informed that we should start with the ebb-tide in the afternoon, so I thought I should have time to pay my respects to Dr. Angelo and other friends, whose extreme courtesy and goodness had made my residence at Cametá so agreeable. After dinner the guests, according to custom at the house of the Correias, walked into the cool verandah which overlooks the river, and there we saw the Santa Rosa, a mere speck in the offing miles away, tacking down river with a fine breeze. I was now in a fix, for it would be useless attempting to overtake the cuberta, and besides the sea ran too high for any montaria. I was then told, that I ought to have been aboard hours before the time fixed for starting, because when a breeze springs up, vessels start before the tide turns; the last hour of the flood not being very strong. All my precious collections, my clothes, and other necessaries, were on board, and it was indispensable that I should be at

Pará when the things were disembarked. I tried to hire a montaria and men, but was told that it would be madness to cross the river in a small boat with this breeze. On going to Senhor Laroque, another of my Cametá friends, I was relieved of my embarrassment; I found there an English gentleman, Mr. Patchett of Pernambuco, who was visiting Pará and its neighbourhood on his way to England, and who, as he was going back to Pará in a small boat with four paddles, which would start at midnight, kindly offered me a passage. The evening from seven to ten o'clock was very stormy. About seven, the night became intensely dark, and a terrific squall of wind burst forth, which made the loose tiles fly over the house tops; to this succeeded lightning and stupendous claps of thunder, both nearly simultaneous. We had had several of these short and sharp storms during the past month. At midnight when we embarked, all was as calm as though a ruffle had never disturbed air, forest or river. The boat sped along like an arrow to the rhythmic paddling of the four stout youths we had with us, who enlivened the passage with their wild songs. Mr. Patchett and I tried to get a little sleep, but the cabin was so small and encumbered with boxes placed at all sorts of angles, that we found sleep impossible. I was just dozing when the day dawned, and, on awaking, the first object I saw was the Santa Rosa, at anchor under a green island in mid-river. I preferred to make the remainder of the voyage in the company of my collections, so bade Mr. Patchett good-day. The owner of the Santa Rosa, Senhor Jacinto Machado, whom I had not seen before, received me aboard, and apologised for having started without me. He was a white man, a planter, and was now taking his year's produce of cacao, about twenty tons, to Pará. The canoe was very heavily laden, and I was rather alarmed to see that it was leaking at all points. The crew were all in the water diving about to feel for the holes, which they stopped with pieces of rag and clay, and an old negro was baling the water out of the hold. This was a pleasant prospect for a three days' voyage! Senhor Machado treated it as the most ordinary incident possible. "It was always likely to leak, for it was an old vessel that had been left as worthless high and dry on the beach, and he had bought it very cheap."

When the leaks were stopped, we proceeded on our journey, and at night reached the mouth of the Anapú. I wrapped myself up in an old sail, and fell asleep on the raised deck. The next day we threaded the Igarapé-mirim, and on the 19th descended the Mojú. Senhor Machado and I by this time had become very good friends. At every interesting spot on the banks of the Mojú, he manned the small boat and took me ashore. There are many large houses on this river belonging to what were formerly large and flourishing plantations. Since the revolution of 1835-6, they had been suffered to go to decay. Two of the largest buildings were constructed by the Jesuits in the early part of the last century. We were told that there were formerly eleven large sugar-mills on the banks of the Mojú, but now there are only three. At Burujúba, there is a large monastery in a state of decay; part of the edifice, however, was inhabited by a Brazilian family. The walls are four feet in thickness. The long dark corridors and gloomy cloisters struck me as very inappropriate in the midst of this young and radiant nature. They would be more in place on some barren moor in northern Europe, than here in the midst of perpetual summer. The next turn in the river below Burujúba brought the city of Pará into view. The wind was now against us, and we were obliged to tack about. Towards evening it began to blow stiffly, the vessel heeled over very much, and Senhor Machado, for the first time, trembled for the safety of his cargo; the leaks burst out afresh, when we were yet two miles from the shore. He ordered another sail to be hoisted, in order to run more quickly into port, but soon afterwards an extra puff of wind came, and the old boat lurched alarmingly, the rigging gave way, and down fell boom and sail with a crash, encumbering us with the wreck. We were then obliged to have recourse to oars, and as soon as we were near the land, I begged Senhor Machado to send me ashore in the boat, with the more precious portion of my collections.

CHAPTER V.

CARIPÍ AND THE BAY OF MARAJÓ.

River Pará and Bay of Marajó—Journey to Caripí—Negro observance of Christmas—A German Family—Bats—Ant-eaters—Humming-birds—Excursion to the Murucupí—Domestic Life of the Inhabitants—Hunting Excursion with Indians—Natural History of the Paca and Cutia—Insects.

THAT part of the Pará river which lies in front of the city, as I have already explained, forms a narrow channel; being separated from the main waters of the estuary by a cluster of islands. This channel is about two miles broad, and constitutes part of the minor estuary of Goajará, into which the three rivers Guamá, Mojú, and Acará discharge their waters. The main channel of the Pará lies 10 miles away from the city, directly across the river; at that point, after getting clear of the islands, a great expanse of water is beheld, 10 to 12 miles in width; the opposite shore—the island of Marajó—being visible only in clear weather as a line of tree tops dotting the horizon. A little further upwards, that is to the south-west, the main land on the right or eastern shore appears, this is called Carnapijó; it is rocky, covered with the never-ending forest, and the coast which is fringed with broad sandy beaches, describes a gentle curve inwards. The broad reach of the Pará in front of this coast is called the Bahia, or bay of Marajó. The coast and the interior of the land are peopled by civilised Indians and Mamelucos, with a mixture of free negroes and mulattos. They are poor, for the waters are not abundant in fish, and they are dependent for a livelihood solely on their small plantations, and the scanty supply of game found in the woods. · The district was originally peopled by various tribes of Indians, of whom the principal were the Tupinambás and Nhengahíbas. Like all the coast tribes, whether inhabiting the banks of the Amazons or the sea-shore between Pará and Bahia, they were far more advanced in civilisation than the hordes scattered through the interior of the country, some of which still remain in the wild state, between the Amazons and the Plata. There are three villages on the coast of Carnapijó, and several planters' houses, formerly the centres of flourishing estates, which have now relapsed into forest in consequence of the scarcity of labour and diminished enterprise. One of the largest of these establishments is called Caripí: at the time of which I am speaking it belonged to a Scotch gentleman, Mr. Campbell, who had married the daughter of a large Brazilian proprietor. Most of the occasional English and American visitors to Pará had made some stay at Caripí, and it had obtained quite a reputation for the number and beauty of the birds and insects found there; I therefore applied for and obtained permission to spend two or three months at the place. The distance from Pará was about 23 miles, round by the northern end of the Ilha das onças (Isle of Tigers), which faces the city. I bargained for a passage thither with the cabo of a small trading vessel, which was going past the place, and started on the 7th of December, 1848.

We were 13 persons aboard; the cabo, his pretty mulatto mistress, the pilot and five Indian canoemen, three young mamelucos, tailor-apprentices who were taking a holiday trip to Cametá, a runaway slave heavily chained, and myself. The young mamelucos were pleasant, gentle fellows: they could read and write, and amused themselves on the voyage with a book containing descriptions and statistics of foreign countries, in which they seemed to take great interest; one reading whilst the others listened. At Uirapiranga, a small island behind the Ilha das onças, we had to stop a short time to embark several pipes of cashaça at a sugar estate. The cabo took the montaria and two men; the pipes were rolled into the water and floated to the canoe, the men passing cables round and towing them through a rough sea. Here we slept, and the following morning, continuing our voyage, entered a narrow channel which intersects the land of Carnapijó. At 2 p.m. we emerged from this channel, which is called the Aititúba, or Arrozal, into the broad Bahia, and then saw, two or three miles away to the left, the red-tiled mansion of Caripí, embosomed in woods on the shores of a charming little bay.

The water is very shallow near the shore, and when the wind blows there is a heavy ground swell. A few years previously an English gentleman, Mr. Graham, an amateur naturalist, was capsized here and drowned with his wife and child, whilst passing in a heavily-laden montaria to his large canoe. Remembering their fate, I was rather alarmed to see that I should be obliged to take all my luggage ashore in one trip in a leaky little boat. The pile of chests with two Indians and myself sank the montaria almost to the level of the water. I was kept busy baling all the way. The Indians manage canoes in this condition with admirable skill. They preserve the nicest equilibrium, and paddle so gently that not the slightest oscillation is perceptible. On landing, an old negress named Florinda, the feitora or manageress of the establishment which was kept only as a poultry farm and hospital for sick slaves, gave me the keys, and I forthwith took possession of the rooms I required.

I remained here nine weeks, or until the 12th of February, 1849. The house was very large and most substantially built, but consisted of only one story. I was told it was built by the Jesuits more than a century ago. The front had no verandah, the doors opening on a slightly elevated terrace about a hundred yards distant from the broad sandy beach. Around the residence the ground had been cleared to the extent of two or three acres, and was planted with fruit trees. Well-trodden pathways through the forest led to little colonies of the natives on the banks of retired creeks and rivulets in the interior. I led here a solitary but not unpleasant life ; there was a great charm in the loneliness of the place. The swell of the river beating on the sloping beach caused an unceasing murmur, which lulled me to sleep at night, and seemed appropriate music in those midday hours when all nature was pausing breathless under the rays of a vertical sun. Here I spent my first Christmas-day in a foreign land. The festival was celebrated by the negroes of their own free will and in a very pleasing manner. The room next to the one I had chosen was the capella, or chapel. It had a little altar which was neatly arranged, and the room was furnished with a magnificent brass chandelier. Men, women, and children were busy in the chapel all day on the 24th of December decorating the altar with flowers and strewing the floor with orange-leaves. They invited some of their neighbours to the evening prayers, and when the simple ceremony began an hour before midnight, the chapel was crowded. They were obliged to dispense with the mass, for they had no priest; the service therefore consisted merely of a long litany and a few hymns. There was placed on the altar a small image of the infant Christ, the "Menino Deos" as they called it, or the child-god, which had a long ribbon depending from its waist. An old white-haired negro led off the litany, and the rest of the people joined in the responses. After the service was over they all went up to the altar, one by one, and kissed the end of the ribbon. The gravity and earnestness shown throughout the proceedings were remarkable. Some of the hymns were very simple and beautiful, especially one beginning "Virgem soberana," a trace of whose melody springs to my recollection whenever I think on the dreamy solitude of Caripí.

The next day after I arrived two blue-eyed and red-haired boys came up and spoke to me in English, and presently their father made his appearance. They proved to be a German family named Petzell, who were living in the woods, Indian fashion, about a mile from Caripí. Petzell explained to me how he came here. He said that thirteen years ago he came to Brazil with a number of other Germans under engagement to serve in the Brazilian army. When his time had expired he came to Pará to see the country, but after a few months' rambling left the place to establish himself in the United States. There he married, went to Illinois, and settled as farmer near St. Louis. He remained on his farm seven or eight years, and had a family of five children. He could never forget, however, the free river life and perpetual summer of the banks of the Amazons, so he persuaded his wife to consent to break up their home in North America, and migrate to Pará. No one can imagine the difficulties the poor fellow had to go through before reaching the land of his choice. He first descended the Mississippi, feeling sure that a passage to Pará could be got at New Orleans. He was there told that the only port in North America he could start from was New York, so away he sailed for New York ; but there was no chance of a vessel sailing thence to Pará, so he took a passage to Demerara, as bringing him, at any rate, near to the desired land. There is no communication whatever between Demerara and Pará, and he was forced to remain here with his family four or five months, during which they all caught the yellow fever, and one of his children died. At length he heard of a small coasting vessel going to Cayenne, so he embarked and got thereby another stage nearer the end of his journey. A short time after reaching Cayenne he shipped in a schooner that was going to Pará, or rather the island of Marajó, for a cargo of cattle. He had now fixed himself, after all his wanderings, in a healthy and fertile little nook on the banks of a rivulet near Caripí, built himself a log hut, and planted a large patch of mandiocca and Indian corn. He seemed to be quite happy, but his wife complained much of the want of wholesome food, meat and wheaten bread. I asked the children whether they liked the country; they shook their heads, and said they would rather be in Illinois. Petzell told me that his Indian neighbours treated him very kindly; one or other of them called almost every day to see how he was getting on, and they had helped him in many ways. He had a high opinion of the Tapuyos, and said, "If you treat them well, they will go through fire to serve you."

Petzell and his family were expert insect collectors, so I employed them at this work during my stay at Caripí. The daily occurrences here were after a uniform fashion. I rose with the dawn, took a cup of coffee, and then sallied forth after birds. At ten I breakfasted, and devoted the hours from ten until three to entomology. The evening was occupied in preserving and storing my captures. Petzell and I sometimes undertook long excursions, occupying the whole day. Our neighbours used to bring me all the quadrupeds, birds, reptiles, and shells they met with, and so altogether I was enabled to acquire a good collection of the productions of the district.

The first few nights I was much troubled by bats. The room where I slept had not been used for many months, and the roof was open to the tiles and rafters. The first night I slept soundly and did not perceive anything unusual, but on the next I was aroused about midnight by the rushing noise made by vast hosts of bats sweeping about the room. The air was alive with them; they had put out the lamp, and when I relighted it the place appeared blackened with the impish multitudes that were whirling round and round. After I had laid about well with a stick for a few minutes they disappeared amongst the tiles, but when all was still again they returned, and once more extinguished the light. I took no further notice of them, and went to sleep. The next night several got into my hammock; I seized them as they were crawling over me, and dashed them against the wall. The next morning I found a wound, evidently caused by a bat, on my hip. This was rather unpleasant, so I set to work with the negroes, and tried to exterminate them. I shot a great many as they hung from the rafters, and the negroes having mounted with ladders to the roof outside, routed out from beneath the eaves many hundreds of them, including young broods. There were altogether four species, two belonging to the genus Dysopes, one to Phyllostoma, and the fourth to Glossophaga. By far the greater number belonged to the Dysopes perotis, a species having very large ears, and measuring two feet from tip to tip of the wings. The Phyllostoma was a small kind, of a dark gray colour, streaked with white down the back, and having a leaf-shaped fleshy expansion on the tip of the nose.

I was never attacked by bats except on this occasion. The fact of their sucking the blood of persons sleeping, from wounds which they make in the toes, is now well established; but it is only a few persons who are subject to this blood-letting. According to the negroes, the Phyllostoma is the only kind which attacks man. Those which I caught crawling over me were Dysopes, and I am inclined to think many different kinds of bat have this propensity.

One day I was occupied searching for insects in the bark of a fallen tree, when I saw a large cat-like animal advancing towards the spot. It came within a dozen yards before perceiving me. I had no weapon with me but an old chisel, and was getting ready to defend myself if it should make a spring, when it turned round hastily and trotted off. I did not obtain a very distinct view of it, but I could see its colour was that of the Puma, or American Lion, although it was much too small for that species. The Puma is not a common animal in the Amazons forests. I did not see altogether more than a dozen skins in the possession of the natives. The fur is of a fawn colour. On account of its hue resembling that of a deer common in the forests, the natives call it the Sassú-arána,* or the false deer; that is, an animal which deceives one at first sight by its superficial resemblance to a deer. The hunters are not at all afraid of it, and speak always in disparaging terms of

* The old zoologist Marcgrave, called the Puma the Cuguacuarana, probably (the c's being soft) a misspelling of Sassú-arána; hence the name Cougouar employed by French zoologists, and copied in most works on natural history.

ANT-EATER GRAPPLING WITH DOG.

Vol. I., page 177.

its courage. Of the Jaguar they give a very different account.

The only species of monkey I met with at Caripí was the same dark-coloured little Midas already mentioned as found near Pará. The great Ant-eater, Tamanduâ of the natives (Myrmecophaga jubata), was not uncommon here. After the first few weeks of residence I ran short of fresh provisions. The people of the neighbourhood had sold me all the fowls they could spare ; I had not yet learnt to eat the stale and stringy salt-fish which is the staple food in these places, and for several days I had lived on rice-porridge, roasted bananas, and farinha. Florinda asked me whether I could eat Tamanduá. I told her almost anything in the shape of flesh would be acceptable, so the same day she went with an old negro named Antonio and the dogs, and in the evening brought one of the animals. The meat was stewed and turned out very good, something like goose in flavour. The people at Caripí would not touch a morsel, saying it was not considered fit to eat in these parts ; I had read, however, that it was an article of food in other countries of South America. During the next two or three weeks, when we were short of fresh meat, Antonio was always ready, for a small reward, to get me a Tamanduá. But one day he came to me in great distress with the news that his favourite dog, Atrevido, had been caught in the grip of an ant-eater, and was killed. We hastened to the place, and found the dog was not dead, but severely torn by the claws of the animal, which itself was mortally wounded, and was now relaxing its grasp.

The habits of the Myrmecophaga jubata are now pretty well known. It is not uncommon in the drier forests of the Amazons valley, but is not found, I believe, in the Ygapó, or flooded lands. The Brazilians call the species the Tamanduá bandeira, or the Banner Ant-eater, the term banner being applied in allusion to the curious colouration of the animal, each side of the body having a broad oblique stripe half-gray and half-black, which gives it some resemblance to a heraldic banner. It has an excessively long slender muzzle, and a worm-like extensile tongue. Its jaws are destitute of teeth. The claws are much elongated, and its gait is very awkward. It lives on the ground, and feeds on termites, or white ants, the long claws being employed to pull in pieces the solid hillocks made by the insects, and the long flexible tongue to lick them up from the crevices. All the other species of this singular genus are arboreal. I met with four species altogether. One was the Myrmecophaga tetradactyla; the two others, more curious and less known, were very small kinds, called Tamanduá-i. Both are similar in size—ten inches in length, exclusive of the tail—and in the number of the claws, having two of unequal length to the anterior feet, and four to the hind feet. One species is clothed with grayish-yellow silky hair; this is of rare occurrence. The other has a fur of a dingy brown colour, without silky lustre. One was brought to me alive at Caripí, having been caught by an Indian clinging motionless inside a hollow tree. I kept it in the house about twenty-four hours. It had a moderately long snout, curved downwards, and extremely small eyes. It remained nearly all the time without motion, except when irritated, in which case it reared itself on its hind legs from the back of a chair to which it clung, and clawed out with its fore paws like a cat. Its manner of clinging with its claws, and the sluggishness of its motions, gave it a great resemblance to a sloth. It uttered no sound, and remained all night on the spot where I had placed it in the morning. The next day I put it on a tree in the open air, and at night it escaped. These small Tamanduás are nocturnal in their habits, and feed on those species of termites which construct earthy nests, that look like ugly excrescences on the trunks and branches of trees. The different kinds of ant-eaters are thus adapted to various modes of life, terrestrial and arboreal. Those which live on trees are again either diurnal or nocturnal, for Myrmecophaga tetradactyla is seen moving along the main branches in the daytime. The allied group of the Sloths, which are still more exclusively South American forms than ant-eaters are, at the present time furnish arboreal species only, but formerly terrestrial forms of sloths existed, as the Megatherium, whose mode of life was a puzzle, seeing that it was of too colossal a size to live on trees, until Owen showed how it might have obtained its food from the ground.

In January the orange-trees became covered with blossom—at least to a greater extent than usual, for they flower more or less in this country all the year round—and the flowers attracted a great number of humming-birds. Every day, in the cooler hours of the morning, and in the evening from four o'clock till six, they were to be seen whirring about the trees by scores. Their motions are unlike those of all other birds. They dart to and fro so swiftly that the eye can scarcely follow them, and when they stop before a flower it is only for a few moments. They poise themselves in an unsteady manner, their wings moving with inconceivable rapidity, probe the flower, and then shoot off to another part of the tree. They do not proceed in that methodical manner which bees follow, taking the flowers seriatim, but skip about from one part of the tree to another in the most capricious way. Sometimes two males close with each other and fight, mounting upwards in the struggle as insects are often seen to do when similarly engaged, and then separating hastily and darting back to their work. Now and then they stop to rest, perching on leafless twigs, when they may be sometimes seen probing, from the place where they sit, the flowers within their reach. The brilliant colours with which they are adorned cannot be seen whilst they are fluttering about, nor can the different species be distinguished unless they have a deal of white hue in their plumage, such as Heliothrix auritus, which is wholly white underneath although of a glittering green colour above, and the white-tailed Florisuga mellivora. There is not a great variety of humming-birds in the Amazons region, the number of species being far smaller in these uniform forest plains than in the diversified valleys of the Andes, under the same parallels of latitude. The family is divisible into two groups contrasted in form and habits, one containing species which live entirely in the shade of the forest, and the other comprising those which prefer open sunny places. The forest species (Phaethorninæ) are seldom seen at flowers, flowers being, in the shady places where they abide, of rare occurrence; but they search for insects on leaves, threading the bushes and passing above and beneath each leaf with wonderful rapidity. The other group (Trochilinæ) are not quite confined to cleared places, as they come into the forest wherever a tree is in blossom, and descend into sunny openings where flowers are to be found. But it is only where the woods are less dense than usual that this is the case; in the lofty forests and twilight shades of the low lands and islands they are scarcely ever seen. I searched well at Caripí, expecting to find the Lophornis Gouldii, which I was told had been obtained in the locality. This is one of the most beautiful of all humming-birds, having round its neck a frill of long white feathers tipped with golden green. I was not, however, so fortunate as to meet with it. Several times I shot by mistake a humming-

Humming-bird and Humming-bird Hawk-moth.

bird hawk-moth instead of a bird. This moth (Macroglossa Titan) is somewhat smaller than humming-birds generally are, but its manner of flight, and the way it poises itself before a flower whilst probing it with its proboscis are precisely like the same actions of humming-birds. It was only after many days' experience that I learnt to distinguish one from the other when on the wing. This resemblance has attracted the notice of the natives, all of whom, even educated whites, firmly believe that one is transmutable into the other. They have observed the metamorphosis of caterpillars into butterflies, and think it not at all more wonderful that a moth should change into a humming-bird. The resemblance between this hawk-moth and a humming-bird is certainly very curious, and strikes one even when both are examined in the hand. Holding them sideways, the shape of the head and position of the eyes in the moth are seen to be nearly the same as in the bird, the extended proboscis representing the long beak. At the tip of the moth's body there is a brush of long hair-scales resembling feathers, which, being expanded, looks very much like a bird's tail. But, of course, all these points of resemblance are merely superficial. The negroes and Indians tried to convince me that the two were of the same species. "Look at their feathers," they said; "their eyes are the same, and so are their tails." This belief is so deeply rooted that it was useless to reason with them on the subject. The Macroglossa moths are found in most countries, and have everywhere the same habits; one well-known species is found in England.

Mr. Gould relates that he once had a stormy altercation with an English gentleman, who affirmed that humming-birds were found in England, for he had seen one flying in Devonshire, meaning thereby the moth Macroglossa stellatarum. The analogy between the two creatures has been brought about, probably, by the similarity of their habits, there being no indication of the one having been adapted in outward appearance with reference to the other.

It has been observed that humming-birds are unlike other birds in their mental qualities, resembling in this respect insects rather than warm-blooded vertebrate animals. The want of expression in their eyes, the small degree of versatility in their actions, the quickness and precision of their movements, are all so many points of resemblance between them and insects. In walking along the alleys of the forest a Phaethornis frequently crosses one's path, often stopping suddenly and remaining poised in mid-air, a few feet distant from the face of the intruder. The Phaethorninæ are certainly more numerous in individuals in the Amazons region than the Trochilinæ. They build their nests, which are made of fine vegetable fibres and lichens, densely woven together and thickly lined with silk-cotton from the fruit of the samaüma tree (Eriodendron samaüma), on the inner sides of the tips of palm fronds. They are long and purse-shaped. The young when first hatched have very much shorter bills than their parents. The only species of Trochilinæ which I found at Caripí were the little brassy-green Polytmus viridissimus, the Sapphire and emerald (Thalurania furcata), and the large falcate-winged Campylopterus obscurus.

Snakes were very numerous at Caripí ; many harmless species were found near the house, and these sometimes came into the rooms. I was wandering one day amongst the green bushes of Guajará, a tree which yields a grape-like berry (Chrysobalanus Icaco) and grows along all these sandy shores, when I was startled by what appeared to be the flexuous stem of a creeping plant endowed with life and threading its way amongst the leaves and branches. This animated liana turned out to be a pale-green snake, the Dryophis fulgida. Its whole body is of the same green hue, and it is thus rendered undistinguishable amidst the foliage of the Guajará bushes, where it prowls in search of its prey, tree-frogs and lizards. The forepart of its head is prolonged into a slender pointed beak, and the total length of the reptile was six feet. There was another kind found amongst bushes on the borders of the forest closely allied to this, but much more slender, viz., the Dryophis acuminata. This grows to a length of 4 feet 8 inches, the tail alone being 22 inches ; but the diameter of the thickest part of the body is little more than a quarter of an inch. It is of light brown colour, with iridescent shades variegated with obscurer markings, and looks like a piece of whipcord. One individual which I caught of this species had a protuberance near the middle of the body. On opening it I found a half-digested lizard which was much more bulky than the snake itself. Another kind of serpent found here, a species of Helicops, was amphibious in its habits.

I saw several of this in wet weather on the beach, which, on being approached, always made straightway for the water, where they swam with much grace and dexterity. Florinda one day caught a Helicops whilst angling for fish, it having swallowed the fish-hook with the bait. She and others told me these water-snakes lived on small fishes, but I did not meet with any proof of the fact. In the woods, snakes were constantly occurring: it was not often, however, that I saw poisonous species. There were many arboreal kinds besides the two just mentioned; and it was rather alarming, in entomologising about the trunks of trees, to suddenly encounter, on turning round, as sometimes happened, a pair of glittering eyes and a forked tongue within a few inches of one's head. The last kind I shall mention is the Coral-snake, which is a most beautiful object when seen coiled up on black soil in the woods. The one I saw here was banded with black and vermilion, the black bands having each two clear white rings. The state of specimens preserved in spirits can give no idea of the brilliant colours which adorn the Coral-snake in life.

Petzell and I, as already mentioned, made many excursions of long extent in the neighbouring forest. We sometimes went to Murucupí, a creek which passes through the forest about four miles behind Caripí, the banks of which are inhabited by Indians and half-breeds who have lived there for many generations in perfect seclusion from the rest of the world, the place being little known or frequented. A path from Caripí leads to it through a gloomy tract of virgin forest, where the trees are so closely packed together that the ground beneath is thrown into the deepest shade, under which nothing but fetid fungi and rotting vegetable débris is to be seen. On emerging from this unfriendly solitude near the banks of the Murucupí, a charming contrast is presented. A glorious vegetation, piled up to an immense height, clothes the banks of the creek, which traverses a broad tract of semi-cultivated ground, and the varied masses of greenery are lighted up with the sunny glow. Open palm-thatched huts peep forth here and there from amidst groves of banana, mango, cotton, and papaw trees and palms. On our first excursion, we struck the banks of the river in front of a house of somewhat more substantial architecture than the rest, having finished mud walls, plastered and white-washed, and a covering of red tiles. It seemed to be full of children, and the aspect of the household was improved by a number of good-looking mameluco women, who were busily employed washing, spinning, and making farinha. Two of them, seated on a mat in the open verandah, were engaged sewing dresses, for a festival was going to take place a few days hence at Balcarem, a village eight miles distant from Murucupí, and they intended to be present to hear mass and show their finery. One of the children, a naked boy about seven years of age, crossed over with the montaria to fetch us. We were made welcome at once, and asked to stay for dinner. On our accepting the invitation a couple of fowls were killed, and a wholesome stew of seasoned rice and fowls soon put in preparation. It is not often that the female members of a family in these retired places are familiar with strangers; but these people had lived a long time in the capital, and therefore were more civilised than their neighbours. Their father had been a prosperous tradesman, and had given them the best education the place afforded. After his death the widow with several daughters, married and unmarried, retired to this secluded spot, which had been their sitio, farm or country house, for many years. One of the daughters was married to a handsome young mulatto, who was present and sang us some pretty songs, accompanying himself on the guitar.

After dinner I expressed a wish to see more of the creek, so a lively and polite old man, whom I took to be one of the neighbours, volunteered as guide. We embarked in a little montaria, and paddled some three or four miles up and down the stream. Although I had now become familiarised with beautiful vegetation, all the glow of fresh admiration came again to me in this place. The creek was about 100 yards wide, but narrower in some places. Both banks were masked by lofty walls of green drapery, here and there a break occurring through which, under over-arching trees, glimpses were obtained of the palm-thatched huts of settlers. The projecting boughs of lofty trees, which in some places stretched half-way across the creek, were hung with natural garlands and festoons, and an endless variety of creeping plants clothed the water frontage, some of which, especially the Bignonias, were ornamented with large gaily-coloured flowers. Art could not have assorted together beautiful vegetable forms so harmoniously as was here done by Nature. Palms, as usual, formed a large proportion of the lower trees; some of them, however, shot up their slim stems to a height of sixty feet or more, and waved their bunches of nodding plumes between us and the sky. One kind of palm, the Pashiúba (Iriartea exorhiza), which grows here in greater abundance than elsewhere, was especially attractive. It is not one of the tallest kinds, for when full-grown its height is not more, perhaps, than forty feet; the leaves are somewhat less drooping, and the leaflets much broader than in other species, so that they have not that feathery appearance which those of some palms have, but still they possess their own peculiar beauty. My guide put me ashore in one place to show me the roots of the Pashiúba. These grow above ground, radiating from the trunk many feet above the surface, so that the tree looks as if supported on stilts; and a person can, in old trees, stand upright amongst the roots with the perpendicular stem wholly above his head. It adds to the singularity of their appearance, that these roots, which have the form of straight rods, are studded with stout thorns, whilst the trunk of the tree is quite smooth. The purpose of this curious arrangement is, perhaps, similar to that of the buttress roots already described; namely, to recompense the tree by root-growth above the soil for its inability, in consequence of the competition of neighbouring roots, to extend it underground. The great amount of moisture and nutriment contained in the atmosphere, may also favour these growths.

On returning to the house, I found Petzell had been

well occupied during the hot hours of the day collecting insects in a neighbouring clearing. He had obtained no less than six species new to me of the beautiful family of Longicornes belonging to the order Coleoptera. Our kind hosts gave us a cup of coffee about five o'clock, and we then started for home. The last mile of our walk was performed in the dark. The forest in this part is obscure even in broad daylight, but I was scarcely prepared for the intense opacity of darkness which reigned here on this night, and which prevented us from seeing each other, although walking side by side. Nothing occurred of a nature to alarm us, except that now and then a sudden rush was heard amongst the trees, and once a dismal shriek startled us. Petzell tripped at one place and fell all his length into the thicket. With this exception, we kept well to the pathway, and in due time arrived safely at Caripí.

One of my neighbours at Murucupí was a hunter of reputation in these parts. He was a civilised Indian, married and settled, named Raimundo, whose habit was to sally forth at intervals to certain productive hunting grounds, whose situation he kept secret, and procure fresh provisions for his family. I had found out by this time, that animal food was as much a necessary of life in this exhausting climate as it is in the North of Europe. An attempt which I made to live on vegetable food was quite a failure, and I could not eat the execrable salt fish which Brazilians use. I had been many days without meat of any kind, and nothing more was to be found near Caripí, so I asked as a favour of Senhor Raimundo, permission to accompany him on one of his hunting trips, and shoot a little game for my own use. He consented, and appointed a day on which I was to come over to his house to sleep, so as to be ready for starting with the ebb-tide shortly after midnight.

The locality we were to visit was situated near the extreme point of the land of Carnapijó, where it projects northwardly into the middle of the Pará estuary and is broken into a number of islands. On the afternoon of January 11th, 1849, I walked through the woods to Raimundo's house, taking nothing with me but a double-barrelled gun, a supply of ammunition and a box for the reception of any insects I might capture. Raimundo was a carpenter, and seemed to be a very industrious man; he had two apprentices, Indians like himself, one a young lad, and the other apparently about twenty years of age. His wife was of the same race. The Indian women are not always of a taciturn disposition like their husbands. Senhora Dominga was very talkative; there was another old squaw at the house on a visit, and the tongues of the two were going at a great rate the whole evening, using only the Tupí language. Raimundo and his apprentices were employed building a canoe. Notwithstanding his industry, he seemed to be very poor, and this was the condition of most of the residents on the banks of the Murucupí. They have, nevertheless, considerable plantations of mandioca and Indian corn, besides small plots of cotton, coffee, and sugar cane; the soil is very fertile, they have no rent to pay, and no direct taxes. There is, moreover, always a market in Pará, twenty miles distant, for their surplus produce, and a ready communication with it by water.

Their poverty seemed to be owing chiefly to two causes. The first is, the prevalence amongst them of a kind of communistic mode of regarding property. The Indian and mameluco country people have a fixed notion that their neighbours have no right to be better off than themselves. If any of them have no food, canoe, or weapons, they beg or borrow without scruple of those who are better provided, and it is the custom not to refuse the gift or the loan. There is no inducement, therefore, for one family to strive or attempt to raise itself above the others. There is always a number of lazy people who prefer to live at the cost of their too good-natured neighbours. The other cause is, the entire dependence of the settlers on the precarious yields of hunting and fishing for their supply of animal food; which is here, as already mentioned, as indispensable an article of diet as in cold climates. The young and strong who are able and willing to hunt and fish, are few. Raimundo, like all other hard-working men in these parts, had to neglect his regular labour every four or five days, and devote a day and a night to hunting or fishing. It does not seem to occur to these people, that they could secure a constant supply of meat by keeping cattle, sheep, or hogs, and feeding them with the produce of their plantations. This touches, however, on a fundamental defect of character which has been inherited from their Indian ancestors. The Brazilian aborigines had no notion of domesticating animals for use; and such is the inflexibility of organisation in the red man, and by inheritance from Indians also in half-breeds, that the habit seems impossible to be acquired by them, although they show great aptitude in other respects for civilised life. Is this attributable fundamentally to the absence in South America of indigenous animals suitable for domestication? It would appear so; and this is a great deficiency in a land otherwise so richly endowed by nature. This, however, is a difficult question, and involves many other considerations. The presence or absence of domesticable animals in a country, no doubt, has a very great influence on the character and culture of races. The North American Indians, especially those of Florida, offered many points of similarity in character and social condition to the Indians of the Amazons region; and they were, like them, condemned, probably from the same cause, to depend for existence chiefly on the produce of the chase or fishing. On the other hand, the Indians of Peru, whose more favoured home contained the Llama, were enabled to reach a high degree of civilisation, a great help thereto being this priceless animal, which served as a beast of burthen, and yielded wool for clothing, and milk, cheese, and flesh for nourishment. In the plains of Tropical America there exists no animal comparable to the ox, the horse, the sheep, or the hog. Of the last-mentioned, indeed, there are two wild species; but they are not closely allied to the European domestic hog. Of the other three animals, which have been such important helps to incipient civilisation in Asia and Europe, the genera even are unknown in South America. There is no lack in the Amazonian forests of tameable animals fit for human food ; the tapir, the paca, the cutía, and the curassow turkeys, are often kept in houses and become quite as tame as the domesticated animals of the old world ; but they are useless from not breeding in confinement. Curassow birds are often seen in the houses of Indians ; one fine species, the Mitu tuberosa, becoming so familiar that it follows children about wherever they go ; it will not propagate, however, in captivity. It is shown to be not wholly the fault of the natives in this case, by their valuing the common fowl, which has been imported from Europe and adopted everywhere, even by remote tribes on rivers rarely visited by white men. It is, however, treated with little attention, and increases very slowly. The Indians do not show themselves so sensible of the advantages derivable from the ox, sheep, and hog, all of which have been introduced into their country. They seem unable to acquire a taste for their flesh, and the management of the animals in a domesticated state is evidently unsuited to their confirmed habits. The inferiority of the native animals compared with those of the old world in regard to capability of breeding in confinement, to which, according to this view, is originally owing the defect in the Indian character regarding the domestication of animals, has been brought about, probably, in some way not easily explicable, by the domination of the forest. It has been lately advanced by ethnologists, that where dense forests clothe the surface of a country, the native races of man cannot make any progress in civilisation.

It might be added, that vast and monotonous naked plains produce the same result. The animals which have been so useful in the infancy of human civilisation are such as roamed originally over open or scantily wooded plains, probably of limited extent. The fact of many delicious wild fruits existing in the forest which they have never learnt to cultivate seems to show, contrary to the view here advanced, that it is innate stupidity rather than want of materials, that has deprived the Indians of these helps to civilization. There is a kind of rice, growing wild on the banks of many of the tributaries of the Amazons, which they have never reclaimed, although they have adopted the plant introduced into the country by Europeans.*

In the evening we had more visitors. The sounds of pipe and tabor were heard, and presently a procession of villagers emerged from a pathway through the mandioca fields. They were on a begging expedition for St. Thomé, the patron saint of Indians and Mamelucos. One carried a banner, on which was rudely painted the figure of St. Thomé with a glory round his head. The pipe and tabor were of the simplest description. The pipe was a reed pierced with four holes, by means of

* Many useful vegetable products have been reclaimed, and it is to the credit of the Indians that they have discovered the use of the Mandioca plant, which is highly poisonous in the raw state, and requires a long preparation to fit it for use. It is cultivated throughout the whole of Tropical America, including Mexico and the West India Islands, but only in the plains, not being seen, according to Humboldt, higher than 600 to 800 metres, at which elevation it grows, on the Mexican Andes. I believe it is not known in what region the plant originated; it is not found wild in the Amazons valley.

which a few unmusical notes were produced, and the tabor was a broad hoop with a skin stretched over each end. A deformed young man played both the instruments. Senhor Raimundo received them with the quiet politeness, which comes so naturally to the Indian when occupying the position of host. The visitors, who had come from the Villa de Condé, five miles through the forest, were invited to rest. Raimundo then took the image of St. Thomé from one of the party, and placed it by the side of Nossa Senhora in his own oratorio, a little decorated box in which every family keeps its household gods; finally lighting a couple of wax candles before it. Shortly afterwards a cloth was laid on a mat, and all the guests were invited to supper. The fare was very scanty; a boiled fowl with rice, a slice of roasted pirarucú, farinha, and bananas. Each one partook very sparingly, some of the young men contenting themselves with a plateful of rice. One of the apprentices stood behind with a bowl of water and a towel, with which each guest washed his fingers and rinsed his mouth after the meal. They stayed all night: the large open shed was filled with hammocks, which were slung from pole to pole; and on retiring, Raimundo gave orders for their breakfast in the morning.

Raimundo called me at two o'clock, when we embarked, he, his older apprentice Joaquim, and myself, in a shady place where it was so dark that I could see neither canoe nor water, taking with us five dogs. We glided down a winding creek where huge trunks of trees slanted across close overhead, and presently emerged into the Murucupí. A few yards further on we entered the broader channel of the Aititúba. This we crossed, and entered another narrow creek on the opposite side. Here the ebb tide was against us, and we had great difficulty in making progress. After we had struggled against the powerful current a distance of two miles, we came to a part where the ebb tide ran in the opposite direction, showing that we had crossed the water-shed. The tide flows into this channel or creek at both ends simultaneously, and meets in the middle, although there is apparently no difference of level, and the breadth of the water is the same. The tides are extremely intricate throughout all the infinite channels and creeks which intersect the lands of the Amazons delta. The moon now broke forth and lighted up the trunks of colossal trees, the leaves of monstrous Júpatí palms which arched over the creek, and revealed groups of arborescent arums standing like rows of spectres on its banks. We had a glimpse now and then into the black depths of the forest, where all was silent except the shrill stridulation of wood-crickets. Now and then a sudden plunge in the water ahead would startle us, caused by heavy fruit or some nocturnal animal dropping from the trees. The two Indians here rested on their paddles and allowed the canoe to drift with the tide. A pleasant perfume came from the forest, which Raimundo said proceeded from a cane-field. He told me that all this land was owned by large proprietors at Pará, who had received grants from time to time from the Government for political services. Raimundo was quite in a talkative humour; he related to me many incidents of the time of the "Cabanagem," as the revolutionary days of 1835-6 are popularly called. He said he had been much suspected himself of being a rebel; but declared that the suspicion was unfounded. The only complaint he had to make against the white man was, that he monopolised the land without having any intention or prospect of cultivating it. He had been turned out of one place where he had squatted and cleared a large piece of forest. I believe the law of Brazil at this time was that the new lands should become the property of those who cleared and cultivated them, if their right was not disputed within a given term of years by some one who claimed the proprietorship. This land-law has since been repealed, and a new one adopted founded on that of the United States. Raimundo spoke of his race as the red-skins, "pelle vermelho;" they meant well to the whites, and only begged to be let alone. "God," he said, "had given room enough for us all." It was pleasant to hear the shrewd good-natured fellow talk in this strain. Our companion, Joaquim, had fallen asleep; the night air was cool, and the moonlight lit up the features of Raimundo, revealing a more animated expression than is usually observable in Indian countenances. I always noticed that Indians were more cheerful on a voyage, especially in the cool hours of night and morning, than when ashore. There is something in their constitution of body which makes them feel excessively depressed in the hot hours of the day, especially inside their houses. Their skin is always hot to the touch. They certainly do not endure the heat of their own climate so well as the whites. The negroes are totally different in this respect; the heat of midday has very little effect on them, and they dislike the cold nights on the river.

We arrived at our hunting-ground about half-past four. The channel was here broader and presented several ramifications. It yet wanted an hour and a half to day-break, so Raimundo recommended me to have a nap. We both stretched ourselves on the benches of the canoe and fell asleep, letting the boat drift with the tide, which was now slack. I slept well considering the hardness of our bed, and when I awoke in the middle of a dream about home-scenes the day was beginning to dawn. My clothes were quite wet with the dew. The birds were astir, the cicadas had begun their music, and the Urania Leilus, a strange and beautiful tailed and gilded moth, whose habits are those of a butterfly, commenced to fly in flocks over the tree tops. Raimundo exclaimed "Clareia o dia!" "The day brightens!" The change was rapid: the sky in the east assumed suddenly the loveliest azure colour, across which streaks of thin, white clouds were painted. It is at such moments as this when one feels how beautiful our earth truly is! The channel on whose waters our little boat was floating was about 200 yards wide; others branched off right and left, surrounding the group of lonely islands which terminate the land of Carnapijó. The forest on all sides formed a lofty hedge without a break: below, it was fringed with mangrove bushes, whose small foliage contrasted with the large glossy leaves of the taller trees, or the feather and fan-shaped fronds of palms.

Being now arrived at our destination, Raimundo turned

up his trousers and shirt-sleeves, took his long hunting-knife, and leapt ashore with the dogs. He had to cut a gap in order to enter the forest. We expected to find Pacas and Cutías; and the method adopted to secure them was this: at the present early hour they would be seen feeding on fallen fruits, but would quickly, on hearing a noise, betake themselves to their burrows: Raimundo was then to turn them out by means of the dogs, and Joaquim and I were to remain in the boat with our guns, ready to shoot all that came to the edge of the stream, the habit of both animals, when hard-pressed, being to take to the water. We had not long to wait. The first arrival was a Paca, a reddish, nearly tailless Rodent, spotted with white on the sides, and intermediate in size and appearance between a hog and a hare. My first shot did not take effect; the animal dived into the water and did not re-appear. A second was brought down by my companion as it was rambling about under the mangrove bushes. A Cutía next appeared: this is also a Rodent, about one-third the size of the Paca: it swims, but does not dive, and I was fortunate enough to shoot it. We obtained in this way two more Pacas and another Cutía. All the time the dogs were yelping in the forest. Shortly afterwards Raimundo made his appearance, and told us to paddle to the other side of the island. Arrived there, we landed and prepared for breakfast. It was a pretty spot; a clean, white, sandy beach beneath the shade of wide-spreading trees. Joaquim made a fire. He first scraped fine shavings from the midrib of a Bacaba palm-leaf; these he piled into a little heap in a dry place, and then struck a light in his bamboo tinder-box with a piece of an old file and a flint, the tinder being a felt-like soft substance manufactured by an ant (Polyrhachis bispinosus). By gentle blowing, the shavings ignited, dry sticks were piled on them, and a good fire soon resulted. He then singed and prepared the cutía, finishing by running a spit through the body and fixing one end in the ground in a slanting position over the fire. We had brought with us a bag of farinha and a cup containing a lemon, a dozen or two of fiery red peppers, and a few spoonsful of salt. We breakfasted heartily when our cutía was roasted, and washed the meal down with a calabash full of the pure water of the river.

After breakfast the dogs found another cutía, which was hidden in its burrow two or three feet beneath the roots of a large tree, and took Raimundo nearly an hour to disinter it. Soon afterwards we left this place, crossed the channel, and, paddling past two islands, obtained a glimpse of the broad river between them, with a long sandy spit, on which stood several scarlet ibises and snowy-white egrets. One of the islands was low and sandy, and half of it was covered with gigantic arum-trees, the often-mentioned Caladium arborescens, which presented a strange sight. Most people are acquainted with the little British species, Arum maculatum, which grows in hedge bottoms, and many, doubtless, have admired the larger kinds grown in hot-houses; they can therefore form some idea of a forest of arums. On this islet the woody stems of the plants near the bottom were 8 to 10 inches in diameter, and the trees were 12 to 15 feet high; all growing together in such a manner that there was just room for a man to walk freely between them. There was a canoe inshore, with a man and a woman: the man, who was hooting with all his might, told us in passing that his son was lost in the "aningal" (arum-grove). He had strayed whilst walking ashore, and the father had now been an hour waiting for him in vain.

About one o'clock we again stopped at the mouth of a little creek. It was now intensely hot. Raimundo said deer were found here, so he borrowed my gun, as being a more effective weapon than the wretched arms called Lazarinos, which he, in common with all the native hunters, used, and which sell at Pará for seven or eight shillings apiece. Raimundo and Joaquim now stripped themselves quite naked, and started off in different directions through the forest, going naked in order to move with less noise over the carpet of dead leaves, amongst which they stepped so stealthily that not the slightest rustle could be heard. The dogs remained in the canoe, in the neighbourhood of which I employed myself two hours entomologising. At the end of that time my two companions returned, having met with no game whatever.

We now embarked on our return voyage. Raimundo cut two slender poles, one for a mast and the other for a sprit: to these he rigged a sail we had brought in the boat, for we were to return by the open river, and expected a good wind to carry us to Caripí. As soon as we got out of the channel we began to feel the wind—the sea-breeze, which here makes a clean sweep from the Atlantic. Our boat was very small and heavily laden, and when, after rounding a point, I saw the great breadth we had to traverse, seven miles, I thought the attempt to cross in such a slight vessel foolhardy in the extreme. The waves ran very high: there was no rudder; Raimundo steered with a paddle, and all we had to rely upon to save us from falling into the trough of the sea and being instantly swamped were his nerve and skill. There was just room in the boat for our three selves, the dogs, and the game we had killed, and whenever we fell in the hollow of a sea our instant destruction seemed inevitable; as it was, we shipped a little water now and then. Joaquim assisted with his paddle to steady the boat: my time was fully occupied in baling out the water and watching the dogs, which were crowded together in the prow, yelling with fear; one or other of them occasionally falling over the side and causing great commotion in scrambling in again. Off the point was a ridge of rocks, over which the surge raged furiously. Raimundo sat at the stern, rigid and silent; his eye steadily watching the prow of the boat. It was almost worth the risk and discomfort of the passage to witness the seamanlike ability displayed by Indians on the water. The little boat rode beautifully, rising well with each wave, and in the course of an hour and a half we arrived at Caripí, thoroughly tired and wet through to the skin.

I will here make a few observations regarding the Paca and the Cutía, although there is little to relate of their habits in addition to what is contained in natural history books. The Paca is the Cœlogenys Paca of zoologists, and the Cutía the Dasyprocta Aguti, or a local variety of that species. Both differ much from our hare and rabbit, which belong to the same order of animals, their fur being coarse and bristly, and their ears short and broad. Their flesh is widely different in taste from that of our English Rodents. The meat of the Paca, in colour, grain, and flavour, resembles young pork; it is much drier, however, and less palatable than pork. The skin is thick, and boils down to a jelly, when it makes a capital soup with rice. Both animals live exclusively in the forests, both dry and moist, being found, perhaps, most abundantly in the ygapós and islands. When these are flooded in the wet season, they escape to the drier lands by swimming across the intervening channels. At Murucupí I saw several semi-domesticated individuals of both species, which had been caught when young, and were suffered to run freely about the houses. The Paca was not so familiar as the Cutía, which generally makes use of a hole or a box in a corner for a hiding-place, and comes out readily to be fed by children. I once saw a tame Cutía running about the woods nibbling the fruits fallen from the Inajá palm-tree (Maximiliana regia), and when I tried to catch it, instead of betaking itself to the thicket, it ran off to the house of its owners, which was about two hundred yards off. When feeding, this species sometimes sits upright, and takes its food in the fore paws like a squirrel.

The Paca and the Cutía belong to a peculiar family of the Rodent order which is confined to South America, and which connects the Rodents to the Pachydermata, the order to which the elephant, horse, and hog belong.

One of the principal points of distinction from other families is the strong, blunt form of the claws, which in one of the forms (the Capybara) are very broad, and approximate in shape to the hoofs of the Pachydermata. On this account the family is named by some authors Subungulati; the great division of mammalian animals to which the Pachydermata belong being called, in the classifications of the best authors, Ungulata, after the hoofed feet, which are considered their leading character. It is an interesting fact that the pachydermatous animal most nearly allied to the Rodents is also American, although found only in the fossil state, namely, the Toxodon, which Professor Owen states resembled the Rodentia in its dentition. The Toxodon, on the other hand, was nearly related to the Elephant, of which the same distinguished zoologist says, "Several particulars in its organization indicate an affinity to the Rodentia." These facts impart a high degree of interest to these semi-hoofed American Rodents, because they make it probable that these animals are the living representatives, albeit somewhat modified, of a group which existed at a former distant epoch in the world's history, and which possessed a structure partaking of the characters of the two great orders, Rodentia and Pachydermata, now so widely distinct in the majority of their forms. I believe that no remains of the order Toxodontia, or of the Rodent family Subungulati, have been found fossil in any other part of the world besides America. In this sort of question it is very unsafe to found generalizations on negative evidence; but does not this tend to show that the great section of mammals to which the Pachydermata belong had its origin on that part of the earth's surface where South America now stands?

On the 16th of January the dry season came abruptly to an end. The sea breezes, which had been increasing in force for some days, suddenly ceased, and the atmosphere became misty; at length heavy clouds collected where a uniform blue sky had for many weeks prevailed, and down came a succession of heavy showers, the first of which lasted a whole day and night. This seemed to give a new stimulus to animal life. On the first night there was a tremendous uproar—tree-frogs, crickets, goat-suckers, and owls, all joining to perform a deafening concert. One kind of goat-sucker kept repeating at intervals throughout the night a phrase similar to the Portuguese words, "Joaō corta pao," "John, cut wood;" a phrase which forms the Brazilian name of the bird. An owl in one of the Genipapa trees muttered now and then a succession of syllables resembling the word "Murucututú." Sometimes the croaking and hooting of frogs and toads were so loud that we could not hear one another's voices within doors. Swarms of dragon-flies appeared in the daytime about the pools of water created by the rain, and ants and termites came forth in the winged state in vast numbers. I noticed that the winged termites, or white ants, which came by hundreds to the lamps at night, when alighting on the table, often jerked off their wings by a voluntary movement. On examination I found that the wings were not shed by the roots, for a small portion of the stumps remained attached to the thorax. The edge of the fracture was in all cases straight, not ruptured: there is, in fact, a natural seam crossing the member towards its root, and at this point the long wing naturally drops or is jerked off when the insect has no further use for it. The white ant is endowed with wings simply for the purpose of flying away from the colony peopled by its wingless companions, to pair with individuals of the same or other colonies, and thus propagate and disseminate its kind. The winged individuals are males and females, whilst the great bulk of their wingless fraternity are of no sex, and are restricted to the functions of building the nests, nursing and defending the young brood. The two sexes mate whilst on the ground after the wings are shed, and then the married couples, if they escape the numerous enemies which lie in wait for them, proceed to the task of founding new colonies. Ants and white ants have much that is analogous in their modes of life: they belong, however, to two widely different orders of insects, strongly contrasted in their structure and manner of growth. In some respects the termites are more wonderful than the ants, but I shall reserve an account of them for another chapter.

I amassed at Caripí a very large collection of beautiful and curious insects, amounting altogether to about twelve hundred species. The number of Coleoptera was remarkable, seeing that this order is so poorly represented near Pará. I attributed their abundance to the number of new clearings made in the virgin forest by the native settlers. The felled timber attracts lignivorous insects, and these draw in their train the predacious species of various families. As a general rule the species were smaller and much less brilliant in colours than those of Mexico and South Brazil. The species too, although numerous, were not represented by great numbers of individuals; they were also extremely nimble, and therefore much less easy of capture than insects of the same order in temperate climates. On the sandy beach I found two species of Tetracha, a genus of tiger-beetles, which have remarkably large heads, and are found only in hot climates. They come forth at night, in the daytime remaining hid in their burrows several inches deep in the light soil. Their powers of running exceed everything I witnessed in this style of insect locomotion. They run in a serpentine course over the smooth sand, and when closely pursued by the fingers in the endeavour to seize them, are apt to turn suddenly back, and thus baffle the most practised hand and eye. I afterwards became much interested in these insects on several accounts, one of which was that they afforded an illustration of a curious problem in natural history. One of the Caripí species (T. nocturna of Dejean) was of a pallid hue like the sand over which it ran; the other was a brilliant copper-coloured kind (T. pallipes of Klug). Many insects whose abode is the sandy beaches are white in colour; I found a large earwig and a mole-cricket of this hue very common in these localities. Now it has been often said, when insects, lizards, snakes, and other animals, are coloured so as to resemble the objects on which they live, that such is a provision of nature, the assimilation of colours being given in order to conceal the creatures from the keen eyes of insectivorous birds and other animals. This is no doubt the right view, but some authors have found a difficulty in the explanation on account of this assimilation of colours being exhibited by some kinds and not by others living in company with them; the dress of some species being in striking contrast to the colours of their dwelling-place. One of our Tetrachas is coloured to resemble the sand, whilst its sister species is a conspicuous object on the sand; the white species, it may be mentioned, being much more swift of foot than the copper-coloured one. The margins of these sandy beaches are frequented throughout the fine season by flocks of sandpipers, who search for insects on moonlit nights as well as by day. If one species of insect obtains immunity from their onslaughts by its deceptive resemblance to the sandy surface on which it runs, why is not its sister species endowed in the same way? The answer is, that the dark-coloured kind has means of protection of quite a different nature, and therefore does not need the peculiar mode of disguise enjoyed by its companion. When handled it emits a strong, offensive, putrid and musky odour, a property which the pale kind does not exhibit. Thus we see that the fact of some species not exhibiting the same adaptation of colours to dwelling-places as their companion species does not throw doubt on the explanation given of the adaptation, but is rather confirmatory of it.

The carnivorous beetles at Caripí were, like those of Pará, chiefly arboreal. Some were found under the bark of trees (Coptodera, Goniotropis, Morio, &c.), others running over the slender twigs, branches, and leaves (Ctenostoma, Lebia, Calophæna, Lia, &c.), and many were concealed in the folds of leaves (Calleida, Agra, &c.). Most of them exhibited a beautiful contrivance for enabling them to cling to and run over smooth or flexible surfaces, such as leaves. Their tarsi or feet are broad, and furnished beneath with a brush of short stiff hairs, whilst their claws are toothed in the form of a comb, adapting them for clinging to the smooth edges of leaves, the joint of the foot which precedes the claw being cleft so as to allow free play to the claw in grasping. The common dung-beetles at Caripí, which flew about in the evening like the Geotrupes, the familiar "shard-borne beetle with his drowsy hum" of our English lanes, were of colossal size and beautiful colours. One kind had a long spear-shaped horn projecting from the crown of its head (Phanæus lancifer). A blow from this fellow, as he came heavily flying along, was never very pleasant. All the tribes of beetles which feed on vegetable substances, fresh or decayed, were very numerous. The most beautiful of these, but not the most common, were the Longicornes; very graceful insects, having slender bodies and long antennæ, often ornamented with fringes and tufts of hair. They were found on flowers, on trunks of trees, or flying about the new clearings. One small species (Coremia hirtipes) has a tuft of hairs on its hind legs, whilst many of its sister species have a similar ornament on the antennæ. It suggests curious reflections when we see an ornament like the feather of a grenadier's cap situated on one part of the body in one species, and in a totally different part in nearly allied ones. I tried in vain to discover the use of these curious brush-like decorations. On the trunk of a living leguminous tree, Petzell found a number of a very rare and handsome species, the Platysternus hebræus, which is of a broad shape, coloured ochreous, but spotted and striped with black, so as to resemble a domino. On the felled trunks of trees, swarms of gilded-green Longicornes occurred, of small size (Chrysoprasis), which looked like miniature musk-beetles, and, indeed, are closely allied to those well-known European insects.

I was interested in the many small kinds of lignivorous or wood-eating insects found at Caripí, a few observations on which may be given in conclusion. It is curious to observe how some small groups of insects exhibit the most diversified forms and habits—one set of species being adapted by their structure for one set of functions in nature, and another set, very closely allied, for an opposite sphere of action. Thus the Histeridæ—small black beetles well known to English entomologists, most of whose species are short and thick in shape and live in the dung of animals—are most diversified in structure and habits in the Amazons region; nevertheless, all the forms preserve in a remarkable degree the essential characters of the family. One set of species live in dung; most of these are somewhat cubical in shape, the head being retractable within the breastplate, as in the tortoise. Another group of Histeridæ are much flatter in form, and live in the moist interior of palm-tree stems; one of these is a veritable colossus, the Hister maximus of Linnæus. A third group (Hololeptæ) are found only under the bark of trees; their heads are not retractable within the breast, and their bodies are excessively depressed, to fit them for living in narrow crevices, some kinds being literally as thin as a wafer. A fourth set of species (Trypanæus) form a perfect contrast to these, being cylindrical in shape. They drill holes into solid wood, and look like tiny animated gimlets when seen at work, their pointed heads being fixed in the wood whilst their smooth glossy bodies work rapidly round, so as to create little streams of sawdust from the holes. Several families of insects show similar diversities of adaptation amongst their species, but none, I think, to the same extent as the Histeridæ, considering the narrow limits of the group. The facts presented by such groups in the animal kingdom must be taken into account in any explanation of the way the almost infinite diversity of the forms of life has been brought about on this wonderful earth.

At length, on the 12th of February, I left Caripí, my Negro and Indian neighbours bidding me a warm "adeos." I had passed a delightful time, notwithstanding the many privations undergone in the way of food. The wet season had now set in; the low lands and islands would soon become flooded daily at high water, and the difficulty of obtaining fresh provisions would increase. I intended, therefore, to spend the next three months at Pará, in whose neighbourhood there was still much to be done in the intervals of fine weather, and then start off on another excursion into the interior.

CHAPTER VI.

THE LOWER AMAZONS—PARÁ TO OBYDOS.

Modes of Travelling on the Amazons—Historical Sketch of the early Explorations of the River—Preparations for Voyage—Life on board a large Trading-vessel—The narrow Channels joining the Pará to the Amazons—First Sight of the great River—Gurupá—The Great Shoal—Flat-topped Mountains—Contraction of the River Valley—Santarem—Obydos—Natural History of Obydos—Origin of Species by Segregation of Local Varieties.

AT the time of my first voyage up the Amazons—namely, in 1849—nearly all communication with the interior was by means of small sailing vessels, owned by traders residing in the remote towns and villages, who seldom came to Pará themselves, but entrusted vessels and cargoes to the care of half-breeds or Portuguese cabos. Sometimes, indeed, they risked all in the hands of the Indian crew, making the pilot, who was also steersman, do duty as supercargo. Now and then, Portuguese and Brazilian merchants at Pará furnished young Portuguese with merchandise, and despatched them to the interior to exchange the goods for produce amongst the scattered population. The means of communication, in fact, with the upper parts of the Amazons had been on the decrease for some time, on account of the augmented difficulty of obtaining hands to navigate vessels. Formerly, when the Government wished to send any important functionary, such as a judge or a military commandant, into the interior, they equipped a swift-sailing galliota, manned with ten or a dozen Indians. These could travel, on the average, in one day further than the ordinary sailing craft could in three. Indian paddlers were now, however, almost impossible to be obtained, and Government officers were obliged to travel as passengers in trading vessels. The voyage made in this way was tedious in the extreme. When the regular east wind blew—the "vento geral," or trade wind, of the Amazons—sailing vessels could get along very well; but when this failed they were obliged to remain, sometimes many days together, anchored near the shore, or progress laboriously by means of the "espia." This latter mode of travelling was as follows. The montaria, with twenty or thirty fathoms of cable, one end of which was attached to the foremast, was sent ahead with a couple of hands, who secured the other end of the rope to some strong bough or tree trunk; the crew then hauled the vessel up to the point, after which the men in the boat re-embarked the cable, and paddled forwards to repeat the process. In the dry season, from August to December, when the trade-wind is strong and the currents slack, a schooner could reach the mouth of the Rio Negro, a thousand miles from Pará, in about forty days; but in the wet season, from January to July, when the east wind no longer blows and the Amazons pours forth its full volume of water, flooding the banks and producing a tearing current, it took three months to travel the same distance. It was a great blessing to the inhabitants when, in 1853, a line of steamers was established, and this same journey could be accomplished with ease and comfort, at all seasons, in eight days!

It is, perhaps, not generally known that the Portuguese, as early as 1710, had a fair knowledge of the Amazons; but the information gathered by their government from various expeditions undertaken on a grand scale, was long withheld from the rest of the world, through the jealous policy which ruled in their colonial affairs. From the foundation of Pará by Caldeira, in 1615, to the settlement of the boundary line between the Spanish and Portuguese possessions, Peru and Brazil, in 1781-91, numbers of these expeditions were in succession undertaken. The largest was the one commanded by Pedro Texeira in 1637-9, who ascended the river to Quito, by way of the Napo, a distance of about 2800 miles, with 45 canoes and 900 men, and returned to Pará without any great misadventure by the same route. The success of this remarkable undertaking amply proved, at that early date, the facility of the river navigation, the practicability of the country, and the good disposition of the aboriginal inhabitants. The river, however, was first discovered by the Spaniards, the mouth having been visited by Pinzon in 1500, and nearly the whole course of the river navigated by Orellana in 1541-2. The voyage of the latter was one of the most remarkable on record. Orellana was a lieutenant of Gonzalo Pizarro, governor of Quito, and accompanied the latter in an adventurous journey which he undertook across the easternmost chain of the Andes, down into the sweltering valley of the Napo, in search of the land of El Dorado, or the Gilded King. They started with 300 soldiers and 4000 Indian porters; but, arrived on the banks of one of the tributaries of the Napo, their followers were so greatly decreased in number by disease and hunger, and the remainder so much weakened, that Pizarro was obliged to despatch Orellana with fifty men, in a vessel they had built, to the Napo, in search of provisions. It can be imagined by those acquainted with the Amazons country how fruitless this errand would be in the wilderness of forest where Orellana and his followers found themselves when they reached the Napo, and how strong their disinclination would be to return against the currents and rapids which they had descended. The idea then seized them to commit themselves to the chances of the stream, although ignorant whither it would lead. So onward they went. From the Napo they emerged into the main Amazons, and, after many and various adventures with the Indians on its banks, reached the Atlantic eight months from the date of their entering the great river.*

Another remarkable voyage was accomplished, in a similar manner, by a Spaniard named Lopez d'Aguirre, from Cusco, in Peru, down the Ucayali, a branch of the Amazons flowing from the south, and therefore from an

* It was during this voyage that the nation of female warriors was said to have been met with; a report which gave rise to the Portuguese name of the river, Amazonas. It is now pretty well known that this is a mere fable, originating in the love of the marvellous which distinguished the early Spanish adventurers, and impaired the credibility of their narratives.

opposite direction to that of the Napo. An account of this journey was sent by D'Aguirre, in a letter to the King of Spain, from which Humboldt has given an extract in his narrative. As it is a good specimen of the quaintness of style and looseness of statement exhibited by these early narrators of adventures in South America, I will give a translation of it. "We constructed rafts, and, leaving behind our horses and baggage, sailed down the river (the Ucayali) with great risk, until we found ourselves in a gulf of fresh water. In this river Marañon we continued more than ten months and a half, down to its mouth, where it falls into the sea. We made one hundred days' journey, and travelled 1500 leagues. It is a great and fearful stream, has 80 leagues of fresh water at its mouth, vast shoals, and 800 leagues of wilderness without any kind of inhabitants,* as your Majesty will see from the true and correct narrative of the journey which we have made. It has more than 6000 islands. God knows how we came out of this fearful sea." Many expeditions were undertaken in the course of the eighteenth century; in fact, the crossing of the continent from the Pacific to the Atlantic, by way of the Amazons, seems to have become by this time a common occurrence. The only voyage, however, which yielded much scientific information to the European public was that of the French astronomer, La Condamine, in 1743-4. The most complete account

* This account disagrees with that of Acunna, the historiographer of Texeira's expedition, who accompanied him, in 1639, on his return voyage from Quito. Acunna speaks of a very numerous population on the banks of the Amazons.

yet published of the river is that given by Von Martius in the third volume of Spix and Martius' Travels. These most accomplished travellers were eleven months in the country—namely, from July, 1819, to June, 1820, and ascended the river to the frontiers of the Brazilian territory. Their accounts of the geography, ethnology, botany, history, and statistics of the Amazons region are the most complete that have ever been given to the world. Their narrative was not published until 1831, and was unfortunately inaccessible to me during the time I travelled in the same country.

Whilst preparing for my voyage it happened fortunately that the half-brother of Dr. Angelo Custodio, a young mestizo named Joaõ da Cunha Correia, was about starting for the Amazons on a trading expedition in his own vessel, a schooner of about forty tons burthen. A passage for me was soon arranged with him through the intervention of Dr. Angelo, and we started on the 5th of September, 1849. I intended to stop at some village on the northern shore of the Lower Amazons, where it would be interesting to make collections, in order to show the relations of the fauna to those of Pará and the coast region of Guiana. As I should have to hire a house or hut wherever I stayed, I took all the materials for housekeeping—cooking utensils, crockery, and so forth. To these were added a stock of such provisions as were difficult to obtain in the interior; also ammunition, chests, store boxes, a small library of natural history books, and a hundredweight of copper money. I engaged, after some trouble, a Mameluco youth to accompany me as servant—a short, fat, yellow-faced boy named Luco, whom I had already employed at Pará in collecting. We weighed anchor at night, and on the following day found ourselves gliding along the dark-brown waters of the Mojú.

Joaō da Cunha, like most of his fellow-countrymen, took matters very easily. He was going to be absent in the interior several years, and therefore intended to diverge from his route to visit his native place, Cametá, and spend a few days with his friends. It seemed not to matter to him that he had a cargo of merchandise, vessel, and crew of twelve persons, which required an economical use of time; "pleasure first and business afterwards" appeared to be his maxim. We stayed at Cametá twelve days. The chief motive for prolonging the stay to this extent was a festival at the Aldeia, two miles below Cametá, which was to commence on the 21st, and which my friend wished to take part in. On the day of the festival the schooner was sent down to anchor off the Aldeia, and master and men gave themselves up to revelry. In the evening a strong breeze sprang up, and orders were given to embark. We scrambled down in the dark through the thickets of cacao, orange, and coffee trees which clothed the high bank, and, after running great risk of being swamped by the heavy sea in the crowded montaria, got all aboard by nine o'clock. We made all sail amidst the "adeos" shouted to us by Indian and mulatto sweethearts from the top of the bank, and, tide and wind being favourable, were soon miles away.

Our crew consisted, as already mentioned, of twelve

persons. One was a young Portuguese from the province of Traz os Montes, a pretty sample of the kind of emigrants which Portugal sends to Brazil. He was two or three and twenty years of age, and had been about two years in the country, dressing and living like the Indians, to whom he was certainly inferior in manners. He could not read and write, whereas one at least of our Tapuyos had both accomplishments. He had a little wooden image of Nossa Senhora in his rough wooden clothes chest, and to this he always had recourse when any squall arose, or when we got aground on a shoal. Another of our sailors was a tawny white of Cametá; the rest were Indians, except the cook, who was a Cafuzo, or half-breed between the Indian and negro. It is often said that this class of mestizos is the most evilly-disposed of all the numerous crosses between the races inhabiting Brazil; but Luiz was a simple, good-hearted fellow, always ready to do one a service. The pilot was an old Tapuyo of Pará, with regular oval face and well-shaped features. I was astonished at his endurance. He never quitted the helm night or day, except for two or three hours in the morning. The other Indians used to bring him his coffee and meals, and after breakfast one of them relieved him for a time, when he used to lie down on the quarter-deck and get his two hours' nap. The Indians forward had things pretty much their own way. No system of watches was followed; when any one was so disposed, he lay down on the deck and went to sleep; but a feeling of good fellowship seemed always to exist amongst them. One of them was a fine specimen of the Indian race: a man very little short of six feet high, with remarkable breadth of shoulder and full muscular chest. His comrades called him the commandant, on account of his having been one of the rebel leaders when the Indians and others took Santarem in 1835. They related of him that, when the legal authorities arrived with an armed flotilla to recapture the town, he was one of the last to quit, remaining in the little fortress which commands the place to make a show of loading the guns, although the ammunition had given out long ago. Such were our travelling companions. We lived almost the same as on board ship. Our meals were cooked in the galley; but, where practicable, and during our numerous stoppages, the men went in the montaria to fish near the shore, so that our breakfasts and dinners of salt piracucu were sometimes varied with fresh food.

Sept. 24th.—We passed Entre-as-Ilhas with the morning tide yesterday, and then made across to the eastern shore—the starting-point for all canoes which have to traverse the broad mouth of the Tocantins going west. Early this morning we commenced the passage. The navigation is attended with danger on account of the extensive shoals in the middle of the river, which are covered only by a small depth of water at this season of the year. The wind was fresh, and the schooner rolled and pitched like a ship at sea. The distance was about fifteen miles. In the middle, the river view was very imposing. Towards the north-east there was a long sweep of horizon clear of land, and on the south-west stretched a similar boundless expanse, but varied with islets clothed with fan-leaved palms, which, however, were visible only as isolated groups of columns, tufted at the top, rising here and there amidst the waste of waters. In the afternoon we rounded the westernmost point ; the land, which is not terra firma, but simply a group of large islands forming a portion of the Tocantins delta, was then about three miles distant.

On the following day (25th) we sailed towards the west, along the upper portion of the Pará estuary, which extends seventy miles beyond the mouth of the Tocantins. It varies in width from three to five miles, but broadens rapidly near its termination, where it is eight or nine miles wide. The northern shore is formed by the island of Marajó, and is slightly elevated and rocky in some parts. A series of islands conceals the southern shore from view most part of the way. The whole country, mainland and islands is covered with forest. We had a good wind all day, and about 7 p.m. entered the narrow river of Breves, which commences abruptly the extensive labyrinth of channels that connect the Pará with the Amazons. The sudden termination of the Pará at a point where it expands to so great a breadth is remarkable ; the water, however, is very shallow over the greater portion of the expanse. I noticed, both on this and on the three subsequent occasions of passing this place in ascending and descending the river, that the flow of the tide from the east along the estuary, as well as up the Breves, was very strong. This seems sufficient to prove that no considerable volume of water passes by this medium from the Amazons to the Pará, and that the opinion of those geographers is an incorrect one, who believe the Pará to be one of the mouths of the great river. There is, however, another channel connecting the two rivers, which enters the Pará six miles to the south of the Breves. The lower part of its course for eighteen miles is formed by the Uanapú, a large and independent river flowing from the south. The tidal flow is said by the natives to produce little or no current up this river; a fact which seems to afford a little support to the view just stated.

We passed the village of Breves at 3 p.m. on the 26th. It consists of about forty houses, most of which are occupied by Portuguese shopkeepers. A few Indian families reside here, who occupy themselves with the manufacture of ornamental pottery and painted cuyas, which they sell to traders or passing travellers. The cuyas—drinking-cups made from gourds—are sometimes very tastefully painted. The rich black ground-colour is produced by a dye made from the bark of a tree called Comateü, the gummy nature of which imparts a fine polish. The yellow tints are made with the Tabatinga clay; the red with the seeds of the Urucú, or anatto plant; and the blue with indigo, which is planted round the huts. The art is indigenous with the Amazonian Indians, but it is only the settled agricultural tribes belonging to the Tupí stock who practise it.

Sept. 27th-30th.—After passing Breves we continued our way slowly along a channel, or series of channels, of variable width. On the morning of the 27th we had a fair wind, the breadth of the stream varying from about 150 to 400 yards. The forest was not remarkable in appearance; the banks were muddy, and in low marshy places groups of Caladiums fringed the edge of the water. About midday we passed, on the western side, the mouth of the Aturiazal, through which, on account of its swifter current, vessels pass in descending from the Amazons to Pará. Shortly afterwards we entered the narrow channel of the Jaburú, which lies twenty miles above the mouth of the Breves. Here commences the peculiar scenery of this remarkable region. We found ourselves in a narrow and nearly straight canal, not more than eighty to a hundred yards in width, and hemmed in by two walls of forest, which rose quite perpendicularly from the water to a height of seventy or eighty feet. The water was of great and uniform depth, even close to the banks. We seemed to be in a deep gorge, and the strange impression the place produced was augmented by the dull echoes produced by the voices of our Indians and the splash of their paddles. The forest was excessively varied. Some of the trees, the dome-topped giants of the Leguminous and Bombaceous orders, reared their heads far above the average height of the green walls. The fan-leaved Mirití palm was scattered in some numbers amidst the rest, a few solitary specimens shooting up their smooth columns above the other trees. The graceful Assai palm grew in little groups, forming feathery pictures set in the rounder foliage of the mass. The Ubussú, lower in height, showed only its shuttlecock-shaped crowns of huge undivided fronds, which, being of a vivid pale green, contrasted forcibly against the sombre hues of the surrounding foliage. The Ubussú grew here in great numbers; the equally remarkable Jupatí palm (Rhaphia tædigera), which, like the Ubussú, is peculiar to this district, occurred more sparsely, throwing its long shaggy leaves, forty to fifty feet in length, in broad arches over the canal. An infinite diversity of smaller-sized palms decorated the water's edge, such as the Marajá-i (Bactris, many species), the Ubim (Geonoma), and a few stately Bacábas (Œnocarpus Bacaba). The shape of this last is exceedingly elegant, the size of the crown being in proper proportion to the straight smooth stem. The leaves, down even to the bases of the glossy petioles, are of a rich dark-green colour, and free from spines. "The forest wall"—I am extracting from my journal—"under which we are now moving consists, besides palms, of a great variety of ordinary forest-trees. From the highest branches of these down to the water sweep ribbons of climbing plants of the most diverse and ornamental foliage possible. Creeping convolvuli and others have made use of the slender lianas and hanging air-roots as ladders to climb by. Now and then appears a Mimosa or other tree having similar fine pinnate foliage, and thick masses of Ingá border the water, from whose branches hang long bean-pods, of different shape and size according to the species, some of them a yard in length. Flowers there are very few. I see, now and then, a gorgeous crimson blossom on long spikes ornamenting the sombre foliage towards the summits of the forest. I suppose it to belong to a climber of the Combretaceous order. There are also a few yellow and violet Trumpet-flowers (Bignoniæ). The blossoms of the Ingás, although not conspicuous, are delicately beautiful. The forest all along offers so dense a front that one never obtains a glimpse into the interior of the wilderness."

The length of the Jaburú channel is about 35 miles, allowing for the numerous abrupt bends which occur between the middle and the northern end of its course. We were three days and a half accomplishing the passage. The banks on each side seemed to be composed of hard river mud with a thick covering of vegetable mould, so that I should imagine this whole district originated in a gradual accumulation of alluvium, through which the endless labyrinths of channels have worked their deep and narrow beds. The flood tide as we travelled northward became gradually of less assistance to us, as it caused only a feeble current upwards. The pressure of the waters from the Amazons here makes itself felt; as this is not the case lower down, I suppose the currents are diverted through some of the numerous channels which we passed on our right, and which traverse, in their course towards the sea, the north-western part of Marajó. In the evening of the 29th we arrived at a point where another channel joins the Jaburú from the north-east. Up this the tide was flowing; we turned westward, and thus met the flood coming from the Amazons. This point is the object of a strange superstitious observance on the part of the canoemen. It is said to be haunted by a Pajé, or Indian wizard, whom it is necessary to propitiate by depositing some article on the spot, if the voyager wishes to secure a safe return from the "sertaô," as the interior of the country is called. The trees were all hung with rags, shirts, straw hats, bunches of fruit, and so forth. Although the superstition doubtless originated with the aborigines, yet I observed, in both my voyages, that it was only the Portuguese and uneducated Brazilians who deposited anything. The pure Indians gave nothing, and treated the whole affair as a humbug ; but they were all civilised Tapuyos.

On the 30th, at 9 p.m., we reached a broad channel called Macaco, and now left the dark, echoing Jaburú. The Macaco sends off branches towards the north-west coast of Marajó. Whilst waiting for the tide I went ashore in the montaria with Joaō da Cunha. The forest was gloomy and forbidding in the extreme, the densely-packed trees producing a deep shade, under which all was dark and cold. There was no animal life visible—vertebrate, articulate, or molluscous. At its commencement the Macaco is about half a mile wide, and runs from S.S.W. to N.N.E. ; towards the north it expands to a breadth of two or three miles. It is merely a passage amongst a cluster of islands, between which a glimpse is occasionally obtained of the broad waters of the main Amazons. A brisk wind carried us rapidly past its monotonous scenery, and early in the morning of the 1st of October we reached the entrance of the Uituquára, or the Wind-hole, which is 15 miles distant from the end of the Jaburú. This is also a winding channel, 35 miles in length, threading a group of islands, but it is much narrower than the Macaco.

On emerging from the Uituquára on the 2nd, we all

went ashore: the men to fish in a small creek; Joaõ da Cunha and I, to shoot birds. We saw a flock of scarlet and blue macaws (Macrocercus Macao) feeding on the fruits of a Bacaba palm, and looking like a cluster of flaunting banners beneath its dark-green crown. We landed about fifty yards from the place, and crept cautiously through the forest, but before we reached them they flew off with loud harsh screams. At a wild-fruit tree we were more successful, as my companion shot an anacá (Derotypus coronatus), one of the most beautiful of the parrot family. It is of a green colour, and has a hood of feathers, red bordered with blue, at the back of its head, which it can elevate or depress at pleasure. The anacá is the only new-world parrot which nearly resembles the cockatoo of Australia. It is found in all the low lands throughout the Amazons region, but is not a common bird anywhere. Few persons succeed in taming it, and I never saw one that had been taught to speak. The natives are very fond of the bird nevertheless, and keep it in their houses for the sake of seeing the irascible creature expand its beautiful frill of feathers, which it readily does when excited. The men returned with a large quantity of fish. I was surprised at the great variety of species; the prevailing kind was a species of Loricaria, a foot in length, and wholly encased in bony armour. It abounds at certain seasons in shallow water. The flesh is dry, but very palatable. They brought also a small alligator, which they called Jacaré-curúa, and said it was a kind found only in shallow creeks. It was not more than two feet in length,

Q 2

although full grown according to the statement of the Indians, who said it was a "mai d'ovos," or mother of

Acarí Fish (Loricaria duodecimalis).

eggs, as they had pillaged the nest, which they had found near the edge of the water. The eggs were rather larger than a hen's, and regularly oval in shape, presenting a rough hard surface of shell. Unfortunately the alligator was cut up ready for cooking when we returned to the schooner, and I could not therefore make a note of its peculiarities. The pieces were skewered and roasted over the fire, each man being his own cook. I never saw this species of alligator afterwards.

October 3rd.—About midnight the wind, for which we had long been waiting, sprang up, the men weighed anchor, and we were soon fairly embarked on the Amazons. I rose long before sunrise to see the great river by moonlight. There was a spanking breeze, and the vessel was bounding gaily over the waters. The channel along which we were sailing was only a narrow arm of the river, about two miles in width : the total breadth at this point is more than 20 miles, but the stream is divided into three parts by a series of large islands. The river, notwithstanding this limitation of its breadth, had a most majestic appearance. It did not present that lake-like aspect which the waters of the Pará and Tocantins affect, but had all the swing, so to speak, of a vast flowing stream. The ochre-coloured turbid waters offered also a great contrast to the rivers belonging to the Pará system. The channel formed a splendid reach, sweeping from south-west to north-east, with a horizon of water and sky both up stream and down. At 11 a.m. we arrived at Gurupá, a small village situated on a rocky bank 30 or 40 feet high. Here we landed, and I had an opportunity of rambling in the neighbouring woods, which are intersected by numerous pathways, and carpeted with Lycopodia growing to a height of 8 or 10 inches, and enlivened by numbers of glossy blue butterflies of the Theclidæ, or hair-streak family. The land on which Gurupá is built appears an isolated rocky area, for the rest of the country round about lies low, and is subject to inundation in the rainy season. At 5 p.m. we were again under way. Soon after sunset, as we were crossing the mouth of the Xingú, the first of the great tributaries of the Amazons, 1200 miles in length, a black cloud arose suddenly in the north-east. Joaō da Cunha ordered all sails to be taken in, and immediately afterwards a furious squall burst forth, tearing the waters into foam, and producing a frightful uproar in the neighbouring forests. A drenching rain followed: but in half an hour all was again calm, and the full moon appeared sailing in a cloudless sky.

From the mouth of the Xingú the route followed by vessels leads straight across the river, here 10 miles broad. Towards midnight the wind failed us, when we were close to a large shoal called the Baixo Grande. We lay here becalmed in the sickening heat for two days, and when the trade wind recommenced with the rising moon at 10 p.m. on the 6th, we found ourselves on a lee-shore. Notwithstanding all the efforts of our pilot to avoid it, we ran aground. Fortunately the bottom consisted only of soft mud, so that, by casting anchor to windward and hauling in with the whole strength of crew and passengers, we got off after spending an uncomfortable night. We rounded the point of the shoal in two fathoms water; the head of the vessel was then put westward, and by sunrise we were bounding forward before a steady breeze, all sail set and everybody in good humour.

The weather was now delightful for several days in succession: the air transparently clear, and the breeze cool and invigorating. At daylight, on the 6th, a chain of blue hills, the Serra de Almeyrim, appeared in the distance on the north bank of the river. The sight was most exhilarating after so long a sojourn in a flat country. We kept to the southern shore, passing in the course of the day the mouths of the Urucuricáya and the Aquiquí, two channels which communicate with the Xingú. The whole of this southern coast hence to near Santarem, a distance of 130 miles, is low land and quite uninhabited. It is intersected by short arms or back waters of the Amazons, which are called in the Tupí language Paraná-mirims or little rivers. By keeping to these, small canoes can travel great part of the distance without being much exposed to the heavy seas of the main river. The coast throughout has a most desolate aspect : the forest is not so varied as on the higher land ; and the water frontage, which is destitute of the green mantle of climbing plants that form so rich a decoration in other parts, is encumbered at every step with piles of fallen trees, peopled by white egrets, ghostly storks, and solitary herons. In the evening we passed Almeyrim. The hills, according to Von Martius, who landed here, are about 800 feet above the level of the river and are thickly wooded to the summit. They commence on the east by a few low isolated and rounded elevations ; but towards the west of the village they assume the appearance of elongated ridges, which seem to have been planed down to a uniform height by some external force. The next day we passed in succession a series of similar flat-topped hills, some isolated and of a truncated-pyramidal shape, others prolonged to a length of several miles. There is an interval of low country between these and the Almeyrim range, which has a total length of about 25 miles : then commences abruptly the Serra de Marauaquá, which is succeeded in a similar way by the Velha Pobre range, the Serras de Tapaiuna-quára, and Parauáquára. All these form a striking contrast to the Serra de Almeyrim in being quite destitute of trees. They have steep, rugged sides, apparently clothed with short herbage, but here and there exposing bare white patches. Their total length is about 40 miles. In the rear, towards the interior, they are succeeded by other ranges of hills communicating with the central mountain chain of Guiana, which divides Brazil from Cayenne.

As we sailed along the southern shore, during the 6th and two following days, the table-topped hills on the opposite side occupied most of our attention. The river is from four to five miles broad, and in some places long, low wooded islands intervene in mid-stream, whose light-green, vivid verdure formed a strangely beautiful foreground to the glorious landscape of broad stream and grey mountain. Ninety miles beyond Almeyrim stands the village of Monte Alegre, which is built near the summit of the last hill visible of this chain. At this point the river bends a little towards the south, and the hilly country recedes from its shores to re-appear at Obydos, greatly decreased in height, about a hundred miles further west. Twenty-five miles to the south-west of Monte Alegre, high land again appears, but now on the opposite side of the river. This is the northernmost limit of the table-land of Brazil, as the hills of Monte Alegre are the southernmost of that of Guiana. In no other part of the river do the high lands on each side approach each other so closely. Beyond Obydos they gradually recede, and the width of the river valley consequently increases, until in the central parts of the Upper Amazons, near Ega, it is no less than 540 miles. At this point, therefore, the valley or river plain of the Amazons is contracted to its narrowest

FLAT-TOPPED MOUNTAINS OF PARAUÁQUÁRA, LOWER AMAZONS.

Vol. I., page 232.

breadth, reckoning from the places 2000 miles from its mouth, where the river and its earliest tributaries rush forth between walls of rock through the easternmost ridges of the Andes. It is, perhaps, necessary to take this in consideration when studying the geographical distribution of the plants and animals which people these vast wooded plains.

We crossed the river three times between Monte Alegre and the next town, Santarem. In the middle the waves ran very high, and the vessel lurched fearfully, hurling everything that was not well secured from one side of the deck to the other. On the morning of the 9th of October, a gentle wind carried us along a "remanso," or still water, under the southern shore. These tracts of quiet water are frequent on the irregular sides of the stream, and are the effect of counter movements caused by the rapid current of its central parts. At 9 a.m. we passed the mouth of a Paraná-mirim, called Mahicá, and then found a sudden change in the colour of the water and aspect of the banks. Instead of the low and swampy water-frontage which had prevailed from the mouth of the Xingú, we saw before us a broad sloping beach of white sand. The forest, instead of being an entangled mass of irregular and rank vegetation as hitherto, presented a rounded outline, and created an impression of repose that was very pleasing. We now approached, in fact, the mouth of the Tapajos, whose clear olive-green waters here replaced the muddy current against which we had so long been sailing. Although this is a river of great extent—1000 miles in length, and, for the last eighty miles of its course, four to ten in breadth—its contribution to the Amazons is not perceptible in the middle of the stream. The white turbid current of the main river flows disdainfully by, occupying nearly the whole breadth of the channel, whilst the darker water of its tributary seems to creep along the shore, and is no longer distinguishable four or five miles from its mouth.

We reached Santarem at 11 a.m. The town has a clean and cheerful appearance from the river. It consists of three long streets, with a few short ones crossing them at right angles, and contains about 2500 inhabitants. It lies just within the mouth of Tapajos, and is divided into two parts, the town and the aldeia or village. The houses of the white and trading classes are substantially built, many being of two and three stories, and all white-washed and tiled. The aldeia, which contains the Indian portion of the population, or did so formerly, consists mostly of mud huts, thatched with palm leaves. The situation of the town is very beautiful. The land, although but slightly elevated, does not form, strictly speaking, a portion of the alluvial river plains of the Amazons, but is rather a northern prolongation of the Brazilian continental land. It is scantily wooded, and towards the interior consists of undulating campos, which are connected with a series of hills extending southward as far as the eye can reach. I subsequently made this place my head-quarters for three years; an account of its neighbourhood is therefore reserved for another chapter. At the first sight of Santarem, one cannot help being struck with the advantages of its situation. Although 400 miles from the sea, it is accessible to vessels of heavy tonnage coming straight from the Atlantic. The river has only two slight bends between this port and the sea, and for five or six months in the year the Amazonian trade wind blows with very little interruption, so that sailing ships coming from foreign countries could reach the place with little difficulty. We ourselves had accomplished 200 miles, or about half the distance from the sea, in an ill-rigged vessel, in three days and a half. Although the land in the immediate neighbourhood is perhaps ill adapted for agriculture, an immense tract of rich soil, with forest and meadow land, lies on the opposite banks of the river, and the Tapajos leads into the heart of the mining provinces of interior Brazil. But where is the population to come from to develop the resources of this fine country? At present the district within a radius of twenty-five miles contains barely 6500 inhabitants; behind the town, towards the interior, the country is uninhabited, and jaguars roam nightly, at least in the rainy season, close up to the ends of the suburban streets.

From information obtained here, I fixed upon the next town, Obydos, as the best place to stay at a few weeks, in order to investigate the natural productions of the north side of the Lower Amazons. We started at sunrise on the 10th, and being still favoured by wind and weather, made a pleasant passage, reaching Obydos, which is nearly fifty miles distant from Santarem, by midnight. We sailed all day close to the southern shore, and found the banks here and there dotted with houses of settlers, each surrounded by its plantation of cacao, which is the staple product of the district. This coast has an evil reputation for storms and mosquitoes, but we fortunately escaped both. It was remarkable that we had been troubled by mosquitoes only on one night, and then to a small degree, during the whole of our voyage.

I landed at Obydos the next morning, and then bid adieu to my kind friend Joaõ da Cunha, who, after landing my baggage, got up his anchor and continued on his way. The town contains about 1200 inhabitants, and is airily situated on a high bluff, 90 or 100 feet above the level of the river. The coast is precipitous for two or three miles hence to the west. The cliffs consist of the parti-coloured clay, or Tabatinga, which occurs so frequently throughout the Amazons region ; the strong current of the river sets full against them in the season of high water, and annually carries away large portions. The clay in places is stratified alternately pink and yellow, the pink beds being the thickest, and of much harder texture than the others. When I descended the river in 1859, a German Major of Engineers, in the employ of the Government, told me that he had found calcareous layers, thickly studded with marine shells interstratified with the clay. On the top of the Tabatinga lies a bed of sand, in some places several feet thick, and the whole formation rests on strata of sandstone, which are exposed only when the river reaches its lowest level. Behind the town rises a fine rounded hill, and a range of similar elevations extends six miles westward, terminating at the mouth of the Trombetas, a large river flowing through the interior of Guiana. Hills and lowlands alike are covered with a sombre rolling forest. The river here is contracted to a breadth of rather less than a mile (1738 yards), and the entire volume of its waters, the collective product of a score of mighty streams, is poured through the strait with tremendous velocity.* It must be remarked, however, that the river valley itself is not contracted to this breadth, the opposite shore not being continental land, but a low alluvial tract, subject to inundation more or less in the rainy season. Behind it lies an extensive lake, called the Lago Grande da Villa Franca, which communicates with the Amazons, both above and below Obydos, and has therefore the appearance of a by-water or an old channel of the river. This lake is about thirty-five miles in length, and from four to ten in width ; but its waters are of little depth, and in the dry season its dimensions are much lessened. It has no perceptible current, and does not therefore now divert any portion of the waters of the Amazons from their main course past Obydos.

I remained at Obydos from the 11th of October to the 19th of November. I spent three weeks here, also,

* It was formerly believed that the river at the strait of Obydos could not be sounded on account of its great depth and the velocity of the current. Lieut. Herndon, of the United States navy, succeeded in doing so, however, in 1852. He found a depth of 30 to 35 fathoms, but in one place he thought he had not touched the bottom at 40 fathoms. Von Martius, estimating the depth in the middle at 60 fathoms, and on the side at 20, and the velocity of the current at 2·4 feet per second, estimated that 499,584 cubic feet of water passed through the strait in each second of time. The tides are felt here in the dry season, but the flood does not press back the current of the Amazons.

in 1859, when the place was much changed through the influx of Portuguese immigrants and the building of a fortress on the top of the bluff. It is one of the pleasantest towns on the river. The houses are all roofed with tiles, and are mostly of substantial architecture. The inhabitants, at least at the time of my first visit, were naïve in their ways, kind and sociable. Scarcely any palm-thatched huts are to be seen, for very few Indians now reside here. It was one of the early settlements of the Portuguese, and the better class of the population consists of old-established white families, who exhibit however, in some cases, traces of cross with the Indian and negro. Obydos and Santarem have received, during the last eighty years, considerable importations of negro slaves; before that time a cruel traffic was carried on in Indians for the same purpose of forced servitude, but their numbers have gradually dwindled away, and Indians now form an insignificant element in the population of the district. Most of the Obydos townsfolk are owners of cacao plantations, which are situated on the low lands in the vicinity. Some are large cattle proprietors, and possess estates of many square leagues' extent in the campo, or grass-land districts, which border the Lago Grande, and other similar inland lakes, near the villages of Faro and Alemquer. These campos bear a crop of nutritious grass; but in certain seasons, when the rising of the Amazons exceeds the average, they are apt to be flooded, and then the large herds of half-wild cattle suffer great mortality from drowning, hunger, and the alligators. Neither in cattle-keeping nor cacao-growing are any but the laziest and most primitive methods followed, and the consequence is, that the proprietors are generally poor. A few, however, have become rich by applying a moderate amount of industry and skill to the management of their estates. People spoke of several heiresses in the neighbourhood whose wealth was reckoned in oxen and slaves ; a dozen slaves and a few hundred head of cattle being considered a great fortune. Some of them I saw had already been appropriated by enterprising young men, who had come from Pará and Maranham to seek their fortunes in this quarter.

The few weeks I spent here passed away pleasantly. I generally spent the evenings in the society of the townspeople, who associated together (contrary to Brazilian custom) in European fashion ; the different families meeting at one another's houses for social amusement, bachelor friends not being excluded, and the whole company, married and single, joining in simple games. The meetings used to take place in the sitting-rooms, and not in the open verandahs—a fashion almost compulsory on account of the mosquitoes ; but the evenings here are very cool, and the closeness of a room is not so much felt as it is in Pará. Sunday was strictly observed at Obydos ; at least all the shops are closed, and almost the whole population went to church. The vicar, Padre Raimundo do Sanchez Brito, was an excellent old man, and I fancy the friendly manners of the people, and the general purity of morals at Obydos, were owing in great part to the good example he set to his parishioners.

One day the owner of the house in which I occupied

a room, Major Martinho da Fonseca Seixas, came over from his estate on the opposite bank of the river. He was a man of great importance in the district, and the only one who had had enterprise sufficient to establish a sugar-mill. He crossed over soon after sunrise in a small boat, with four dark-skinned paddlers, who made the morning air ring with a wild chorus which their master, I was told, always made them sing, to beguile the way. I found him a tall, wiry, and sharp-featured old gentleman, with a shrewd but good-humoured expression of countenance—quite a typical specimen, in fact, of the old school of Brazilian planters. He landed in dressing-gown and slippers, and came up the beach chattering, scolding, and gesticulating. Several friends joined him, and we soon had the house full of company. After taking coffee and a hot buttered roll, he dressed and went to mass, whilst I slipped off to spend an hour or two in the woods. When I came back I found the Major with his friends seated in hammocks, two by two, slung in the four corners of the room, and all engaged in a lively discussion on political questions. They had a demijohn of cashaça in their midst, and were helping themselves freely, drinking out of little tea-cups. One of the company was a dark-skinned Cametaense, named Senhor Calisto Pantoja, a very agreeable fellow, and as full of talk as the Major. Like most of his townsmen, he was a Santa Luzia, or Liberal, whilst the old gentleman was a rabid Tory. Pantoja rather nettled the old man by saying that the Cametá people had held their town against the rebels in 1835, whilst the whites of Obydos abandoned theirs to be pillaged by them. The Major then launched out into a denunciation of the Cametaenses and the Liberals in general. He said he was a pure white, a "Massagonista;"* the blood of the Fidalguia of Portugal flowed in his veins, whilst the people of Cametá were a mixed breed of whites and Indians. I noticed that this boasting was ill received by the rest; it is generally, in fact, considered bad taste in Brazil to boast of purity of descent. Soon afterwards most of the visitors departed, and we dined in quiet. A few days afterwards I crossed the river to the Major's place, and spent two days with him. The house was a very large two-story building, having a large verandah to the upper floor. There was an appearance of disorder and cheerlessness about the place which was very dispiriting. The old gentleman was a widower. His only son had been brutally massacred by the rebels in 1835, whilst he was crossing the river in a small boat, and his two daughters were now completing their education at a seminary in Pará. The household affairs appeared to be managed by a middle-aged mulatto woman; and a number of dirty negro children were playing about the rooms. Amongst the outbuildings there were several large sheds, containing the cane-mill and sugar factory, and beyond these a curral, or enclosure for cattle. The mill for grinding the sugar-cane was a rude affair, worked by bullocks. The cane was pressed between wooden cylinders, and the juice received in troughs formed of hollowed logs. Sugar-cane here grows to a height of 18

* The Massagonistas are the descendants of the Portuguese colonists of Massagaõ, in Morocco, who forsook this place in a body in 1769, and migrated to the banks of the Amazons.

to 20 feet, the sugar-yielding part of the stem being about 8 feet in length and 3 inches in diameter. The land for miles around the establishment is rich alluvial soil, and as level as a bowling-green. Beyond the belt of forest which runs along the banks of the river, there is a large tract of soft green meadow with patches of woodland and scattered trees, combining to form a landscape like that of an English park. But a meadow on the banks of the Amazons is a very different thing from what it is in a temperate climate: the vegetation is rank and monotonous, and there are absolutely no flowers. The old gentleman had built a pretty little chapel on his estate, on the occasion of a visit from the Bishop of Pará, who sometimes travels through his diocese, and I slept in the Bishop's room attached to the building. The abundance of mosquitoes is a great drawback to the rich agricultural country on this side of the river. A little before night sets in, the inhabitants are obliged to close the doors and windows of their sleeping apartments; and it is singular that this simple means of keeping out the pests seems to be pretty successful. On the Upper Amazons the precaution is of no use, and every one is obliged to sling his hammock under a mosquito tent. The whole of this coast, as well as the banks of the many inlets which intersect it, is inhabited by scattered settlers. The population of the municipal district of Obydos, which comprises about twenty miles of river frontage, is estimated at 12,000 souls.

I made a large collection in the neighbourhood of Obydos, chiefly of insects. The forest is more varied than it is in the Amazons region generally. There is only one path leading into it for any considerable distance. It ascends first the rising ground behind the town, and then leads down through a broad alley where the trees arch overhead, to the sandy margins of a small lake choked up with aquatic plants, on the opposite bank of which rises the wooded hill before mentioned. Passing a swampy tract at the head of the lake, the road continues for three or four miles along the slopes of a ravine, after which it dwindles into a mere picada or hunter's track, and finally ceases altogether. Another shorter road runs along the top of the cliff westward, and terminates at a second small lake, which fills a basin-shaped depression between the hills, and is called Jauareté-paúa, or the Jaguar's Mud-hole. The vegetation on this rising ground is, of course, different from that of the low land. The trees, however, grow to an immense height. Those plants, such as the Heliconiæ and Marantaceæ, which have large, broad, and glossy leaves, and which give so luxuriant a character to the moister areas, are absent; but in their stead is an immense diversity of plants of the Bromeliaceous or pine-apple order, which grow in masses amongst the underwood, and make the forest in many places utterly impenetrable. Cacti also, which are peculiar to the drier soils, are very numerous, some of them growing to an unwieldy size, and resembling in shape huge candelabra.

The forest seemed to abound in monkeys, for I rarely passed a day without seeing several. I noticed four species: the Coaitá (Ateles paniscus), the Chrysothrix sciureus, the Callithrix torquatus, and our old Pará friend, Midas ursulus. The Coaitá is a large black monkey, covered with coarse hair, and having the prominent parts of the face of a tawny flesh-coloured hue. It is the largest of the Amazonian monkeys in stature, but is excelled in bulk by the "Barrigudo" (Lagothrix Humboldtii) of the Upper Amazons. It occurs throughout the low lands of the Lower and Upper Amazons, but does not range to the south beyond the limits of the river plains. At that point an allied species, the White-whiskered Coaitá (Ateles marginatus) takes its place. The Coaitás are called by some French zoologists spider monkeys, on account of the length and slenderness of their body and limbs. In these apes the tail, as a prehensile organ, reaches its highest degree of perfection; and on this account it would, perhaps, be correct to consider the Coaitás as the extreme development of the American type of apes. As far as we know, from living and fossil species, the New World has progressed no farther than the Coaitá towards the production of a higher form of the Quadrumanous order. The tendency of Nature here has been, to all appearance, simply to perfect those organs which adapt the species more and more completely to a purely arboreal life; and no nearer approach has been made towards the more advanced forms of anthropoid apes, which are the products of the Old World solely. The tail of the Coaitá is endowed with a wonderful degree of flexibility. It is always in motion, coiling and uncoiling like the trunk of an elephant, and grasping whatever comes within reach. Another remarkable character of the Coaitá is the absence of a thumb to the anterior hands. It is worthy of note that this strange deficiency occurs again in the Quadrumanous order only in the Colobi, a genus of apes peculiar to Africa. The Colobi, however, are not furnished with prehensile tails, and belong, in all their essential characters, to the Catarhinæ, or Old World monkeys, a group entirely distinct from the Platyrhinæ, or South American sub-order. The want of the thumb, therefore, is not a sign of near relationship between the Colobi and the Coaitás, but is a mere analogical character, which must have originated, in each case, through independent, although perhaps similar, causes. One species of Coaitá has a rudiment of thumb, without a nail. The flesh of this monkey is much esteemed by the natives in this part of the country, and the Military Commandant of Obydos, Major Gama, every week sent a negro hunter to shoot one for his table. One day I went on a Coaitá hunt, borrowing a negro slave of a friend to show me the way. On the road I was much amused by the conversation of my companion. He was a tall, handsome negro, about forty years of age, with a staid, courteous demeanour and a deliberate manner of speaking. Strangely enough in a negro, he was a total abstainer from liquors and tobacco. He told me he was a native of Congo, and the son of a great chief or king. He narrated the events of a great battle between his father's and some other tribe, in which he was taken prisoner and sold to the Portuguese slave-dealers. When in the deepest part of the ravine we heard a rustling sound in the trees overhead, and Manoel soon pointed out a Coaitá to me. There was something human-like in its appearance, as the lean, dark, shaggy creature moved deliberately amongst the branches at a great height. I fired, but unfortunately only wounded it in the belly. It fell with a crash headlong about twenty or thirty feet, and then caught a bough with its tail, which grasped it instantaneously, and then the animal remained suspended in mid-air. Before I could reload it recovered itself, and mounted nimbly to the topmost branches out of the reach of a fowling-piece, where we could perceive the poor thing apparently probing the wound with its fingers. Coaitás are more frequently kept in a tame state than any other kind of monkey. The Indians are very fond of them as pets, and the women often suckle them when young at their breasts. They become attached to their masters, and will sometimes follow them on the ground to considerable distances. I once saw a most ridiculously tame Coaitá. It was an old female, which accompanied its owner, a trader on the river, in all his voyages. By way of giving me a specimen of its intelligence and feeling, its master set to and rated it soundly, calling it scamp, heathen, thief, and so forth, all through the copious Portuguese vocabulary of vituperation. The poor monkey, quietly seated on the ground, seemed to be in sore trouble at this display of anger. It began by looking earnestly at him, then it whined, and lastly rocked its body to and fro with emotion, crying piteously, and passing its long, gaunt arms continually over its forehead; for this was its habit when excited, and the front of the head was worn quite bald in consequence. At length its master altered his tone. "It's all a lie, my old woman; you're an angel, a flower, a good affectionate old creature," and so forth. Immediately the poor monkey ceased its wailing, and soon after came over to where the man sat. The disposition of the Coaitá is mild in the extreme: it has none of the painful, restless vivacity of its kindred, the Cebi, and no trace of the surly, untameable temper of its still nearer relatives, the Mycetes, or howling monkeys. It is, however, an arrant thief, and shows considerable cunning in pilfering small articles of clothing, which it conceals in its sleeping place. The natives of the Upper Amazons procure the Coaitá, when full grown, by shooting it with the blowpipe and poisoned darts, and restoring life by putting a little salt (the antidote to the Urarí poison with which the darts are tipped) in its mouth. The animals thus caught become tame forthwith. Two females were once kept at the Jardin des Plantes of Paris, and Geoffroy St. Hilaire relates of them that they rarely quitted each other, remaining most part of the time in close embrace, folding their tails round one another's bodies. They took their meals together; and it was remarked on such occasions, when the friendship of animals is put to a hard test, that they never quarrelled or disputed the possession of a favourite fruit with each other.

The neighbourhood of Obydos was rich in insects. In the broad alleys of the forest a magnificent butterfly of the genus Morpho, six to eight inches in expanse, the Morpho Hecuba, was seen daily gliding along at a height of twenty feet or more from the ground. Amongst the lower trees and bushes numerous kinds of

Heliconii, a group of butterflies peculiar to tropical America, having long narrow wings, were very abundant. The prevailing ground colour of the wings of these insects is a deep black, and on this are depicted spots and streaks of crimson, white, and bright yellow, in different patterns according to the species. Their elegant shape, showy colours, and slow, sailing mode of flight make them very attractive objects, and their numbers are so great that they form quite a feature in the physiognomy of the forest, compensating for the scarcity of flowers. Next to the Heliconii the Catagrammas (C. astarte and C. peristera) were the most conspicuous. These have a very rapid and short flight, settling frequently and remaining stationary for a long time on the trunks of trees. The colours of their wings are vermilion and black, the surface having a rich velvety appearance. The genus owes its Greek name Catagramma (signifying "a letter beneath") to the curious markings of the underside of the wings, resembling Arabic numerals. The species and varieties are of almost endless diversity, but the majority inhabit the hot valleys of the eastern parts of the Andes. Another butterfly nearly allied to these, Callithea Leprieurii, was also very abundant here at the marshy head of the pool before mentioned. The wings are of a rich dark-blue colour, with a broad border of silvery green. These two groups of Callithea and Catagramma are found only in tropical America, chiefly near the equator, and are certainly amongst the most beautiful productions of a region where the animals and plants seem to have been fashioned in nature's choicest moulds. A great variety of other beautiful and curious insects adorned these pleasant woods. Others were seen only in the sunshine in open places. As the waters retreated from the beach, vast numbers of sulphur-yellow and orange coloured butterflies congregated on the moist sand. The greater portion of them belonged to the genus Callidryas.* They assembled in densely packed masses, sometimes two or three yards in circumference, their wings all held in an upright position, so that the beach looked as though variegated with beds of crocuses. These Callidryades seem to be migratory insects, and have large powers of dissemination. During the last two days of our voyage the great numbers constantly passing over the river attracted the attention of every one on board. They all crossed in one direction, namely, from north to south, and the processions were uninterrupted from an early hour in the morning until sunset. All the individuals which resort to the margins of sandy beaches are of the male sex. The females are much more rare, and are seen only on the borders of the forest, wandering from tree to tree, and depositing their eggs on low mimosas which grow in the shade. The migrating hordes, as far as I could ascertain, are composed only of males, and on this account I believe their wanderings do not extend very far. In confirmation of this is the fact that, although the same species generally

* More than three-fourths of the individuals in these congregations of butterflies consisted of a pale sulphur-coloured species, C. Statira : two yellow kinds, C. Eubule and C. Trite, and one orange-coloured, C. Argante, were less numerous. A few of a much larger species (C. Leachiana), sulphur-coloured with orange tips to the wings, now and then occurred amidst the masses.

has a very wide range, some being found from the central parts of the United States down to 32° S. lat., yet each distant region has its tolerably distinct local variety. But the effect of this general wandering habit of the group is, in the long run, a wide dissemination of the species; the formation of local varieties showing that the process is, nevertheless, a slow one. None of the species are found much beyond the tropics, but the genus is well represented within the tropical zone throughout the world; and an East Indian kind (C. Alcmeone) is so nearly allied to a South American one (C. Statira), as to have been mistaken for it by some authors.

A strange kind of wood-cricket is found in this neighbourhood. The males produce a very loud and not unmusical noise by rubbing together the overlapping edges of their wing-cases. The notes are certainly the loudest and most extraordinary that I ever heard produced by an orthopterous insect. The natives call it the Tananá, in allusion to its music, which is a sharp, resonant stridulation resembling the syllables ta-na-ná, ta-na-ná, succeeding each other with little intermission. It seems to be rare in the neighbourhood. When the natives capture one they keep it in a wicker-work cage for the sake of hearing it sing. A friend of mine kept one six days. It was lively only for two or three, and then its loud note could be heard from one end of the village to the other. When it died he gave me the specimen, the only one I was able to procure. It is a member of the family Locustidæ, a group intermediate between the Crickets (Achetidæ) and the Grasshoppers

(Acridiidæ). The total length of the body is two inches and a quarter; when the wings are closed the insect has an inflated vesicular or bladder-like shape, owing to the great convexity of the thin but firm parchmenty wing-cases, and the colour is wholly pale-green. The instru-

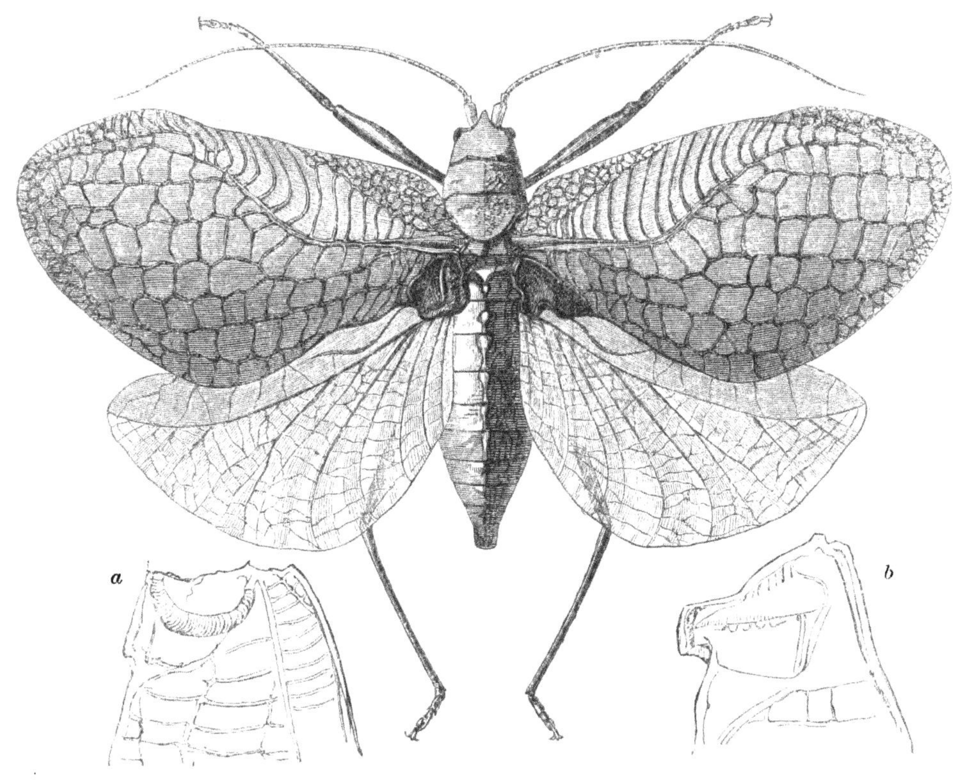

Musical Cricket (Chlorocœlus Tananá).
a. b. Lobes of wing-cases transformed into a musical instrument.

ment by which the Tananá produces its music is curiously contrived out of the ordinary nervures of the wing-cases. In each wing-case the inner edge, near its origin, has a horny expansion or lobe; on one wing (*b*) this lobe has sharp raised margins; on the other (*a*), the strong nervure which traverses the lobe on the under side is crossed by a number of fine sharp furrows like those of a file. When the insect rapidly moves its wings, the file of the one lobe is scraped sharply across the horny margin of the other, thus producing the sounds; the parchmenty wing-cases and the hollow drum-like space which they enclose assisting to give resonance to the tones. The projecting portions of both wing-cases are traversed by a similar strong nervure, but this is scored like a file only in one of them, in the other remaining perfectly smooth. Other species of the family to which the Tananá belongs have similar stridulating organs, but in none are these so highly developed as in this insect; they exist always in the males only, the other sex having the edges of the wing-cases quite straight and simple. The mode of producing the sounds and their object have been investigated by several authors with regard to certain European species. They are the call-notes of the males. In the common field-cricket of Europe the male has been observed to place itself, in the evening, at the entrance of its burrow, and stridulate until a female approaches, when the louder notes are succeeded by a more subdued tone, whilst the successful musician caresses with his antennæ the mate he has won. Any one who will take the trouble may observe a similar proceeding in the common house-cricket. The nature and object of this insect music are more uniform than the structure and situation of the instrument by which it is produced. This differs in each of the three allied families above mentioned. In the crickets the wing-cases are symmetrical; both have straight edges and sharply-scored nervures adapted to produce the stridulation. A distinct portion of their edges is not, therefore, set apart for the elaboration of a sound-producing instrument. In this family the wing-cases lie flat on the back of the insect, and overlap each other for a considerable portion of their extent. In the Locustidæ the same members have a sloping position on each side of the body, and do not overlap, except to a small extent near their bases ; it is out of this small portion that the stridulating organ is contrived. Greater resonance is given in most species by a thin transparent plate, covered by a membrane, in the centre of the overlapping lobes. In the Grasshoppers (Acridiidæ) the wing-cases meet in a straight suture, and the friction of portions of their edges is no longer possible. But Nature exhibits the same fertility of resource here as elsewhere; and in contriving other methods of supplying the males with an instrument for the production of call-notes indicates the great importance which she attaches to this function. The music in the males of the Acridiidæ is produced by the scraping of the long hind thighs against the horny nervures of the outer edges of the wing-cases ; a drum-shaped organ placed in a cavity near the insertion of the thighs being adapted to give resonance to the tones.

I obtained very few birds at Obydos. There was no scarcity of birds, but they were mostly common Cayenne species. In early morning the woods near my house were quite animated with their songs—an unusual thing in this country. I heard here for the first time the pleasing wild notes of the Carashué, a species of thrush, probably the Mimus lividus of ornithologists. I found it afterwards to be a common bird in the scattered woods of the campo district near Santarem. It is a much smaller and plainer-coloured bird than our thrush, and its song is not so loud, varied, or so long sustained; but the tone is of a sweet and plaintive quality, which harmonizes well with the wild and silent woodlands, where alone it is heard in the mornings and evenings of sultry tropical days. In course of time the song of this humble thrush stirred up pleasing associations in my mind, in the same way as those of its more highly endowed sisters formerly did at home. There are several allied species in Brazil; in the southern provinces they are called Sabiahs. The Brazilians are not insensible to the charms of this their best songster, for I often heard some pretty verses in praise of the Sabiah sung by young people to the accompaniment of the guitar. I found several times the nest of the Carashué, which is built of dried grass and slender twigs, and lined with mud; the eggs are coloured and spotted like those of our blackbird, but they are considerably smaller. I was much pleased with a brilliant little red-headed manikin, which I shot here (Pipra cornuta). There were three males seated on a low branch, and hopping slowly backwards and forwards, near to one another, as though engaged in a kind of dance. In the pleasant airy woods surrounding the sandy shores of the pool behind the town, the yellow-bellied Trogon (T. viridis) was very common. Its back is of a brilliant metallic-green colour, and the breast steel blue. The natives call it the Suruquá do Ygapó, or Trogon of the flooded lands, in contradistinction to the red-breasted species, which are named Suruquás da terra firma. I often saw small companies of half a dozen individuals quietly seated on the lower branches of trees. They remained almost motionless for an hour or two at a time, simply moving their heads, on the watch for passing insects; or, as seemed more generally to be the case, scanning the neighbouring trees for fruit; which they darted off now and then, at long intervals, to secure, returning always to the same perch.

The species of mammals, birds, and insects found at Obydos are, to a great extent, the same as those inhabiting the well-explored tract of country lying along the seacoast of Guiana. No other locality visited in the Amazons region supplied, among its productions, so large a proportion of Guiana forms. The four monkeys already mentioned all recur at Cayenne. A general resemblance of the species to those of Guiana is one of the principal features in the zoology of the Amazons valley; but in the low lands a great number exist only in the form of strongly modified local varieties; indeed, many of them are so much transformed that they pass for distinct species; and so they truly are, according to the received definitions of species. In the somewhat drier district of Obydos, the forms are more constant to their Guiana types. We seem to obtain here a glimpse of the manufacture of new species in nature. The way in which these modifications occur merits a few remarks. I will therefore give an account of one very instructive case which presented itself in this neighbourhood.

The case was furnished by certain kinds of handsome butterflies belonging to the genus Heliconius,* a group peculiar to Tropical America, abounding in individuals everywhere in the shades of its luxuriant forests, and presenting clusters of varieties and closely allied species, as well as many distinct, better marked forms. The closely allied species and varieties are a great puzzle to classifiers; in fact, the group is one of those wherein great changes seem to be now going on. A conspicuous mem-

Heliconius Melpomene.

ber of the group is the H. Melpomene of Linnæus. This elegant form is found throughout Guiana, Venezuela, and some parts of New Granada. It is very common at Obydos, and reappears on the south side of the river in the dry forests behind Santarem, at the mouth of the Tapajos. In all other parts of the Amazons valley, east-

* This genus has long been known under the name of Heliconia: a most inconvenient term, as a botanical genus bears the same name. An author has lately proposed to revert to the masculine termination of the words as first employed by Linnæus (Felder, in the "Wiener Entomologische Monatschrift," March, 1862), and, as I think the correction a good one, I adopt it.

ward to Pará and westward to Peru, it is entirely absent. This absence at first appeared to me very strange; for the local conditions of these regions did not appear so strongly contrasted as to check, in this abrupt manner, the range of so prolific a species; especially as at Obydos and Santarem it occurred in moist woods close to the edge of the river. Another and nearly allied species, however, takes its place in the forest plains; namely, the H. Thelxiope of Hübner. It is of the same size and shape as its sister kind, but differs very strikingly in colours: H. Melpomene being simply black with a large crimson spot on its wings, whilst H. Thelxiope has these beautifully rayed with black and crimson, and is further adorned with a number of bright yellow spots. Both have the same habits. H. Melpomene ornaments the

Heliconius Thelxiope.

sandy alleys in the forests of Obydos, floating lazily in great numbers over the lower trees; whilst H. Thelxiope, in a similar manner and in equal numbers, adorns the moister forests which constitute its domain. No one who has studied the group has doubted for a moment that the two are perfectly and originally distinct species, like the hare and rabbit, for instance, or any other two allied species of one and the same genus. The following facts, however, led me to conclude that the one is simply a modification of the other. There are, as might be supposed, districts of forest intermediate in character between the drier areas of Obydos, &c., and the moister tracts which compose the rest of the immense river valley. At two places in these intermediate districts, namely, Serpa, 180 miles west of Obydos, on the same side of the river, and Aveyros, on the lower Tapajos, most of the individuals of these Heliconii which occurred were transition forms between the two species. Already, at Obydos, H. Melpomene showed some slight variation amongst its individuals in the direction of H. Thelxiope, but not anything nearly approaching it. It might be said that these transition forms were hybrids, produced by the intercrossing of two originally distinct species; but the two come in contact in several places where these intermediate examples are unknown, and I never observed them to pair with each other. Besides which, many of them occur also on the coast of Guiana, where H. Thelxiope has never been found. These hybrid-looking specimens are connected together by so complete a chain of gradations that it is difficult to separate them even into varieties, and they are incomparably more rare than the two extreme forms. They link together gradually the wide interval between the two species. One is driven to conclude, from these facts, that the two were originally one and the same: the

Transition forms between Heliconius Melpomene and H. Thelxiope.

mode in which they occur and their relative geographical positions being in favour of the supposition that H. Thelxiope has been derived from H. Melpomene. Both are nevertheless good and true species in all the essential characters of species; for, as already observed, they do not pair together when existing side by side, nor is their any appearance of reversion to an original common form under the same circumstances.

In the controversy which is being waged amongst Naturalists, since the publication of the Darwinian theory of the origin of species, it has been rightly said that no proof at present existed of the production of a physiological species,—that is, a form which will not interbreed with the one from which it was derived, although given ample opportunities of doing so, and does not exhibit signs of reverting to its parent form when placed under the same conditions with it. Morphological species,—that is, forms which differ to an amount that would justify their being considered good species, have been produced in plenty through selection by man out of variations arising under domestication or cultivation. The facts just given are, therefore, of some scientific importance; for they tend to show that a physiological species can be and is produced in nature out of the varieties of a pre-existing closely allied one. This is not an isolated case; for I observed, in the course of my travels, a number of similar instances. But in very few has it happened that the species which clearly appears to be the parent coexists with one that has been evidently derived from it. Generally the supposed parent also seems to have been modified, and then the demonstration is not so clear, for some of the links in the chain of variation are wanting. The process of origination of a species in nature, as it takes place successively, must be ever perhaps beyond man's power to trace, on account of the great lapse of time it requires. But we can obtain a fair view of it by tracing a variable and far-spreading species over the wide area of its present distribution; and a long observation of such will lead to the conclusion that new species in all cases must have arisen out of variable and widely-disseminated forms. It sometimes happens, as in the present instance, that we find in one locality a species under a certain form which is constant to all the individuals concerned; in another exhibiting numerous varieties; and in a third presenting itself as a constant form, quite distinct from the one we set out with. If we meet with any two of these modifications living side by side, and maintaining their distinctive characters under such circumstances, the proof of the natural origination of a species is complete: it could not be much more so were we able to watch the process step by step. It might be objected that the difference between our two species is but slight, and that by classing them as varieties nothing further would be proved by them. But the differences between them are such as obtain between allied species generally. Large genera are composed, in great part, of such species; and it is interesting to show how the great and beautiful diversity within a large genus is brought about by the working of laws within our comprehension.

A few remarks on the way races are produced will be here in place. Naturalists have been generally inclined to attribute the formation of local varieties or races of a species to the direct action of physical conditions on individuals belonging to it which have migrated into new localities. It might be said, therefore, that our Heliconius Thelxiope of the moist forests has resulted from such operation of the local conditions on H. Melpomene, especially as intermediate varieties are found in districts of intermediate character and position. It is true that external agencies—such as food and climate, causing delayed or quickened growth, —have great effect on insects, acting on their adolescent states, and so by correlation of growth on the shape and colours of the adult forms.* But there is no proof that a complete local variety or race has been produced wholly by this means, modifications acquired by individuals not being generally transmissible to offspring. The examination of these races or closely allied species of Heliconii, with reference to their geographical distribution, throws light also on this subject. Thus Heliconius Thelxiope is disseminated over a district 2000 miles in length from east to west, from the mouth of the Amazons to the eastern slopes of the Andes, but shows no remarkable modification throughout all that area; some slight variations only occurring at the extreme points of it. If local conditions

* M. Bellier de la Chavignerie, in the "Annales de la Société Entomologique de France, 1858," p. 299, relates experiments on the effect of retardation of the pupa development through exposure to unusual cold, showing that striking varieties of the adult insects are producible by this means.

acting directly on individuals had originally produced this race or species, they certainly would have caused much modification of it in different parts of this region; for the upper Amazons country differs greatly from the district near the Atlantic in climate, sequence of seasons, soil, forest clothing, periodical inundations, and so forth. These differences moreover graduate away, so that the species is subjected to a great diversity of physical conditions from locality to locality, and ought in consequence to present an endless series of local varieties, on the view mentioned, instead of one constant form throughout. Besides, how should we explain the fact of H. Thelxiope and H. Melpomene both existing under the same local conditions; and how account for the diversified modifications presented in one and the same locality as at Serpa and on the Tapajos?*

There is evidently therefore some more subtle agency at work in the segregation of a race than the direct operation of external conditions. The principle of natural selection, as lately propounded by Darwin, seems to offer an intelligible explanation of the facts. According to this theory, the variable state of the species exhibited in the districts above mentioned would be owing to Heliconius Melpomene having been rendered vaguely instable by the indirect action of local conditions dis-

* As the action of external influences would be on the early states of the insects and not on the adults, it is well to mention that the broods of the Heliconii appear to be social; the larvæ feeding together and undergoing their last transformation on the same tree. This I observed with regard to the H. Erato, a species closely allied to H. Thelxiope.

similar to those where it exists under a constant normal form. In these districts selection has not operated, or it is suitable to the conditions of life there prevailing, that the species should exist under an instable form. But in the adjoining moister forests, as the result shows, the local conditions were originally more favourable to one of these varieties than to the others. The selected one, therefore, increased more rapidly than its relatives ; and the fact of the entire absence of these latter from an area whence they are now separated only by a few miles, points to the conclusion that they could not there maintain their ground. Those individuals of successive broods which were still better suited to the new conditions would for the same reasons be preferred over their relatives ; and this process going forward for a few generations, the extreme form of H. Thelxiope would be reached. At this point the race became well adapted to the new area, which we may suppose to have been at that epoch in process of formation as the river plains became dry land, at the last geological changes in the level of the country. In the higher and drier areas of Guiana and the neighbouring countries, H. Melpomene has been the selected form ; in the lower and more humid regions of the Amazons, H. Thelxiope has been preferred. An existing proof of this perfect adaptation is shown by the swarming abundance of the species ; the derivation of H. Thelxiope from H. Melpomene is made extremely probable by the existence of a complete series of connecting links ; and lastly, its permanent establishment is made evident by its refusal to intercross with its parent form, or revert to its former shape when brought by natural redistribution into contact with it.*

* If this explanation of the derivation of Heliconius Thelxiope be true, the origination by natural process of a host of now distinct allied species of this genus, as well as, in fact, all other genera containing numerous closely related species, will have to be admitted. A species allied to H. Thelxiope, namely, H. Vesta, seems to have been derived also from H. Melpomene, for amongst the numerous varieties already mentioned are many examples intermediate between the two. There is this difference, however, between H. Thelxiope and H. Vesta: the former is confined in its range to the Amazons valley, whilst H. Vesta extends beyond this region over Guiana and the central valleys of the Andes; it seems, therefore, to have acquired a power of adaptation to a much wider diversity of local conditions. Insects seem to be well adapted to furnish data in illustration of this interesting but difficult subject. This arises chiefly from the ease with which ample suites of specimens can be obtained for comparison from many points in the areas of distribution, both of species and varieties. It is scarcely necessary to add that the conclusions thus arrived at will apply to all organic beings.

CHAPTER VII.

THE LOWER AMAZONS—OBYDOS TO MANAOS, OR THE BARRA OF THE RIO NEGRO.

Departure from Obydos—River banks and by-channels—Cacao planters—Daily life on board our vessel—Great storm—Sand-island and its birds—Hill of Parentins—Negro trader and Mauhés Indians—Villa Nova, its inhabitants, climate, forest, and animal productions—Cararaucú—A rustic festival—Lake of Cararaucú—Motúca flies—Serpa—Christmas holidays—River Madeira—A mameluco farmer—Mura Indians—Rio Negro—Description of Barra—Descent to Pará—Yellow fever.

A TRADER of Obydos, named Penna, was about proceeding in a cuberta laden with merchandise to the Rio Negro, intending to stop frequently on the road; so I bargained with him for a passage. He gave up a part of the toldo, or fore-cabin as it may be called, and here I slung my hammock and arranged my boxes, so as to be able to work as we went along. The stoppages I thought would be an advantage, as I could collect in the woods whilst he traded, and thus acquire a knowledge of the productions of many places on the river which in a direct voyage it would be impossible to do. I provided a stock of groceries for two months' consumption; and, after the usual amount of unnecessary fuss and delay on the part of the owner, we started on the 19th of November. Penna took his family with him; this comprised a smart, lively mameluco woman, named Catarina, whom we called Senhora Katita, and two children. The crew consisted of three men, one a sturdy Indian, another a Cafuzo, godson of Penna, and the third, our best hand, a steady, good-natured mulatto, named Joaquim. My boy Luco was to assist in rowing and so forth. Penna was a timid middle-aged man, a white with a slight cross of Indian; when he was surly and obstinate, he used to ask me to excuse him on account of the Tapuyo blood in his veins. He tried to make me as comfortable as the circumstances admitted, and provided a large stock of eatables and drinkables; so that altogether the voyage promised to be a pleasant one.

On leaving the port of Obydos we crossed over to the right bank, and sailed with a light wind all day, passing numerous houses, each surrounded by its grove of cacao trees. On the 20th we made slow progress. After passing the high land at the mouth of the Trombetas, the banks were low, clayey, or earthy on both sides. The breadth of the river varies hereabout from two and a half to three miles, but neither coast is the true terra firma. On the northern side a by-channel runs for a long distance inland, communicating with the extensive lake of Faro; on the south, three channels lead to the similar fresh-water sea of Villa Franca; these are in part arms of the river, so that the land they surround consists, properly speaking, of islands. When this description of land is not formed wholly of river deposit, as sometimes happens, or is raised above the level of the highest floods, it is called *Ygapó alto,* and is distinguished by the natives from the true islands of mid-river, as well as from the terra firma. We landed at one of the cacao plantations. The house was substantially built; the walls formed of strong upright posts, lathed across, plastered with mud and whitewashed, and the roof tiled. The family were mamelucos, and seemed to be an average sample of the poorer class of cacao growers. All were loosely dressed and bare-footed. A broad verandah extended along one side of the house, the floor of which was simply the well-trodden earth; and here hammocks were slung between the bare upright supports, a large rush mat being spread on the ground, upon which the stout matron-like mistress, with a tame parrot perched upon her shoulder, sat sewing with two pretty little mulatto girls. The master, coolly clad in shirt and drawers, the former loose about the neck, lay in his hammock smoking a long gaudily-painted wooden pipe. The household utensils, earthenware jars, water-pots and saucepans, lay at one end, near which was a wood fire, with the ever-ready coffee-pot simmering on the top of a clay tripod. A large shed stood a short distance off, embowered in a grove of banana, papaw, and mango trees; and under it were the ovens, troughs, sieves, and all other apparatus for the preparation of mandioca. The cleared space around the house was only a few yards in extent; beyond it lay the cacao plantations, which stretched on each side parallel to the banks of the river. There was a path through the forest which led to the mandioca fields, and several miles beyond to other houses on the banks of an interior channel. We were kindly received, as is always the case when a stranger visits these out-of-the-way habitations; the people being invariably civil and hospitable. We had a long chat, took coffee, and on departing one of the daughters sent a basket full of oranges for our use down to the canoe.

The cost of a cacao plantation in the Obydos district is after the rate of 240 reis or sixpence per tree, which is much higher than at Cametá, where I believe the yield is not so great. The forest here is cleared before planting, and the trees are grown in rows. The smaller cultivators are all very poor. Labour is scarce; one family generally manages its own small plantation of 10,000 to 15,000 trees, but at the harvest time neighbours assist each other. It appeared to me to be an easy, pleasant life; the work is all done under shade, and occupies only a few weeks in the year. The incorrigible nonchalance and laziness of the people alone prevent them from surrounding themselves with all the luxuries of a tropical country. They might plant orchards of the choicest fruit-trees around their houses, grow Indian corn, and rear cattle and hogs, as intelligent settlers from Europe would certainly do, instead of indolently relying solely on the produce of their small plantations, and living on a meagre diet of fish and farinha. In preparing the cacao they have not devised any means of separating the seeds well from the pulp, or drying it in a systematic way; the consequence is that, although naturally of good quality, it moulds before reaching the merchants' stores, and does not fetch more than half the price of the same article grown in other parts of tropical America. The Amazons region is the original home of the principal species of chocolate tree, the Theobroma cacao; and it grows in abundance in the forests of the upper river. The cultivated crop appears to be a precarious one; little or no care, however, is bestowed on the trees, and even weeding is done very inefficiently. The plantations are generally old, and have been made on the low ground near the river, which renders them liable to inundation when this rises a few inches more than the average. There is plenty of higher land quite suitable to the tree, but it is uncleared, and the want of labour and enterprise prevents the establishment of new plantations.*

We passed the last houses in the Obydos district on the 20th, and the river scenery then resumed its usual wild and solitary character, which the scattered human habitations relieved, although in a small degree. We soon fell into a regular mode of life on board our little ark. Penna would not travel by night; indeed, our small crew, wearied by the day's labour, required rest, and we very rarely had wind in the night. We used to moor the vessel to a tree, giving out plenty of cable, so as to sleep at a distance from the banks and free of mosquitoes, which although swarming in the forest, rarely came

* Next to india-rubber, cacao is the chief article of exportation from Pará. The yield, however, varies greatly in different years. The price also fluctuates considerably, and does not follow the abundance or scarcity of the crop. The following valuation of exports of the article is taken from an official statement of exports, given me by Mr. Bailey, U. S. Consul at Pará. In 1856, £99,247 7*s.* 9*d.*; 1857, £208,926; 1858, £133,013 8*s.* The quantity in weight exported was in 1856, 4,343,136lb.; in 1857, 7,428,480lb.

many yards out into the river at this season of the year. The strong current at a distance of thirty or forty yards from the coast steadied the cuberta head to stream, and kept us from drifting ashore. We all slept in the open air, as the heat of the cabins was stifling in the early part of the night. Penna, Senhora Katita, and I slung our hammocks in triangle between the mainmast and two stout poles fixed in the raised deck. A sheet was the only covering required, besides our regular clothing; for the decrease of temperature at night on the Amazons is never so great as to be felt otherwise than as a delightful coolness after the sweltering heat of the afternoons. We used to rise when the first gleam of dawn showed itself above the long, dark line of forest. Our clothes and hammocks were then generally soaked with dew, but this was not felt to be an inconvenience. The Indian Manoel used to revive himself by a plunge in the river, under the bows of the vessel. It is the habit of all Indians, male and female, to bathe early in the morning; they do it sometimes for warmth's sake, the temperature of the water being often considerably higher than that of the air. Penna and I lolled in our hammocks, whilst Katita prepared the indispensable cup of strong coffee, which she did with wonderful celerity, smoking meanwhile her early morning pipe of tobacco. Liberal owners of river craft allow a cup of coffee sweetened with molasses, or a ration of cashaça, to each man of their crews; Penna gave them coffee. When all were served, the day's work began. There was seldom any wind at this early hour; so if there was a remanso along the shore the men rowed, if not there was no way of progressing but by espia. In some places the currents ran with great force close to the banks, especially where these receded to form long bays or *enseadas,* as they are called, and then we made very little headway. In such places the banks consist of loose earth, a rich crumbly vegetable mould, supporting a growth of most luxuriant forest, of which the currents almost daily carry away large portions, so that the stream for several yards out is encumbered with fallen trees, whose branches quiver in the current. When projecting points of land were encountered, it was impossible, with our weak crew, to pull the cuberta against the whirling torrents which set round them; and in such cases we had to cross the river, drifting often with the current, a mile or two lower down on the opposite shore. There generally sprung a light wind as the day advanced, and then we took down our hammocks, hoisted all sail, and bowled away merrily. Penna generally preferred to cook the dinner ashore, when there was little or no wind. About midday on these calm days we used to look out for a nice shady nook in the forest, with cleared space sufficient to make a fire upon. I then had an hour's hunting in the neighbouring wilderness, and was always rewarded by the discovery of some new species. During the greater part of our voyage, however, we stopped at the house of some settler, and made our fire in the port. Just before dinner it was our habit to take a bath in the river, and then, according to the universal custom on the Amazons, where it seems to be suitable on account of the weak fish diet, we each took half a tea-cup full of neat cashaça, the "abre" or "opening," as it is called, and set to on our mess of stewed pirarucú, beans, and bacon. Once or twice a week we had fowls and rice; at supper, after sunset, we often had fresh fish caught by our men in the evening. The mornings were cool and pleasant until towards midday; but in the afternoons the heat became almost intolerable, especially in gleamy, squally weather, such as generally prevailed. We then crouched in the shade of the sails, or went down to our hammocks in the cabin, choosing to be half stifled rather than expose ourselves on deck to the sickening heat of the sun. We generally ceased travelling about nine o'clock, fixing upon a safe spot wherein to secure the vessel for the night. The cool evening hours were delicious; flocks of whistling ducks (Anas autumnalis), parrots, and hoarsely-screaming macaws, pair by pair, flew over from their feeding to their resting places, as the glowing sun plunged abruptly beneath the horizon. The brief evening chorus of animals then began, the chief performers being the howling monkeys, whose frightful unearthly roar deepened the feeling of solitude which crept on as darkness closed around us. Soon after, the fireflies in great diversity of species came forth and flitted about the trees. As night advanced, all became silent in the forest, save the occasional hooting of tree-frogs, or the monotonous chirping of wood-crickets and grasshoppers.

We made but little progress on the 20th and two following days, on account of the unsteadiness of the wind. The dry season had been of very brief duration this year; it generally lasts in this part of the Amazons from July to January, with a short interval of showery weather in November. The river ought to sink thirty or thirty-five feet below its highest point; this year it had declined only about twenty-five feet, and the November rains threatened to be continuous. The drier the weather the stronger blows the east wind; it now failed us altogether, or blew gently for a few hours merely in the afternoons. I had hitherto seen the great river only in its sunniest aspect; I was now about to witness what it could furnish in the way of storms.

On the night of the 22nd the moon appeared with a misty halo. As we went to rest, a fresh watery wind was blowing, and a dark pile of clouds gathering up river in a direction opposite to that of the wind. I thought this betokened nothing more than a heavy rain which would send us all in a hurry to our cabins. The men moored the vessel to a tree alongside a hard clayey bank, and after supper all were soon fast asleep, scattered about the raised deck. About eleven o'clock I was awakened by a horrible uproar, as a hurricane of wind suddenly swept over from the opposite shore. The cuberta was hurled with force against the clayey bank; Penna shouted out, as he started to his legs, that a trovoada de cima, or a squall from up river, was upon us. We took down our hammocks, and then all hands were required to save the vessel from being dashed to pieces. The moon set, and a black pall of clouds spread itself over the dark forests and river; a frightful crack of thunder now burst over our heads, and down fell the drenching rain. Joaquim leapt ashore through the drowning spray with a strong pole, and tried to pass the cuberta round a small projecting point, whilst we on deck aided in keeping her off and lengthened the cable. We succeeded in getting free, and the stout-built boat fell off into the strong current further away from the shore, Joaquim swinging himself dexterously aboard by the bowsprit as it passed the point. It was fortunate for us that we happened to be on a sloping clayey bank, where there was no fear of falling trees; a few yards further on, where the shore was perpendicular and formed of crumbly earth, large portions of loose soil, with all their superincumbent mass of forest, were being washed away; the uproar thus occasioned adding to the horrors of the storm.

The violence of the wind abated in the course of an hour, but the deluge of rain continued until about three o'clock in the morning; the sky being lighted up by almost incessant flashes of pallid lightning, and the thunder pealing from side to side without interruption. Our clothing, hammocks, and goods were thoroughly soaked by the streams of water which trickled through between the planks. In the morning all was quiet; but an opaque, leaden mass of clouds overspread the sky, throwing a gloom over the wild landscape that had a most dispiriting effect. These squalls from the west are always expected about the time of the breaking up of the dry season in these central parts of the Lower Amazons. They generally take place about the beginning of February, so that this year they had commenced much earlier than usual. The soil and climate are much drier in this part of the country than in the region lying further to the west, where the denser forests and more clayey, humid soil produce a considerably cooler

T 2

atmosphere. The storms may be therefore attributed to the rush of cold moist air from up river, when the regular trade-wind coming from the sea has slackened or ceased to blow.

On the 26th we arrived at a large sand bank connected with an island in midriver, in front of an inlet called Maracá-uassú. Here we anchored and spent half a day ashore. Penna's object in stopping was simply to enjoy a ramble on the sands with the children, and give Senhora Katita an opportunity to wash the linen. The sandbank was now fast going under water with the rise of the river; in the middle of the dry season it is about a mile long and half a mile in width. The canoe-men delight in these open spaces, which are a great relief to the monotony of the forest that clothes the land in every other part of the river. Further westward they are much more frequent, and of larger extent. They lie generally at the upper end of islands; in fact, the latter originate in accretions of vegetable matter formed by plants and trees growing on a shoal. The island was wooded chiefly with the trumpet tree (Cecropia peltata), which has a hollow stem and smooth pale bark. The leaves are similar in shape to those of the horse-chestnut, but immensely larger; beneath they are white, and when the welcome trade-wind blows they show their silvery undersides,—a pleasant signal to the weary canoe traveller. The mode of growth of this tree is curious: the branches are emitted at nearly right angles with the stem, the branchlets in minor whorls around these, and so forth, the leaves growing at their extremities; so that the total appearance is that of a huge candelabrum. Cecropiæ of different species are characteristic of Brazilian forest scenery; the kind of which I am speaking grows in great numbers everywhere on the banks of the Amazons where the land is low. In the same places the curious Monguba tree (Bombax ceiba) is also plentiful; the dark green bark of its huge tapering trunk, scored with gray, forming a conspicuous object. The principal palm-tree on the low lands is the Jauarí (Astryocaryum Jauarí), whose stem, surrounded by whorls of spines, shoots up to a great height. On the borders of the island were large tracts of arrow-grass (Gynerium saccharoides), which bears elegant plumes of flowers, like those of the reed, and grows to a height of twenty feet, the leaves arranged in a fan-shaped figure near the middle of the stem. I was surprised to find on the higher parts of the sandbank the familiar foliage of a willow (Salix Humboldtiana). It is a dwarf species, and grows in patches resembling beds of osiers; as in the English willows, the leaves were peopled by small chrysomelideous beetles. In wandering about, many features reminded me of the seashore. Flocks of white gulls were flying overhead, uttering their well-known cry, and sandpipers coursed along the edge of the water. Here and there lonely wading-birds were stalking about; one of these, the Curicáca (Ibis melanopis), flew up with a loud cackling noise, and was soon joined by a unicorn bird (Palamedea cornuta), which I startled up from amidst the bushes, whose harsh screams, resembling the bray of a jackass, but shriller, disturbed unpleasantly the solitude of the place. Amongst the willow bushes were flocks of a handsome bird belonging to the Icteridæ or troupial family, adorned with a rich plumage of black and saffron-yellow. I spent some time watching an assemblage of a species of bird called by the natives Tamburí-pará, on the Cecropia trees. It is the Monasa nigrifrons of ornithologists, and has a plain slate-coloured plumage with the beak of an orange hue. It belongs to the family of Barbets, most of whose members are remarkable for their dull inactive temperament. Those species which are ranged by ornithologists under the genus Bucco are called by the Indians, in the Tupí language, Tai-assú uirá, or pig-birds. They remain seated sometimes for hours together on low branches in the shade, and are stimulated to exertion only when attracted by passing insects. This flock of Tamburí-pará were the reverse of dull; they were gambolling and chasing each other amongst the branches. As they sported about, each emitted a few short tuneful notes, which altogether produced a ringing, musical chorus that quite surprised me.

On the 27th we reached an elevated wooded promontory, called Parentins, which now forms the boundary between the provinces of Pará and the Amazons. Here we met a small canoe descending to Santarem. The owner was a free negro named Lima, who, with his wife, was going down the river to exchange his year's crop of tobacco for European merchandise. The long shallow canoe was laden nearly to the water level. He resided on the banks of the Abacaxí, a river which discharges its waters into the Canomá, a broad interior channel which extends from the river Madeira to the Parentins, a distance of 180 miles. Penna offered him advantageous terms, so a bargain was struck, and the man saved his long journey. The negro seemed a frank, straightforward fellow; he was a native of Pernambuco, but had settled many years ago in this part of the country. He had with him a little Indian girl belonging to the Mauhés tribe, whose native seat is the district of country lying in the rear of the Canomá, between the Madeira and the Tapajos. The Maúhés are considered, I think with truth, to be a branch of the great Mundurucú nation, having segregated from them at a remote period, and by long isolation acquired different customs and a totally different language, in a manner which seems to have been general with the Brazilian aborigines. The Mundurucús seem to have retained more of the general characteristics of the original Tupí stock than the Mauhés. Senhor Lima told me, what I afterwards found to be correct, that there were scarcely two words alike in the languages of the two peoples, although there are words closely allied to Tupí in both.* The little girl had not the slightest trace of the savage in her appearance. Her features were finely shaped, the cheek-bones not at all prominent, the lips thin, and the expression of her countenance frank and smiling. She had been brought only a few weeks previously from a remote settlement of her tribe on the banks of the Abacaxí, and did not yet know five words of Portuguese. The Indians, as a general rule, are very manageable when they are young, but it is a general complaint that

* Thus the word Woman, in Mauhé, is Unihá; in Tupí, Cunhá; in Mundurucú, Taishí. Fire in Mauhé, is Ariá; in Tupí, Tatá; in Mundurucú, Idashá or Tashá.

when they reach the age of puberty they become restless and discontented. The rooted impatience of all restraint then shows itself, and the kindest treatment will not prevent them running away from their masters; they do not return to the malocas of their tribes, but join parties who go out to collect the produce of the forests and rivers, and lead a wandering semi-savage kind of life.

We remained under the Serra dos Parentins all night. Early the next morning a light mist hung about the tree-tops, and the forest resounded with the yelping of Whaiápu-sai monkeys. I went ashore with my gun and got a glimpse of the flock, but did not succeed in obtaining a specimen. They were of small size and covered with long fur of a uniform gray colour. I think the species was the Callithrix donacophilus. The rock composing the elevated ridge of the Parentins is the same coarse iron-cemented conglomerate which I have spoken of as occurring near Pará and in several other places. Many loose blocks were scattered about. The forest was extremely varied, and inextricable coils of woody climbers stretched from tree to tree. Thongs of cacti were spread over the rocks and tree-trunks. The variety of small, beautifully-shaped ferns, lichens, and boleti made the place quite a museum of cryptogamic plants. I found here two exquisite species of Longicorn beetles, and a large kind of grasshopper (Pterochroza) whose broad fore-wings resembled the leaf of a plant, providing the insect with a perfect disguise when they were closed; whilst the hind-wings were decorated with gaily-coloured eye-like spots.

The negro left us and turned up a narrow channel, the Paraná-mirim dos Ramos (the little river of the branches, *i.e.*, having many ramifications), on the road to his home, 130 miles distant. We then continued our voyage, and in the evening arrived at Villa Nova, a straggling village containing about seventy houses, many of which scarcely deserve the name, being mere mud-huts roofed with palm-leaves. We stayed here four days. The village is built on a rocky bank, composed of the same coarse conglomerate as that already so often mentioned. In some places a bed of Tabatinga clay rests on the conglomerate. The soil in the neighbourhood is sandy, and the forest, most of which appears to be of second growth, is traversed by broad alleys which terminate to the south and east on the banks of pools and lakes, a chain of which extends through the interior of the land. As soon as we anchored I set off with Luco to explore the district. We walked about a mile along the marly shore, on which was a thick carpet of flowering shrubs, enlivened by a great variety of lovely little butterflies, and then entered the forest by a dry watercourse. About a furlong inland this opened on a broad placid pool, whose banks, clothed with grass of the softest green hue, sloped gently from the water's edge to the compact wall of forest which encompassed the whole. The pool swarmed with water-fowl; snowy egrets, dark-coloured striped herons, and storks of various species standing in rows around its margins. Small flocks of Macaws were stirring about the topmost branches of the trees. Long-legged piosócas (Parra Jacana) stalked over the water-plants on the surface of the pool, and in the bushes on its margin were great numbers of a kind of canary (Sycalis brasiliensis) of a greenish-yellow colour, which has a short and not very melodious song. We had advanced but a few steps when we startled a pair of the Jaburú-moleque (Mycteria Americana), a powerful bird of the stork family, four and a half feet in height, which flew up and alarmed the rest, so that I got only one bird out of the tumultuous flocks which passed over our heads. Passing towards the farther end of the pool I saw, resting on the surface of the water, a number of large round leaves, turned up at their edges; they belonged to the Victoria water-lily. The leaves were just beginning to expand (December 3rd), some were still under water, and the largest of those which had reached the surface measured not quite three feet in diameter. We found a montaria with a paddle in it, drawn up on the bank, which I took leave to borrow of the unknown owner, and Luco paddled me amongst the noble plants to search for flowers, meeting, however, with no success. I learnt afterwards that the plant is common in nearly all the lakes of this neighbourhood. The natives call it the furno do Piosoca, or oven of the Jacana, the shape of the leaves being like that of the ovens on which Mandioca meal is roasted. We saw many kinds of hawks and eagles, one of which, a black species, the Caracára-í (Milvago nudicollis), sat on the top of a tall naked stump, uttering its hypocritical whining notes. This eagle is considered a bird of ill omen by the Indians; it often perches on the tops of trees in the neighbourhood of their huts, and is then said to bring a warning of death to some member of the household. Others say that its whining cry is intended to attract other defenceless birds within its reach. The little courageous flycatcher Bem-ti-vi (Saurophagus sulphuratus) assembles in companies of four or five, and attacks it boldly, driving it from the perch where it would otherwise sit for hours. I shot three hawks of as many different species; and these, with a Magoary stork, two beautiful gilded-green jacamars (Galbula chalcocephala), and half-a-dozen leaves of the water-lily made a heavy load, with which we trudged off back to the canoe.

A few years after this visit, namely, in 1854-5, I passed eight months at Villa Nova. The district of which it is the chief town is very extensive, for it has about forty miles of linear extent along the banks of the river; but the whole does not contain more than 4000 inhabitants. More than half of these are pure-blood Indians, who live in a semi-civilized condition on the banks of the numerous channels and lakes. The trade of the place is chiefly in India-rubber, balsam of Copaiba (which are collected on the banks of the Madeira and the numerous rivers that enter the Canomá channel), and salt fish, prepared in the dry season, nearer home. These articles are sent to Pará in exchange for European goods. The few Indian and half-breed families who reside in the town, are many shades inferior in personal qualities and social condition to those I lived amongst near Pará and Cametá. They live in wretched dilapidated mud-hovels; the women cultivate small patches of mandioca; the men spend most of their time in fishing, selling what they do not require themselves and getting drunk with the most exemplary regularity on cashaça, purchased with the proceeds.

The configuration of the district of country in which Villa Nova is situated, is remarkable. About a mile inland, there commences a chain of lakes of greater or lesser extent, which are connected together by narrow channels, and extend to the interior by-water of the Ramos. This latter communicates with the channel of Canomá, already mentioned as connected with the river Madeira. The whole tract of land, therefore, forms an island, or group of islands, which extends from a little below Villa Nova, to the mouth of the Madeira, a distance of 180 miles; the breadth varying from ten to twenty miles. The district is known by the name of the Island of Tupinambarána. The Canomá is an outlet to the waters of the Madeira when this river is fuller than the main Amazons, which is the case from November to February. But it also receives the contributions of eight other independent rivers, most of which have broad, lake-like expansions of water near their junction with the Canomá. One of them, the Andirá-mirim, I was told, is a league broad for some distance from its mouth. The country bordering these interior waters is extremely fertile, and the broad lakes have clear waters and sandy shores. They abound in fish and turtle. The country is healthy along the banks of the Canomá, and for some distance up its tributary streams. In certain places on the banks of these, intermittent fevers prevail, as they do on all those affluents of the Amazons which have clear, dark waters and slow currents. The incidence of this endemic is somewhat remarkable, for it exists on one side of the Andirá-mirim, where the land is high and rocky, and not on the other which is low and swampy. The old historians relate that the island of Tupinambarána was colonised by a portion of the great Tupí or Tupinámba nation, who were driven from the sea-coast near Pernambuco, by the early Portuguese settlers in the 16th century. I think, however, there is reason to conclude, that different tribes, having more or less affinity with the Tupís, originally existed in many places on the banks of the Amazons, and that they had frequent communication with each other, before the time of the Portuguese. Much partial migration probably occurred when the aborigines had the navigation of the main Amazons all to themselves. It seems to me very unlikely, that a compact body of Indians wandered at once from the sea-coast near Pernambuco to the central parts of the Amazons. However this may be, no trace of the aboriginal Tupís now exists in this quarter. The district is thinly populated, and the Indians who now reside here, are scattered hordes of the Mundurucú, Múra, and Mauhés tribes : semi-civilised families of the two latter live in or near the town.

I found some very friendly and intelligent people amongst the white and mameluco families residing at Villa Nova. The vicar, Father Torquato de Souza, is not quite unknown to the European public, having been the guide of Prince Adalbert of Prussia when he visited the Jurúna Indians on the Xingú, and mentioned in the published narrative of the journey. He is now a distinguished citizen of the new Province of the Amazons, having been elected, several times in succession, President of the Provincial Chamber. Together with many other natives of the Amazons region, he affords a proof that an equatorial climate in the new world has not necessarily a deteriorating effect on the white race. He is a well-built man: above the middle height, with handsome features, and a fine, healthy, ruddy complexion. He is a most lively and energetic fellow. When we first landed at Villa Nova, in 1849, the church was being repaired, and as carpenters were scarce, he had buckled to the work himself, and I found him, with sleeves turned up, sawing and planing as though he was well used to the trade. Next to Padre Torquato, Senhor Meirelles, well deserves mention; a more sensible, intelligent and kind-hearted man I never met with in Brazil. He also held some appointment under Government, but his time was chiefly taken up with the management of his plantations situated three miles below the village. Both these worthy men were fond of reading, and subscribed regularly to Rio Janeiro daily newspapers. Senhor Meirelles spent a deal of money on dear books, which he sent for by a parcel at a time from the metropolis, 2000 miles off. Some of these were Portuguese periodicals, on the plan of the English Penny Magazine; most of them, however, were translations of romances chiefly French. They circulated freely amongst the many readers at Villa Nova. At the time of my visit "Uncle Tom's Cabin," translated into Portuguese, was a great favourite. I found a love of reading not at all uncommon amongst the better sort of people in the towns and villages on the Amazons; it seems natural to the climate, and is promoted by the occupation being well suited to the hot and lazy hours of mid-day. It is a pity the Portuguese language, on account of the poverty of its modern literature, is so poor a medium for acquiring knowledge, and that books are so scarce in Northern Brazil, otherwise the Amazonian people would not be condemned to the wretchedly narrow range of information which is now generally their lot. A system of popular education supported entirely by the Government, has been established for some time in Brazil, and a primary school for boys exists in every small town from Pará to the frontiers of the Empire. Padre Torquato was the schoolmaster, as well as the priest at Villa Nova. He had about thirty scholars, who were of all shades of colour, from the negro and Indian to the pure white. The schoolmasters, as mentioned in a former chapter, receive the same amount of salary as the priests, namely, 600 milreis, or about 70*l.* a year; but they are entitled to a bonus if the number of scholars exceeds a certain limit. In some of the larger villages, schools for girls have also been established. It is very desirable that these should be well supported, for the future advancement of the Brazilian people towards a better social condition depends in a great measure on the improvement in the education of their women.

Villa Nova, like most places on the main Amazons, is very healthy; it is considerably more so than Santarem, where the climate is much drier and hotter, or the regions further west, where the air is sultry and stagnant. The cool and invigorating east wind becomes neutralised before reaching the Rio Negro, but at Villa Nova, in average seasons, it blows daily, with the exception of a few weeks' interval in November, from the beginning of September to the end of January. The river, here about two and a half miles broad, makes a bold sweep of ten or twelve miles free of islands, the blue ridge of the Parentins terminating the prospect down stream. The broad, rapidly-flowing current, with the brisk counter-movement of the atmosphere, are no doubt the chief causes of the salubrity of the district. The seasons vary very considerably. Thus, in 1849, as already mentioned, the period of dryness and strong breezes was unusually short, and the river, in consequence, did not sink to its usual level. In 1854 I witnessed the opposite extreme. The wet season, from February to June, had been very severe, and the waters had risen to their highest point. It took us, in the months of June and July, in a well-manned vessel, fourteen days to ascend from Santarem, a distance of only 110 miles. The currents were very strong; all the low lands were flooded, and great portions of land planted with cacao on the coast of Obydos were swept away. At Villa Nova it was very hot, gleamy, and showery up to the end of August. The welcome dry winds then set in, and lasted until the 20th of November, by which time the river had receded to its lowest level. At that date commenced a series of heavy rains, which continued, however, only nine days; but the weather remained showery to the end of the year. On the 3rd of January a kind of second summer began, and this was a most delightful time. The vegetation which had become parched up in November had been freshened by the showery weather of December, and the open places were covered with a carpet of the brightest verdure. The marly and sandy terrace-formed beaches were clothed with a great diversity of flowering shrubs. Birds and insects were far more numerous and active than they had been before. A species of swallow of a brown colour, with a short square tail (Cotyle), then made its appearance in great numbers, and built its nests in holes of the bank on which the village is built, trilling forth in the mornings and evenings a short but sweet song. The east wind recommenced. It blew at first gently, but increased in strength daily as the dryness augmented: and with it came a dense fog, a rare phenomenon in this country, but which I found to be of regular occurrence in the central parts of the Lower Amazons when the dry season was much prolonged. For three successive weeks the daily order of the weather was almost uniform. The mornings dawned with a clear sky, a stiff breeze blowing and tossing the waters into billows, searching through our dwellings, and communicating a healthful exhilarating glow to the body. As the sun ascended, a light mistiness crept along the lower strata of the atmosphere; after midday this increased in density, until an hour before sunset the sun became obscured, and no longer produced that sickening heat which at all other times was so depressing in the late hours of afternoon. An hour or two after sunset the mist cleared away, and the nights were starlit and deliciously cool. Every day the fog increased in amount, until at the beginning of February a thick moist veil enveloped the whole landscape both night and day. The wind then increased to a gale; every sailing craft on the river was obliged to seek shelter; and when the monthly river steamer, a vessel of 400 tons burthen, anchored in the port, it pitched up and down as I have seen ships do in breezy weather in the Southampton water. This lasted three days, at the end of which the wind suddenly lulled, black clouds gathered in the east, the fog lifted up like a curtain, and down came the deluging rain which inaugurates the wet season.

I made, in this second visit to Villa Nova, an extensive collection of the natural productions of the neighbourhood. A few remarks on some of the more interesting of these must suffice. The forests are very different in their general character from those of Pará, and in fact those of humid districts generally throughout the Amazons. The same scarcity of large-leaved Musaceous and Marantaceous plants was noticeable here as at Obydos. The low-lying areas of forest or Ygapós, which alternate everywhere with the more elevated districts, did not furnish the same luxuriant vegetation as they do in the Delta region of the Amazons. They are flooded during three or four months in the year, and when the waters retire, the soil—to which the very thin coating of alluvial deposit imparts little fertility—remains bare, or covered with a matted bed of dead leaves, until the next flood season. These tracts have then a barren appearance; the trunks and lower branches of the trees are coated with dried slime, and disfigured by rounded masses of fresh-water sponges, whose long horny spiculæ and dingy colours give them the appearance of hedgehogs. Dense bushes of a harsh, cutting grass, called Tiriríca, form almost the only fresh vegetation in the dry season. Perhaps the dense shade, the long period during which the land remains under water, and the excessively rapid desiccation when the waters retire, all contribute to the barrenness of these Ygapós. The higher and drier land is everywhere sandy, and tall coarse grasses line the borders of the broad alleys which have been cut through the second-growth woods. These places swarm with carapátos, ugly ticks belonging to the genus Ixodes, which mount to the tips of blades of grass, and attach themselves to the clothes of passers by. They are a great annoyance. It occupied me a full hour daily to pick them off my flesh after my diurnal ramble. There are two species; both are much flattened in shape, have four pairs of legs, a thick short proboscis and a horny integument. Their habit is to attach themselves to the skin by plunging their proboscides into it, and then suck the blood until their flat bodies are distended into a globular form. The whole proceeding, however, is very slow, and it takes them several days to pump their fill. No pain or itching is felt, but serious sores are caused if care is not taken in removing them, as the proboscis is liable to break off and remain in the wound. A little tobacco juice is generally applied to make them loosen their hold. They do not cling firmly to the skin by their legs, although each of these has a pair of sharp

U 2

and fine claws connected with the tips of the member by means of a flexible pedicle. When they mount to the summits of slender blades of grass, or the tips of leaves, they hold on by their fore legs only, the other three pairs being stretched out so as to fasten on any animal which comes in their way. The smaller of the two species is of a yellowish colour; it is much the most abundant, and sometimes falls upon one by scores. When distended it is about the size of a No. 8 shot; the larger kind, which fortunately comes only singly to the work, swells to the size of a pea.

Peuririma Palm (Bactris).

In some parts of the interior the soil is composed of very coarse sand and small angular fragments of quartz; in these places no trees grow. I visited, in company with Padre Torquato, one of these treeless spaces or campos, as they are called, situated five miles from the village. The road thither led through a varied and beautiful forest, containing many gigantic trees. I missed the Assai, Mirití, Paxiúba, and other palms which are all found only on rich moist soils, but the noble Bacába was not uncommon, and there was a great diversity of dwarf species of Marajá palms (Bactris), one of which, called the Peuriríma, was very elegant, growing to a height of twelve or fifteen feet, with a stem no thicker than a man's finger. On arriving at the campo all this beautiful forest abruptly ceased, and we saw before us an oval tract of land, three or four miles in circumference, destitute even of the smallest bush. The only vegetation was a crop of coarse hairy grass growing in patches. The forest formed a hedge all round the isolated field, and its borders were composed in great part of trees which do not grow in the dense virgin forest, such as a great variety of bushy Melastomas, low Byrsomina trees, myrtles, and Lacre-trees, whose berries exude globules of wax resembling gamboge. On the margins of the campo wild pine-apples also grew in great quantity. The fruit was of the same shape as our cultivated kind, but much smaller, the size being that of a moderately large apple. We gathered several quite ripe; they were pleasant to the taste, of the true pine-apple flavour, but had an abundance of fully developed seeds, and only a small quantity of eatable pulp. There was no path beyond this campo; in fact all beyond is terra incognita to the inhabitants of Villa Nova.

The only interesting Mammalian animal which I saw

at Villa Nova was a monkey of a species new to me ; it was not, however, a native of the district, having been brought by a trader from the river Madeira, a few miles above Borba. It was a howler, probably the Mycetes stramineus of Geoffroy St. Hilaire. The howlers are the only kinds of monkey which the natives have not succeeded in taming. They are often caught, but they do not survive captivity many weeks. The one of which I am speaking was not quite full grown. It measured sixteen inches in length, exclusive of the tail; the whole body was covered with rather long and shining dingy-white hair, the whiskers and beard only being of a tawny hue. It was kept in a house, together with a Coaitá and a Caiarára monkey (Cebus albifrons). Both these lively members of the monkey order seemed rather to court attention, but the Mycetes slunk away when any one approached it. When it first arrived, it occasionally made a gruff subdued howling noise early in the morning. The deep volume of sound in the voice of the howling monkeys, as is well known, is produced by a drum-shaped expansion of the larynx. It was curious to watch the animal whilst venting its hollow cavernous roar, and observe how small was the muscular exertion employed. When howlers are seen in the forest there are generally three or four of them mounted on the topmost branches of a tree. It does not appear that their harrowing roar is emitted from sudden alarm ; at least, it was not so in captive individuals. It is probable, however, that the noise serves to intimidate their enemies. I did not meet with the Mycetes stramineus in any other part of the Amazons region ; in the neighbourhood of Pará a reddish-coloured species prevails (M. Belzebuth); in the narrow channels near Breves I shot a large, entirely black kind; another yellow-handed species, according to the report of the natives, inhabits the island of Macajó, which is probably the M. flavimanus of Kuhl; some distance up the Tapajos the only howler found is a brownish-black species; and on the Upper Amazons the sole species seen was the Mycetes ursinus, whose fur is of a shining yellowish-red colour.

In the dry forests of Villa Nova I saw a rattlesnake for the first time. I was returning home one day through a narrow alley, when I heard a pattering noise close to me. Hard by was a tall palm tree, whose head was heavily weighted with parasitic plants, and I thought the noise was a warning that it was about to fall. The wind lulled for a few moments, and then there was no doubt that the noise proceeded from the ground. On turning my head in that direction, a sudden plunge startled me, and a heavy gliding motion betrayed a large serpent making off almost from beneath my feet. The ground is always so encumbered with rotting leaves and branches that one only discovers snakes when they are in the act of moving away. The residents of Villa Nova would not believe that I had seen a rattlesnake in their neighbourhood; in fact, it is not known to occur in the forests at all, its place being the open campos, where, near Santarem, I killed several. On my second visit to Villa Nova I saw another. I had then a favourite little dog, named Diamante, who used to accompany me in my rambles. One day he rushed into the thicket, and made a dead set at a large snake, whose head I saw raised above the herbage. The foolish little brute approached quite close, and then the serpent reared its tail slightly in a horizontal position and shook its terrible rattle. It was many minutes before I could get the dog away; and this incident, as well as the one already related, shows how slow the reptile is to make the fatal spring.

I was much annoyed, and at the same time amused, with the Urubú vultures. The Portuguese call them corvos or crows; in colour and general appearance, they somewhat resemble rooks, but they are much larger, and have naked, black, wrinkled skin about their face and throat. They assemble in great numbers in the villages about the end of the wet season, and are then ravenous with hunger. My cook could not leave the open kitchen at the back of the house for a moment, whilst the dinner was cooking, on account of their thievish propensities. Some of them were always loitering about, watching their opportunity, and the instant the kitchen was left unguarded, the bold marauders marched in and lifted the lids of the saucepans with their beaks to rob them of their contents. The boys of the village lie in wait and shoot them with bow and arrow; and vultures have consequently acquired such a dread of these weapons, that they may be often kept off by hanging a bow from the rafters of the kitchen. As the dry season advances, the hosts of Urubús follow the fishermen to the lakes, where they gorge themselves with the offal of the fisheries. Towards February, they return to the villages, and are then not nearly so ravenous as before their summer trips.

The insects of Villa Nova are, to a great extent, the same as those of Santarem and the Tapajos. A few species of all orders, however, are found here, which occurred nowhere else on the Amazons, besides several others which are properly considered local varieties or races of others found at Pará, on the Northern shore of the Amazons or in other parts of Tropical America. The Hymenoptera were especially numerous, as they always are in districts which possess a sandy soil; but the many interesting facts which I gleaned relative to their habits will be more conveniently introduced when I treat of the same or similar species found in the localities above-named. One of the most conspicuous insects peculiar to Villa Nova is an exceedingly handsome butterfly, which has been named Agrias Phalcidon. It is of large size, and the colours of the upper surface of its wings, resemble those of the Callithea Leprieurii, already described, namely, dark blue, with a broad silvery-green border. When it settles on leaves of trees, fifteen or twenty feet from the ground, it closes its wings and then exhibits a row of brilliant pale-blue eye-like spots with white pupils, which adorns their under surface. Its flight is exceedingly swift, but when at rest it is not easily made to budge from its place; or if driven off, returns soon after to the same spot. Its superficial resemblance to Callithea Leprieurii, which is a very abundant species in the same locality, is very close. The likeness might be considered a mere accidental coincidence, especially as it refers chiefly to the upper surface of the wings, if similar parallel resemblances did not occur between other species of the same two genera. Thus, on the Upper Amazons, another totally distinct kind of Agrias mimicks still more closely another Callithea; both insects being peculiar to the district where they are found flying together. Resemblances of this nature are very numerous in the insect world. I was much struck with them in the course of my travels, especially when, on removing from one district to another, local varieties of certain species were found accompanied by local varieties of the species which counterfeited them in the former locality, under a dress changed to correspond with the altered liveries of the species they mimicked. One cannot help concluding these imitations to be intentional, and that nature has some motive in their production. In many cases, the reason of the imitation is sufficiently plain. For instance, when a fly or parasitic bee has a deceptive resemblance to the species of working bee, in whose nest it deposits the egg it has otherwise no means of providing for, or when a leaping-spider, as it crouches in the axil of a leaf waiting for its prey, presents an exact imitation of a flower-bud; it is evident that the benefit of the imitating species is the object had in view. When, however, an insect mimicks another species of its own order where predaceous or parasitic habits are out of the question, it is not so easy to divine the precise motive of the adaptation. We may be sure, nevertheless, that one of the two is assimilated in external appearance to the other for some purpose useful,—perhaps of life and death importance—to the species. I believe these imitations are of the same nature as those in which an insect or lizard is coloured and marked so as to resemble the soil, leaf, or bark on which it lives; the resemblance serving to conceal the creatures from the prying eyes of their enemies; or, if they are predaceous species, serving them as a disguise to enable them to approach their prey. When an insect, instead of a dead or inorganic substance, mimicks another species of its own order, and does not prey, or is not parasitic, may it not be inferred that the mimicker is subject to a persecution by insectivorous animals from which its model is free? Many species of insects have a most deceptive resemblance to living or dead leaves; it is generally admitted, that this serves to protect them from the onslaughts of insect-feeding animals who would devour the insect, but refuse the leaf. The same might be said of a species mimicking another of the same order; one may be as repugnant to the tastes of insect persecutors, as a leaf or a piece of bark would be, and its imitator not enjoying this advantage would escape by being deceptively assimilated to it in external appearances. In the present instance, it is not very clear what property the Callithea possesses to render it less liable to persecution than the Agrias, except it be that it has a strong odour somewhat resembling Vanilla, which the Agrias is destitute of. This odour becomes very powerful when the insect is roughly handled or pinched, and if it serves as a protection to the Callithea, it would explain why the Agrias is assimilated to it in colours. The resemblance, as before remarked, applies chiefly to the upper side; in other species* it is equally close on both surfaces of the wings. Some birds, and the

* Agrias Hewitsonius and Callithea Markii.

great Æschnæ dragon-flies, take their insect prey whilst on the wing, when the upper surface of the wings is the side most conspicuous.

In the broad alleys of the forest where these beautiful insects are found, several species of Morpho were common. One of these is a sister form to the Morpho Hecuba, which I have mentioned as occurring at Obydos. The Villa Nova kind differs from Hecuba sufficiently to be considered a distinct species, and has been described under the name of M. Cisseis; but it is clearly only a local variety of it, the range of the two being limited by the barrier of the broad Amazons. It is a grand sight to see these colossal butterflies by twos and threes floating at a great height in the still air of a tropical morning. They flap their wings only at long intervals, for I have noticed them to sail a very considerable distance without a stroke. Their wing-muscles and the thorax to which they are attached, are very feeble in comparison with the wide extent and weight of the wings: but the large expanse of these members doubtless assists the insects in maintaining their aërial course. Morphos are amongst the most conspicuous of the insect denizens of Tropical American forests, and the broad glades of the Villa Nova woods seemed especially suited to them, for I noticed here six species. The largest specimens of Morpho Cisseis measure seven inches and a half in expanse. Another smaller kind, which I could not capture, was of a pale silvery-blue colour, and the polished surface of its wings flashed like a silver speculum, as the insect flapped its wings at a great elevation in the sunlight.

To resume our voyage. We left Villa Nova on the 4th of December. A light wind on the 5th carried us across to the opposite shore and past the mouth of the Paranámirím do arco, or the little river of the bow, so called on account of its being a short arm of the main river of a curved shape, rejoining the Amazons a little below Villa Nova. On the 6th, after passing a large island in mid-river, we arrived at a place where a line of perpendicular clay cliffs, called the Barreiros de Cararaucú, diverts slightly the course of the main stream, as at Obydos. A little below these cliffs were a few settlers' houses ; here Penna remained ten days to trade, a delay which I turned to good account in augmenting very considerably my collections.

At the first house a festival was going forward. We anchored at some distance from the shore, on account of the water being shoaly, and early in the morning three canoes put off laden with salt fish, oil of manatee, fowls and bananas, wares which the owners wished to exchange for different articles required for the festa. Soon after I went ashore. The head man was a tall, well-made, civilised Tapuyo named Marcellino, who, with his wife, a thin, active, wiry old squaw, did the honours of their house, I thought, admirably. The company consisted of 50 or 60 Indians and Mamelucos ; some of them knew Portuguese, but the Tupí language was the only one used amongst themselves. The festival was in honour of our Lady of Conception ; and when the people learnt that Penna had on board an image of the saint handsomer than their own, they put off in their canoes to borrow it ; Marcellino taking charge of the doll, covering it carefully with a neatly-bordered white towel. On landing with the image, a procession was formed from the port to the house, and salutes fired from a couple of lazarino guns, the saint being afterwards carefully deposited in the family oratorio. After a litany and hymn were sung in the evening, all assembled to supper around a large mat spread on a smooth terrace-like space in front of the house. The meal consisted of a large boiled Pirarucú, which had been harpooned for the purpose in the morning, stewed and roasted turtle, piles of mandioca-meal and bananas. The old lady, with two young girls, showed the greatest activity in waiting on the guests, Marcellino standing gravely by, observing what was wanted and giving the necessary orders to his wife. When all was done hard drinking began, and soon after there was a dance, to which Penna and I were invited. The liquor served was chiefly a spirit distilled by the people themselves from mandioca cakes. The dances were all of the same class, namely, different varieties of the "Landum," an erotic dance similar to the fandango originally learnt from the Portuguese. The music was supplied by a couple of wire-stringed guitars, played alternately by the young men. All passed off very quietly considering the amount of strong liquor drunk, and the ball was kept up until sunrise the next morning.

We visited all the houses one after the other. One of them was situated in a charming spot, with a broad sandy beach before it, at the entrance to the Paraná-miri̇́m do Mucambo, a channel leading to an interior lake peopled by savages of the Múra tribe. This seemed to be the abode of an industrious family, but all the men were absent, salting Pirarucú on the lakes. The house, like its neighbours, was simply a framework of poles thatched with palm-leaves, the walls roughly latticed and plastered with mud: but it was larger, and much cleaner inside than the others. It was full of women and children, who were busy all day with their various employments; some weaving hammocks in a large clumsy frame, which held the warp whilst the shuttle was passed by the hand slowly across the six feet breadth of web; others spinning cotton, and others again scraping, pressing, and roasting mandioca. The family had cleared and cultivated a large piece of ground; the soil was of extraordinary richness, the perpendicular banks of the river, near the house, revealing a depth of many feet of crumbling vegetable mould. There was a large plantation of tobacco, besides the usual patches of Indian-corn, sugar-cane, and mandioca; and a grove of cotton, cacao, coffee and fruit-trees surrounded the house. We passed two nights at anchor in shoaly water off the beach. The weather was most beautiful; scores of Dolphins rolled and snorted about the canoe all night. I saw here, for the first time, the flesh-coloured species (Delphinus pallidus of Gervais?), which rolled always in pairs, both individuals being of the same colour. In the day-time the margin of the beach abounded with a small tiger-beetle (Cicindela hebræa of Klug), which flew up like a swarm of house-flies before our steps as we walked along. It is not easily detected, for its colour is assimilated to that of the moist sand over which it runs. I have a pleasant recollection of this sand-bank, from having here observed, for the first time, in ascending the river, one of the handsomest of the many handsome butterflies which are found exclusively in the interior parts of the South American continent, namely the Papilio Columbus. It is of a cream-white colour bordered with black, and has a patch of crimson near the commencement of its long slender tails. In the forest, amongst a host of other beautiful and curious insects, I found another species of the same genus, which was new to me, namely, the Papilio Lysander, remarkable for the contrasted colours of its livery—crimson and blue-green spots on a black ground. This conspicuous insect may be cited as affording another illustration of the way in which species so very commonly become modified according to the different localities they inhabit. P. Lysander is found throughout the interior of the Amazons country, from Villa Nova to Peru, and also in Dutch and British Guiana. In the Delta region of the Amazons it is replaced by a form which has been treated as a distinct species, namely, the P. Parsodes of Gray. In French Guiana, however, numerous varieties intermediate between the two are found, so that we are compelled to consider them as local modifications of one and the same species. The difference between the two local forms is of a slight nature, and many naturalists on this account alone would consider them to belong to the same species; but the numerous existing intermediate shades of variation show how many grades are possible between even two local varieties of a species. In fact, the steps of modification are found to be exceedingly small and numerous in all cases where the filiation of races or species can be traced; and this circumstance may be held as confirming the truth of the axiom, "Natura non facit saltum," which has been impugned by some writers.

About two miles beyond this sand-bank was the miserable abode of a family of Mura Indians, the most degraded tribe inhabiting the banks of the Amazons. It was situated on a low terrace on the shores of a pretty little bay at the commencement of the high barreiros. With the exception of a cluster of bananas there were no fruit-trees or plantation of any description near the house. We saw in the bay several large alligators, with head and shoulders just reared above the level of the water. The house was a mere hovel; a thatch of palm-leaves supported on a slender framework of upright posts and rafters, bound with flexible lianas, and the walls were partially plastered up with mud. A low doorway led into the dark chamber; the bare earth floor was filthy in the extreme; and in a damp corner I espied two large toads whose eyes glittered in the darkness. The furniture consisted of a few low stools; there was no mat, and the hammock was a rudely woven web of ragged strips of the inner bark of the Mongúba tree. Bows and arrows hung from the smoke-blackened rafters. An ugly woman, clad in a coarse petticoat, and holding a child astride across her hip, sat crouched over a fire roasting the head of a large fish. Her husband was occupied in notching pieces of bamboo for arrow-heads. Both of them seemed rather disconcerted at our sudden entrance; we could get nothing but curt and surly answers to our questions, and so were glad to depart.

We crossed the river at this point, and entered a narrow channel which penetrates the interior of the island of Tupinambarána, and leads to a chain of lakes called the Lagos de Cararaucú. A furious current swept along the coast, eating into the crumbling earthy banks, and strewing the river with débris of the forest. The mouth of the channel lies about twenty-five miles from Villa Nova; the entrance is only about forty yards broad, but it expands, a short distance inland, into a large sheet of water. We suffered terribly from insect pests during the twenty-four hours we remained here. At night it was quite impossible to sleep for mosquitos; they fell upon us by myriads, and without much piping came straight at our faces as thick as raindrops in a shower. The men crowded into the cabins, and then tried to expel the pests by the smoke from burnt rags, but it was of little avail, although we were half suffocated during the operation. In the daytime the Motúca, a much larger and more formidable fly than the mosquito, insisted upon levying his tax of blood. We had been tormented by it for many days past, but this place seemed to be its metropolis. The species has been described by Perty, the author of the Entomological portion of Spix and Martius' travels, under the name of Hadaus lepidotus. It is a member of the Tabanidæ family, and indeed is closely related to the Hæmatopota pluvialis, a brown fly which haunts the borders of woods in summer time in England. The Motúca is of a bronzed-black colour; its proboscis is formed of a bundle of horny lancets, which are shorter and broader than is usually the case in the family to which it belongs. Its puncture does not produce much pain, but it makes such a large gash in the flesh that the blood trickles forth in little streams. Many scores of them were flying about the canoe all day, and sometimes eight or ten would settle on one's ancles at the same time. It is sluggish in its motions, and may be easily killed with the fingers when it settles. Penna went forward in the montaria to the Pirarucú fishing stations, on a lake lying further inland; but he did not succeed in reaching them on account of the length and intricacy of the channels; so after wasting a day, during which, however, I had a profitable ramble in the forest, we again crossed the river, and on the 16th continued our voyage along the northern shore.

The clay cliffs of Cararaucú are several miles in length. The hard pink and red-coloured beds are here extremely thick, and in some places present a compact stony texture. The total height of the cliff is from thirty to sixty feet above the mean level of the river, and the clay rests on strata of the same coarse iron-cemented conglomerate which has already been so often mentioned. Large blocks of this latter have been detached and rolled by the force of currents up parts of the cliff where they are seen resting on terraces of the clay. On the top of all lies a bed of sand and vegetable mould which supports a lofty forest growing up to the very brink of the precipice. After passing these barreiros we continued our way along a low uninhabited coast, clothed, wherever it was elevated above high water-mark, with the usual vividly-coloured forests of the higher Ygapó lands, to which the broad and regular fronds of the Murumurú palm, here extremely abundant, served as a great decoration. Wherever the land was lower than the flood height of the Amazons, Cecropia trees prevailed, sometimes scattered over meadows of tall broad-leaved grasses, which surrounded shallow pools swarming with water-fowl. Alligators were common on most parts of the coast; in some places we saw also small herds of Capybaras (a large Rodent animal, like a colossal Guinea-pig) amongst the rank herbage on muddy banks, and now and then flocks of the graceful squirrel monkey (Chrysothrix sciureus), and the vivacious Caiarára (Cebus albifrons) were seen taking flying leaps from tree to tree. On the 22nd we passed the mouth of the most easterly of the numerous channels which lead to the large interior lake of Saracá, and on the 23rd threaded a series of passages between islands, where we again saw human habitations, ninety miles distant from the last house at Cararaucú. On the 24th we arrived at Serpa.

Serpa is a small village consisting of about eighty houses, built on a bank elevated twenty-five feet above the level of the river. The beds of Tabatínga clay, which are here intermingled with scoria-looking conglomerate, are in some parts of the declivity prettily variegated in colour; the name of the town in the Tupí language, Ita-coatiára, takes its origin from this circumstance, signifying striped or painted rock. It is an old settlement, and was once the seat of the district government, which had authority over the Barra of the Rio

Negro. It was in 1849 a wretched-looking village, but it has since revived, on account of having being chosen by the Steamboat Company of the Amazons as a station for steam saw-mills and tile manufactories. We arrived on Christmas-eve, when the village presented an animated appearance from the number of people congregated for the holidays. The port was full of canoes, large and small—from the montaria, with its arched awning of woven lianas and Maranta leaves, to the two-masted cuberta of the peddling trader, who had resorted to the place in the hope of trafficking with settlers coming from remote sitios to attend the festival. We anchored close to an igarité, whose owner was an old Jurí Indian, disfigured by a large black tatooed patch in the middle of his face, and by his hair being close cropped, except a fringe in front of the head. In the afternoon we went ashore. The population seemed to consist chiefly of semi-civilised Indians, living as usual in half-finished mud hovels. The streets were irregularly laid out and overrun with weeds and bushes swarming with "mocuim," a very minute scarlet acarus, which sweeps off to one's clothes in passing, and attaching itself in great numbers to the skin causes a most disagreeable itching. The few whites and better class of mameluco residents live in more substantial dwellings, white-washed and tiled. All, both men and women, seemed to me much more cordial, and at the same time more brusque in their manners than any Brazilians I had yet met with. One of them, Captain Manoel Joaquim, I knew for a long time afterwards; a lively, intelligent, and thoroughly good-hearted man, who had quite a reputation throughout the interior of the country for generosity, and for being a firm friend of foreign residents and stray travellers. Some of these excellent people were men of substance, being owners of trading vessels, slaves, and extensive plantations of cacao and tobacco.

We stayed at Serpa five days. Some of the ceremonies observed at Christmas were interesting, inasmuch as they were the same, with little modification, as those taught by the Jesuit missionaries more than a century ago to the aboriginal tribes whom they had induced to settle on this spot. In the morning all the women and girls, dressed in white gauze chemises and showy calico print petticoats, went in procession to church, first going the round of the town to take up the different "mordomos" or stewards, whose office is to assist the Juiz of the festa. These stewards carried each a long white reed, decorated with coloured ribbons; several children also accompanied, grotesquely decked with finery. Three old squaws went in front, holding the "sairé,", a large semicircular frame, clothed with cotton and studded with ornaments, bits of looking-glass, and so forth. This they danced up and down, singing all the time a monotonous whining hymn in the Tupí language, and at frequent intervals turning round to face the followers, who then all stopped for a few moments. I was told that this sairé was a device adopted by the Jesuits to attract the savages to church, for these everywhere followed the mirrors, in which they saw as it were magically reflected their own persons. In the evening, good-humoured revelry prevailed on all sides. The negroes, who had a saint of their own colour—St. Benedito—had their holiday apart from the rest, and spent the whole night singing and dancing to the music of a long drum (gambá) and the caracashá. The drum was a hollow log, having one end covered with skin, and was played by the performer sitting astride upon it and drumming with his knuckles. The caracashá is a notched bamboo tube, which produces a harsh rattling noise by passing a hard stick over the notches. Nothing could exceed in dreary monotony this music and the singing and dancing, which were kept up with unflagging vigour all night long. The Indians did not get up a dance; for the whites and mamelucos had monopolised all the pretty coloured girls for their own ball, and the older squaws preferred looking on to taking a part themselves. Some of their husbands joined the negroes, and got drunk very quickly. It was amusing to notice how voluble the usually taciturn red-skins became under the influence of liquor. The negroes and Indians excused their own intemperance by saying the whites were getting drunk at the other end of the town, which was quite true.

The forest which encroaches on the ends of the weed-grown streets yielded me a large number of interesting insects, some of which have been described in the preceding chapter. The elevated land on which Serpa is built appears to be a detached portion of the terra firma; behind, lies the great interior lake of Saracá, to the banks of which there is a foot-road through the forest, but I could not ascertain what was the distance. Outlets from the lake enter the Amazons both above and below the village. The woods were remarkably dense, and the profoundest solitude reigned at the distance of a few minutes' walk from the settlement. The first mile or two of the forest road was very pleasant; the path was broad, shady, and clean; the lower trees presented the most beautiful and varied foliage imaginable, and a compact border of fern-like selaginellas lined the road on each side. The only birds I saw were ant-thrushes in the denser thickets, and two species of Ceræba, a group allied to the creepers. These were feeding on the red gummy seeds of Clusia trees, which were here very numerous, their thick oval leaves, and large, white, wax-like flowers making them very conspicuous objects in the crowded woods. The only insect I will name amongst the numbers of species which sported about these shady places is the Papilio Ergeteles, and this for the purpose of again showing how much may be learned by noting the geographical relations of races and closely-allied species. The Papilio Ergeteles is of a velvety black colour, with two large spots of green and two belts of crimson on its wings. Its range is limited to the North side of the lower Amazons from Obydos to the Rio Negro; on the south side of the river it is replaced by a distinct kind called the Papilio Echelus. The two might be considered, as they have been hitherto, perfectly distinct species, had not an intermediate variety been found to inhabit Cayenne, where neither extreme form occurs. The two forms are as distinct as any two allied species can well be, and they are different in both sexes. They are found in no other part of America than the districts mentioned. The intermediate varieties, however, link the two together, so that they cannot be considered otherwise than as modifications of one and the same species; one produced on the North, the other on the South side of the Amazons. It is worthy of especial mention that here as well as in the cases of P. Lysander and the Heliconii, described in the preceding chapter, the connecting links are found inhabiting distinct localities, and not mingled with the extreme forms which they connect.

We left Serpa on the 29th of December, in company of an old planter named Senhor Joaõ Trinidade; at whose sitio, situated opposite the mouth of the Madeira, Penna intended to spend a few days. Our course on the 29th and 30th lay through narrow channels between islands. On the 31st we passed the last of these, and then beheld to the south a sea-like expanse of water, where the Madeira, the greatest tributary of the Amazons, after 2000 miles of course, blends its waters with those of the king of rivers. I was hardly prepared for a junction of waters on so vast a scale as this, now nearly 900 miles from the sea. Whilst travelling week after week along the somewhat monotonous stream, often hemmed in between islands, and becoming thoroughly familiar with it, my sense of the magnitude of this vast water system had become gradually deadened; but this noble sight renewed the first feelings of wonder. One is inclined, in such places as these, to think the Paraenses do not exaggerate much when they call the Amazons the Mediterranean of South America. Beyond the mouth of the Madeira, the Amazons sweeps down in a majestic reach, to all appearance not a whit less in breadth before than after this enormous addition to its waters. The Madeira does not ebb and flow simultaneously with the Amazons; it rises and sinks about two months earlier, so that it was now fuller than the main river. Its current therefore poured forth freely from its mouth, carrying with it a long line of floating trees and patches of grass, which had been torn from its crumbly banks in the lower part of its course. The current, however, did not reach the middle of the main stream, but swept along nearer to the southern shore.

A few items of information which I gleaned relative to this river may find a place here. The Madeira is navigable for about 480 miles from its mouth; a series of cataracts and rapids then commences, which extends with some intervals of quiet water, about 160 miles, beyond which is another long stretch of navigable stream. Canoes sometimes descend from Villa Bella, in the interior province of Matto Grosso, but not so frequently as formerly, and I could hear of very few persons who had attempted of late years to ascend the river to that point. It was explored by the Portuguese in the early part of the eighteenth century; the chief and now the only town on its banks, Borba, 150 miles from its mouth, being founded in 1756. Up to the year 1853, the lower part of the river, as far as about 100 miles beyond Borba, was regularly visited by traders from Villa Nova, Serpa, and Barra, to collect salsaparilla, copaüba balsam, turtle-oil, and to trade with the Indians, with whom their relations were generally on a friendly footing. In that year many India-rubber collectors resorted to this region, stimulated by the high price (2*s.* 6*d.* a pound) which the article was at that time fetching at Pará; and then the Aráras, a fierce and intractable tribe of Indians, began to be troublesome. They attacked several canoes and massacred every one on board, the Indian crews as well as the white traders. Their plan was to lurk in ambush near the sandy beaches where canoes stop for the night, and then fall upon the people whilst asleep. Sometimes they came under pretence of wishing to trade, and then as soon as they could get the trader at a disadvantage shot him and his crew from behind trees. Their arms were clubs, bows, and Taquára arrows, the latter a formidable weapon tipped with a piece of flinty bamboo shaped like a spear-head; they could propel it with such force as to pierce a man completely through the body. The whites of Borba made reprisals, inducing the warlike Mundurucús, who had an old feud with the Aráras, to assist them. This state of things lasted two or three years, and made a journey up the Madeira a risky undertaking, as the savages attacked all comers. Besides the Aráras and the Mundurucús, the latter a tribe friendly to the whites, attached to agriculture, and inhabiting the interior of the country from the Madeira to beyond the Tapajos, two other tribes of Indians now inhabit the lower Madeira, namely, the Parentintins and the Muras. Of the former I did not hear much; the Muras lead a lazy quiet life on the banks of the labyrinths of lakes and channels which intersect the low country on both sides of the river below Borba. The Aráras are one of those tribes which do not plant mandioca; and indeed have no settled habitations. They are very similar in stature and other physical features to the Mundurucús, although differing from them so widely in habits and social condition. They paint their chins red with Urucú (Anatto), and have usually a black tatooed streak on each side of the face, running from the corner of the mouth to the temple. They have not yet learnt the use of firearms, have no canoes, and spend their lives roaming over the interior of the country, living on game and wild fruits. When they wish to cross a river they make a temporary canoe with the thick bark of trees, which they secure in the required shape of a boat by means of lianas. I heard it stated by a trader of Santarem, who narrowly escaped being butchered by them in 1854, that the Aráras numbered two thousand fighting men. The number I think must be exaggerated, as it generally is with regard to Brazilian tribes. When the Indians show a hostile disposition to the whites, I believe it is most frequently owing to some provocation they have received at their hands; for the first impulse of the Brazilian red-man is to respect Europeans; they have a strong dislike to be forced into their service, but if strangers visit them with a friendly intention they are well treated. It is related, however, that the Indians of the Madeira were hostile to the Portuguese from the first; it was then the tribes of Muras and Torazes who attacked travellers. In 1855 I met with an American, an odd character, named Kemp, who had lived for many years amongst the Indians on the Madeira, near the abandoned settlement of Crato. He told me his neighbours were a kindly-disposed and cheerful people, and that the onslaught of the Aráras was provoked by a trader from Barra, who wantonly fired into a family of them, killing the parents, and carrying off their children to be employed as domestic servants.

We remained nine days at the sitio of Senhor Joaõ Trinidade. It is situated on a tract of high Ygapó land, which is raised, however, only a few inches above high-water mark. This skirts the northern shore for a long distance; the soil consisting of alluvium and rich vegetable mould, and exhibiting the most exuberant fertility. Such districts are the first to be settled on in this country, and the whole coast for many miles was dotted with pleasant-looking sitios like that of our friend. The establishment was a large one, the house and outbuildings covering a large space of ground. The industrious proprietor seemed to be Jack-of-all-trades; he was planter, trader, fisherman, and canoe-builder, and a large igarité was now on the stocks under a large shed. There was greater pleasure in contemplating this prosperous farm from its being worked almost entirely by free labour; in fact, by one family, and its dependents. Joaõ Trinidade had only one female slave; his other workpeople were a brother and sister-in-law, two godsons, a free negro, one or two Indians, and a family of Muras. Both he and his wife were mamelucos; the negro children called them always father and mother. The order, abundance, and comfort about the place, showed what industry and good management could effect in this country without slave-labour. But the surplus produce of such small plantations is very trifling.

All we saw, had been done since the disorders of 1835-6, during which Joaõ Trinidade was a great sufferer; he was obliged to fly, and the Mura Indians destroyed his house and plantations. There was a large, well-weeded grove of cacao along the banks of the river, comprising about 8000 trees, and further inland, considerable plantations of tobacco, mandioca, Indian corn, fields of rice, melons, and water-melons. Near the house was a kitchen garden, in which grew cabbages and onions introduced from Europe, besides a wonderful variety of tropical vegetables. It must not be supposed that these plantations and gardens were enclosed or neatly kept, such is never the case in this country where labour is so scarce; but it was an unusual thing to see vegetables grown at all, and the ground tolerably well weeded. The space around the house was plentifully planted with fruit-trees, some, belonging to the Anonaceous order, yielding delicious fruits large as a child's head, and full of custardy pulp which it is necessary to eat with a spoon; besides oranges, lemons, guavas, alligator pears, Abíus (Achras cainito), Genipapas and bananas. In the shade of these, coffee trees grew in great luxuriance. The table was always well supplied with fish, which the Mura, who was attached to the household as fisherman, caught every morning a few hundred yards from the port. The chief kinds were the Surubim, Pira-peëua and Piramutába, three species of Siluridæ, belonging to the genus Pimelodus. To these we used a sauce in the form of a yellow paste, quite new to me, called Arubé, which is made of the poisonous juice of the mandioca root, boiled down before the starch or tapioca is precipitated, and seasoned with capsicum peppers. It is kept in stone bottles several weeks before using, and is a most appetising relish to fish. Tucupí, another sauce made also from mandioca juice, is much more common in the interior of the country than Arubé. This is made by boiling or heating the pure liquid after the tapioca has been separated, daily for several days in succession, and seasoning it with peppers and small fishes; when old it has the taste of essence of anchovies. It is generally made as a liquid, but the Jurí and Miranha tribes on the Japurá, make it up in the form of a black paste by a mode of preparation I could not learn; it is then called Tucupí-pixúna, or black Tucupí. I have seen the Indians on the Tapajos, where fish is scarce, season Tucupí with Saüba ants. It is there used chiefly as a sauce to Tacacá, another preparation from mandioca, consisting of the starch beaten up in boiling water.

I thoroughly enjoyed the nine days we spent at this place. Our host and hostess took an interest in my pursuit; one of the best chambers in the house was given up to me, and the young men took me long rambles in the neighbouring forests. I saw very little hard work going forward. Everyone rose with the dawn, and went down to the river to bathe; then came the never-failing cup of rich and strong coffee, after which all proceeded to their avocations. At this time, nothing was being done at the plantations; the cacao and tobacco crops were not ripe; weeding time was over, and the only work on foot was the preparation of a little farinha by the women. The men dawdled about; went shooting and fishing, or did trifling jobs about the house.

The only laborious work done during the year in these establishments is the felling of timber for new clearings; this happens at the beginning of the dry season, namely, from July to September. Whatever employment the people were engaged in, they did not intermit it during the hot hours of the day. Those who went into the woods took their dinners with them—a small bag of farinha, and a slice of salt fish. About sunset all returned to the house; they then had their frugal suppers, and towards 8 o'clock, after coming to ask a blessing of the patriarchal head of the household, went off to their hammocks to sleep.

There was another visitor besides ourselves, a negro, whom Joaō Trinidade introduced to me as his oldest and dearest friend, who had saved his life during the revolt of 1835. I have, unfortunately, forgotten his name; he was a freeman, and had a sitio of his own, situated about a day's journey from this. There was the same manly bearing about him that I had noticed with pleasure in many other free negroes; but his quiet, earnest manner, and the thoughtful and benevolent expression of his countenance showed him to be a superior man of his class. He told me he had been intimate with our host for thirty years, and that a wry word had never passed between them. At the commencement of the disorders of 1835 he got into the secret of a plot for assassinating his friend, hatched by some villains whose only cause of enmity was their owing him money and envying his prosperity. It was such as these who aroused the stupid and brutal animosity of the Múras against the whites. The negro, on obtaining this news, set off alone in a montaria on a six hours' journey in the dead of night, to warn his "compadre" of the fate in store for him, and thus gave him time to fly. It was a pleasing sight to notice the cordiality of feeling and respect for each other shown by these two old men. They used to spend hours together enjoying the cool breeze, seated under a shed which overlooked the broad river, and talking of old times. Joaõ Trinidade was famous for his tobacco and Tauarí cigarettes. He took particular pains in preparing the Tauarí, the envelope of the cigarettes. It is the inner bark of a tree, which separates into thin papery layers. Many trees yield it, amongst them the Courataria Guianensis and the Sapucaya nut-tree (Lecythis ollaria), both belonging to the same natural order. The bark is cut in long strips, of a breadth suitable for folding the tobacco; the inner portion is then separated, boiled, hammered with a wooden mallet, and exposed to the air for a few hours. Some kinds have a reddish colour and an astringent taste, but the sort prepared by our host was of a beautiful satiny-white hue, and perfectly tasteless. He obtained sixty, eighty, and sometimes a hundred layers from the same strip of bark. The best tobacco in Brazil is grown in the neighbourhood of Borba, on the Madeira, where the soil is a rich black loam; but tobacco of very good quality was grown by Joaõ Trinidade and his neighbours along this coast, on similar soil. It is made up into slender rolls, an inch and a half in diameter and six feet in length, tapering at each end. When the leaves are gathered and partially dried, layers of them, after the mid-ribs are plucked out, are placed on a mat and rolled up into the required shape. This is done by the women and children, who also manage the planting, weeding, and gathering of the tobacco. The process of tightening the rolls is a long and heavy task, and can be done only by men. The cords used for this purpose are of very great strength. They are made of the inner bark of a peculiar light-wooded and slender tree, called Uaissíma, which yields, when beaten out, a great quantity of most beautiful silky fibre, many feet in length. I think this might be turned to some use by English manufacturers, if they could obtain it in large quantity. The tree is abundant on light soils on the southern side of the Lower Amazons, and grows very rapidly. When the rolls are sufficiently well pressed they are bound round with narrow thongs of remarkable toughness, cut from the bark of the climbing Jacitára palm tree (Desmoncus macracanthus), and are then ready for sale or use.

A narrow channel runs close by this house, which communicates at a distance of six hours' journey (about eighteen miles) with the Urubú, a large and almost unknown river, flowing through the interior of Guiana. Our host told me the Urubú presented an expanse of clear dark water, in some places a league in width, and was surrounded by an undulating country, partly forest and partly campo. Its banks are fringed with white sandy beaches, and peopled only by a few families of Múra savages. The family now in his employ, and who were living gipsy fashion, the only way they can be induced to live, under a wretched shed on his grounds, were brought from this river six months previously. The channel was navigable by montarias only in the rainy season ; it was now a half-dry watercourse, the mouth lying about eight feet above the present level of the Amazons. The principal mouth of the Urubú lies between this place and Serpa. The river communicates with the lake of Saracá, but I could make out nothing clearly as to its precise geographical relations with that large sheet of water, which is ten or twelve leagues in length and one to two in breadth, and has an old-established Brazilian settlement, called Silves, on its banks.

It was very pleasant to roam in our host's cacaoal. The ground was clear of underwood, the trees were about thirty feet in height, and formed a dense shade. Two species of monkey frequented the trees, and I was told committed great depredations when the fruit was ripe. One of these, the macaco prego (Cebus cirrhifer ?), is a most impudent thief ; it destroys more than it eats by its random, hasty way of plucking and breaking the fruits, and when about to return to the forest, carries away all it can in its hands or under its arms. The other species, the pretty little Chrysothrix sciureus, contents itself with devouring what it can on the spot. A variety of beautiful insects basked on the foliage where stray gleams of sunlight glanced through the canopy of broad soft-green leaves. Numbers of an elegant, long-legged tiger-beetle (Odontocheila egregia) ran and flew about over the herbage. It belongs to a sub-genus peculiar to the warmest parts of America, the species of which are found only in the shade of the forest, and are seen quite as frequently pursuing their prey on trees and herbage as on the ground. The typical tiger-beetles, or Cicindelæ, inhabit only open and sunny situations, and are wholly terrestrial in their habits. They are the sole forms of the family which occur in the Northern and Central parts of Europe and North America. In the Amazons region, the shade-loving and semi-arboreal Odontocheilæ outnumber in species the Cicindelæ as twenty-two to six; all but one of this number are exclusively peculiar to the Amazonian forests, and this affords another proof of the adaptation of the Fauna to a forest-clad country, pointing to a long and uninterrupted existence of land covered by forests on this part of the earth's surface.

We left this place on the 8th of January, and on the afternoon of the 9th, arrived at Matarí, a miserable little settlement of Múra Indians. Here we again anchored and went ashore. The place consisted of about twenty slightly-built mud-hovels, and had a most forlorn appearance, notwithstanding the luxuriant forest in its rear. A horde of these Indians settled here many years ago, on the site of an abandoned missionary station, and the government had lately placed a resident director over them, with the intention of bringing the hitherto intractable savages under authority. This, however, seemed to promise no other result than that of driving them to their old solitary haunts on the banks of the interior waters, for many families had already withdrawn themselves. The absence of the usual cultivated trees and plants, gave the place a naked and poverty-stricken aspect. I entered one of the hovels, where several women were employed cooking a meal. Portions of a large fish were roasting over a fire made in the middle of the low chamber, and the entrails were scattered about the floor, on which the women with their children were squatted. These had a timid, distrustful expression of countenance, and their bodies were begrimed with black mud, which is smeared over the skin as a protection against musquitoes. The children were naked, the women wore petticoats of coarse cloth, ragged round the edges, and stained in blotches with murixí, a dye made from the bark of a tree. One of them wore a necklace of monkey's teeth. There were scarcely any household utensils; the place was bare with the exception of two dirty grass hammocks hung in the corners. I missed the usual mandioca sheds behind the house, with their surrounding cotton, cacao, coffee, and lemon trees. Two or three young men of the tribe were lounging about the low open doorway. They were stoutly-built fellows, but less well-proportioned than the semi-civilised Indians of the Lower Amazons generally are. Their breadth of chest was remarkable, and their arms were wonderfully thick and muscular. The legs appeared short in proportion to the trunk; the expression of their countenances was unmistakeably more sullen and brutal, and the skin of a darker hue than is common in the Brazilian red man. Before we left the hut, an old couple came in; the husband carrying his paddle, bow, arrows, and harpoon, the woman bent beneath the weight of a large basket filled with palm fruits. The man was of low stature and had a wild appearance from the long coarse hair which hung over his forehead. Both his lips were pierced with holes, as is usual with the older Múras seen on the river. They used formerly to wear tusks of the wild hog in these holes whenever they went out to encounter strangers or their enemies in war. The gloomy savagery, filth, and poverty of the people in this place, made me feel quite melancholy, and I was glad to return to the canoe. They offered us no civilities; they did not even pass the ordinary salutes, which all the semi-civilised and many savage Indians proffer on a first meeting. The men persecuted Penna for cashaça, which they seemed to consider the only good thing the white man brings with him. As they had nothing whatever to give in exchange, Penna declined to supply them. They followed us as we descended to the port, becoming very troublesome when about a dozen had collected together. They brought their empty bottles with them and promised fish and turtle, if we would only trust them first with the coveted aguardente, or cau-im, as they called it. Penna was inexorable: he ordered the crew to weigh anchor, and the disappointed savages remained hooting after us with all their might from the top of the bank as we glided away.

The Múras have a bad reputation all over this part of the Amazons, the semi-civilised Indians being quite as severe upon them as the white settlers. Every one spoke of them as lazy, thievish, untrustworthy, and cruel. They have a greater repugnance than any other class of Indians to settled habits, regular labour, and the service of the whites; their distaste, in fact, to any approximation towards civilised life is invincible. Yet most of these faults are only an exaggeration of the fundamental defects of character in the Brazilian red man. There is nothing, I think, to show that the Múras had a different origin from the nobler agricultural tribes belonging to the Tupí nation, to some of whom they are close neighbours, although the very striking contrast in their characters and habits would suggest the conclusion that they had, in the same way as the Semangs of Malacca, for instance, with regard to the Malays. They are merely an offshoot from them, a number of segregated hordes becoming degraded by a residence most likely of very many centuries in Ygapó lands, confined to a fish diet, and obliged to wander constantly in search of food. Those tribes which are supposed to be more nearly related to the Tupís are distinguished by their settled agricultural habits, their living in well-constructed houses, their practice of many arts, such as the manufacture of painted earthenware, weaving, and their general custom of tattooing, social organisation, obedience to chiefs, and so forth. The Múras have become a nation of nomade fishermen, ignorant of agriculture and all other arts practised by their neighbours. They do not build substantial and fixed dwellings, but live in separate families or small hordes, wandering from place to place along the margins of those rivers and lakes which most abound in fish and turtle. At each resting-place they construct temporary huts at the edge of the stream, shifting them higher or lower on the banks, as the waters advance or recede. Their canoes originally were made simply of the thick bark of trees, bound up into a semi-cylindrical shape by means of woody lianas; these are now rarely seen, as most families possess montarias, which they have contrived to steal from the settlers from time to time. Their food is chiefly fish and turtle, which they are very expert in capturing. It is said by their neighbours that they dive after turtles, and succeed in catching them by the legs, which I believe is true in the shallow lakes where turtles are imprisoned in the dry season. They shoot fish with bow and arrow, and have no notion of any other method of cooking it than by roasting. It is not quite clear whether the whole tribe were originally quite ignorant of agriculture; as some families on the banks of the streams behind Villa Nova, who could scarcely have acquired the art in recent times, plant mandioca; but, as a general rule, the only vegetable food used by the Múras is bananas and wild fruits. The original home of this tribe was the banks of the Lower Madeira. It appears they were hostile to the European settlers from the beginning; plundering their sitios, waylaying their canoes, and massacreing all who fell into their power. About fifty years ago the Portuguese succeeded in turning the warlike propensities of the Mundurucús against them; and these, in the course of many years' persecution, greatly weakened the power of the tribe, and drove a great part of them from their seats on the banks of the Madeira. The Múras are now scattered in single hordes and families over a wide extent of country bordering the main river from Villa Nova to Catuá, near Ega, a distance of 800 miles. Since the disorders of 1835-6, when they committed great havoc amongst the peaceable settlements from Santarem to the Rio Negro, and were pursued and slaughtered in great numbers by the Mundurucús in alliance with the Brazilians, they have given no serious trouble.

The reasons which lead me to think the Múras are merely an offshoot from the Mundurucús, or some other allied section of the widely-spread Tupí nation, and not an originally distinct people, are founded on a general comparison of the different tribes of Amazonian Indians. In the first place, there is no sharply-defined difference between sections of the Indian race, either in physical or moral qualities. They are all very much alike in bodily structure; and, although some are much lower in the scale of culture than others, yet the numerous tribes in this respect form a graduated link from the lowest to the highest. The same customs reappear in tribes who are strongly contrasted in other respects and live very wide apart. The Mauhés, who live in the neighbourhood of the Mundurucús and Múras, have much in common with both; but, according to tradition, they once formed part of the Mundurucú nation. The language of the Múras is entirely different from that of the tribes mentioned; but language is not a sure guide in the filiation of Brazilian tribes; seven or eight different languages being sometimes spoken on the same river, within a distance of 200 or 300 miles. There are certain peculiarities in Indian habits which lead to a quick corruption of language and segregation of dialects. When Indians, men or women, are conversing amongst themselves, they seem to take pleasure in inventing new modes of pronunciation, or in distorting words. It is amusing to notice how the whole party will laugh when the wit of the circle perpetrates a new slang term, and these new words are very often retained. I have noticed this during long voyages made with Indian crews. When such alterations occur amongst a family or horde, which often live many years without communication with the rest of their tribe, the local corruption of language becomes perpetuated. Single hordes belonging to the same tribe and inhabiting the banks of the same river thus become, in the course of many years' isolation, unintelligible to other hordes, as happens with the Collínas on the Jurúa. I think it, therefore, very probable that the disposition to invent new words and new modes of pronunciation, added to the small population and habits of isolation of hordes and tribes, are the causes of the wonderful diversity of languages in South America.

There is one curious custom of the Múras which requires noticing before concluding this digression ; this is the practice of snuff-taking with peculiar ceremonies. The snuff is called Paricá, and is a highly stimulating powder, made from the seeds of a species of Ingá, belonging to the Leguminous order of plants. The seeds are dried in the sun, pounded in wooden mortars, and kept in bamboo tubes. When they are ripe, and the snuff-making season sets in, they have a fuddling-bout, lasting many days, which the Brazilians call a *Quarentena*, and which forms a kind of festival of a semi-religious character. They begin by drinking large quantities of caysúma and cashirí, fermented drinks made of various fruits and mandioca, but they prefer cashaça, or rum, when they can get it. In a short time they drink themselves into a soddened semi-intoxicated state, and then commence taking the Paricá. For this purpose they pair off, and each of the partners, taking a reed containing a quantity of the snuff, after going through a deal of unintelligible mummery, blows the contents with all his force into the nostrils of his companion. The effect on the usually dull and taciturn savages is wonderful; they become exceedingly talkative, sing, shout, and leap about in the wildest excitement. A re-action soon follows; more drinking is then necessary to rouse them from their stupor, and thus they carry on for many days in succession. The Mauhés also use the Paricá, although it is not known amongst their neighbours the Mundurucús. Their manner of taking it is very different from that of the swinish Múras, it being kept in the form of a paste, and employed chiefly as a preventive against ague in the months between the dry and wet seasons, when the disease prevails. When a dose is required, a small quantity of the paste is dried and pulverised on a flat shell, and the powder then drawn up into both nostrils at once through two vulture quills secured together by cotton thread. The use of Paricá was found by the early travellers amongst the Omaguas, a section of the Tupís, who formerly lived on the Upper Amazons, a thousand miles distant from the homes of the Mauhés and Múras. This community of habits is one of those facts which support the view of the common origin and near relationship of the Amazonian Indians.

After leaving Matarí, we continued our voyage along the northern shore. The banks of the river were of moderate elevation during several days' journey; the terra firma lying far in the interior, and the coast being either low land or masked with islands of alluvial formation. On the 14th we passed the upper mouth of the Parana-mirim de Eva, an arm of the river of small breadth, formed by a straggling island some ten miles in length, lying parallel to the northern bank. On passing the western end of this, the main land again appeared; a rather high rocky coast, clothed with a magnificent forest of rounded outline, which continues hence for twenty miles to the mouth of the Rio Negro, and forms the eastern shore of that river. Many houses of settlers, built at a considerable elevation on the wooded heights, now enlivened the river banks. One of the first objects which here greeted us was a beautiful bird we had not hitherto met with, namely, the scarlet and black tanager (Ramphocœlus nigrogularis), flocks of which were seen sporting about the trees on the edge of the water, their flame-coloured liveries lighting up the masses of dark-green foliage.

The weather, from the 14th to the 18th, was wretched; it rained sometimes for twelve hours in succession, not heavily, but in a steady drizzle, such as we are familiar with in our English climate. We landed at several places on the coast, Penna to trade as usual, and I to ramble in the forest in search of birds and insects. In one spot the wooded slope enclosed a very picturesque scene: a brook, flowing through a ravine in the high bank, fell in many little cascades to the broad river beneath, its margins decked out with an infinite variety of beautiful plants. Wild bananas arched over the watercourse, and the trunks of the trees in its vicinity were clothed with ferns, large-leaved species belonging to the genus Lygodium, which, like Osmunda, have their spore-cases collected together on contracted leaves. On the 18th, we arrived at a large fazenda (plantation and cattle-farm), called Jatuarána. A rocky point here projects into the stream, and as we found it impossible to stem the strong current which whirled round it, we crossed over to the southern shore. Canoes, in approaching the Rio Negro, generally prefer the southern side on account of the slackness of the current near the banks. Our progress, however, was most tediously slow, for the regular east wind had now entirely ceased, and the vento de cima or wind from up river, having taken its place, blew daily for a few hours, dead against us. The weather was oppressively close, and every afternoon a squall arose, which, however, as it came from the right quarter and blew for an hour or two, was very welcome. We made acquaintance on this coast with a new insect pest, the Piúm, a minute fly, two-thirds of a line in length, which here commences its reign, and continues henceforward as a terrible scourge along the upper river, or Solimoens, to the end of the navigation on the Amazons. It comes forth only by day, relieving the mosquito at sunrise with the greatest punctuality, and occurs only near the muddy shores of the stream, not one ever being found in the shade of the forest. In places where it is abundant it accompanies canoes in such dense swarms as to resemble thin clouds of smoke. It made its appearance in this way the first day after we crossed the river. Before I was aware of the presence of flies, I felt a slight itching on my neck, wrist, and ankles, and on looking for the cause, saw a number of tiny objects having a disgusting resemblance to lice, adhering to the skin. This was my introduction to the much-talked of Piúm, On close examination, they are seen to be minute two-winged insects, with dark-coloured body and pale legs and wings, the latter closed lengthwise over the back. They alight imperceptibly, and squatting close, fall at once to work; stretching forward their long front legs, which are in constant motion and seem to act as feelers, and then applying their short, broad snouts to the skin. Their abdomens soon become distended and red with blood, and then, their thirst satisfied, they slowly move off, sometimes so stupefied with their potations that they can scarcely fly. No pain is felt whilst they are at work, but they each leave a small circular raised spot on the skin and a disagreeable irritation. The latter may be avoided in great measure by pressing out the blood which remains in the spot; but this is a troublesome task when one has several hundred punctures in the course of a day. I took the trouble to dissect specimens to ascertain the way in which the little pests operate. The mouth consists of a pair of thick fleshy lips, and two triangular horny lancets, answering to the upper lip and tongue of other insects. This is applied closely to the skin, a puncture is made with the lancets, and the blood then sucked through between these into the œsophagus, the circular spot which results coinciding with the shape of the lips. In the course of a few days the red spots dry up, and the skin in time becomes blackened with the endless number of discoloured punctures that are crowded together. The irritation they produce is more acutely felt by some persons than others. I once travelled with a middle-aged Portuguese, who was laid up for three weeks from the attacks of Piúm; his legs being swelled to an enormous size, and the punctures aggravated into spreading sores.*

A brisk wind from the east sprang up early in the morning of the 22nd: we then hoisted all sail, and made for the mouth of the Rio Negro. This noble stream at its junction with the Amazons, seems, from its position, to be a direct continuation of the main river, whilst the Solimoens which joins at an angle and is somewhat narrower than its tributary, appears to be a branch instead of the main trunk of the vast water-system. One sees therefore at once, how the early explorers came to give a separate name to this upper part of the Amazons. The Brazilians have lately taken to applying the convenient term Alto Amazonas (High or Upper Amazons), to the Solimoens, and it is probable that this will gradually prevail over the old name. The Rio Negro broadens considerably from its mouth upwards, and presents the appearance of a great lake; its black-dyed waters having no current, and seeming to be dammed up by the impetuous flow of the yellow, turbid Solimoens, which here belches forth a continuous line of uprooted trees

* The Piúm belongs probably to the same species as the Mosquito of the Orinoco, described by Humboldt, and which he referred to the genus Simulium, several kinds of which inhabit Europe. Our insect is nearly allied to Simulium, but differs from the genus in several points, chiefly in the nervures of the wings.

and patches of grass, and forms a striking contrast with its tributary. In crossing, we passed the line, a little more than half-way over, where the waters of the two rivers meet and are sharply demarcated from each other. On reaching the opposite shore, we found a remarkable change. All our insect pests had disappeared, as if by magic, even from the hold of the canoe: the turmoil of an agitated, swiftly flowing river, and its torn, perpendicular, earthy banks, had given place to tranquil water and a coast indented with snug little bays, fringed with sloping sandy beaches. The low shore and vivid light green endlessly-varied foliage, which prevailed on the south side of the Amazons, were exchanged for a hilly country, clothed with a sombre, rounded, and monotonous forest. Our tedious voyage now approached its termination; a light wind carried us gently along the coast to the city of Barra, which lies about seven or eight miles within the mouth of the river. We stopped for an hour in a clean little bay, to bathe and dress, before showing ourselves again among civilised people. The bottom was visible at a depth of six feet, the white sand taking a brownish tinge from the stained but clear water. In the evening I went ashore, and was kindly received by Senhor Henriques Antony, a warm-hearted Italian, established here in a high position as merchant, who was the never-failing friend of stray travellers. He placed a couple of rooms at my disposal, and in a few hours I was comfortably settled in my new quarters, sixty-four days after leaving Obydos.

The town of Barra is built on a tract of elevated, but

very uneven land, on the left bank of the Rio Negro, and contained in 1850, about 3000 inhabitants. There was originally a small fort here, erected by the Portuguese to protect their slave-hunting expeditions amongst the numerous tribes of Indians which peopled the banks of the river. The most distinguished and warlike of these were the Manáos, who had many traits in common with the Omaguas, or Cambevas, of the Upper Amazons, the Mundurucús of the Tapajos, the Jurúnas of the Xingú, and other sections of the Tupí nation. The Manáos were continually at war with the neighbouring tribes, and had the custom of enslaving the prisoners made during their predatory expeditions. The Portuguese disguised their slave-dealing motives under the pretext of ransoming (*resgatando*), these captives; indeed, the term *resgatar* (to ransom) is still applied by the traders on the Upper Amazons to the very general, but illegal, practice of purchasing Indian children of the wild tribes. The older inhabitants of the place remember the time when many hundreds of these captives were brought down by a single expedition. In 1809, Barra became the chief town of the Rio Negro district; many Portuguese and Brazilians from other provinces then settled here; spacious houses and warehouses were built, and it grew, in the course of thirty or forty years, to be, next to Santarem, the principal settlement on the banks of the Amazons. At the time of my visit it was on the decline, in consequence of the growing distrust, or increased cunning, of the Indians, who once formed a numerous and the sole labouring class, but having got to know that the laws protected them against forced servitude, were rapidly withdrawing themselves from the place. When the new province of the Amazons was established, in 1852, Barra was chosen as the capital, and was then invested with the appropriate name of the city of Manáos.

The situation of the town has many advantages; the climate is healthy; there are no insect pests; the soil is fertile and capable of growing all kinds of tropical produce (the coffee of the Rio Negro, especially, being of very superior quality), and it is near the fork of two great navigable rivers. The imagination becomes excited when one reflects on the possible future of this place, situated near the centre of the equatorial part of South America, in the midst of a region almost as large as Europe, every inch of whose soil is of the most exuberant fertility, and having water communication on one side with the Atlantic, and on the other with the Spanish republics of Venezuela, New Granada, Ecuador, Peru, and Bolivia. Barra is now the principal station for the lines of steamers which were established in 1853, and passengers and goods are transhipped here for the Solimoens and Peru. A steamer runs once a fortnight between Pará and Barra, and a bi-monthly one plies between this place and Nauta in the Peruvian territory. The steam-boat company is supported by a large annual grant, about 50,000*l.* sterling, from the imperial government. Barra was formerly a pleasant place of residence, but it is now in a most wretched plight, suffering from a chronic scarcity of the most necessary articles of food. The attention of the settlers was formerly devoted almost entirely to the collection of the spontaneous produce of the forests and rivers; agriculture was consequently neglected, and now the neighbourhood does not produce even mandioca-meal sufficient for its own consumption. Many of the most necessary articles of food, besides all luxuries, come from Portugal, England, and North America. A few bullocks are brought now and then from Obydos, 500 miles off, the nearest place where cattle are reared in any numbers, and these furnish at long intervals a supply of fresh beef, but this is generally monopolised by the families of government officials. Fowls, eggs, fresh fish, turtles, vegetables, and fruit, were excessively scarce and dear in 1859, when I again visited the place; for instance, six or seven shilings were asked for a poor lean fowl, and eggs were twopence half-penny a piece. In fact, the neighbourhood produces scarcely anything; the provincial government is supplied with the greater part of its funds from the treasury of Pará; its revenue, which amounts to about 50 contos of reis (5,600*l.*), derived from export taxes on the produce of the entire province, not sufficing for more than about one-fifth of its expenditure. The population of the province of the Amazons, according to a census taken in 1858, is 55,000 souls; the municipal district of Barra, which comprises a large area around the capital, containing only 4500 inhabitants. For the government, however, of this small number of people, an immense staff of officials is gathered together in the capital, and, notwithstanding the endless number of trivial formalities which Brazilians employ in every small detail of administration, these have nothing to do the greater part of their time. None of the people who flocked to Barra on the establishment of the new government, seemed to care about the cultivation of the soil and the raising of food, although these would have been most profitable speculations. The class of Portuguese who emigrate to Brazil seem to prefer petty trading to the honourable pursuit of agriculture. If the English are a nation of shopkeepers, what are we to say of the Portuguese ? I counted in Barra, one store for every five dwelling-houses. These stores, or *tavernas*, have often not more than fifty pounds' worth of goods for their whole stock, and the Portuguese owners, big lusty fellows, stand all day behind their dirty counters for the sake of selling a few coppers' worth of liquors, or small wares. These men all give the same excuse for not applying themselves to agriculture, namely, that no hands can be obtained to work on the soil. Nothing can be done with Indians ; indeed, they are fast leaving the neighbourhood altogether, and the importation of negro slaves, in the present praiseworthy temper of the Brazilian mind, is out of the question. The problem, how to obtain a labouring class for a new and tropical country, without slavery, has to be solved before this glorious region can become what its delightful climate and exuberant fertility fit it for—the abode of a numerous, civilised, and happy people.

I found at Barra my companion, Mr. Wallace, who, since our joint Tocantins expedition, had been exploring, partly with his brother, lately arrived from England, the north-eastern coast of Marajó, the river Capim (a branch of the Guamá, near Pará), Monte Alegre, and Santarem.

He had passed us by night below Serpa, on his way to Barra, and so had arrived about three weeks before me. Besides ourselves, there were half-a-dozen other foreigners here congregated,—Englishmen, Germans, and Americans; one of them a Natural History collector, the rest traders on the rivers. In the pleasant society of these, and of the family of Senhor Henriques, we passed a delightful time; the miseries of our long river voyages were soon forgotten, and in two or three weeks we began to talk of further explorations. Meantime we had almost daily rambles in the neighbouring forest. The country around Barra is undulating and furrowed by ravines, through which flow rivulets of clear cold water, along whose banks many picturesque nooks occur. The whole surface of the land down to the water's edge is covered by the uniform dark-green rolling forest, the *caá-apoam* (convex woods) of the Indians, characteristic of the Rio Negro. This clothes also the extensive areas of low land, which are flooded by the river in the rainy season. The olive-brown tinge of the water seems to be derived from the saturation in it of the dark green foliage during these annual inundations. The great contrast in form and colour between the forests of the Rio Negro and those of the Amazons arises from the predominance in each of different families of plants. On the main river, palms of twenty or thirty different species form a great proportion of the mass of trees; whilst on the Rio Negro they play a very subordinate part. The characteristic kind in the latter region is the Jará (Leopoldinia pulchra), a species not found on the margins of the Amazons, which has a scanty head of fronds with narrow leaflets of the same dark green hue as the rest of the forest. The stem is smooth, and about two inches in diameter; its height is not more than twelve to fifteen feet; it does not, therefore, rise amongst the masses of foliage of the exogenous trees, so as to form a feature in the landscape, like the broad-leaved Murumurú and Urucurí, the slender Assaí, the tall Jauarí, and the fan-leaved Muritì of the banks of the Amazons. On the shores of the main river the mass of the forest is composed, besides palms, of Leguminosæ, or trees of the bean family, in endless variety as to height, shape of foliage, flowers, and fruit; of silk-cotton trees, colossal nut-trees (Lecythideæ), and Cecropiæ, the underwood and water-frontage consisting in great part of broad-leaved Musaceæ, Marantaceæ, and succulent grasses: all of which are of light shades of green. The forests of the Rio Negro are almost destitute of these large-leaved plants and grasses, which give so rich an appearance to the vegetation wherever they grow; the margins of the stream being clothed with bushes or low trees, having the same gloomy monotonous aspect as the mangroves of the shores of creeks near the Atlantic. The uniformly small but elegantly-leaved exogenous trees, which constitute the mass of the forest, consist in great part of members of the Laurel, Myrtle, Bignoniaceous, and Rubiaceous orders. The soil is generally a stiff loam, whose chief component part is the Tabatinga clay, which also forms low cliffs on the coast in some places, where it overlies strata of coarse sandstone. This kind of soil and the same geological formation prevail, as we have seen, in many places on the banks of the Amazons, so that the great contrast in the forest-clothing of the two rivers cannot arise from this cause.

I did not stay long enough at Barra to make a large collection of the animal productions of the neighbourhood. I obtained one species of monkey; not more than a dozen birds, and about 300 species of insects. Judging from these materials, the fauna appears to have much in common with that of the sea-coast of Guiana; but, at the same time, it contains a considerable number of species not hitherto found in Guiana, or in any other part of South America. The resemblance between the eastern shore of the Rio Negro and the distant coast of Guiana, in this respect, appears to be greater than that between the Rio Negro and the banks of the Upper Amazons.*

The species of monkey mentioned above was rather

* My own material is perhaps not sufficient to establish this view of the relations of the fauna, for it requires the comparison of an extensive series of species to obtain sound results on such subjects. A few conspicuous instances, however, pointed to the conclusion above mentioned. For example: in birds, the beautiful seven-coloured Tanager, Calliste tatao, the "sete cores" of the Brazilians, a Cayenne bird, is common to Guiana and the neighbourhood of Barra, but does not range further westward to the banks of the Solimoens; where, from Ega to Tabatinga, the allied form of Calliste Yeni takes its place. The Ramphastos Toco, or Tocano pacova (so named from its beak resembling a banana or pacova), a well-known Guianian bird, is found also at Barra, but not further west at Ega. In Coleopterous insects such species as Aniara sepulchralis, Agra ænea, Stenocheila Lacordairei, and others, confirm this view, being common to Cayenne and the Rio Negro, but not found further west on the banks of the Solimoens. Mr. Wallace discovered that the Rio Negro served as a barrier to the distribution of many species of mammals and birds, certain kinds being peculiar to the east, and others to the west bank (Travels on the Amazons and Rio Negro, p. 471). The Upper Amazons Fauna, nevertheless, contains a very large proportion of Guiana species.

common in the forest; it is the Midas bicolor of Spix, a kind I had not before met with, and peculiar, as far as at present known, to the eastern bank of the Rio Negro. The colour is brown, with the neck and arms white. Like its congeners, it keeps together in small troops, and runs along the main boughs of the loftier trees, climbing perpendicular trunks, but never taking flying leaps. The locality seemed to be a poor one for birds and insects. I do not know how far this apparent scarcity is attributable to the rainy weather which prevailed, and to the unfavourable time of the year. The months spent here (from January to March) I always found to be the best for collecting Coleopterous insects in this climate, but they are not so well for other orders of insects or for birds, which abound most from July to October. The forest was very pleasant for rambling. In some directions broad pathways led down gentle slopes, through what one might fancy were interminable shrubberies of evergreens, to moist hollows where springs of water bubbled up, or shallow brooks ran over their beds of clean white sand. But the most beautiful road was one that ran through the heart of the forest to a waterfall, which the citizens of Barra consider as the chief natural curiosity of their neighbourhood. The waters of one of the larger rivulets which traverse the gloomy wilderness, here fall over a ledge of rock about ten feet high. It is not the cascade itself, but the noiseless solitude, and the marvellous diversity and richness of trees, foliage, and flowers, encircling the water basin, that form the attraction of the place. Families make picnic excursions to this spot; and the gentlemen—it is said the ladies also—spend the sultry hours of midday bathing in the cold and bracing waters. This place is classic ground to the Naturalist, from having been a favourite spot with the celebrated travellers Spix and Martius, during their stay at Barra in 1820. Von Martius was so much impressed by its magical beauty that he commemorated the visit by making a sketch of the scenery serve as background in one of the plates of his great work on the palms.

Birds and insects, however, were scarce amidst these charming sylvan scenes. I often traversed the whole distance from Barra to the waterfall, about two miles by the forest road, without seeing or hearing a bird, or meeting with so many as a score of Lepidopterous and Coleopterous insects. In the thinner woods near the borders of the forest many pretty little blue and green creepers of the Dacnidæ group, were daily seen feeding on berries; and a few very handsome birds occurred in the forest. But the latter were so rare that we could obtain them only by employing a native hunter; who used to spend a whole day, and go a great distance, to obtain two or three specimens. In this way I obtained, amongst others, specimens of the Trogon pavoninus (the Suruquá grande of the natives), a most beautiful creature, having soft golden green plumage, red breast, and an orange-coloured beak; also the Ampelis Pompadoura, a rich glossy-purple chatterer with wings of a snowy-white hue. The borders of the forest yielded me more insects also than the shady central pathways. A few species occurred which I had previously found at Obydos and Serpa, but certain kinds were met with which are not known in any other part of South America. The small-leaved bushes and low trees on the skirts of the forest and along the more open pathways were sparingly tenanted by a variety of curious phytophagous beetles. None of these offered any remarkable feature, except perhaps the species of Chlamys. These are small beetles of a cubical shape and grotesque appearance, the upper surface of their bodies being studded with tubercles. They look like anything rather than insects; some of them are an exact imitation of the dung of caterpillars on leaves; others have a deceptive likeness to small flower-buds, galls, and other vegetable excrescences, whilst some large kinds are like fragments of metallic ore. They are very sluggish in their motions, and live in the most exposed situations on the surface of leaves; their curious shapes are therefore no doubt so many disguises to protect them from the keen eyes of insectivorous birds and lizards. A nearly allied group, the Lamprosomas, of which several species occurred in the same places, have perfectly smooth convex bodies; these glitter like precious stones on the foliage, and seem to be protected by the excessive hardness of their integuments. The Eumolpidæ and Galerucidæ, two groups belonging also to the leaf-eating family, were much more numerous than the Chlamydes and Lamprosomas, although having neither the disguised appearance of the one nor the hard integuments of the other; but many of them secrete a foul liquor when handled, which may perhaps serve the same purpose of passive defence. The Chlamydes are almost confined to the warmer parts of America, and the species, although extremely numerous (about 300 are known in collections), are nearly all very rare. It is worthy of note that mimicking insects are very generally of great scarcity; that is, few examples of each species occur in the places where they are found, and they constitute groups which are remarkable for the strongly-marked diversity and limited ranges of their species.

After we had rested some weeks in Barra, we arranged our plans for further explorations in the interior of the country. Mr. Wallace chose the Rio Negro for his next trip, and I agreed to take the Solimoens. My colleague has already given to the world an account of his journey on the Rio Negro, and his adventurous ascent of its great tributary the Uapés. I left Barra for Ega, the first town of any importance on the Solimoens, on the 26th of March, 1850. The distance is nearly 400 miles, which we accomplished in a small cuberta, manned by ten stout Cucama Indians, in thirty-five days. On this occasion, I spent twelve months in the upper region of the Amazons; circumstances then compelled me to return to Pará. I revisited the same country in 1855, and devoted three years and a-half to a fuller exploration of its natural productions. The results of both journies will be given together in subsequent chapters of this work; in the meantime, I will proceed to give an account of Santarem and the river Tapajos, whose neighbourhoods I investigated in the years 1851-4.

A few words on my visit to Pará in 1851, may be

here introduced. I descended the river from Ega to the capital, a distance of 1400 miles, in a heavily-laden schooner belonging to a trader of the former place. The voyage occupied no less than twenty-nine days, although we were favoured by the powerful currents of the rainy season. The hold of the vessel was filled with turtle oil contained in large jars, the cabin was crammed with Brazil nuts, and a great pile of salsaparilla, covered with a thatch of palm leaves, occupied the middle of the deck. We had, therefore (the master and two passengers), but rough accommodation, having to sleep on deck exposed to the wet and stormy weather under little toldos or arched shelters, arranged with mats of woven lianas and maranta leaves. I awoke many a morning, with clothes and bedding soaked through with the rain. With the exception, however, of a slight cold at the commencement I never enjoyed better health than during this journey. When the wind blew from up river or off the land, we sped away at a great rate; but it was often squally from those quarters, and then it was not safe to hoist the sails. The weather was generally calm, a motionless mass of leaden clouds covering the sky and the broad expanse of waters flowing smoothly down with no other motion than the ripple of the current. When the wind came from below, we tacked down the stream; sometimes it blew very strong, and then the schooner, having the wind abeam, laboured through the waves, shipping often heavy seas which washed everything that was loose from one side of the deck to the other.

On arriving at Pará, I found the once cheerful and

healthy city desolated by two terrible epidemics. The yellow fever, which visited the place the previous year (1850) for the first time since the discovery of the country, still lingered, after having carried off nearly 5 per cent. of the population. The number of persons who were attacked, namely, three-fourths of the entire population, showed how general is the onslaught of an epidemic on its first appearance in a place. At the heels of this plague came the smallpox. The yellow fever had fallen most severely on the whites and mamelucos, the negroes wholly escaping ; but the smallpox attacked more especially the Indians, negroes, and people of mixed colour, sparing the whites almost entirely, and taking off about a twentieth part of the population in the course of the four months of its stay. I heard many strange accounts of the yellow fever. I believe Pará was the second port in Brazil attacked by it. The news of its ravages in Bahia, where the epidemic first appeared, arrived some few days before the disease broke out. The government took all the sanitary precautions that could be thought of ; amongst the rest was the singular one of firing cannon at the street corners, to purify the air. Mr. Norris, the American consul, told me, the first cases of fever occurred near the port, and that it spread rapidly and regularly from house to house, along the streets which run from the waterside to the suburbs, taking about twenty-four hours to reach the end. Some persons related that for several successive evenings before the fever broke out the atmosphere was thick, and that a body of murky vapour accompanied by a strong stench, travelled from street to street. This moving vapour was called the "Maî da peste," "the mother or spirit of the plague"; and it was useless to attempt to reason them out of the belief that this was the forerunner of the pestilence. The progress of the disease was very rapid. It commenced in April, in the middle of the wet season. In a few days, thousands of persons lay sick, dying or dead. The state of the city during the time the fever lasted, may be easily imagined. Towards the end of June it abated, and very few cases occurred during the dry season from July to December.

As I said before, the yellow fever still lingered in the place when I arrived from the interior in April. I was in hopes I should escape it, but was not so fortunate; it seemed to spare no new comer. At the time I fell ill, every medical man in the place was worked to the utmost in attending the victims of the other epidemic; it was quite useless to think of obtaining their aid, so I was obliged to be my own doctor, as I had been in many former smart attacks of fever. I was seized with shivering and vomit at 9 o'clock in the morning. Whilst the people of the house went down to the town for the medicines I ordered, I wrapped myself in a blanket and walked sharply to and fro along the verandah, drinking at intervals a cup of warm tea, made of a bitter herb in use amongst the natives, called Pajémarióba, a leguminous plant growing in all waste places. About an hour afterwards, I took a good draught of a decoction of elder blossoms as a sudorific, and soon after fell insensible into my hammock. Mr. Philipps, an English resident with whom I was then lodging, came home in the afternoon and found me sound asleep and perspiring famously. I did not wake till towards midnight, when I felt very weak and aching in every bone of my body. I then took as a purgative, a small dose of Epsom salts and manna. In forty-eight hours the fever left me, and in eight days from the first attack, I was able to get about my work. Little else happened during my stay, which need be recorded here. I shipped off all my collections to England, and received thence a fresh supply of funds. It took me several weeks to prepare for my second and longest journey into the interior. My plan now was first to make Santarem head-quarters for some time, and ascend from that place the river Tapajos, as far as practicable. Afterwards I intended to revisit the marvellous country of the Upper Amazons, and work well its natural history at various stations I had fixed upon, from Ega to the foot of the Andes.

END OF VOL. I.

BRADBURY AND EVANS, PRINTERS, WHITEFRIARS.

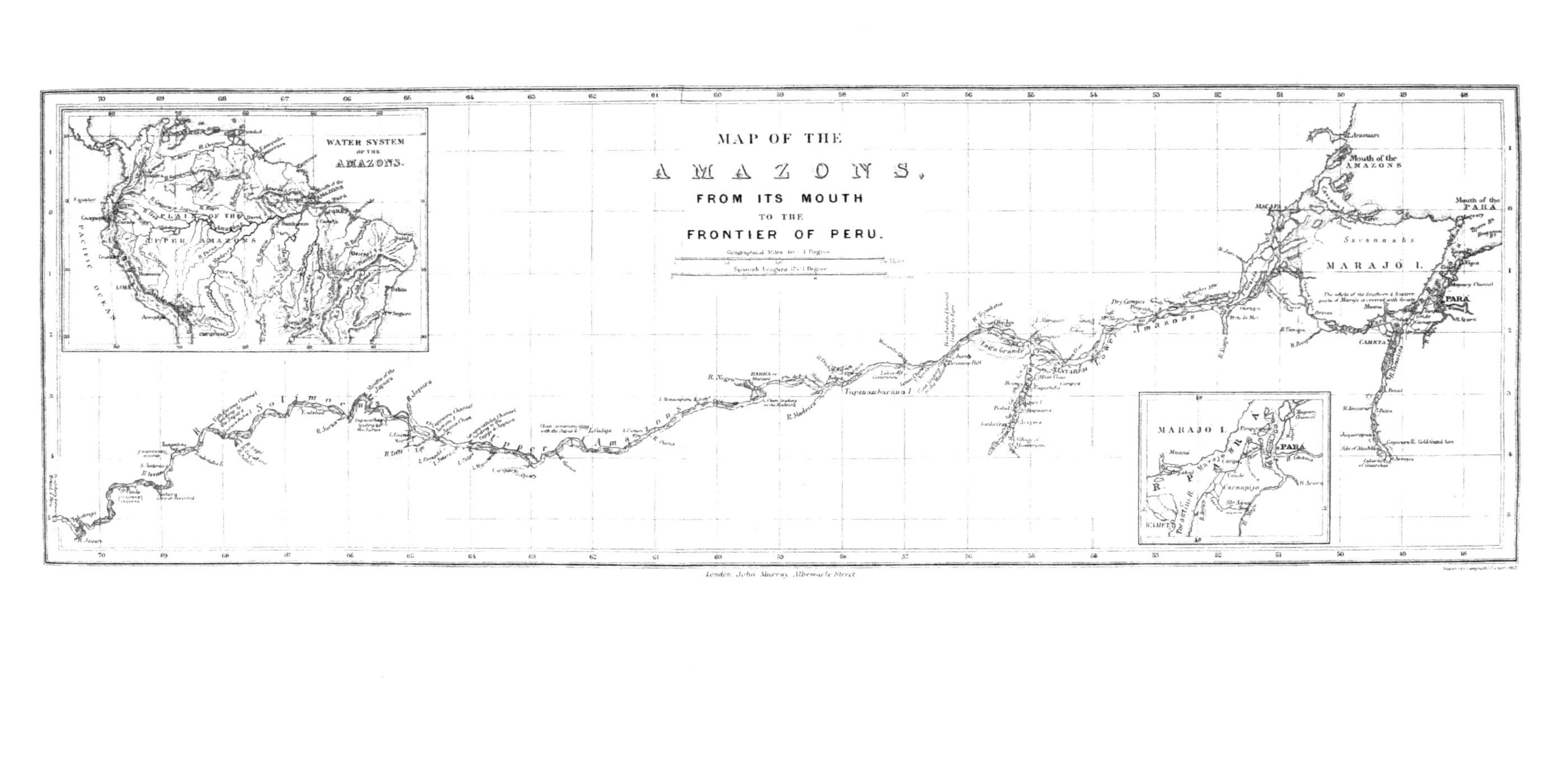
MAP OF THE
AMAZONS,
FROM ITS MOUTH
TO THE
FRONTIER OF PERU.
WATER SYSTEM
OF THE
AMAZONS.
PACIFIC OCEAN
Mouth of the AMAZONS
Mouth of the PARA
MACAPA
Savannahs
MARAJO I.
PARA
CAMETA
SANTAREM
Lago Grande
Tupinambarana I.
R. Negro
R. Madeira
R. Purus
R. Teffe
R. Japura
R. Jurua
Lower Amazons
Upper Amazons
Solimoens
London John Murray, Albemarle Street

MR. MURRAY'S
GENERAL LIST OF WORKS.

ABBOTT'S (REV. J.) Philip Musgrave; or, Memoirs of a Church of England Missionary in the North American Colonies. Post 8vo. 2*s.*

ABERCROMBIE'S (JOHN) Enquiries concerning the Intellectual Powers and the Investigation of Truth. *Sixteenth Edition.* Fcap. 8vo. 6*s.* 6*d.*

——— Philosophy of the Moral Feelings. *Twelfth Edition.* Fcap. 8vo. 4*s.*

——— Pathological and Practical Researches on the Diseases of the Stomach, &c. *Third Edition.* Fcap. 8vo. 6*s.*

ACLAND'S (REV. CHARLES) Popular Account of the Manners and Customs of India. Post 8vo. 2*s.*

ADOLPHUS'S (J. L.) Letters from Spain, in 1856 and 1857. Post 8vo. 10*s.* 6*d.*

ÆSOP'S FABLES. A New Translation. With Historical Preface. By Rev. THOMAS JAMES. With 100 Woodcuts, by TENNIEL and WOLF. 38*th Thousand.* Post 8vo. 2*s.* 6*d.*

AGRICULTURAL (THE) JOURNAL. Of the Royal Agricultural Society of England. 8vo. 10*s.* *Published half-yearly.*

AIDS TO FAITH: a Series of Essays. By various Writers. Edited by WILLIAM THOMSON, D.D., Lord Archbishop of York. 8vo. 9*s.*

CONTENTS.

Rev. H. L. MANSEL—*On Miracles.*
BISHOP FITZGERALD—*Christian Evidences.*
REV. DR. MCCAUL—*On Prophecy.*
Rev. F. C. COOK—*Ideology and Subscription.*
Rev. DR. MCCAUL—*Mosaic Record of Creation.*
Rev. GEORGE RAWLINSON—*The Pentateuch.*
ARCHBISHOP THOMSON—*Doctrine of the Atonement.*
Rev. HAROLD BROWNE—*On Inspiration.*
BISHOP ELLICOTT—*Scripture and its Interpretation.*

AMBER-WITCH (THE). The most interesting Trial for Witchcraft ever known. Translated from the German by LADY DUFF GORDON. Post 8vo. 2*s.*

ARTHUR'S (LITTLE) History of England. By LADY CALLCOTT. 120*th Thousand.* With 20 Woodcuts. Fcap. 8vo. 2*s.* 6*d.*

ATKINSON'S (MRS.) Recollections of Tartar Steppes and their Inhabitants. With Illustrations. Post 8vo. 12*s.*

AUNT IDA'S Walks and Talks; a Story Book for Children. By a LADY. Woodcuts. 16mo. 5*s.*

AUSTIN'S (JOHN) PROVINCE OF JURISPRUDENCE DETERMINED; or, Philosophy of Positive Law. *Second Edition.* 8vo. 15*s.*

——— Lectures on Jurisprudence. Being a Continuation of the "Province of Jurisprudence Determined." 2 vols. 8vo.

——— (SARAH) Fragments from German Prose Writers. With Biographical Notes. Post 8vo. 10*s.*

B

ADMIRALTY PUBLICATIONS; Issued by direction of the Lords Commissioners of the Admiralty:—

A MANUAL OF SCIENTIFIC ENQUIRY, for the Use of Travellers. Edited by Sir JOHN F. HERSCHEL, and Rev. ROBERT MAIN. *Third Edition.* Woodcuts. Post 8vo. 9*s.*

AIRY'S ASTRONOMICAL OBSERVATIONS MADE AT GREENWICH. 1836 to 1847. Royal 4to. 50*s.* each.

—— ASTRONOMICAL RESULTS. 1848 to 1858. 4to. 8*s.* each.

—— APPENDICES TO THE ASTRONOMICAL OBSERVATIONS.

1836.—I. Bessel's Refraction Tables.
II. Tables for converting Errors of R.A. and N.P.D. into Errors of Longitude and Ecliptic P.D. } 8*s.*
1837.—I. Logarithms of Sines and Cosines to every Ten Seconds of Time.
II. Table for converting Sidereal into Mean Solar Time. } 8*s.*
1842.—Catalogue of 1439 Stars. 8*s.*
1845.—Longitude of Valentia. 8*s.*
1847.—Twelve Years' Catalogue of Stars. 14*s.*
1851.—Maskelyne's Ledger of Stars. 6*s.*
1852.—I. Description of the Transit Circle. 5*s.*
II. Regulations of the Royal Observatory. 2*s.*
1853.—Bessel's Refraction Tables. 3*s.*
1854.—I. Description of the Zenith Tube. 3*s.*
II. Six Years' Catalogue of Stars. 10*s.*
1856.—Description of the Galvanic Apparatus at Greenwich Observatory. 8s.

—— MAGNETICAL AND METEOROLOGICAL OBSERVATIONS. 1840 to 1847. Royal 4to. 50*s.* each.

—— ASTRONOMICAL, MAGNETICAL, AND METEOROLOGICAL OBSERVATIONS, 1848 to 1860. Royal 4to. 50*s.* each.

—— ASTRONOMICAL RESULTS. 1859. 4to.

—— MAGNETICAL AND METEOROLOGICAL RESULTS. 1848 to 1859. 4to. 8*s.* each.

—— REDUCTION OF THE OBSERVATIONS OF PLANETS. 1750 to 1830. Royal 4to. 50*s.*

—————— LUNAR OBSERVATIONS. 1750 to 1830. 2 Vols. Royal 4to. 50*s.* each.

—————— 1831 to 1851. 4to. 20*s.*

BERNOULLI'S SEXCENTENARY TABLE. *London*, 1779. 4to.

BESSEL'S AUXILIARY TABLES FOR HIS METHOD OF CLEARING LUNAR DISTANCES. 8vo.

—— FUNDAMENTA ASTRONOMIÆ: *Regiomontii*, 1818. Folio. 60*s.*

BIRD'S METHOD OF CONSTRUCTING MURAL QUADRANTS. *London*, 1768. 4to. 2*s.* 6*d.*

—— METHOD OF DIVIDING ASTRONOMICAL INSTRUMENTS. *London*, 1767. 4to. 2*s.* 6*d.*

COOK, KING, AND BAYLY'S ASTRONOMICAL OBSERVATIONS. *London* 1782. 4to. 21*s.*

EIFFE'S ACCOUNT OF IMPROVEMENTS IN CHRONOMETERS. 4to. 2*s.*

ENCKE'S BERLINER JAHRBUCH, for 1830. *Berlin*, 1828. 8vo. 9*s.*

GROOMBRIDGE'S CATALOGUE OF CIRCUMPOLAR STARS. 4to. 10*s.*

HANSEN'S TABLES DE LA LUNE. 4to. 20*s.*

HARRISON'S PRINCIPLES OF HIS TIME-KEEPER. PLATES. 1797. 4to. 5*s.*

HUTTON'S TABLES OF THE PRODUCTS AND POWERS OF NUMBERS. 1781. Folio. 7*s.* 6*d.*

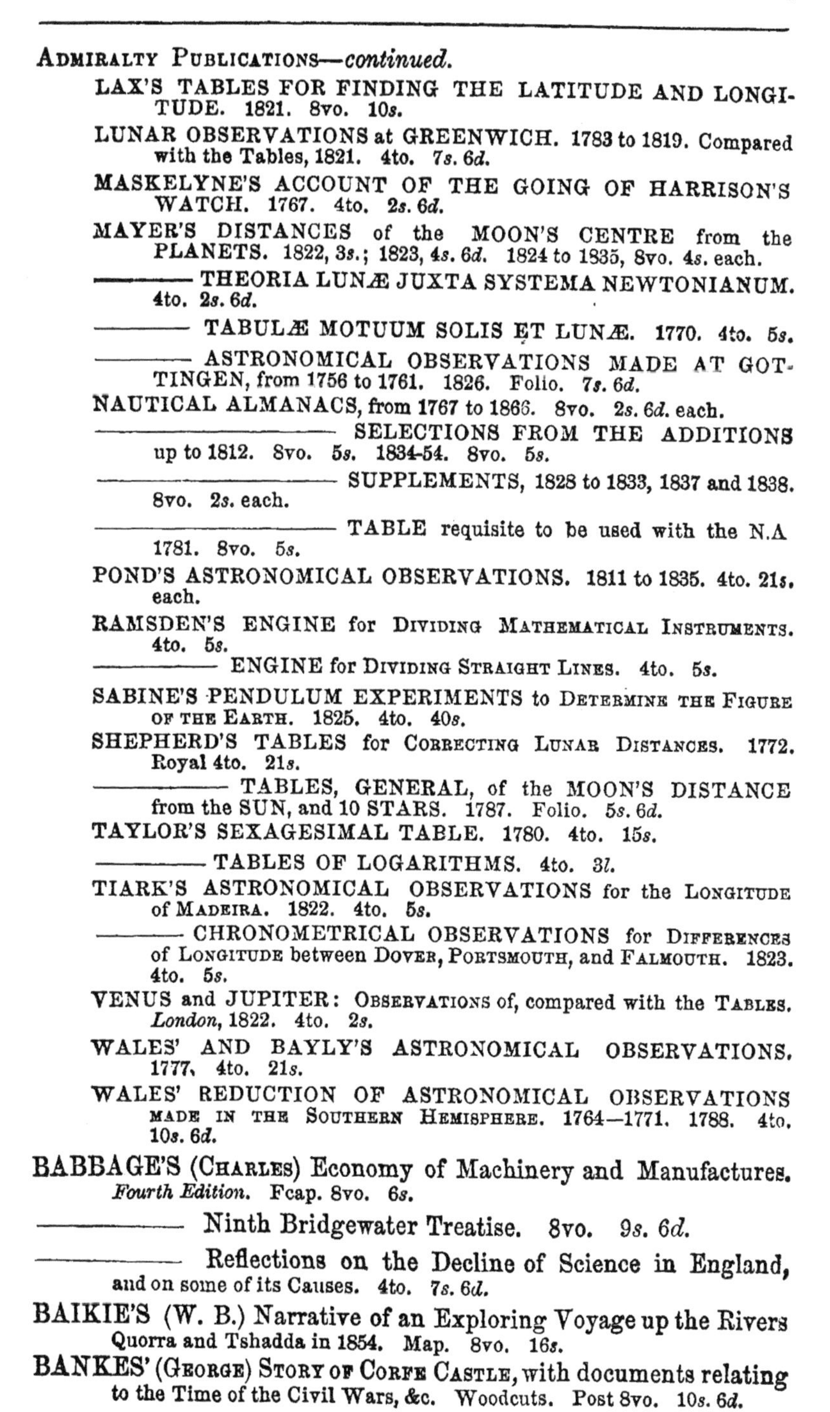

ADMIRALTY PUBLICATIONS—*continued.*

LAX'S TABLES FOR FINDING THE LATITUDE AND LONGITUDE. 1821. 8vo. 10*s.*

LUNAR OBSERVATIONS at GREENWICH. 1783 to 1819. Compared with the Tables, 1821. 4to. 7*s.* 6*d.*

MASKELYNE'S ACCOUNT OF THE GOING OF HARRISON'S WATCH. 1767. 4to. 2*s.* 6*d.*

MAYER'S DISTANCES of the MOON'S CENTRE from the PLANETS. 1822, 3*s.*; 1823, 4*s.* 6*d.* 1824 to 1835, 8vo. 4*s.* each.

——— THEORIA LUNÆ JUXTA SYSTEMA NEWTONIANUM. 4to. 2*s.* 6*d.*

——— TABULÆ MOTUUM SOLIS ET LUNÆ. 1770. 4to. 5*s.*

——— ASTRONOMICAL OBSERVATIONS MADE AT GOTTINGEN, from 1756 to 1761. 1826. Folio. 7*s.* 6*d.*

NAUTICAL ALMANACS, from 1767 to 1866. 8vo. 2*s.* 6*d.* each.

——— SELECTIONS FROM THE ADDITIONS up to 1812. 8vo. 5*s.* 1834-54. 8vo. 5*s.*

——— SUPPLEMENTS, 1828 to 1833, 1837 and 1838. 8vo. 2*s.* each.

——— TABLE requisite to be used with the N.A 1781. 8vo. 5*s.*

POND'S ASTRONOMICAL OBSERVATIONS. 1811 to 1835. 4to. 21*s.* each.

RAMSDEN'S ENGINE for DIVIDING MATHEMATICAL INSTRUMENTS. 4to. 5*s.*

——— ENGINE for DIVIDING STRAIGHT LINES. 4to. 5*s.*

SABINE'S PENDULUM EXPERIMENTS to DETERMINE THE FIGURE OF THE EARTH. 1825. 4to. 40*s.*

SHEPHERD'S TABLES for CORRECTING LUNAR DISTANCES. 1772. Royal 4to. 21*s.*

——— TABLES, GENERAL, of the MOON'S DISTANCE from the SUN, and 10 STARS. 1787. Folio. 5*s.* 6*d.*

TAYLOR'S SEXAGESIMAL TABLE. 1780. 4to. 15*s.*

——— TABLES OF LOGARITHMS. 4to. 3*l.*

TIARK'S ASTRONOMICAL OBSERVATIONS for the LONGITUDE of MADEIRA. 1822. 4to. 5*s.*

——— CHRONOMETRICAL OBSERVATIONS for DIFFERENCES of LONGITUDE between DOVER, PORTSMOUTH, and FALMOUTH. 1823. 4to. 5*s.*

VENUS and JUPITER: OBSERVATIONS of, compared with the TABLES. *London*, 1822. 4to. 2*s.*

WALES' AND BAYLY'S ASTRONOMICAL OBSERVATIONS. 1777. 4to. 21*s.*

WALES' REDUCTION OF ASTRONOMICAL OBSERVATIONS MADE IN THE SOUTHERN HEMISPHERE. 1764—1771. 1788. 4to. 10*s.* 6*d.*

BABBAGE'S (CHARLES) Economy of Machinery and Manufactures. *Fourth Edition.* Fcap. 8vo. 6*s.*

——— Ninth Bridgewater Treatise. 8vo. 9*s.* 6*d.*

——— Reflections on the Decline of Science in England, and on some of its Causes. 4to. 7*s.* 6*d.*

BAIKIE'S (W. B.) Narrative of an Exploring Voyage up the Rivers Quorra and Tshadda in 1854. Map. 8vo. 16*s.*

BANKES' (GEORGE) STORY OF CORFE CASTLE, with documents relating to the Time of the Civil Wars, &c. Woodcuts. Post 8vo. 10*s.* 6*d.*

B 2

BARROW'S (SIR JOHN) Autobiographical Memoir, including Reflections, Observations, and Reminiscences at Home and Abroad. From Early Life to Advanced Age. Portrait. 8vo. 16*s*.

——— Voyages of Discovery and Research within the Arctic Regions, from 1818 to the present time. Abridged and arranged from the Official Narratives. 8vo. 15*s*.

——— (SIR GEORGE) Ceylon; Past and Present. Map. Post 8vo. 6*s*. 6*d*.

——— (JOHN) Naval Worthies of Queen Elizabeth's Reign, their Gallant Deeds, Daring Adventures, and Services in the infant state of the British Navy. 8vo. 14*s*.

——— Life and Voyages of Sir Francis Drake. With numerous Original Letters. Post 8vo. 2*s*.

BASSOMPIERRE'S Memoirs of his Embassy to the Court of England in 1626. Translated with Notes. 8vo. 9*s*. 6*d*.

BASTIAT'S (FREDERIC) Harmonies of Political Economy. Translated, with a Notice of his Life and Writings, by P. J. STIRLING. 8vo. 7*s*. 6*d*.

BATES' (H. W.) Naturalist on the Amazon; Adventures, Social Sketches, Native Life, Habits of Animals and Features of Nature in the Tropics during eleven years of Travel. Illustrations. 2 Vols. Post 8vo.

BEES AND FLOWERS. Two Essays. By Rev. Thomas James. Reprinted from the "Quarterly Review." Fcap. 8vo. 1*s*. each.

BELL'S (SIR CHARLES) Mechanism and Vital Endowments of the Hand as evincing Design. *Sixth Edition.* Woodcuts. Post 8vo. 6*s*.

BENEDICT'S (JULES) Sketch of the Life and Works of Felix Mendelssohn-Bartholdy. *Second Edition.* 8vo. 2*s*. 6*d*.

BERTHA'S Journal during a Visit to her Uncle in England. Containing a Variety of Interesting and Instructive Information. *Seventh Edition.* Woodcuts. 12mo.

BIRCH'S (SAMUEL) History of Ancient Pottery and Porcelain: Egyptian, Assyrian, Greek, Roman, and Etruscan. With 200 Illustrations. 2 Vols. Medium 8vo. 42*s*.

BLUNT'S (REV. J. J.) Principles for the proper understanding of the Mosaic Writings, stated and applied, together with an Incidental Argument for the truth of the Resurrection of our Lord. Being the HULSEAN LECTURES for 1832. Post 8vo. 6*s*. 6*d*.

——— Undesigned Coincidences in the Writings of the Old and New Testament, an Argument of their Veracity: with an Appendix containing Undesigned Coincidences between the Gospels, Acts, and Josephus. *7th Edition.* Post 8vo.

——— History of the Church in the First Three Centuries. *Third Edition.* Post 8vo. 7*s*. 6*d*.

——— Parish Priest; His Duties, Acquirements and Obligations. *Third Edition.* Post 8vo. 7*s*. 6*d*.

——— Lectures on the Right Use of the Early Fathers. *Second Edition.* 8vo. 15*s*.

——— Plain Sermons Preached to a Country Congregation. *Second Edition.* 3 Vols. Post 8vo. 7*s*. 6*d*. each.

——— Literary Essays, from the Quarterly Review. 8vo. 12*s*.

BLACKSTONE'S COMMENTARIES on the Laws of England. Adapted to the present state of the law. By R. MALCOLM KERR, LL.D. *Third Edition*, corrected to 1861. 4 Vols. 8vo. 63*s*.

——— For STUDENTS. Being those Portions which relate to the BRITISH CONSTITUTION and the RIGHTS OF PERSONS. Post 8vo. 9*s*.

BLAKISTON'S (CAPT.) Five Months on the Yang-Tsze, with a Narrative of the Expedition sent to explore its Upper Waters. Maps and 24 Illustrations. 8vo. 18*s*.

BLOMFIELD'S (REV. A.) Memoir of the late Bishop Blomfield, D.D., with Selections from his Correspondence. Portrait, 2 Vols. post 8vo.

BOOK OF COMMON PRAYER. Illustrated with Borders, Initials, Letters, and Woodcuts. A new and carefully printed edition. 8vo.

BOSWELL'S (JAMES) Life of Samuel Johnson, LL.D. Including the Tour to the Hebrides. Edited by Mr. CROKER. Portraits. Royal 8vo. 10*s*.

BORROW'S (GEORGE) Bible in Spain; or the Journeys, Adventures, and Imprisonments of an Englishman in an Attempt to circulate the Scriptures in the Peninsula. 3 Vols. Post 8vo. 27*s*.; or *Popular Edition*, 16mo, 3*s*. 6*d*.

——— Zincali, or the Gipsies of Spain; their Manners, Customs, Religion, and Language. 2 Vols. Post 8vo. 18*s*.; or *Popular Edition*, 16mo, 3*s*. 6*d*.

——— Lavengro; The Scholar—The Gipsy—and the Priest. Portrait. 3 Vols. Post 8vo. 30*s*.

——— Romany Rye; a Sequel to Lavengro. *Second Edition*. 2 Vols. Post 8vo. 21*s*.

——— Wild Wales: its People, Language, and Scenery. 3 Vols. Post 8vo. 30*s*.

BRAY'S (MRS.) Life of Thomas Stothard, R.A. With Personal Reminiscences. Illustrated with Portrait and 60 Woodcuts of his chief works. 4to.

BREWSTER'S (SIR DAVID) Martyrs of Science, or the Lives of Galileo, Tycho Brahe, and Kepler. *Fourth Edition*. Fcap. 8vo. 4*s*. 6*d*.

——— More Worlds than One. The Creed of the Philosopher and the Hope of the Christian. *Eighth Edition*. Post 8vo. 6*s*.

——— Stereoscope: its History, Theory, Construction, and Application to the Arts and to Education. Woodcuts. 12mo. 5*s*. 6*d*.

——— Kaleidoscope: its History, Theory, and Construction, with its application to the Fine and Useful Arts. *Second Edition*. Woodcuts. Post 8vo. 5*s*. 6*d*.

BRINE'S (L.) Narrative of the Rise and Progress of the Taeping Rebellion in China. Maps and Plans. Post 8vo. 10*s*. 6*d*.

BRITISH ASSOCIATION REPORTS. 8vo. York and Oxford, 1831-32, 13*s*. 6*d*. Cambridge, 1833, 12*s*. Edinburgh, 1834, 15*s*. Dublin, 1835, 13*s*. 6*d*. Bristol, 1836, 12*s*. Liverpool, 1837, 16*s*. 6*d*. Newcastle, 1838, 15*s*. Birmingham, 1839, 13*s*. 6*d*. Glasgow, 1840, 15*s*. Plymouth, 1841, 13*s*. 6*d*. Manchester, 1842, 10*s*. 6*d*. Cork, 1843, 12*s*. York, 1844, 20*s*. Cambridge, 1845, 12*s*. Southampton, 1846, 15*s*. Oxford, 1847, 18*s*. Swansea, 1848, 9*s*. Birmingham, 1849, 10*s*. Edinburgh, 1850, 15*s*. Ipswich, 1851, 16*s*. 6*d*. Belfast, 1852, 15*s*. Hull, 1853, 10*s*. 6*d*. Liverpool, 1854, 18*s*. Glasgow, 1855, 15*s*.; Cheltenham, 1856, 18*s*.; Dublin, 1857, 15*s*.; Leeds, 1858, 20*s*. Aberdeen, 1859, 15*s*. Oxford, 1860. Manchester, 1861. 15*s*.

BRITISH CLASSICS. A New Series of Standard English Authors, printed from the most correct text, and edited with elucidatory notes. Published occasionally in demy 8vo. Volumes, varying in price.

Already Published.

GOLDSMITH'S WORKS. Edited by PETER CUNNINGHAM, F.S.A. Vignettes. 4 Vols. 30*s.*

GIBBON'S DECLINE AND FALL OF THE ROMAN EMPIRE. Edited by WILLIAM SMITH, LL.D. Portrait and Maps. 8 Vols. 60*s.*

JOHNSON'S LIVES OF THE ENGLISH POETS. Edited by PETER CUNNINGHAM, F.S.A. 3 Vols. 22*s.* 6*d.*

BYRON'S POETICAL WORKS. Edited, with Notes. 6 vols. 45*s.*

In Preparation.

WORKS OF POPE. With Life, Introductions, and Notes, by REV. WHITWELL ELWIN. Portrait.

HUME'S HISTORY OF ENGLAND. Edited, with Notes.

LIFE AND WORKS OF SWIFT. Edited by JOHN FORSTER.

BROUGHAM'S (LORD) Address at the Social Science Association, Dublin. August, 1861. Revised, with Notes. 8vo. 1*s.*

BROUGHTON'S (LORD) Journey through Albania and other Provinces of Turkey in Europe and Asia, to Constantinople, 1809—10. *Third Edition.* Maps and Woodcuts. 2 Vols. 8vo. 30*s.*

——— Visits to Italy. *Third Edition.* 2 vols. Post 8vo. 18*s.*

BUBBLES FROM THE BRUNNEN OF NASSAU. By an Old MAN. *Sixth Edition.* 16mo. 5*s.*

BUNBURY'S (C. J. F.) Journal of a Residence at the Cape of Good Hope; with Excursions into the Interior, and Notes on the Natural History and Native Tribes of the Country. Woodcuts. Post 8vo. 9*s.*

BUNYAN (JOHN) and Oliver Cromwell. Select Biographies. By ROBERT SOUTHEY. Post 8vo. 2*s.*

BUONAPARTE'S (NAPOLEON) Confidential Correspondence with his Brother Joseph, sometime King of Spain. *Second Edition.* 2 vols. 8vo. 26*s.*

BURGHERSH'S (LORD) Memoir of the Operations of the Allied Armies under Prince Schwarzenberg and Marshal Blucher during the latter end of 1813—14. 8vo. 21*s.*

——— Early Campaigns of the Duke of Wellington in Portugal and Spain. 8vo. 8*s.* 6*d.*

BURGON'S (Rev. J. W.) Memoir of Patrick Fraser Tytler. *Second Edition.* Post 8vo. 9*s.*

——— Letters from Rome, written to Friends at Home. Illustrations. Post 8vo. 12*s.*

BURN'S (LIEUT.-COL.) French and English Dictionary of Naval and Military Technical Terms. *Fourth Edition.* Crown 8vo. 15*s.*

BURNS' (ROBERT) Life. By JOHN GIBSON LOCKHART. Fifth *Edition.* Fcap. 8vo. 3*s.*

BURR'S (G. D.) Instructions in Practical Surveying, Topographical Plan Drawing, and on sketching ground without Instruments. *Third Edition.* Woodcuts. Post 8vo. 7*s.* 6*d.*

BUTTMAN'S LEXILOGUS; a Critical Examination of the Meaning of numerous Greek Words, chiefly in Homer and Hesiod. Translated by Rev. J. R. FISHLAKE. *Fifth Edition.* 8vo. 12*s.*

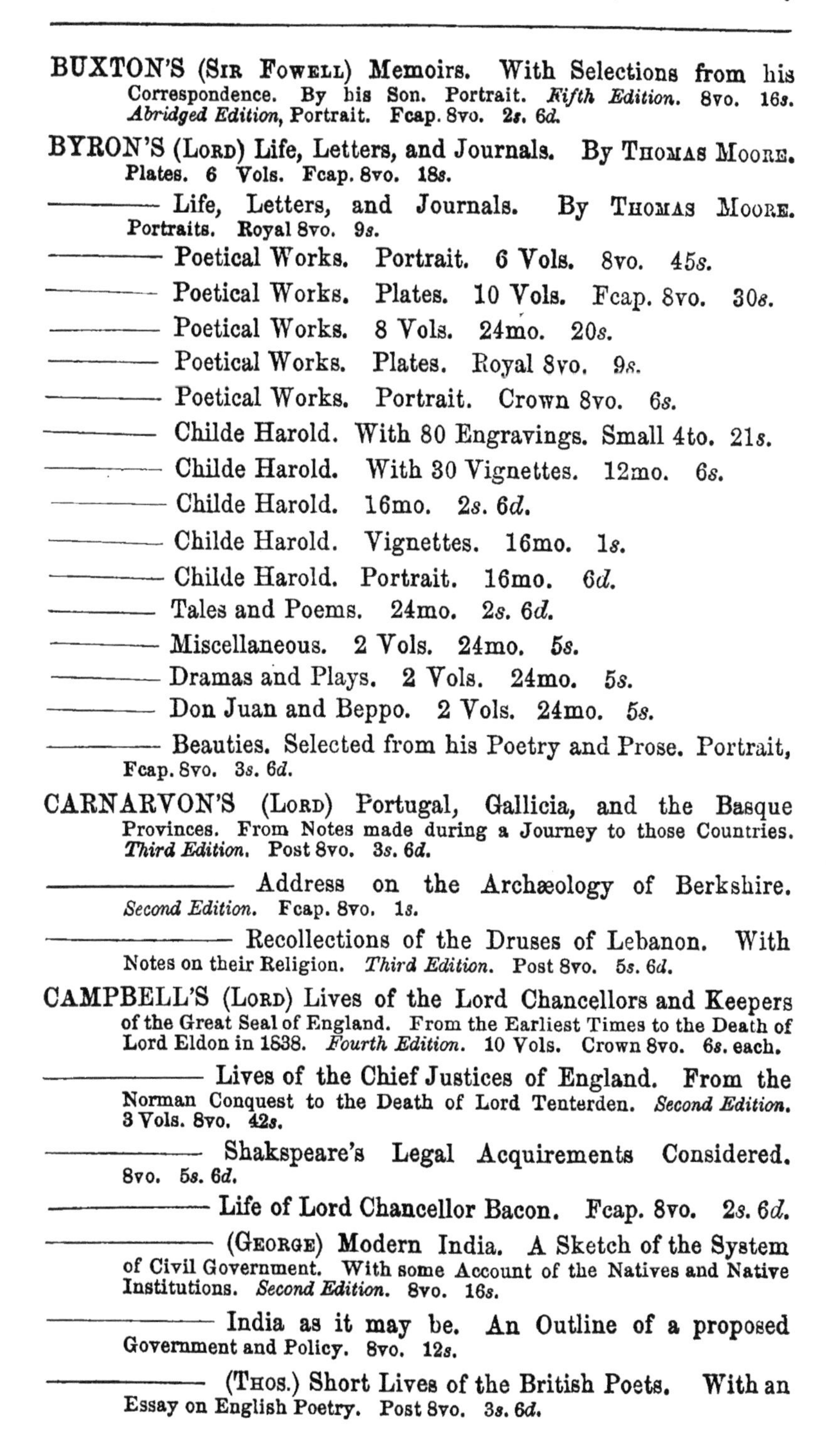

BUXTON'S (SIR FOWELL) Memoirs. With Selections from his Correspondence. By his Son. Portrait. *Fifth Edition.* 8vo. 16*s.* *Abridged Edition,* Portrait. Fcap. 8vo. 2*s.* 6*d.*

BYRON'S (LORD) Life, Letters, and Journals. By THOMAS MOORE. Plates. 6 Vols. Fcap. 8vo. 18*s.*

——— Life, Letters, and Journals. By THOMAS MOORE. Portraits. Royal 8vo. 9*s.*

——— Poetical Works. Portrait. 6 Vols. 8vo. 45*s.*

——— Poetical Works. Plates. 10 Vols. Fcap. 8vo. 30*s.*

——— Poetical Works. 8 Vols. 24mo. 20*s.*

——— Poetical Works. Plates. Royal 8vo. 9*s.*

——— Poetical Works. Portrait. Crown 8vo. 6*s.*

——— Childe Harold. With 80 Engravings. Small 4to. 21*s.*

——— Childe Harold. With 30 Vignettes. 12mo. 6*s.*

——— Childe Harold. 16mo. 2*s.* 6*d.*

——— Childe Harold. Vignettes. 16mo. 1*s.*

——— Childe Harold. Portrait. 16mo. 6*d.*

——— Tales and Poems. 24mo. 2*s.* 6*d.*

——— Miscellaneous. 2 Vols. 24mo. 5*s.*

——— Dramas and Plays. 2 Vols. 24mo. 5*s.*

——— Don Juan and Beppo. 2 Vols. 24mo. 5*s.*

——— Beauties. Selected from his Poetry and Prose. Portrait, Fcap. 8vo. 3*s.* 6*d.*

CARNARVON'S (LORD) Portugal, Gallicia, and the Basque Provinces. From Notes made during a Journey to those Countries. *Third Edition.* Post 8vo. 3*s.* 6*d.*

——— Address on the Archæology of Berkshire. *Second Edition.* Fcap. 8vo. 1*s.*

——— Recollections of the Druses of Lebanon. With Notes on their Religion. *Third Edition.* Post 8vo. 5*s.* 6*d.*

CAMPBELL'S (LORD) Lives of the Lord Chancellors and Keepers of the Great Seal of England. From the Earliest Times to the Death of Lord Eldon in 1838. *Fourth Edition.* 10 Vols. Crown 8vo. 6*s.* each.

——— Lives of the Chief Justices of England. From the Norman Conquest to the Death of Lord Tenterden. *Second Edition.* 3 Vols. 8vo. 42*s.*

——— Shakspeare's Legal Acquirements Considered. 8vo. 5*s.* 6*d.*

——— Life of Lord Chancellor Bacon. Fcap. 8vo. 2*s.* 6*d.*

——— (GEORGE) Modern India. A Sketch of the System of Civil Government. With some Account of the Natives and Native Institutions. *Second Edition.* 8vo. 16*s.*

——— India as it may be. An Outline of a proposed Government and Policy. 8vo. 12*s.*

——— (THOS.) Short Lives of the British Poets. With an Essay on English Poetry. Post 8vo. 3*s.* 6*d.*

CALVIN'S (JOHN) Life. With Extracts from his Correspondence. By THOMAS H. DYER. Portrait. 8vo. 15*s.*

CALLCOTT'S (LADY) Little Arthur's History of England. 100*th Thousand.* With 20 Woodcuts. Fcap. 8vo. 2*s.* 6*d.*

CARMICHAEL'S (A. N.) Greek Verbs. Their Formations, Irregularities, and Defects. *Second Edition.* Post 8vo. 8*s.* 6*d.*

CASTLEREAGH (THE) DESPATCHES, from the commencement of the official career of the late Viscount Castlereagh to the close of his life. Edited by the MARQUIS OF LONDONDERRY. 12 Vols. 8vo. 14*s.* each.

CATHCART'S (SIR GEORGE) Commentaries on the War in Russia and Germany, 1812-13. Plans. 8vo. 14*s.*

———— Military Operations in Kaffraria, which led to the Termination of the Kaffir War. *Second Edition.* 8vo. 12*s.*

CAVALCASELLE (G. B.). Notices of the Early Flemish Painters; Their Lives and Works. Woodcuts. Post 8vo. 12*s.*

CHAMBERS' (G. F.) Handbook of Descriptive and Practical Astronomy. Illustrations. Post 8vo. 12*s.*

CHANTREY (SIR FRANCIS). Winged Words on Chantrey's Woodcocks. Edited by JAS. P. MUIRHEAD. Etchings. Square 8vo. 10*s.* 6*d.*

CHARMED ROE (THE); or, The Story of the Little Brother and Sister. By OTTO SPECKTER. Plates. 16mo. 5*s.*

CHURTON'S (ARCHDEACON) Gongora. An Historical Essay on the Age of Philip III. and IV. of Spain. With Translations. Portrait. 2 Vols. Small 8vo. 15*s.*

CLAUSEWITZ'S (CARL VON) Campaign of 1812, in Russia. Translated from the German by LORD ELLESMERE. Map. 8vo. 10*s.* 6*d.*

CLIVE'S (LORD) Life. By REV. G. R. GLEIG, M.A. Post 8vo. 3*s.* 6*d.*

COBBOLD'S (REV. R. H.) Pictures of the Chinese drawn by themselves. With 24 Plates. Crown 8vo. 9*s.*

COLCHESTER (THE) PAPERS. The Diary and Correspondence of Charles Abbott, Lord Colchester, Speaker of the House of Commons, 1802-1817. Edited by HIS SON. Portrait. 3 Vols. 8vo. 42*s.*

COLERIDGE'S (SAMUEL TAYLOR) Table-Talk. *Fourth Edition.* Portrait. Fcap. 8vo. 6*s.*

———— (HENRY NELSON) Introductions to the Greek Classic Poets. *Third Edition.* Fcap. 8vo. 5*s.* 6*d.*

———— (SIR JOHN) on Public School Education, with especial reference to Eton. *Third Edition.* Fcap. 8vo. 2*s.*

COLONIAL LIBRARY. [See Home and Colonial Library.]

COOKERY (MODERN DOMESTIC). Founded on Principles of Economy and Practical Knowledge, and adapted for Private Families. By a Lady *New Edition.* Woodcuts. Fcap. 8vo. 5*s.*

CORNWALLIS (THE) Papers and Correspondence during the American War,—Administrations in India,—Union with Ireland, and Peace of Amiens. Edited by CHARLES ROSS. *Second Edition.* 3 Vols. 8vo. 63*s.*

CRABBE'S (REV. GEORGE) Life, Letters, and Journals. By his SON. Portrait. Fcap. 8vo. 3*s.*

———— Poetical Works. With his Life. Plates. 8 Vols. Fcap. 8vo. 24*s.*

———— Life and Poetical Works. Plates. Royal 8vo. 7*s.*

CROKER'S (J. W.) Progressive Geography for Children. *Fifth Edition.* 18mo. 1*s.* 6*d.*

——— Stories for Children, Selected from the History of England. *Fifteenth Edition.* Woodcuts. 16mo. 2*s.* 6*d.*

——— Boswell's Life of Johnson. Including the Tour to the Hebrides. Portraits. Royal 8vo. 10*s.*

——— Lord Hervey's Memoirs of the Reign of George the Second, from his Accession to the death of Queen Caroline. Edited with Notes. *Second Edition.* Portrait. 2 Vols. 8vo. 21*s.*

——— Essays on the Early Period of the French Revolution. 8vo. 15*s.*

——— Historical Essay on the Guillotine. Fcap. 8vo. 1*s.*

CROMWELL (Oliver) and John Bunyan. By Robert Southey. Post 8vo. 2*s.*

CROWE'S (J. A.) Notices of the Early Flemish Painters; their Lives and Works. Woodcuts. Post 8vo. 12*s.*

CUNNINGHAM'S (Allan) Life of Sir David Wilkie. With his Journals and Critical Remarks on Works of Art. Portrait. 3 Vols. 8vo. 42*s.*

——— Poems and Songs. Now first collected and arranged, with Biographical Notice. 24mo. 2*s.* 6*d.*

——— (Capt. J. D.) History of the Sikhs. From the Origin of the Nation to the Battle of the Sutlej. *Second Edition.* Maps. 8vo. 15*s.*

CURETON (Rev. W.) Remains of a very Ancient Recension of the Four Gospels in Syriac, hitherto unknown in Europe. Discovered, Edited, and Translated. 4to. 24*s.*

CURTIUS' (Professor) Student's Greek Grammar, for the use of Colleges and the Upper Forms. Translated from the German. Edited by Dr. Wm. Smith. Post 8vo.

——— Smaller Greek Grammar, abridged from the above, 12mo.

CURZON'S (Hon. Robert) Visits to the Monasteries of the Levant. *Fourth Edition.* Woodcuts. Post 8vo. 15*s.*

——— Armenia and Erzeroum. A Year on the Frontiers of Russia, Turkey, and Persia. *Third Edition.* Woodcuts. Post 8vo. 7*s.* 6*d.*

CUST'S (General) Annals of the Wars of the Nineteenth Century —1800-15. 4 Vols. Fcap. 8vo. 5*s.* each.

——— Annals of the Wars of the Eighteenth Century. 5 Vols. Fcap. 8vo. 5*s.* each.

DARWIN'S (Charles) Journal of Researches into the Natural History and Geology of the Countries visited during a Voyage round the World. *Tenth Thousand.* Post 8vo. 9*s.*

——— Origin of Species by Means of Natural Selection; or, the Preservation of Favoured Races in the Struggle for Life. *Seventh Thousand.* Post 8vo. 14*s.*

——— Various Contrivances by which Orchids are Fertilised through Insect Agency, and as to the good of Intercrossing. Woodcuts. Post 8vo. 9*s.*

DAVIS' (Nathan) Ruined Cities within Numidian and Carthaginian Territories. Map and Illustrations. 8vo. 16*s.*

DAVY'S (Sir Humphry) Consolations in Travel; or, Last Days of a Philosopher. *Fifth Edition.* Woodcuts. Fcap. 8vo. 6*s.*

——— Salmonia; or, Days of Fly Fishing. With some Account of the Habits of Fishes belonging to the genus Salmo. *Fourth Edition.* Woodcuts. Fcap. 8vo. 6*s.*

DELEPIERRE'S (OCTAVE) History of Flemish Literature and its celebrated Authors. From the Twelfth Century to the present Day. 8vo. *9s.*

DENNIS' (GEORGE) Cities and Cemeteries of Etruria. Plates. 2 Vols. 8vo. *42s.*

DIXON'S (HEPWORTH) Story of the Life of Lord Bacon. Portrait. Fcap. 8vo. *7s. 6d.*

DOG-BREAKING; the Most Expeditious, Certain, and Easy Method, whether great excellence or only mediocrity be required. By LIEUT.-COL. HUTCHINSON. *Third Edition.* Woodcuts. Post 8vo. *9s.*

DOMESTIC MODERN COOKERY. Founded on Principles of Economy and Practical Knowledge, and adapted for Private Families. *New Edition.* Woodcuts. Fcap. 8vo. *5s.*

DOUGLAS'S (GENERAL SIR HOWARD) Treatise on the Theory and Practice of Gunnery. *Fifth Edition.* Plates. 8vo. *21s.*

———— Treatise on Military Bridges, and the Passages of Rivers in Military Operations. *Third Edition.* Plates. 8vo. *21s.*

———— Naval Warfare with Steam.. *Second Edition.* 8vo. *8s. 6d.*

———— Modern Systems of Fortification, with special reference to the Naval, Littoral, and Internal Defence of England. Plans. 8vo. *12s.*

———— Life and Adventures; from his Notes, Conversations, and Correspondence. By S. W. FULLOM. Portrait. 8vo.

DRAKE'S (SIR FRANCIS) Life, Voyages, and Exploits, by Sea and Land. By JOHN BARROW. *Third Edition.* Post 8vo. *2s.*

DRINKWATER'S (JOHN) History of the Siege of Gibraltar, 1779-1783. With a Description and Account of that Garrison from the Earliest Periods. Post 8vo. *2s.*

DU CHAILLU'S (PAUL B.) EQUATORIAL AFRICA, with Accounts of the Manners and Customs of the People, and of the Chase of the Gorilla, the Nest-building Ape, Chimpanzee, Crocodile, &c. *Tenth Thousand.* Illustrations. 8vo. *21s.*

DUDLEY'S (EARL OF) Letters to the late Bishop of Llandaff. *Second Edition.* Portrait. 8vo. *10s. 6d.*

DUFFERIN'S (LORD) Letters from High Latitudes, being some Account of a Yacht Voyage to Iceland, &c., in 1856. *Fourth Edition.* Woodcuts. Post 8vo. *9s.*

DURHAM'S (ADMIRAL SIR PHILIP) Naval Life and Services. By CAPT. ALEXANDER MURRAY. 8vo. *5s. 6d.*

DYER'S (THOMAS H.) Life and Letters of John Calvin. Compiled from authentic Sources. Portrait. 8vo. *15s.*

———— History of Modern Europe, from the taking of Constantinople by the Turks to the close of the War in the Crimea. Vols. 1 & 2. 8vo. *30s.*

EASTLAKE'S (SIR CHARLES) Italian Schools of Painting. From the German of KUGLER. Edited, with Notes. *Third Edition.* Illustrated from the Old Masters. 2 Vols. Post 8vo. *30s.*

EASTWICK'S (E. B.) Handbook for Bombay and Madras, with Directions for Travellers, Officers, &c. Map. 2 Vols. Post 8vo. *24s.*

EDWARDS' (W. H.) Voyage up the River Amazon, including a Visit to Para. Post 8vo. *2s.*

EGERTON'S (HON. CAPT. FRANCIS) Journal of a Winter's Tour in India; with a Visit to Nepaul. Woodcuts. 2 Vols. Post 8vo. *18s.*

ELDON'S (LORD) Public and Private Life, with Selections from his Correspondence and Diaries. By HORACE TWISS. *Third Edition.* Portrait. 2 Vols. Post 8vo. 21*s.*

ELIOT'S (HON. W. G. C.) Khans of the Crimea. Being a Narrative of an Embassy from Frederick the Great to the Court of Krim Gerai. Translated from the German. Post 8vo. 6*s.*

ELLIS (REV. W.) Visits to Madagascar, including a Journey to the Capital, with notices of Natural History, and Present Civilisation of the People. *Fifth Thousand.* Map and Woodcuts. 8vo. 16*s.*

——— (MRS.) Education of Character, with Hints on Moral Training. Post 8vo. 7*s.* 6*d.*

ELLESMERE'S (LORD) Two Sieges of Vienna by the Turks. Translated from the German. Post 8vo. 2*s.*

——— Second Campaign of Radetzky in Piedmont. The Defence of Temeswar and the Camp of the Ban. From the German. Post 8vo. 6*s.* 6*d.*

——— Campaign of 1812 in Russia, from the German of General Carl Von Clausewitz. Map. 8vo. 10*s.* 6*d.*

——— Pilgrimage, and other Poems. Crown 4to. 24*s.*

——— Essays on History, Biography, Geography, and Engineering. 8vo. 12*s.*

ELPHINSTONE'S (HON. MOUNTSTUART) History of India—the Hindoo and Mahomedan Periods. *Fourth Edition.* Map. 8vo. 18*s.*

ENGLAND (HISTORY OF) from the Peace of Utrecht to the Peace of Versailles, 1713–83. By LORD MAHON. *Library Edition,* 7 Vols. 8vo. 93*s.*; or *Popular Edition,* 7 Vols. Post 8vo. 35*s.*

——— From the First Invasion by the Romans, down to the 14th year of Queen Victoria's Reign. By MRS. MARKHAM. 118*th Edition.* Woodcuts. 12mo. 6*s.*

——— Social, Political, and Industrial, in the 19th Century. By W. JOHNSTON. 2 Vols. Post 8vo. 18*s.*

ENGLISHWOMAN IN AMERICA. Post 8vo. 10*s.* 6*d.*

——— RUSSIA. Woodcuts. Post 8vo. 10*s.* 6*d.*

EOTHEN; or, Traces of Travel brought Home from the East. *A New Edition.* Post 8vo. 7*s.* 6*d.*

ERSKINE'S (ADMIRAL) Journal of a Cruise among the Islands of the Western Pacific, including the Fejees, and others inhabited by the Polynesian Negro Races. Plates. 8vo. 16*s.*

ESKIMAUX and English Vocabulary, for Travellers in the Arctic Regions. 16mo. 3*s.* 6*d.*

ESSAYS FROM "THE TIMES." Being a Selection from the LITERARY PAPERS which have appeared in that Journal. *Seventh Thousand.* 2 vols. Fcap. 8vo. 8*s.*

EXETER'S (BISHOP OF) Letters to the late Charles Butler, on the Theological parts of his Book of the Roman Catholic Church; with Remarks on certain Works of Dr. Milner and Dr. Lingard, and on some parts of the Evidence of Dr. Doyle. *Second Edition.* 8vo. 16*s.*

FAIRY RING; A Collection of TALES and STORIES. From the German. By J. E. TAYLOR. Illustrated by RICHARD DOYLE. *Second Edition.* Fcap. 8vo.

FALKNER'S (FRED.) Muck Manual for the Use of Farmers. A Treatise on the Nature and Value of Manures. *Second Edition.* Fcap. 8vo. 5*s.*

FAMILY RECEIPT-BOOK. A Collection of a Thousand Valuable and Useful Receipts. Fcap. 8vo. *5s. 6d.*

FANCOURT'S (Col.) History of Yucatan, from its Discovery to the Close of the 17th Century. With Map. 8vo. *10s. 6d.*

FARRAR'S (Rev. A. S.) Science in Theology. Sermons Preached before the University of Oxford. 8vo. *9s.*

——— Bampton Lectures, 1862. History of Free Thought in reference to the Christian Religion. 8vo.

——— (F. W.) Origin of Language, based on Modern Researches. Fcap. 8vo. *5s.*

FEATHERSTONHAUGH'S (G. W.) Tour through the Slave States of North America, from the River Potomac to Texas and the Frontiers of Mexico. Plates. 2 Vols. 8vo. *26s.*

FELLOWS' (Sir Charles) Travels and Researches in Asia Minor, more particularly in the Province of Lycia. *New Edition.* Plates. Post 8vo. *9s.*

FERGUSSON'S (James) Palaces of Nineveh and Persepolis Restored: an Essay on Ancient Assyrian and Persian Architecture. Woodcuts. 8vo. *16s.*

——— History of Architecture. Being a Concise and Popular Account of the Different Styles prevailing in all Ages and Countries in the World. With a Description of the most remarkable Buildings. With 850 Illustrations. 8vo. *26s.*

——— Modern Styles of Architecture. With 30 Illustrations. 8vo. *31s. 6d.*

FERRIER'S (T. P.) Caravan Journeys in Persia, Afghanistan, Herat, Turkistan, and Beloochistan, with Descriptions of Meshed, Balk, and Candahar, &c. *Second Edition.* Map. 8vo. *21s.*

——— History of the Afghans. Map. 8vo. *21s.*

FISHER'S (Rev. George) Elements of Geometry, for the Use of Schools. *Fifth Edition.* 18mo. *1s. 6d.*

——— First Principles of Algebra, for the Use of Schools. *Fifth Edition.* 18mo. *1s. 6d.*

FLOWER GARDEN (The). An Essay. By Rev. Thos. James. Reprinted from the "Quarterly Review." Fcap. 8vo. *1s.*

FORBES' (C. S.) Iceland; its Volcanoes, Geysers, and Glaciers. Illustrations. Post 8vo. *14s.*

FORD'S (Richard) Handbook for Spain, Andalusia, Ronda, Valencia, Catalonia, Granada, Gallicia, Arragon, Navarre, &c. *Third Edition.* 2 Vols. Post 8vo. *30s.*

——— Gatherings from Spain. Post 8vo. *3s. 6d.*

FORSTER'S (John) Arrest of the Five Members by Charles the First. A Chapter of English History re-written. Post 8vo. *12s.*

——— Debates on the Grand Remonstrance, 1641. With an Introductory Essay on English freedom under the Plantagenet and Tudor Sovereigns. *Second Edition.* Post 8vo. *12s.*

——— Oliver Cromwell, Daniel De Foe, Sir Richard Steele, Charles Churchill, Samuel Foote. Biographical Essays. *Third Edition.* Post 8vo. *12s.*

FORSYTH'S (William) Hortensius, or the Advocate: an Historical Essay on the Office and Duties of an Advocate. Post 8vo. *12s.*

——— History of Napoleon at St. Helena. From the Letters and Journals of Sir Hudson Lowe. Portrait and Maps. 3 Vols. 8vo. *45s.*

FORTUNE'S (ROBERT) Narrative of Two Visits to the Tea Countries of China, between the years 1843-52, with full Descriptions of the Tea Plant. *Third Edition.* Woodcuts. 2 Vols. Post 8vo. 18*s.*

———— Chinese, Inland, on the Coast, and at Sea. A Narrative of a Third Visit in 1853-56. Woodcuts. 8vo. 16*s.*

FRANCE (HISTORY OF). From the Conquest by the Gauls to the Death of Louis Philippe. By Mrs. MARKHAM. *56th Thousand.* Woodcuts. 12mo. 6*s.*

FRENCH (THE) in Algiers; The Soldier of the Foreign Legion—and the Prisoners of Abd-el-Kadir. Translated by Lady DUFF GORDON. Post 8vo. 2*s.*

GALTON'S (FRANCIS) Art of Travel; or, Hints on the Shifts and Contrivances available in Wild Countries. *Third Edition.* Woodcuts. Post 8vo. 7*s.* 6*d.*

GEOGRAPHICAL (THE) Journal. Published by the Royal Geographical Society of London. 8vo.

GERMANY (HISTORY OF). From the Invasion by Marius, to the present time. By Mrs. MARKHAM. *Fifteenth Thousand.* Woodcuts. 12mo. 6*s.*

GIBBON'S (EDWARD) History of the Decline and Fall of the Roman Empire. *A New Edition.* Preceded by his Autobiography. Edited, with Notes, by Dr. WM. SMITH. Maps. 8 Vols. 8vo. 60*s.*

———— (The Student's Gibbon); Being an Epitome of the above work, incorporating the Researches of Recent Commentators. By Dr. WM. SMITH. *Ninth Thousand.* Woodcuts. Post 8vo. 7*s.* 6*d.*

GIFFARD'S (EDWARD) Deeds of Naval Daring; or, Anecdotes of the British Navy. New Edition. Fcap. 8vo.

GOLDSMITH'S (OLIVER) Works. A New Edition. Printed from the last editions revised by the Author. Edited by PETER CUNNINGHAM. Vignettes. 4 Vols. 8vo. 30*s.* (Murray's British Classics.)

GLEIG'S (REV. G. R.) Campaigns of the British Army at Washington and New Orleans. Post 8vo. 2*s.*

———— Story of the Battle of Waterloo. Compiled from Public and Authentic Sources. Post 8vo. 3*s.* 6*d.*

———— Narrative of Sir Robert Sale's Brigade in Affghanistan, with an Account of the Seizure and Defence of Jellalabad. Post 8vo. 2*s.*

———— Life of Robert Lord Clive. Post 8vo. 3*s.* 6*d.*

———— Life and Letters of General Sir Thomas Munro. Post 8vo. 3*s.* 6*d.*

GORDON'S (SIR ALEX. DUFF) Sketches of German Life, and Scenes from the War of Liberation. From the German. Post 8vo. 3*s.* 6*d.*

———— (LADY DUFF) Amber-Witch: the most interesting Trial for Witchcraft ever known. From the German. Post 8vo. 2*s.*

———— French in Algiers. 1. The Soldier of the Foreign Legion. 2. The Prisoners of Abd-el-Kadir. From the French. Post 8vo. 2*s.*

GOUGER'S (HENRY) Personal Narrative of Two Years' Imprisonment in Burmah. *Second Edition.* Woodcuts. Post 8vo. 12*s.*

GRANT'S (ASAHEL) Nestorians, or the Lost Tribes; containing Evidence of their Identity, their Manners, Customs, and Ceremonies; with Sketches of Travel in Ancient Assyria, Armenia, and Mesopotamia; and Illustrations of Scripture Prophecy. *Third Edition.* Fcap. 8vo. 6*s.*

GRENVILLE (THE) PAPERS. Being the Public and Private Correspondence of George Grenville, including his PRIVATE DIARY. Edited by W. J. SMITH. 4 Vols. 8vo. 16*s.* each.

GREEK GRAMMAR FOR SCHOOLS. Abridged from Matthiæ. By the BISHOP OF LONDON. *Ninth Edition,* revised by Rev. J. EDWARDS. 12mo. 3*s.*

GREY'S (SIR GEORGE) Polynesian Mythology, and Ancient Traditional History of the New Zealand Race. Woodcuts. Post 8vo. 10*s.* 6*d.*

GROTE'S (GEORGE) History of Greece. From the Earliest Times to the close of the generation contemporary with the death of Alexander the Great. *Fourth Edition.* Portrait and Maps. 8 vols. 8vo. 112*s.*

——— (MRS.) Memoir of the Life of the late Ary Scheffer. *Second Edition.* Portrait. Post 8vo. 8*s.* 6*d.*

——— Collected Papers in Prose and Verse (Original and Reprinted.) 8vo. 10*s.* 6*d.*

HALLAM'S (HENRY) Constitutional History of England, from the Accession of Henry the Seventh to the Death of George the Second. *Seventh Edition.* 3 Vols. 8vo. 30*s.*

——— History of Europe during the Middle Ages. *Tenth Edition.* 3 Vols. 8vo. 30*s.*

——— Literary History of Europe, during the 15th, 16th and 17th Centuries. *Fourth Edition.* 3 Vols. 8vo. 36*s.*

——— Literary Essays and Characters. Selected from the last work. Fcap. 8vo. 2*s.*

——— Historical Works. History of England,—Middle Ages of Europe,—Literary History of Europe. 10 Vols. Post 8vo. 6*s.* each.

——— (ARTHUR) Remains; in Verse and Prose. With Preface, Memoir, and Portrait. (*Now first Published.*) Fcap. 8vo. 7*s.* 6*d.*

HAMILTON'S (JAMES) Wanderings in Northern Africa, Benghazi, Cyrene, the Oasis of Siwah, &c. Woodcuts. Post 8vo. 12*s.*

HAMPDEN'S (BISHOP) Philosophical Evidence of Christianity, or the Credibility obtained to a Scripture Revelation from its Coincidence with the Facts of Nature. 8vo. 9*s.* 6*d.*

HARCOURT'S (EDWARD VERNON) Sketch of Madeira; with Map and Plates. Post 8vo. 8*s.* 6*d.*

HART'S ARMY LIST. (*Quarterly and Annually.*) 8vo. 10*s.* 6*d.* and 21*s.*

HAY'S (J. H. DRUMMOND) Western Barbary, its wild Tribes and savage Animals. Post 8vo. 2*s.*

HEBER'S (BISHOP) Journey through the Upper Provinces of India, From Calcutta to Bombay, with a Journey to Madras and the Southern Provinces. *Twelfth Edition.* 2 Vols. Post 8vo. 7*s.*

——— Poetical Works. *Sixth Edition.* Portrait. Fcap. 8vo. 6*s.*

——— Parish Sermons. *Sixth Edition.* 2 Vols. Post 8vo. 16*s.*

——— Sermons Preached in England. *Second Edition.* 8vo. 9*s.* 6*d.*

——— Hymns for the Weekly Church Service of the Year. *Twelfth Edition.* 16mo. 2*s.*

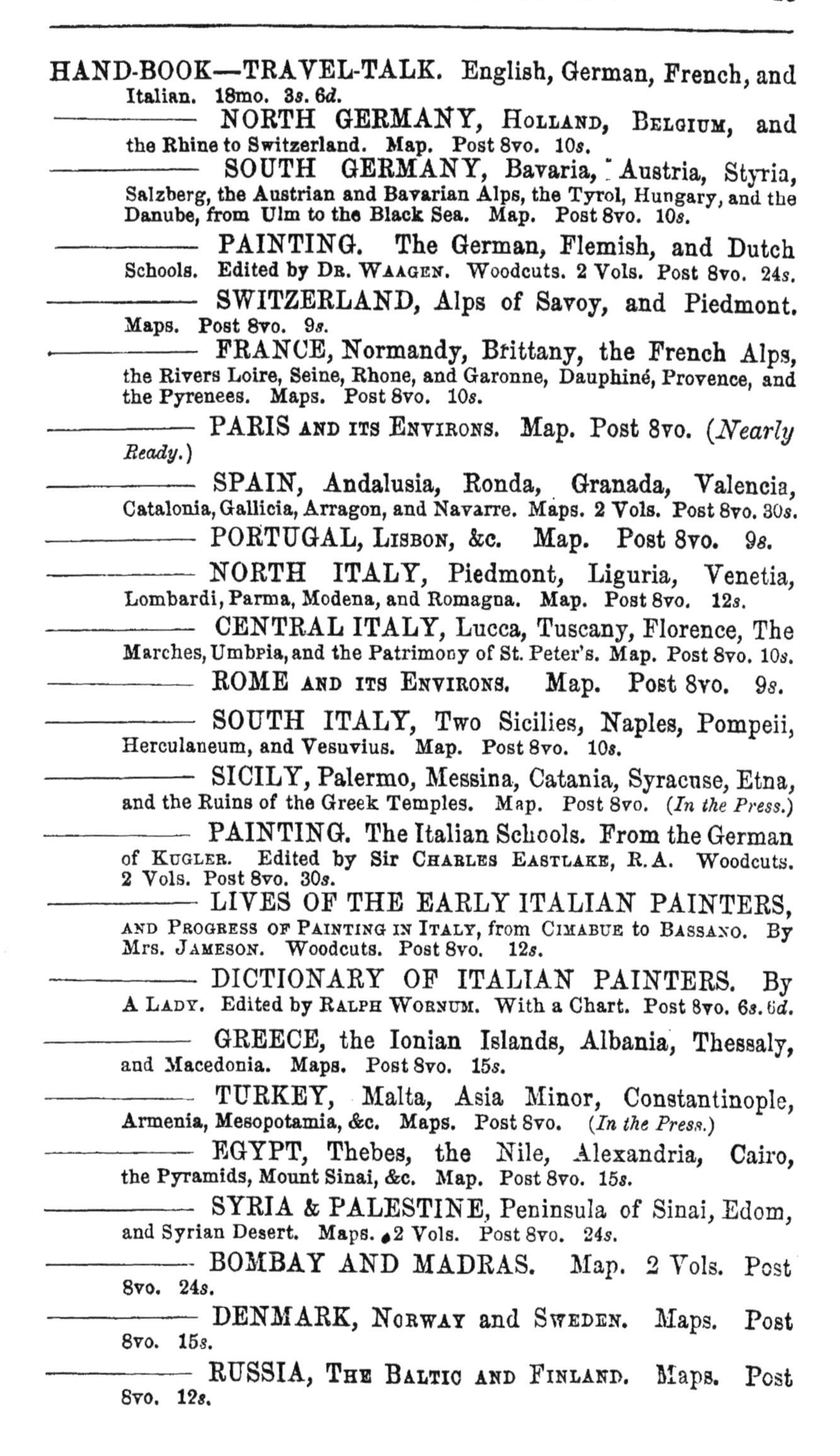

HAND-BOOK—TRAVEL-TALK. English, German, French, and Italian. 18mo. 3*s*. 6*d*.

——— NORTH GERMANY, HOLLAND, BELGIUM, and the Rhine to Switzerland. Map. Post 8vo. 10*s*.

——— SOUTH GERMANY, Bavaria, Austria, Styria, Salzberg, the Austrian and Bavarian Alps, the Tyrol, Hungary, and the Danube, from Ulm to the Black Sea. Map. Post 8vo. 10*s*.

——— PAINTING. The German, Flemish, and Dutch Schools. Edited by DR. WAAGEN. Woodcuts. 2 Vols. Post 8vo. 24*s*.

——— SWITZERLAND, Alps of Savoy, and Piedmont. Maps. Post 8vo. 9*s*.

——— FRANCE, Normandy, Brittany, the French Alps, the Rivers Loire, Seine, Rhone, and Garonne, Dauphiné, Provence, and the Pyrenees. Maps. Post 8vo. 10*s*.

——— PARIS AND ITS ENVIRONS. Map. Post 8vo. (*Nearly Ready.*)

——— SPAIN, Andalusia, Ronda, Granada, Valencia, Catalonia, Gallicia, Arragon, and Navarre. Maps. 2 Vols. Post 8vo. 30*s*.

——— PORTUGAL, LISBON, &c. Map. Post 8vo. 9*s*.

——— NORTH ITALY, Piedmont, Liguria, Venetia, Lombardi, Parma, Modena, and Romagna. Map. Post 8vo. 12*s*.

——— CENTRAL ITALY, Lucca, Tuscany, Florence, The Marches, Umbria, and the Patrimony of St. Peter's. Map. Post 8vo. 10*s*.

——— ROME AND ITS ENVIRONS. Map. Post 8vo. 9*s*.

——— SOUTH ITALY, Two Sicilies, Naples, Pompeii, Herculaneum, and Vesuvius. Map. Post 8vo. 10*s*.

——— SICILY, Palermo, Messina, Catania, Syracuse, Etna, and the Ruins of the Greek Temples. Map. Post 8vo. (*In the Press.*)

——— PAINTING. The Italian Schools. From the German of KUGLER. Edited by Sir CHARLES EASTLAKE, R.A. Woodcuts. 2 Vols. Post 8vo. 30*s*.

——— LIVES OF THE EARLY ITALIAN PAINTERS, AND PROGRESS OF PAINTING IN ITALY, from CIMABUE to BASSANO. By Mrs. JAMESON. Woodcuts. Post 8vo. 12*s*.

——— DICTIONARY OF ITALIAN PAINTERS. By A LADY. Edited by RALPH WORNUM. With a Chart. Post 8vo. 6*s*. 6*d*.

——— GREECE, the Ionian Islands, Albania, Thessaly, and Macedonia. Maps. Post 8vo. 15*s*.

——— TURKEY, Malta, Asia Minor, Constantinople, Armenia, Mesopotamia, &c. Maps. Post 8vo. (*In the Press.*)

——— EGYPT, Thebes, the Nile, Alexandria, Cairo, the Pyramids, Mount Sinai, &c. Map. Post 8vo. 15*s*.

——— SYRIA & PALESTINE, Peninsula of Sinai, Edom, and Syrian Desert. Maps. 2 Vols. Post 8vo. 24*s*.

——— BOMBAY AND MADRAS. Map. 2 Vols. Post 8vo. 24*s*.

——— DENMARK, NORWAY and SWEDEN. Maps. Post 8vo. 15*s*.

——— RUSSIA, THE BALTIC AND FINLAND. Maps. Post 8vo. 12*s*.

HAND-BOOK—KENT AND SUSSEX, Canterbury, Dover, Ramsgate, Sheerness, Rochester, Chatham, Woolwich, Brighton, Chichester, Worthing, Hastings, Lewes, Arundel, &c. Map. Post 8vo. 10*s.*

——— SURREY, HANTS, Kingston, Croydon, Reigate, Guildford, Winchester, Southampton, Portsmouth, and Isle of Wight. Maps. Post 8vo. 7*s.* 6*d.*

——— MODERN LONDON. A Complete Guide to all the Sights and Objects of Interest in the Metropolis. Map. 16mo.

——— LONDON, PAST AND PRESENT. *Second Edition.* Post 8vo. 16*s.*

——— WESTMINSTER ABBEY. Woodcuts. 16mo. 1*s.*

——— ENVIRONS OF LONDON. Maps. Post 8vo. (*In preparation.*)

——— BERKS, BUCKS, AND OXON, Windsor, Eton, Reading, Aylesbury, Uxbridge, Wycombe, Henley, the City and University of Oxford, and the Descent of the Thames to Maidenhead and Windsor. Map. Post 8vo. 7*s.* 6*d.*

——— WILTS, DORSET, AND SOMERSET, Salisbury, Chippenham, Weymouth, Sherborne, Wells, Bath, Bristol, Taunton, &c. Map. Post 8vo. 7*s.* 6*d.*

——— DEVON AND CORNWALL, Exeter, Ilfracombe, Linton, Sidmouth, Dawlish, Teignmouth, Plymouth, Devonport, Torquay, Launceston, Truro, Penzance, Falmouth, &c. Maps. Post 8vo. 7*s.* 6*d.*

——— CATHEDRALS OF ENGLAND—Southern Division, Winchester, Salisbury, Exeter, Wells, Chichester, Rochester, Canterbury. With 110 Illustrations. 2 Vols. Crown 8vo. 24*s.*

——— CATHEDRALS OF ENGLAND—Eastern Division, Oxford, Peterborough, Norwich, Ely, and Lincoln. With 90 Illustrations. Crown 8vo. 18*s.*

——— NORTH AND SOUTH WALES, Bangor, Carnarvon, Beaumaris, Snowdon, Conway, Menai Straits, Carmarthen, Pembroke, Tenby, Swansea, The Wye, &c. Maps. 2 Vols. Post 8vo. 12*s.*

——— FAMILIAR QUOTATIONS. From English Authors. *Third Edition.* Fcap. 8vo. 5*s.*

HEAD'S (SIR FRANCIS) Horse and his Rider. Woodcuts. Post 8vo. 5*s.*

——— Rapid Journeys across the Pampas and over the Andes. Post 8vo. 2*s.*

——— Descriptive Essays. 2 Vols. Post 8vo. 18*s.*

——— Bubbles from the Brunnen of Nassau. By an OLD MAN. 16mo. 5*s.*

——— Emigrant. Fcap. 8vo. 2*s.* 6*d.*

——— Stokers and Pokers; or, the North-Western Railway. Post 8vo. 2*s.*

——— Defenceless State of Great Britain. Post 8vo. 12*s.*

——— Faggot of French Sticks; or, Sketches of Paris. 2 Vols. Post 8vo. 12*s.*

——— Fortnight in Ireland. Map. 8vo. 12*s.*

——— (SIR GEORGE) Forest Scenes and Incidents in Canada. Post 8vo. 10*s.*

——— Home Tour through the Manufacturing Districts of England. 2 Vols. Post 8vo. 12*s.*

——— (SIR EDMUND) Shall and Will; or, Two Chapters on Future Auxiliary Verbs. Fcap. 8vo. 4*s.*

HEIRESS (The) in Her Minority; or, The Progress of Character. By the Author of "Bertha's Journal." 2 Vols. 12mo. 18*s*.

HERODOTUS. A New English Version. Edited with Notes and Essays, historical, ethnographical, and geographical. By Rev. G. Rawlinson, assisted by Sir Henry Rawlinson and Sir J. G. Wilkinson. *Second Edition*. Maps and Woodcuts. 4 Vols. 8vo. 48*s*.

HERVEY'S (Lord) Memoirs of the Reign of George the Second, from his Accession to the Death of Queen Caroline. Edited, with Notes, by Mr. Croker. *Second Edition*. Portrait. 2 Vols. 8vo. 21*s*.

HESSEY (Rev. Dr.). Sunday—Its Origin, History, and Present Obligations. Being the Bampton Lectures for 1860. *Second Edition*. 8vo. 16*s*.

HICKMAN'S (Wm.) Treatise on the Law and Practice of Naval Courts-Martial. 8vo. 10*s*. 6*d*.

HILLARD'S (G. S.) Six Months in Italy. 2 Vols. Post 8vo. 16*s*.

HOLLWAY'S (J. G.) Month in Norway. Fcap. 8vo. 2*s*.

HONEY BEE (The). An Essay. By Rev. Thomas James. Reprinted from the "Quarterly Review." Fcap. 8vo. 1*s*.

HOOK'S (Dean) Church Dictionary. *Eighth Edition*. 8vo. 16*s*.

——— Discourses on the Religious Controversies of the Day. 8vo. 9*s*.

——— (Theodore) Life. By J. G. Lockhart. Reprinted from the "Quarterly Review." Fcap. 8vo. 1*s*.

HOOKER'S (Dr. J. D.) Himalayan Journals; or, Notes of an Oriental Naturalist in Bengal, the Sikkim and Nepal Himalayas, the Khasia Mountains, &c. *Second Edition*. Woodcuts. 2 Vols. Post 8vo.

HOOPER'S (Lieut.) Ten Months among the Tents of the Tuski; with Incidents of an Arctic Boat Expedition in Search of Sir John Franklin. Plates. 8vo. 14*s*.

HOPE'S (A. J. Beresford) English Cathedral of the Nineteenth Century. With Illustrations. 8vo. 12*s*.

HORACE (Works of). Edited by Dean Milman. With 300 Woodcuts. Crown 8vo. 21*s*.

——— (Life of). By Dean Milman. Woodcuts, and coloured Borders. 8vo. 9*s*.

HOSPITALS AND SISTERHOODS. By a Lady. Fcap. 8vo. 3*s*. 6*d*.

HUME'S (David) History of England, from the Invasion of Julius Cæsar to the Revolution of 1688. Abridged for Students. Correcting his errors, and continued to 1858. *Twenty-fifth Thousand*. Woodcuts. Post 8vo. 7*s*. 6*d*.

HUTCHINSON (Col.) on the most expeditious, certain, and easy Method of Dog-Breaking. *Third Edition*. Woodcuts. Post 8vo. 9*s*.

HUTTON'S (H. E.) Principia Græca; an Introduction to the Study of Greek. Comprehending Grammar, Delectus and Exercise-book, with Vocabularies. *Second Edition*. 12mo. 3*s*.

c

HOME AND COLONIAL LIBRARY. A Series of Works adapted for all circles and classes of Readers, having been selected for their acknowledged interest and ability of the Authors. Post 8vo. Published at 2*s.* and 3*s.* 6*d.* each, and arranged under two distinctive heads as follows :—

CLASS A.

HISTORY, BIOGRAPHY, AND HISTORIC TALES.

1. SIEGE OF GIBRALTAR. By JOHN DRINKWATER. 2*s.*
2. THE AMBER-WITCH. By LADY DUFF GORDON. 2*s.*
3. CROMWELL AND BUNYAN. By ROBERT SOUTHEY. 2*s.*
4. LIFE OF SIR FRANCIS DRAKE. By JOHN BARROW,
5. CAMPAIGNS AT WASHINGTON. By REV. G. R. GLEIG. 2*s.*
6. THE FRENCH IN ALGIERS. By LADY DUFF GORDON. 2*s.*
7. THE FALL OF THE JESUITS. 2*s.*
8. LIVONIAN TALES. 2*s.*
9. LIFE OF CONDE. By LORD MAHON. 3*s.* 6*d.*
10. SALE'S BRIGADE. By REV. G. R. GLEIG. 2*s.*
11. THE SIEGES OF VIENNA. By LORD ELLESMERE. 2*s.*
12. THE WAYSIDE CROSS. By CAPT. MILMAN. 2*s.*
13. SKETCHES OF GERMAN LIFE. By SIR A. GORDON. 3*s.* 6*d.*
14. THE BATTLE OF WATERLOO. By REV. G. R. GLEIG. 3*s.* 6*d.*
15. AUTOBIOGRAPHY OF STEFFENS. 2*s.*
16. THE BRITISH POETS. By THOMAS CAMPBELL. 3*s.* 6*d.*
17. HISTORICAL ESSAYS. By LORD MAHON. 3*s.* 6*d.*
18. LIFE OF LORD CLIVE. By REV. G. R. GLEIG. 3*s.* 6*d.*
19. NORTH - WESTERN RAILWAY. By SIR F. B. HEAD. 2*s.*
20. LIFE OF MUNRO. By REV. G. R. GLEIG. 3*s.* 6*d.*

CLASS B.

VOYAGES, TRAVELS, AND ADVENTURES.

1. BIBLE IN SPAIN. By GEORGE BORROW. 3*s.* 6*d.*
2. GIPSIES OF SPAIN. By GEORGE BORROW. 3*s.* 6*d.*

3 & 4. JOURNALS IN INDIA. By BISHOP HEBER. 2 Vols. 7*s.*

5. TRAVELS IN THE HOLY LAND. By IRBY and MANGLES. 2*s.*
6. MOROCCO AND THE MOORS. By J. DRUMMOND HAY. 2*s.*
7. LETTERS FROM THE BALTIC. By a LADY. 2*s.*
8. NEW SOUTH WALES. By MRS. MEREDITH. 2*s.*
9. THE WEST INDIES. By M. G. LEWIS. 2*s.*
10. SKETCHES OF PERSIA. By SIR JOHN MALCOLM. 3*s.* 6*d.*
11. MEMOIRS OF FATHER RIPA. 2*s.*

12 & 13. TYPEE AND OMOO. By HERMANN MELVILLE. 2 Vols. 7*s.*

14. MISSIONARY LIFE IN CANADA. By REV. J. ABBOTT. 2*s.*
15. LETTERS FROM MADRAS. By a LADY. 2*s.*
16. HIGHLAND SPORTS. By CHARLES ST. JOHN. 3*s.* 6*d.*
17. PAMPAS JOURNEYS. By SIR F. B. HEAD. 2*s.*
18. GATHERINGS FROM SPAIN. By RICHARD FORD. 3*s.* 6*d.*
19. THE RIVER AMAZON. By W. H. EDWARDS. 2*s.*
20. MANNERS & CUSTOMS OF INDIA. By REV. C. ACLAND. 2*s.*
21. ADVENTURES IN MEXICO. By G. F. RUXTON. 3*s.* 6*d.*
22. PORTUGAL AND GALLICIA. By LORD CARNARVON. 3*s.* 6*d.*
23. BUSH LIFE IN AUSTRALIA. By Rev. H. W. HAYGARTH. 2*s.*
24. THE LIBYAN DESERT. By BAYLE ST. JOHN. 2*s.*
25. SIERRA LEONE. By a LADY. 3*s.* 6*d.*

*** Each work may be had separately.

IRBY AND MANGLES' Travels in Egypt, Nubia, Syria, and the Holy Land. Post 8vo. 2*s.*

JAMES' (REV. THOMAS) Fables of Æsop. A New Translation, with Historical Preface. With 100 Woodcuts by TENNIEL and WOLF. *Thirty-eighth Thousand.* Post 8vo. 2*s.* 6*d.*

JAMESON'S (MRS.) Lives of the Early Italian Painters, from Cimabue to Bassano, and the Progress of Painting in Italy. *New Edition.* With Woodcuts. Post 8vo. 12*s.*

JERVIS'S (CAPT.) Manual of Operations in the Field. Post 8vo. 9*s.* 6*d.*

JESSE'S (EDWARD) Scenes and Occupations of Country Life. *Third Edition.* Woodcuts. Fcap. 8vo. 6*s.*

——— Gleanings in Natural History. *Eighth Edition.* Fcap 8vo. 6*s.*

JOHNSON'S (DR. SAMUEL) Life. By James Boswell. Including the Tour to the Hebrides. Edited by the late MR. CROKER. Portraits. Royal 8vo. 10*s.*

——— Lives of the most eminent English Poets. Edited by PETER CUNNINGHAM. 3 vols. 8vo. 22*s.* 6*d.* (Murray's British Classics.)

JOHNSTON'S (WM.) England: Social, Political, and Industrial, in 19th Century. 2 Vols. Post 8vo. 18*s.*

JOURNAL OF A NATURALIST. *Fourth Edition.* Woodcuts. Post 8vo. 9*s.* 6*d.*

JOWETT (Rev. B.) on St. Paul's Epistles to the Thessalonians, Galatians, and Romans. *Second Edition.* 2 Vols. 8vo. 30*s.*

JONES' (Rev. R.) Literary Remains. With a Prefatory Notice. By Rev. W. WHEWELL, D.D. Portrait. 8vo. 14*s.*

KEN'S (BISHOP) Life. By A LAYMAN. *Second Edition.* Portrait. 2 Vols. 8vo. 18*s.*

——— Exposition of the Apostles' Creed. Extracted from his "Practice of Divine Love." *New Edition.* Fcap. 1*s.* 6*d.*

——— Approach to the Holy Altar. Extracted from his "Manual of Prayer" and "Practice of Divine Love." *New Edition.* Fcap. 8vo. 1*s.* 6*d.*

KING'S (REV. S. W.) Italian Valleys of the Alps; a Tour through all the Romantic and less-frequented "Vals" of Northern Piedmont. Illustrations. Crown 8vo. 18*s.*

——— (REV. C. W.) Antique Gems; their Origin, Use, and Value, as Interpreters of Ancient History, and as illustrative of Ancient Art. Illustrations. 8vo. 42*s.*

KING EDWARD VITH'S Latin Grammar; or, an Introduction to the Latin Tongue, for the Use of Schools. *Sixteenth Edition.* 12mo. 3*s.* 6*d.*

——— First Latin Book; or, the Accidence, Syntax, and Prosody, with an English Translation for the Use of Junior Classes. *Fourth Edition.* 12mo. 2*s.* 6*d.*

c 2

KNAPP'S (J. A.) English Roots and Ramifications; or, the Derivation and Meaning of Divers Words. Fcap. 8vo. 4*s.*

KUGLER'S Italian Schools of Painting. Edited, with Notes, by SIR CHARLES EASTLAKE. *Third Edition.* Woodcuts. 2 Vols. Post 8vo. 30*s.*

——— German, Dutch, and Flemish Schools of Painting. Edited, with Notes, by DR. WAAGEN. *Second Edition.* Woodcuts. 2 Vols. Post 8vo. 24*s.*

LABARTE'S (M. JULES) Handbook of the Arts of the Middle Ages and Renaissance. With 200 Woodcuts. 8vo. 18*s.*

LABORDE'S (LEON DE) Journey through Arabia Petræa, to Mount Sinai, and the Excavated City of Petræa,—the Edom of the Prophecies. *Second Edition.* With Plates. 8vo. 18*s.*

LANE'S (E. W.) Manners and Customs of the Modern Egyptians. *Fifth Edition.* Edited by E. STANLEY POOLE. Woodcuts. 8vo. 18*s.*

LATIN GRAMMAR (KING EDWARD VITH'S). For the Use of Schools. *Fifteenth Edition.* 12mo. 3*s.* 6*d.*

——— First Book (KING EDWARD VITH'S); or, the Accidence, Syntax, and Prosody, with English Translation for Junior Classes. *Fourth Edition.* 12mo. 2*s.* 6*d.*

LAYARD'S (A. H.) Nineveh and its Remains. Being a Narrative of Researches and Discoveries amidst the Ruins of Assyria. With an Account of the Chaldean Christians of Kurdistan; the Yezedis, or Devil-worshippers; and an Enquiry into the Manners and Arts of the Ancient Assyrians. *Sixth Edition.* Plates and Woodcuts. 2 Vols. 8vo. 36*s.*

——— Nineveh and Babylon; being the Result of a Second Expedition to Assyria. *Fourteenth Thousand.* Plates. 8vo. 21*s.* Or *Fine Paper,* 2 Vols. 8vo. 30*s.*

——— Popular Account of Nineveh. 15*th Edition.* With Woodcuts. Post 8vo. 5*s.*

LESLIE'S (C. R.) Handbook for Young Painters. With Illustrations. Post 8vo. 10*s.* 6*d.*

——— Autobiographical Recollections, with Selections from his Correspondence. Edited by TOM TAYLOR. Portrait. 2 Vols. Post 8vo. 18*s.*

——— Life of Sir Joshua Reynolds. With an Account of his Works, and a Sketch of his Cotemporaries. By TOM TAYLOR. Fcap. 4to. (*In the Press.*)

LEAKE'S (COL.) Topography of Athens, with Remarks on its Antiquities. *Second Edition.* Plates. 2 Vols. 8vo. 30*s.*

——— Travels in Northern Greece. Maps. 4 Vols. 8vo. 60*s.*

——— Disputed Questions of Ancient Geography. Map. 8vo. 6*s.* 6*d.*

——— Numismata Hellenica, and Supplement. Completing a descriptive Catalogue of Twelve Thousand Greek Coins, with Notes Geographical and Historical. With Map and Appendix. 4to. 63*s.*

——— Peloponnesiaca. 8vo. 15*s.*

——— On the Degradation of Science in England. 8vo. 3*s.* 6*d.*

LETTERS FROM THE BALTIC. By a LADY. Post 8vo. 2*s.*

——— MADRAS; or, Life and Manners in India. By a LADY. Post 8vo. 2*s.*

LETTERS from SIERRA LEONE, written to Friends at Home. By a LADY. Edited by Mrs. NORTON. Post 8vo. *3s. 6d.*

——— Head Quarters; or, The Realities of the War in the Crimea. By a STAFF OFFICER. Plans. Post 8vo. *6s.*

LEXINGTON (THE) PAPERS; or, Some Account of the Courts of London and Vienna at the end of the 17th Century. Edited by HON. H. MANNERS SUTTON. 8vo. *14s.*

LEWIS' (SIR G. C.) Essay on the Government of Dependencies. 8vo. *12s.*

——— Glossary of Provincial Words used in Herefordshire and some of the adjoining Counties. 12mo. *4s. 6d.*

——— (LADY THERESA) Friends and Contemporaries of the Lord Chancellor Clarendon, illustrative of Portraits in his Gallery. With a Descriptive Account of the Pictures, and Origin of the Collection. Portraits. 3 Vols. 8vo. *42s.*

——— (M. G.) Journal of a Residence among the Negroes in the West Indies. Post 8vo. *2s.*

LIDDELL'S (DEAN) History of Rome. From the Earliest Times to the Establishment of the Empire. With the History of Literature and Art. 2 Vols. 8vo. *28s.*

——— Student's History of Rome. Abridged from the above Work. *Twentieth Thousand.* With Woodcuts. Post 8vo. *7s. 6d.*

LINDSAY'S (LORD) Lives of the Lindsays; or, a Memoir of the Houses of Crawfurd and Balcarres. With Extracts from Official Papers and Personal Narratives. *Second Edition.* 3 Vols. 8vo. *24s.*

———Report of the Claim of James, Earl of Crawfurd and Balcarres, to the Original Dukedom of Montrose, created in 1488. Folio. *15s.*

——— Scepticism; a Retrogressive Movement in Theology and Philosophy. 8vo. *9s.*

LISPINGS from LOW LATITUDES; or, the Journal of the Hon. Impulsia Gushington. With 24 Plates, 4to.

LITTLE ARTHUR'S HISTORY OF ENGLAND. By LADY CALLCOTT. *120th Thousand.* With 20 Woodcuts. Fcap. 8vo. *2s. 6d.*

LIVINGSTONE'S (REV. DR.) Missionary Travels and Researches in South Africa; including a Sketch of Sixteen Years' Residence in the Interior of Africa, and a Journey from the Cape of Good Hope to Loanda on the West Coast; thence across the Continent, down the River Zambesi, to the Eastern Ocean. *Thirtieth Thousand.* Map, Plates, and Index. 8vo. *21s.*

——— Popular Account of Travels in South Africa. Condensed from the above. Map and Illustrations. Post 8vo. *6s.*

LIVONIAN TALES. By the Author of "Letters from the Baltic." Post 8vo. *2s.*

LOCKHART'S (J. G.) Ancient Spanish Ballads. Historical and Romantic. Translated, with Notes. *Illustrated Edition.* 4to. *21s.* Or, *Popular Edition*, Post 8vo. *2s. 6d.*

——— Life of Robert Burns. *Fifth Edition.* Fcap. 8vo. *3s.*

LONDON'S (BISHOP OF). Dangers and Safeguards of Modern Theology. Containing Suggestions to the Theological Student under present difficulties. 8vo. *9s.*

LOUDON'S (Mrs.) Instructions in Gardening for Ladies. With Directions and Calendar of Operations for Every Month. *Eighth Edition.* Woodcuts. Fcap. 8vo. 5s.

——— Modern Botany; a Popular Introduction to the Natural System of Plants. *Second Edition.* Woodcuts. Fcap. 8vo. 6s.

LOWE'S (Sir Hudson) Letters and Journals, during the Captivity of Napoleon at St. Helena. By William Forsyth. Portrait. 3 Vols. 8vo. 45s.

LUCKNOW: A Lady's Diary of the Siege. *Fourth Thousand.* Fcap. 8vo. 4s. 6d.

LYELL'S (Sir Charles) Principles of Geology; or, the Modern Changes of the Earth and its Inhabitants considered as illustrative of Geology. *Ninth Edition.* Woodcuts. 8vo. 18s.

———Visits to the United States, 1841-46. *Second Edition.* Plates. 4 Vols. Post 8vo. 24s.

——— Geological Evidences of the Antiquity of Man. With 50 Illustrations. 8vo. 14s.

MAHON'S (Lord) History of England, from the Peace of Utrecht to the Peace of Versailles, 1713—83. *Library Edition,* 7 Vols. 8vo. 93s. *Popular Edition,* 7 Vols. Post 8vo. 35s.

——— Life of Right Hon. William Pitt, with Extracts from his MS. Papers. *Second Edition.* Portraits. 4 Vols. Post 8vo. 42s.

——— "Forty-Five;" a Narrative of the Rebellion in Scotland. Post 8vo. 3s.

——— History of British India from its Origin till the Peace of 1783. Post 8vo. 3s. 6d.

——— History of the War of the Succession in Spain. *Second Edition.* Map. 8vo. 15s.

——— Spain under Charles the Second; or, Extracts from the Correspondence of the Hon. Alexander Stanhope, British Minister at Madrid from 1690 to 1700. *Second Edition.* Post 8vo. 6s. 6d.

——— Life of Louis, Prince of Condé, surnamed the Great. Post 8vo. 3s. 6d.

——— Life of Belisarius. *Second Edition.* Post 8vo. 10s. 6d.

——— Historical and Critical Essays. Post 8vo. 3s. 6d.

——— Story of Joan of Arc. Fcap. 8vo. 1s.

——— Addresses Delivered at Manchester, Leeds, and Birmingham. Fcap. 8vo. 1s.

McCLINTOCK'S (Capt. Sir F. L.) Narrative of the Discovery of the Fate of Sir John Franklin and his Companions in the Arctic Seas. *Twelfth Thousand.* Illustrations. 8vo. 16s.

McCOSH (Rev. Dr.) on the Intuitive Convictions of the Mind inductively investigated. 8vo. 12s.

McCULLOCH'S (J. R.) Collected Edition of Ricardo's Political Works. With Notes and Memoir. *Second Edition.* 8vo. 16s.

MAINE (H. Sumner) on Ancient Law: its Connection with the Early History of Society, and its Relation to Modern Ideas. *Second Edition.* 8vo. 12s.

MALCOLM'S (Sir John) Sketches of Persia. *Third Edition.* Post 8vo. 3s. 6d.

MANSEL (Rev. H. L.) Limits of Religious Thought Examined. Being the Bampton Lectures for 1858. *Fourth Edition.* Post 8vo. 7s. 6d.

MANTELL'S (Gideon A.) Thoughts on Animalcules; or, the Invisible World, as revealed by the Microscope. *Second Edition.* Plates. 16mo. 6s.

MANUAL OF SCIENTIFIC ENQUIRY, Prepared for the Use of Officers and Travellers. By various Writers. Edited by Sir J. F. HERSCHEL and Rev. R. MAIN. *Third Edition.* Maps. Post 8vo. 9*s.* (*Published by order of the Lords of the Admiralty.*)

MARKHAM'S (MRS.) History of England. From the First Invasion by the Romans, down to the fourteenth year of Queen Victoria's Reign. 156*th Edition.* Woodcuts. 12mo. 6*s.*

——— History of France. From the Conquest by the Gauls, to the Death of Louis Philippe. *Sixtieth Edition.* Woodcuts. 12mo. 6*s.*

——— History of Germany. From the Invasion by Marius, to the present time. *Fifteenth Edition.* Woodcuts. 12mo. 6*s.*

——— History of Greece. From the Earliest Times to the Roman Conquest. By Dr. WM. SMITH. Woodcuts. 16 mo. 3*s.* 6*d.*

——— History of Rome, from the Earliest Times to the Establishment of the Empire. By DR. WM. SMITH. Woodcuts. 16mo. 3*s.* 6*d.*

——— (CLEMENTS, R.) Travels in Peru and India, for the purpose of collecting Cinchona Plants, and introducing Bark into India. Maps and Illustrations. 8vo. 16*s.*

MARKLAND'S (J. H.) Reverence due to Holy Places. *Third Edition.* Fcap. 8vo. 2*s.*

MARRYAT'S (JOSEPH) History of Modern and Mediæval Pottery and Porcelain. With a Description of the Manufacture. *Second Edition.* Plates and Woodcuts. 8vo. 31*s.* 6*d.*

——— (HORACE) Residence in Jutland, the Danish Isles, and Copenhagen. Illustrations. 2 Vols. Post 8vo. 24*s.*

——— Year in Sweden, including a Visit to the Isle of Gothland. Illustrations. 2 Vols. Post 8vo. 28*s.*

MATTHIÆ'S (AUGUSTUS) Greek Grammar for Schools. Abridged from the Larger Grammar. By Blomfield. *Ninth Edition.* Revised by EDWARDS. 12mo. 3*s.*

MAUREL'S (JULES) Essay on the Character, Actions, and Writings of the Duke of Wellington. *Second Edition.* Fcap. 8vo. 1*s.* 6*d.*

MAWE'S (H. L.) Journal of a Passage from the Pacific to the Atlantic. 8vo. 12*s.*

MAXIMS AND HINTS on Angling and Chess. To which is added the Miseries of Fishing. By RICHARD PENN. *New Edition.* Woodcuts. 12mo. 1*s.*

MAYNE'S (R. C.) Four Years in British Columbia and Vancouver Island. Its Forests, Rivers, Coasts, and Gold Fields, and its Resources for Colonisation. Map and Illustrations. 8vo. 16*s.*

MAYO'S (DR.) Pathology of the Human Mind. Fcap. 8vo. 5*s.* 6*d.*

MELVILLE'S (HERMANN) Typee and Omoo; or, Adventures amongst the Marquesas and South Sea Islands. 2 Vols. Post 8vo. 7*s.*

MENDELSSOHN'S Life. By JULES BENEDICT. 8vo. 2*s.* 6*d.*

MEREDITH'S (MRS. CHARLES) Notes and Sketches of New South Wales, during a Residence from 1839 to 1844. Post 8vo. 2*s.*

——— Tasmania, during a Residence of Nine Years. With Illustrations. 2 Vols. Post 8vo. 18*s.*

MERRIFIELD (MRS.) on the Arts of Painting in Oil, Miniature, Mosaic, and Glass; Gilding, Dyeing, and the Preparation of Colours and Artificial Gems, described in several old Manuscripts. 2 Vols. 8vo. 30*s.*

MESSIAH (THE). By Author of the "Life of Bishop Ken." Map. 8vo. 18*s*.

MILLS' (ARTHUR) India in 1858; A Summary of the Existing Administration—Political, Fiscal, and Judicial; with Laws and Public Documents, from the earliest to the present time. *Second Edition.* With Coloured Revenue Map. 8vo. 10*s*. 6*d*.

MITCHELL'S (THOMAS) Plays of Aristophanes. With English Notes. FROGS. 8vo. 15*s*.

MILMAN'S (DEAN) History of Latin Christianity; including that of the Popes to the Pontificate of Nicholas V. *Second Edition.* 6 Vols. 8vo. 72*s*.

——— History of the Jews, brought down to Modern Times. 3 Vols. 8vo.

——— Character and Conduct of the Apostles considered as an Evidence of Christianity. 8vo. 10*s*. 6*d*.

——— Life and Works of Horace. With 300 Woodcuts. *New Edition.* 2 Vols. Crown 8vo. 30*s*.

——— Poetical Works. Plates. 3 Vols. Fcap. 8vo. 18*s*.

——— Fall of Jerusalem. Fcap. 8vo. 1*s*.

——— (CAPT. E. A.) Wayside Cross; or, the Raid of Gomez. A Tale of the Carlist War. Post 8vo. 2*s*.

MODERN DOMESTIC COOKERY. Founded on Principles of Economy and Practical Knowledge, and adapted for Private Families. *New Edition.* Woodcuts. Fcap. 8vo. 5*s*.

MOLTKE'S (BARON) Russian Campaigns on the Danube and the Passage of the Balkan, 1828-9. Plans. 8vo. 14*s*.

MONASTERY AND THE MOUNTAIN CHURCH. By Author of "Sunlight through the Mist." Woodcuts. 16mo. 4*s*.

MOORE'S (THOMAS) Life and Letters of Lord Byron. *Cabinet Edition.* Plates. 6 Vols. Fcap. 8vo. 18*s*.

——— Life and Letters of Lord Byron. Portraits. Royal 8vo. 9*s*.

MOTLEY'S (J. L.) History of the United Netherlands: from the Death of William the Silent to the Synod of Dort. Embracing the English-Dutch struggle against Spain; and a detailed Account of the Spanish Armada. *Fourth Thousand.* Portraits. 2 Vols. 8vo. 30*s*.

MOZLEY'S (REV. J. B.) Treatise on the Augustinian Doctrine of Predestination. 8vo. 14*s*.

——— Primitive Doctrine of Baptismal Regeneration. 8vo. 7*s*. 6*d*.

MUCK MANUAL (The) for the Use of Farmers. A Practical Treatise on the Chemical Properties, Management, and Application of Manures. By FREDERICK FALKNER. *Second Edition.* Fcap. 8vo. 5*s*.

MUNDY'S (GEN.) Pen and Pencil Sketches during a Tour in India. *Third Edition.* Plates. Post 8vo. 7*s*. 6*d*.

MUNRO'S (GENERAL SIR THOMAS) Life and Letters. By the REV. G. R. GLEIG. Post 8vo. 3*s*. 6*d*.

MURCHISON'S (SIR RODERICK) Russia in Europe and the Ural Mountains; Geologically Illustrated. With Coloured Maps, Plates, Sections, &c. 2 Vols. Royal 4to.

——— Siluria; or, a History of the Oldest Rocks containing Organic Remains. *Third Edition.* Map and Plates. 8vo. 42*s*.

MURRAY'S RAILWAY READING. For all classes of Readers.

[*The following are published:*]

Wellington. By Lord Ellesmere. 6*d*.
Nimrod on the Chase, 1*s*.
Essays from "The Times." 2 Vols. 8*s*.
Music and Dress. 1*s*.
Layard's Account of Nineveh. 5*s*.
Milman's Fall of Jerusalem. 1*s*.
Mahon's "Forty-Five." 3*s*.
Life of Theodore Hook. 1*s*.
Deeds of Naval Daring. 2 Vols. 5*s*.
The Honey Bee. 1*s*.
James' Æsop's Fables. 2*s*. 6*d*.
Nimrod on the Turf. 1*s*. 6*d*.
Oliphant's Nepaul. 2*s*. 6*d*.
Art of Dining. 1*s*. 6*d*.
Hallam's Literary Essays. 2*s*.
Mahon's Joan of Arc. 1*s*.
Head's Emigrant. 2*s*. 6*d*.
Nimrod on the Road. 1*s*.
Wilkinson's Ancient Egyptians. 12*s*
Croker on the Guillotine. 1*s*.
Hollway's Norway. 2*s*.
Maurel's Wellington. 1*s*. 6*d*.
Campbell's Life of Bacon. 2*s*. 6*d*.
The Flower Garden. 1*s*.
Lockhart's Spanish Ballads. 2*s*. 6*d*.
Lucas on History. 6*d*.
Beauties of Byron. 3*s*.
Taylor's Notes from Life. 2*s*.
Rejected Addresses. 1*s*.
Penn's Hints on Angling. 1*s*.

MURRAY'S (Capt. A.) Naval Life and Services of Admiral Sir Philip Durham. 8vo. 5*s*. 6*d*.

MUSIC AND DRESS. Two Essays, by a Lady. Reprinted from the "Quarterly Review." Fcap. 8vo. 1*s*.

NAPIER'S (Sir Wm.) English Battles and Sieges of the Peninsular War. *Third Edition*. Portrait. Post 8vo. 10*s*. 6*d*.

——— Life and Letters. Edited by H. A. Bruce, M.P. Portraits. 2 Vols. Crown 8vo.

——— Life of General Sir Charles Napier; chiefly derived from his Journals, Letters, and Familiar Correspondence. *Second Edition*. Portraits. 4 Vols. Post 8vo. 48*s*.

NAUTICAL ALMANACK (The). Royal 8vo. 2*s*. 6*d*. (*Published by Authority*.)

NAVY LIST (The Quarterly). (*Published by Authority*.) Post 8vo. 2*s*. 6*d*.

NELSON (Robert), Memoir of his Life and Times. By Rev. C. T. Secretan, M.A. Portrait. 8vo. 10*s*. 6*d*.

NEWBOLD'S (Lieut.) Straits of Malacca, Penang, and Singapore. 2 Vols. 8vo. 26*s*.

NEWDEGATE'S (C. N.) Customs' Tariffs of all Nations; collected and arranged up to the year 1855. 4to. 30*s*.

NICHOLLS' (Sir George) History of the English Poor-Laws. 2 Vols. 8vo. 28*s*.

——— History of the Irish Poor-Law. 8vo. 14*s*.

——— History of the Scotch Poor-Law. 8vo. 12*s*.

——— (Rev. H. G.) Historical and Descriptive Account of the Forest of Dean: from Sources Public, Private, Legendary, and Local. Woodcuts, &c. Post 8vo. 10*s*. 6*d*.

NICOLAS' (Sir Harris) Historic Peerage of England. Exhibiting the Origin, Descent, and Present State of every Title of Peerage which has existed in this Country since the Conquest. Being a New Edition of the "Synopsis of the Peerage." Revised and Continued to the Present Time. By William Courthope, Somerset Herald. 8vo. 30*s*.

NIMROD On the Chace—The Turf—and The Road. Reprinted from the "Quarterly Review." Woodcuts. Fcap. 8vo. 3*s*. 6*d*.

O'CONNOR'S (R.) Field Sports of France; or, Hunting, Shooting, and Fishing on the Continent. Woodcuts. 12mo. 7*s*. 6*d*.

OXENHAM'S (Rev. W.) English Notes for Latin Elegiacs; designed for early Proficients in the Art of Latin Versification, with Prefatory Rules of Composition in Elegiac Metre. *Fourth Edition*. 12mo. 3*s*. 6*d*.

PAGET'S (John) Hungary and Transylvania. With Remarks on their Condition, Social, Political, and Economical. *Third Edition.* Woodcuts. 2 Vols. 8vo. 18*s.*

PARIS' (Dr.) Philosophy in Sport made Science in Earnest; or, the First Principles of Natural Philosophy inculcated by aid of the Toys and Sports of Youth. *Eighth Edition.* Woodcuts. Post 8vo. 7*s.* 6*d.*

PEEL'S (Sir Robert) Memoirs. Left in MSS. Edited by Earl Stanhope and the Right Hon. Edward Cardwell. 2 Vols. Post 8vo. 7*s.* 6*d.* each.

PEILE'S (Rev. Dr.) Agamemnon and Choephorœ of Æschylus. A New Edition of the Text, with Notes. *Second Edition.* 2 Vols. 8vo. 9*s.* each.

PENN'S (Richard) Maxims and Hints for an Angler and Chess-player. *New Edition.* Woodcuts. Fcap. 8vo. 1*s.*

PENROSE'S (F. C.) Principles of Athenian Architecture, and the Optical Refinements exhibited in the Construction of the Ancient Buildings at Athens, from a Survey. With 40 Plates. Folio. 5*l.* 5*s.*

PERCY'S (John, M.D.) Metallurgy; or, the Art of Extracting Metals from their Ores and adapting them to various purposes of Manufacture. *First Division* — Slags, Fire-Clays, Fuel-Copper, Zinc, and Brass. Illustrations. 8vo. 21*s.*

PERRY'S (Sir Erskine) Bird's-Eye View of India. With Extracts from a Journal kept in the Provinces, Nepaul, &c. Fcap. 8vo. 5*s.*

PHILLIPS' (John) Memoirs of William Smith, LL.D. (the Geologist). Portrait. 8vo. 7*s.* 6*d.*

——— Geology of Yorkshire, The Yorkshire Coast, and the Mountain-Limestone District. Plates. 4to. Part I., 20*s.* –Part II., 30*s.*

——— Rivers, Mountains, and Sea Coast of Yorkshire. With Essays on the Climate, Scenery, and Ancient Inhabitants of the Country. *Second Edition,* with 36 Plates. 8vo. 15*s.*

——— (March.) Jurisprudence. 8vo.

PHILPOTT'S (Bishop) Letters to the late Charles Butler, on the Theological parts of his "Book of the Roman Catholic Church;" with Remarks on certain Works of Dr. Milner and Dr. Lingard, and on some parts of the Evidence of Dr. Doyle. *Second Edition.* 8vo. 16*s.*

PHIPPS' (Hon. Edmund) Memoir, Correspondence, Literary and Unpublished Diaries of Robert Plumer Ward. Portrait. 2 Vols. 8vo. 28*s.*

POPE'S (Alexander) Life and Works. *A New Edition.* Containing nearly 500 unpublished Letters. Edited with a New Life, Introductions and Notes. By Rev. Whitwell Elwin. Portraits. Vol. I. 8vo. (*In the Press.*)

PORTER'S (Rev. J. L.) Five Years in Damascus. With Travels to Palmyra, Lebanon, and other Scripture Sites. Map and Woodcuts. 2 Vols. Post 8vo. 21*s.*

——— Handbook for Syria and Palestine: including an Account of the Geography, History, Antiquities, and Inhabitants of these Countries, the Peninsula of Sinai, Edom, and the Syrian Desert. Maps. 2 Vols. Post 8vo. 24*s.*

——— (Mrs.) Rational Arithmetic for Schools and for Private Instruction. 12mo. 3*s.* 6*d.*

PRAYER-BOOK (The Illustrated), with 1000 Illustrations of Borders, Initials, Vignettes, &c. Medium 8vo. 21*s.*

PRECEPTS FOR THE CONDUCT OF LIFE. Extracted from the Scriptures. *Second Edition.* Fcap. 8vo. 1*s.*

PRINSEP'S (Jas.) Essays on Indian Antiquities, Historic, Numismatic, and Palæographic, with Tables. Edited by Edward Thomas. Illustrations. 2 Vols. 8vo. 52*s*. 6*d*.

PROGRESS OF RUSSIA IN THE EAST. An Historical Summary, continued to the Present Time. *Third Edition.* Map. 8vo. 6*s*. 6*d*.

PUSS IN BOOTS. With 12 Illustrations; for Old and Young. By Otto Speckter. 16mo. 1*s*. 6*d*.; or Coloured, 2*s*. 6*d*.

QUARTERLY REVIEW (The). 8vo. 6*s*.

RAWLINSON'S (Rev. George) Herodotus. A New English Version. Edited with Notes and Essays. Assisted by Sir Henry Rawlinson and Sir J. G. Wilkinson. *Second Edition.* Maps and Woodcut. 4 Vols. 8vo. 48*s*.

——— Historical Evidences of the truth of the Scripture Records stated anew, with special reference to the Doubts and Discoveries of Modern Times; the Bampton Lectures for 1859. *Second Edition.* 8vo. 14*s*.

——— Five Great Monarchies of the Ancient World. Or the History, Geography, and Antiquities of Chaldæa, Assyria, Babylonia, Media, and Persia. Illustrations. Vol. I. 8vo. 16*s*.

REJECTED ADDRESSES (The). By James and Horace Smith. *New Edition.* Fcap. 8vo. 1*s*., or *Fine Paper*, with Portrait, fcap. 8vo, 5*s*.

RICARDO'S (David) Political Works. With a Notice of his Life and Writings. By J. R. M'Culloch. *New Edition.* 8vo. 16*s*.

RIPA'S (Father) Memoirs during Thirteen Years' Residence at the Court of Peking. From the Italian. Post 8vo. 2*s*.

ROBERTSON'S (Canon) History of the Christian Church, From the Apostolic Age to the Concordat of Worms, A.D. 1123. *Second Edition.* 3 Vols. 8vo. 38*s*.

——— Life of Becket. Illustrations. Post 8vo. 9*s*.

ROBINSON'S (Rev. Dr.) Biblical Researches in the Holy Land. Being a Journal of Travels in 1838, and of Later Researches in 1852. Maps. 3 Vols. 8vo. 36*s*.

ROMILLY'S (Sir Samuel) Memoirs and Political Diary. By his Sons. *Third Edition.* Portrait. 2 Vols. Fcap. 8vo. 12*s*.

ROSS'S (Sir James) Voyage of Discovery and Research in the Southern and Antarctic Regions, 1839-43. Plates. 2 Vols. 8vo. 36*s*.

ROWLAND'S (David) Manual of the English Constitution; a Review of its Rise, Growth, and Present State. Post 8vo. 10*s*. 6*d*.

RUNDELL'S (Mrs.) Domestic Cookery, founded on Principles of Economy and Practice, and adapted for Private Families. *New and Revised Edition.* Woodcuts. Fcap. 8vo. 5*s*.

RUSSELL'S (J. Rutherfurd, M.D.) Art of Medicine—Its History and its Heroes. Portraits. 8vo. 14*s*.

RUSSIA; A Memoir of the Remarkable Events which attended the Accession of the Emperor Nicholas. By Baron M. Korff, Secretary of State. 8vo. 10*s*. 6*d*. (*Published by Imperial Command.*)

RUXTON'S (George F.) Travels in Mexico; with Adventures among the Wild Tribes and Animals of the Prairies and Rocky Mountains. Post 8vo. 3*s*. 6*d*.

SALE'S (Lady) Journal of the Disasters in Affghanistan. Post 8vo. 12*s*.

——— (Sir Robert) Brigade in Affghanistan. With an Account of the Defence of Jellalabad. By Rev. G. R. Gleig. Post 8vo. 2*s*.

SANDWITH'S (HUMPHRY) Siege of Kars and Resistance by the Turkish Garrison under General Williams. Post 8vo. 3*s.* 6*d.*

SCOTT'S (G. GILBERT) Secular and Domestic Architecture, Present and Future. *Second Edition.* 8vo. 9*s.*

——— (Master of Baliol) Sermons Preached before the University of Oxford. Post 8vo. 8*s.* 6*d.*

SCROPE'S (WILLIAM) Days of Deer-Stalking; with some Account of the Red Deer. *Third Edition.* Woodcuts. Crown 8vo. 20*s.*

——— Days and Nights of Salmon Fishing in the Tweed; with a short Account of the Salmon. *Second Edition.* Woodcuts. Royal 8vo. 31*s.* 6*d.*

——— (G. P.) Memoir of Lord Sydenham, and his Administration in Canada. *Second Edition.* Portrait. 8vo. 9*s.* 6*d.*

——— Geology and Extinct Volcanoes of Central France. *Second Edition.* Illustrations. Medium 8vo. 30*s.*

SELF-HELP. With Illustrations of Character and Conduct. By SAMUEL SMILES. *Fifty-fifth Thousand.* Post 8vo. 6*s.*

SENIOR'S (N. W.) Suggestions on Popular Education. 8vo. 9*s.*

SHAFTESBURY (LORD CHANCELLOR); Memoirs of his Early Life. With his Letters, &c. By W. D. CHRISTIE. Portrait. 8vo. 10*s.* 6*d.*

SHAW'S (J. F.) Outlines of English Literature for Students. *Second Edition.* Revised. Post 8vo. (*In the Press.*)

SIERRA LEONE; Described in Letters to Friends at Home. By A LADY. Post 8vo. 3*s.* 6*d.*

SIMMONS on Courts-Martial. *5th Edition.* Adapted to the New Mutiny Act and Articles of War, the Naval Discipline Act, and the Criminal Law Consolidation Acts. 8vo.

SMILES' (SAMUEL) Lives of Engineers; from the Earliest Period to the Death of Telford; with an account of their Principal Works, and a History of Inland Communication in Britain. Portraits and numerous Woodcuts. 2 Vols. 8vo. 42*s.*

——— George and Robert Stephenson. Forming the Third Volume of "Lives of the Engineers." With 2 Portraits and 70 Illustrations. Medium 8vo. 21*s.*

——— Story of the Life of George Stephenson. Woodcuts. *Eighteenth Thousand.* Post 8vo. 6*s.*

——— Self-Help. With Illustrations of Character and Conduct. *Fifty-fifth Thousand.* Post 8vo. 6*s.*

———Workmen's Earnings, Savings, and Strikes. *Fifth Thousand.* Fcap. 8vo. 1*s.* 6*d.*

SOMERVILLE'S (MARY) Physical Geography. *Fifth Edition.* Portrait. Post 8vo. 9*s.*

——— Connexion of the Physical Sciences. *Ninth Edition.* Woodcuts. Post 8vo. 9*s.*

SOUTH'S (JOHN F.) Household Surgery; or, Hints on Emergencies. *Seventeenth Thousand.* Woodcuts. Fcp. 8vo. 4*s.* 6*d.*

SOUTHEY'S (ROBERT) Book of the Church. *Seventh Edition.* Post 8vo. 7*s.* 6*d.*

——— Lives of Bunyan and Cromwell. Post 8vo. 2*s.*

SPECKTER'S (OTTO) Puss in Boots. With 12 Woodcuts. Square 12mo. 1*s.* 6*d.* plain, or 2*s.* 6*d.* coloured.

——— Charmed Roe; or, the Story of the Little Brother and Sister. Illustrated. 16mo.

SMITH'S (Dr. Wm.) Dictionary of the Bible; its Antiquities, Biography, Geography, and Natural History. *Second Edition.* Woodcuts. Vol. I. 8vo. 42*s.*

——— Greek and Roman Antiquities. *2nd Edition.* Woodcuts. 8vo. 42*s.*

——— Biography and Mythology. Woodcuts. 3 Vols. 8vo. 5*l.* 15*s.* 6*d.*

——— Geography. Woodcuts. 2 Vols. 8vo. 80*s.*

——— Latin-English Dictionary. Based upon the Works of Forcellini and Freund. *Ninth Thousand.* 8vo. 21*s.*

——— Classical Dictionary. *6th Edition.* 750 Woodcuts. 8vo. 18*s.*

——— Smaller Classical Dictionary. *Twentieth Thousand.* 200 Woodcuts. Crown 8vo. 7*s.* 6*d.*

——— Smaller Dictionary of Antiquities. *Twentieth Thousand.* 200 Woodcuts. Crown 8vo. 7*s.* 6*d.*

——— Smaller Latin-English Dictionary. *Twenty-fifth Thousand.* Square 12mo. 7*s.* 6*d.*

——— Principia Latina—Part I. A Grammar, Delectus, and Exercise Book, with Vocabularies. *3rd Edition.* 12mo. 3*s.* 6*d.*

——— Principia Latina—Part II. A Reading-book, Mythology, Geography, Roman Antiquities, and History. With Notes and Dictionary. *Second Edition.* 12mo. 3*s.* 6*d.*

——— Principia Latina.—Part III. A Latin Poetry Book. Containing:—Easy Hexameters and Pentameters; Eclogæ Ovidianæ; Latin Prosody. 12mo. 3*s.* 6*d.*

——— Latin-English Vocabulary; applicable for those reading Phædrus, Cornelius Nepos, and Cæsar. *Second Edition.* 12mo. 3*s.* 6*d.*

——— Principia Græca; a First Greek Course. A Grammar, Delectus, and Exercise-book with Vocabularies. By H. E. Hutton, M.A. *3rd Edition.* 12mo. 3*s.*

——— Student's Greek Grammar. Translated from the German of Professor Curtius. Post 8vo.

——— Student's Latin Grammar, for the use of Colleges and the Upper Forms in Schools. Post 8vo.

——— Smaller Greek Grammar, for the use of the Middle and Lower Forms. Abridged from the above work. 12mo.

——— Smaller Latin Grammar, for the use of the Middle and Lower Forms. Abridged from the above work. 12mo.

STANLEY'S (Canon) Lectures on the History of the Eastern Church. *Second Edition.* Plans. 8vo. 16*s.*

——— Lectures on the History of the Jewish Church. From Abraham to Samuel. Plans. 8vo. 16*s.*

——— Sermons on the Unity of Evangelical and Apostolical Teaching. *Second Edition.* Post 8vo. 7*s.* 6*d.*

——— St. Paul's Epistles to the Corinthians, with Notes and Dissertations. *Second Edition.* 8vo. 18*s.*

——— Historical Memorials of Canterbury. *Third Edition.* Woodcuts. Post 8vo. 7*s.* 6*d.*

——— Sinai and Palestine, in Connexion with their History. *Sixth Edition.* Map. 8vo. 16*s.*

——— Bible in the Holy Land. Being Extracts from the above work. Woodcuts. Fcp. 8vo. 2*s.* 6*d.*

——— Addresses and Charges of Bishop Stanley. With Memoir. *Second Edition.* 8vo. 10*s.* 6*d.*

ST. JOHN'S (Charles) Wild Sports and Natural History of the Highlands. Post 8vo. 3*s.* 6*d.*

——— (Bayle) Adventures in the Libyan Desert and the Oasis of Jupiter Ammon. Woodcuts. Post 8vo. 2*s.*

STEPHENSON (George) The Railway Engineer. The Story of his Life. By Samuel Smiles. *Eighteenth Thousand.* Woodcuts. Post 8vo. 6*s.*

STOTHARD'S (Thos.) Life. With Personal Reminiscences. By Mrs. Bray. With Portrait and 60 Woodcuts. 4to.

STREET'S (G. E.) Brick and Marble Architecture of Italy in the Middle Ages. Plates. 8vo. 21*s.*

STRIFE FOR THE MASTERY. Two Allegories. With Illustrations. Crown 8vo. 6*s.*

STUDENT'S HUME. A History of England from the Invasion of Julius Cæsar to the Revolution of 1688. Based on the Work by David Hume. Continued to 1858. *Twenty-fifth Thousand.* Woodcuts. Post 8vo. 7*s.* 6*d.*

*** A Smaller History of England. 12mo. 3*s.* 6*d.*

——— HISTORY OF FRANCE; From the Earliest Times to the Establishment of the Second Empire, 1852. Woodcuts. Post 8vo. 7*s.* 6*d.*

——— HISTORY OF GREECE; from the Earliest Times to the Roman Conquest. With the History of Literature and Art. By Wm. Smith, LL.D. 20*th Thousand.* Woodcuts. Crown 8vo. 7*s.* 6*d.* (Questions. 2*s.*)

*** A Smaller History of Greece, for Junior Classes. 12mo. 3*s.* 6*d.*

——— HISTORY OF ROME; from the Earliest Times to the Establishment of the Empire. With the History of Literature and Art. By H. G. Liddell, D.D. 20*th Thousand.* Woodcuts. Crown 8vo. 7*s.* 6*d.*

*** A Smaller History of Rome, for Junior Classes. By Dr. Wm. Smith. 12mo. 3*s.* 6*d.*

——— GIBBON; an Epitome of the History of the Decline and Fall of the Roman Empire. Incorporating the Researches of Recent Commentators. 9*th Thousand.* Woodcuts. Post 8vo. 7*s.* 6*d.*

——— MANUAL OF ANCIENT GEOGRAPHY. Based on the larger Dictionary of Greek and Roman Geography. Woodcuts. Post 8vo. 9*s.*

——— MODERN GEOGRAPHY. Post 8vo.

——— ENGLISH LANGUAGE. By George P. Marsh. Post 8vo. 7*s.* 6*d.*

SWIFT'S (Jonathan) Life, Letters and Journals. By John Forster. 8vo. (*In Preparation.*)

——— Works. Edited, with Notes. By John Forster. 8vo. (*In Preparation.*)

SYME'S (Jas.) Principles of Surgery. *Fourth Edition.* 8vo. 14*s.*

TAIT'S (Bishop) Dangers and Safeguards of Modern Theology. 8vo. 9*s.*

TAYLOR'S (Henry) Notes from Life. Fcap. 8vo. 2*s.*

THOMSON'S (Archbishop) Sermons Preached in Lincoln's Inn Chapel. 8vo. 10*s.* 6*d.*

——— (Dr.) Story of New Zealand; Past and Present—Savage and Civilised. *Second Edition.* Illustrations. 2 Vols. Post 8vo. 24*s.*

THREE-LEAVED MANUAL OF FAMILY PRAYER; arranged so as to save the trouble of turning the Pages backwards and forwards. Royal 8vo. 2*s.*

TICKNOR'S (GEORGE) History of Spanish Literature. With Criticisms on particular Works, and Biographical Notices of Prominent Writers. *Second Edition.* 3 Vols. 8vo. 24*s.*

TOCQUEVILLE'S (M. DE) State of France before the Revolution, 1789, and on the Causes of that Event. Translated by HENRY REEVE, ESQ. 8vo. 14*s.*

TREMENHEERE'S (H. S.) Political Experience of the Ancients, in its bearing on Modern Times. Fcap. 8vo. 2*s.* 6*d.*

——— Notes on Public Subjects, made during a Tour in the United States and Canada. Post 8vo. 10*s.* 6*d.*

——— Constitution of the United States compared with our own. Post 8vo. 9*s.* 6*d.*

TRISTRAM'S (H. B.) Great Sahara; or, Wanderings South of the Atlas Mountains. Illustrations. Post 8vo. 15*s.*

TWISS' (HORACE) Public and Private Life of Lord Chancellor Eldon, with Selections from his Correspondence. Portrait. *Third Edition.* 2 Vols. Post 8vo. 21*s.*

TYNDALL'S (JOHN) Glaciers of the Alps. Being a Narrative of various Excursions among them, and an Account of Three Years' Observations and Experiments on their Motion, Structure, and General Phenomena. Woodcuts. Post 8vo. 14*s.*

TYTLER'S (PATRICK FRASER) Memoirs. By REV. J. W. BURGON, M.A. *Second Edition.* 8vo. 9*s.*

UBICINI'S (M. A.) Letters on Turkey and its Inhabitants—the Moslems, Greeks, Armenians, &c. Translated by LADY EASTHOPE. 2 Vols. Post 8vo. 21*s.*

VAUGHAN'S (REV. DR.) Sermons preached in Harrow School. 8vo. 10*s.* 6*d.*

VENABLES' (REV. R. L.) Domestic Scenes in Russia during a Year's Residence, chiefly in the Interior. *Second Edition.* Post 8vo. 5*s.*

VOYAGE to the Mauritius and back, touching at the Cape of Good Hope and St. Helena. By Author of "PADDIANA." Post 8vo. 9*s.* 6*d.*

WAAGEN'S (DR.) Treasures of Art in Great Britain. Being an Account of the Chief Collections of Paintings, Sculpture, Manuscripts, Miniatures, &c. &c., in this Country. Obtained from Personal Inspection during Visits to England. 3 Vols. 8vo. 36*s.*

——— Galleries and Cabinets of Art in England. Being an Account of more than Forty Collections, visited in 1854-56. With Index. 8vo. 18*s.*

WADDINGTON'S (DEAN) Condition and Prospects of the Greek Church. *New Edition.* Fcap. 8vo. 3*s.* 6*d.*

WAKEFIELD'S (E. J.) Adventures in New Zealand. With some Account of the Beginning of the British Colonisation of the Island. Map. 2 Vols. 8vo. 28*s.*

WALKS AND TALKS. A Story-book for Young Children. By AUNT IDA. With Woodcuts. 16mo. 5*s.*

WALSH'S (SIR JOHN) Practical Results of the Reform Bill of 1832. 8vo. 5*s.* 6*d.*

WARD'S (ROBERT PLUMER) Memoir, Correspondence, Literary and Unpublished Diaries and Remains. By the HON. EDMUND PHIPPS. Portrait. 2 Vols. 8vo. 28*s.*

WATT'S (JAMES) Life. Incorporating the most interesting passages from his Private and Public Correspondence. By JAMES P. MUIRHEAD, M.A. *Second Edition.* Portrait. 8vo. 16*s.*

———— Origin and Progress of his Mechanical Inventions. Illustrated by his Correspondence. By J. P. MUIRHEAD. Plates. 3 Vols. 8vo. 45*s.*

WILKIE'S (SIR DAVID) Life, Journals, Tours, and Critical Remarks on Works of Art, with a Selection from his Correspondence. By ALLAN CUNNINGHAM. Portrait. 3 Vols. 8vo. 42*s.*

WOOD'S (LIEUT.) Voyage up the Indus to the Source of the River Oxus, by Kabul and Badakhshan. Map. 8vo. 14*s.*

WELLINGTON'S (THE DUKE OF) Despatches during his various Campaigns. Compiled from Official and other Authentic Documents. By COL. GURWOOD, C.B. *New Enlarged Edition.* 8 Vols. 8vo. 21*s.* each.

———— Supplementary Despatches, and other Papers. Edited by his SON. Vols. I. to IX. 8vo. 20*s.* each.

———— Selections from his Despatches and General Orders. By COLONEL GURWOOD. 8vo. 18*s.*

———— Speeches in Parliament. 2 Vols. 8vo. 42*s.*

WILKINSON'S (SIR J. G.) Popular Account of the Private Life, Manners, and Customs of the Ancient Egyptians. *New Edition.* Revised and Condensed. With 500 Woodcuts. 2 Vols. Post 8vo. 12*s.*

———— Dalmatia and Montenegro; with a Journey to Mostar in Hertzegovina, and Remarks on the Slavonic Nations. Plates and Woodcuts. 2 Vols. 8vo. 42*s.*

———— Handbook for Egypt.—Thebes, the Nile, Alexandria, Cairo, the Pyramids, Mount Sinai, &c. Map. Post 8vo. 15*s.*

———— On Colour, and on the Necessity for a General Diffusion of Taste among all Classes; with Remarks on laying out Dressed or Geometrical Gardens. With Coloured Illustrations and Woodcuts. 8vo. 18*s.*

———— (G. B.) Working Man's Handbook to South Australia; with Advice to the Farmer, and Detailed Information for the several Classes of Labourers and Artisans. Map. 18mo. 1*s.* 6*d.*

WILSON'S (DANIEL, D.D., BISHOP OF CALCUTTA) Life, with Extracts from his Letters and Journals. By Rev. JOSIAH BATEMAN. *New and Condensed Edition.* Illustrations. Post 8vo. 9*s.*

———— (GENL. SIR ROBERT) Secret History of the French Invasion of Russia, and Retreat of the French Army, 1812. *Second Edition.* 8vo. 15*s.*

———— Private Diary of Travels, Personal Services, and Public Events, during Missions and Employments in Spain, Sicily, Turkey, Russia, Poland, Germany, &c. 1812-14. 2 Vols. 8vo. 26*s.*

———— Life. Edited from Autobiographical Memoirs. Portrait. 2 Vols. 8vo. 26*s.*

WORDSWORTH'S (CANON) Journal of a Tour in Athens and Attica. *Third Edition.* Plates. Post 8vo. 8*s.* 6*d.*

———— Pictorial, Descriptive, and Historical Account of Greece, with a History of Greek Art, by G. SCHARF, F.S.A. *New Edition.* With 600 Woodcuts. Royal 8vo. 28*s.*

WORNUM (RALPH). A Biographical Dictionary of Italian Painters: with a Table of the Contemporary Schools of Italy. By a LADY. Post 8vo. 6*s.* 6*d.*

WROTTESLEY'S (LORD) Thoughts on Government and Legislation. Post 8vo. 7*s.* 6*d.*

YOUNG'S (DR. THOS.) Life and Miscellaneous Works, edited by DEAN PEACOCK and JOHN LEITCH. Portrait and Plates. 4 Vols. 8vo. 15*s.* each.

BRADBURY AND EVANS, PRINTERS, WHITEFRIARS.

MASKED-DANCE AND WEDDING-FEAST OF TUCÚNA INDIANS.

Frontispiece to Vol. II.

THE

NATURALIST ON THE RIVER AMAZONS,

A RECORD OF ADVENTURES, HABITS OF ANIMALS, SKETCHES OF BRAZILIAN AND INDIAN LIFE, AND ASPECTS OF NATURE UNDER THE EQUATOR, DURING ELEVEN YEARS OF TRAVEL.

BY HENRY WALTER BATES.

Pelopæus Wasp building nest.

IN TWO VOLUMES.—VOL. II.

LONDON:
JOHN MURRAY, ALBEMARLE STREET.
1863.

[*The Right of Translation is Reserved.*]

CONTENTS OF VOL. II.

CHAPTER I.

SANTAREM.

CHAPTER II.

VOYAGE UP THE TAPAJOS.

CHAPTER III.

THE UPPER AMAZONS—VOYAGE TO EGA.

CHAPTER IV.

EXCURSIONS IN THE NEIGHBOURHOOD OF EGA.

CHAPTER V.

ANIMALS OF THE NEIGHBOURHOOD OF EGA.

CHAPTER VI.

EXCURSIONS BEYOND EGA.

LIST OF ILLUSTRATIONS.

VOL. II.

THE

NATURALIST ON THE AMAZONS.

CHAPTER I.

SANTAREM.

Situation of Santarem—Manners and customs of the inhabitants—Trade—Climate—Leprosy—Historical sketch—Grassy campos and woods—Excursions to Mapirí, Mahicá, and Irurá, with sketches of their Natural History ; Palms, wild fruit-trees, Mining Wasps, Mason Wasps, Bees, Sloths, and Marmoset Monkeys—Natural History of Termites or White Ants.

I HAVE already given a short account of the size, situation, and general appearance of Santarem. Although containing not more than 2500 inhabitants, it is the most civilised and important settlement on the banks of the main river from Peru to the Atlantic. The pretty little town, or city as it is called, with its rows of tolerably uniform, white-washed and red-tiled houses surrounded by green gardens and woods, stands on gently sloping ground on the eastern side of the Tapajos, close to its point of junction with the Amazons. A small eminence on which a fort has been erected, but which is now in a dilapidated condition, overlooks the streets, and forms the eastern limit of the mouth of the tributary. The Tapajos at Santarem is contracted to a breadth of about a mile-and-a-half by an accretion of low alluvial land, which forms a kind of delta on the western side; fifteen miles further up the river is seen at its full width of ten or a dozen miles, and the magnificent hilly country through which it flows from the south, is then visible on both shores. This high land, which appears to be a continuation of the central table-lands of Brazil, stretches almost without interruption on the eastern side of the river down to its mouth at Santarem. The scenery as well as the soil, vegetation and animal tenants of this region, are widely different from those of the flat and uniform country which borders the Amazons along most part of its course. After travelling week after week on the main river, the aspect of Santarem with its broad white sandy beach, limpid dark-green waters, and line of picturesque hills rising behind over the fringe of green forest, affords an agreeable surprise. On the main Amazons, the prospect is monotonous unless the vessel runs near the shore, when the wonderful diversity and beauty of the vegetation afford constant entertainment. Otherwise, the unvaried, broad yellow stream, and the long low line of forest, which dwindles away in a broken line of trees on the sea-like horizon and is renewed, reach after reach, as the voyager advances; weary by their uniformity.

I arrived at Santarem on my second journey into the interior, in November, 1851, and made it my head quarters for a period, as it turned out, of three years and a half. During this time I made, in pursuance of the plan I had framed, many excursions up the Tapajos, and to other places of interest in the surrounding region. On landing, I found no difficulty in hiring a suitable house on the outskirts of the place. It was pleasantly situated near the beach, going towards the aldeia or Indian part of the town. The ground sloped from the back premises down to the waterside, and my little raised verandah overlooked a beautiful flower-garden, a great rarity in this country, which belonged to the neighbours. The house contained only three rooms, one with brick and two with boarded floors. It was substantially built, like all the better sort of houses in Santarem, and had a stuccoed front. The kitchen, as is usual, formed an outhouse placed a few yards distant from the other rooms. The rent was 12,000 reis, or about twenty-seven shillings a month. In this country, a tenant has no extra payments to make ; the owners of house property pay a dizimo or tithe, to the "collectoria geral," or general treasury, but with this the occupier of course has nothing to do. In engaging servants, I had the good fortune to meet with a free mulatto, an industrious and trustworthy young fellow, named José, willing to arrange with me ; the people of his family cooking for us, whilst he assisted me in collecting ; he proved of the greatest service in the different excursions we subsequently made. Servants of any kind were almost impossible to be obtained at Santarem, free people being too proud to hire themselves, and slaves too few and valuable to their masters, to be let out to others. These matters arranged, the house put in order, and a rude table, with a few chairs, bought or borrowed to furnish the house with, I was ready in three or four

B 2

days to commence my Natural History explorations in the neighbourhood.

I found Santarem quite a different sort of place from the other settlements on the Amazons. At Cametá, the lively, good-humoured, and plain-living Mamelucos formed the bulk of the population, the white immigrants there, as on the Rio Negro and Upper Amazons, seeming to have fraternised well with the aborigines. In the neighbourhood of Santarem the Indians, I believe, were originally hostile to the Portuguese; at any rate, the blending of the two races has not been here on a large scale. I did not find the inhabitants the pleasant, easy-going, and blunt-spoken country folk that are met with in other small towns of the interior. The whites, Portuguese and Brazilians, are a relatively more numerous class here than in other settlements, and make great pretensions to civilisation; they are the merchants and shopkeepers of the place; owners of slaves, cattle estates, and cacao plantations. Amongst the principal residents must also be mentioned the civil and military authorities, who are generally well-bred and intelligent people from other provinces. Few Indians live in the place; it is too civilised for them, and the lower class is made up (besides the few slaves) of half-breeds, in whose composition negro blood predominates. Coloured people also exercise the different handicrafts; the town supports two goldsmiths, who are mulattoes and have each several apprentices; the blacksmiths are chiefly Indians, as is the case generally throughout the province. The manners of the upper class (copied from those of Pará), are very stiff and formal, and the absence of the hearty hospitality met with in other places, produces a disagreeable impression at first. Much ceremony is observed in the intercourse of the principal people with each other, and with strangers. The best room in each house is set apart for receptions, and visitors are expected to present themselves in black dress coats, regardless of the furious heat which rages in the sandy streets of Santarem towards mid-day, the hour when visits are generally made. In the room a cane-bottomed sofa and chairs, all lacquered and gilded, are arranged in quadrangular form, and here the visitors are invited to seat themselves, whilst the compliments are passed, or the business arranged. In taking leave, the host backs out his guests with repeated bows, finishing at the front door. Smoking is not in vogue amongst this class, but snuff-taking is largely indulged in, and great luxury is displayed in gold and silver snuff-boxes. All the gentlemen, and indeed most of the ladies also, wear gold watches and guard chains. Social parties are not very frequent; the principal men being fully occupied with their business and families, and the rest spending their leisure in billiard and gambling rooms, leaving wives and daughters shut up at home. Occasionally, however, one of the principal citizens gives a ball. In the first that I attended, the gentlemen were seated all the evening on one side of the room, and the ladies on the other, and partners were allotted by means of numbered cards, distributed by a master of the ceremonies. But the customs changed rapidly in these matters after steamers began to run on the Amazons (in 1853), bringing a flood of new ideas and fashions into the country. The old, bigoted, Portuguese system of treating women, which stifled social intercourse and wrought endless evils in the private life of the Brazilians, is now being gradually, although slowly, abandoned.

When a stranger arrives at an interior town in Brazil, with the intention of making some stay, he is obliged within three days to present himself at the Police office, to show his passport. He is then expected to call on the different magistrates, the military commander, and the principal private residents. This done, he has to remain at home a day or two to receive return visits, after which he is considered to be admitted into the best society. Santarem being the head of a comarca or county, as well as a borough, has a resident high judge (Juiz de Direito), besides a municipal judge (Juiz Municipal) and recorder (Promotor publico). The head of the police is also a magistrate, having jurisdiction in minor cases; he is called the delegado or delegate of police, from being appointed by and subordinate to the chief of police in the capital: all these officers are nominated by the Central Government. In a pretentious place like Santarem, the people attach great importance to these matters, and I had to go a round of visiting before I finally settled down to work. Notwithstanding the ceremonious manners of the principal inhabitants, I found several most worthy and agreeable people amongst them. Some of the older families, who spend most of their time on their plantations or cattle estates, were as kind-hearted and simple in their ways as the Obydos townsfolk. But these are rarely in town, coming only for a few days during the festivals. They have, however, spacious town-houses, some of them two stories high, with massive walls of stone or adobe. The principal citizen, Senhor Miguel Pinto de Guimaraens, is a native of the place, and is an example of the readiness with which talent and industry meet with their reward under the wise government of Brazil. He began life in a very humble way; I was told he was once a fisherman, and retailed the produce of his hook and line or nets in the port. He is now the chief merchant of the district; a large cattle and landed proprietor; and owner of a sugar estate and mills. When the new National Guard was formed in Brazil in 1853, he received from the Emperor the commission of colonel. He is a pale, grave, and white-haired, though only middle-aged, man. I saw a good deal of him, and liked his sincerity and the uprightness of his dealings. When I arrived in Santarem he was the delegado of police. He is rather unmerciful both in and out of office towards the shortcomings, in private and public morality, of his fellow-countrymen; but he is very much respected. The nation cannot be a despicable one, whose best men are thus able to work themselves up to positions of trust and influence.

The religious festivals were not so numerous here as in other towns, and such as did take place were very poor and ill attended. There is a handsome church, but the vicar showed remarkably little zeal for religion, except for a few days now and then when the Bishop came from Pará, on his rounds through the diocese. The people are as fond of holiday making here as in other parts of the province; but it seemed to be a growing fashion to substitute rational amusements for the processions and mummeries of the saints' days. The young folks are very musical, the principal instruments in use being the flute, violin, Spanish guitar, and a small four-stringed viola, called cavaquinho. During the early part of my stay at Santarem, a little party of instrumentalists, led by a tall, thin, ragged mulatto, who was quite an enthusiast in his art, used frequently to serenade their friends in the cool and brilliant moonlit evenings of the dry season, playing French and Italian marches and dance music with very good effect. The guitar was the favourite instrument with both sexes, as at Pará; the piano, however, is now fast superseding it. The ballads sung to the accompaniment of the guitar were not learnt from written or printed music, but communicated orally from one friend to another. They were never spoken of as songs, but *modinhas*, or "little fashions," each of which had its day, giving way to the next favourite brought by some young fellow from the capital. At festival times there was a great deal of masquerading, in which all the people, old and young, white, negro, and Indian, took great delight. The best things of this kind used to come off during the Carnival, in Easter week, and on St. John's eve; the negroes having a grand semi-dramatic display in the streets at Christmas time. The more select affairs were got up by the young whites, and coloured men associating with whites. A party of thirty or forty of these used to dress themselves in uniform style, and in very good taste, as cavaliers and dames, each disguised with a peculiar kind of light gauze mask. The troop, with a party of musicians, went the round of their friends' houses in the evening, and treated the large and gaily-dressed companies which were there assembled to a variety of dances. The principal citizens, in the large rooms of whose houses these entertainments were given, seemed quite to enjoy them; great preparations were made at each place; and, after the dance, guests and masqueraders were regaled with pale ale and sweet-meats. Once a year the Indians, with whom masked dances and acting are indigenous, had their turn, and on one occasion they gave us a great treat. They assembled from different parts of the neighbourhood at night, on the outskirts of the town, and then marched through the streets by torchlight towards the quarter inhabited by the whites, to perform their hunting and devil dances before the doors of the principal inhabitants. There were about a hundred men, women, and children in the procession. Many of the men were dressed in the magnificent feather crowns, tunics, and belts, manufactured by the Mundurucús, and worn by them on festive occasions, but the women were naked to the waist, and the children quite naked, and all were painted and smeared red with anatto. The ringleader enacted the part of the Tushaua, or chief, and carried a sceptre, richly decorated with the orange, red, and green feathers of toucans and parrots. The pajé or medicine-man came along, puffing at a long tauarí cigar, the instrument by which he professes to make his wonderful cures. Others blew harsh jarring blasts with the turé, a horn made of long and thick bamboo, with a split reed in the mouthpiece. This is the war trumpet of many tribes of Indians, with which the sentinels of predatory hordes, mounted on a lofty tree, give the signal for attack to their comrades. Those Brazilians who are old enough to remember the times of warfare between Indians and settlers, retain a great horror of the turé, its loud harsh note heard in the dead of the night having been often the prelude to an onslaught of bloodthirsty Múras on the outlying settlements. The rest of the men in the procession carried bows and arrows, bunches of javelins, clubs, and paddles. The older children brought with them the household pets; some had monkeys or coatis on their shoulders, and others bore tortoises on their heads. The squaws carried their babies in aturás, or large baskets, slung on their backs, and secured with a broad belt of bast over their foreheads. The whole thing was accurate in its representation of Indian life, and showed more ingenuity than some people give the Brazilian red man credit for. It was got up spontaneously by the Indians, and simply to amuse the people of the place.

The entire produce in cacao, salt fish, and other articles of a very large district, passes through the hands of the Santarem merchants, and a large trade, for this country, is done with the Indians on the Tapajos in salsaparilla, balsam of copaüba, India-rubber, farinha, and other productions. I was told the average annual yield of the Tapajos in salsaparilla, was about 2000 arrobas (of 32 lbs. each). The quality of the drug found in the forests of the Tapajos, is much superior to that of the Upper Amazons, and always fetches double the price at Pará. The merchants send out young Brazilians and Portuguese in small canoes to trade on the rivers and collect the produce, and the cargoes are shipped to the capital in large cubertas and schooners, of from twenty to eighty tons burthen. The risk and profits must be great, or capital scarce, for the rate of interest on lent money or overdue accounts is two-and-a-half to three per cent. per month ; this is the same, however, as that which rules at Pará. The shops are numerous, and well-stocked with English, French, German, and North American wares ; the retail prices of which are very little above those of the capital. There is much competition amongst the traders and shopkeepers, yet they all seem to thrive, if one may judge from external appearances; but it is said, that most of them are over head and ears in debt to rich Portuguese merchants of Pará, who act as their correspondents.

The people seem to be thoroughly alive to the advantages of education for their children. Besides the usual primary schools, one for girls, and another for boys, there is a third of a higher class, where Latin and French, amongst other accomplishments, are taught by professors, who, like the common schoolmasters, are paid by the provincial government. This is used as a preparatory school to the Lyceum and Bishop's seminary, well-endowed institutions at Pará, whither it is the ambition of traders and planters to send their sons to finish their studies. The rudiments of education only are taught in the primary schools, and it is surprising how quickly and well the little lads, both coloured and white, learn reading, writing, and arithmetic. But the simplicity of the Portuguese language, which is written as it is pronounced, or according to unvarying rules, and the use of the decimal system of accounts, make these acquirements much easier than they are with us. Students in the superior school have to pass an examination before they can be admitted at the colleges in Pará, and the managers once did me the honour to make me one of the examiners for the year. The performances of the youths, most of whom were under fourteen years of age, were very creditable, especially in grammar; there was a quickness of apprehension displayed which would have gladdened the heart of a northern schoolmaster. The course of study followed at the colleges of Pará must be very deficient; for it is rare to meet with an educated Paraense who has the slightest knowledge of the physical sciences, or even of geography, if he has not travelled out of the province. The young men all become smart rhetoricians and lawyers; any of them is ready to plead in a law case at an hour's notice; they are also great at statistics, for the gratification of which taste there is ample field in Brazil, where every public officer has to furnish volumes of dry reports annually to the government; but they are wofully ignorant on most other subjects. I do not recollect seeing a map of any kind at Santarem. The quick-witted people have a suspicion of their deficiencies in this respect, and it is difficult to draw them out on geography; but one day a man holding an important office betrayed himself by asking me, "on what side of the river was Paris situated?" This question did not arise, as might be supposed, from a desire for accurate topographical knowledge of the Seine, but from the idea, that all the world was a great river, and that the different places he had heard of must lie on one shore or the other. The fact of the Amazons being a limited stream, having its origin in narrow rivulets, its beginning and its ending, has never entered the heads of most of the people who have passed their whole lives on its banks.

Santarem is a pleasant place to live in, irrespective of its society. There are no insect pests, mosquito, pium, sand-fly, or motuca. The climate is glorious ; during six months of the year, from August to February, very little rain falls, and the sky is cloudless for weeks together, the fresh breezes from the sea, nearly 400 miles distant, moderating the great heat of the sun. The wind is sometimes so strong for days together, that it is difficult to make way against it in walking along the streets, and it enters the open windows and doors of houses, scattering loose clothing and papers in all directions. The place is considered healthy ; but at the changes of season, severe colds and ophthalmia are prevalent. I found three Englishmen living here, who had resided many years in the town or its neighbourhood, and who still retained their florid complexions ; the plump and fresh appearance of many of the middle-aged Santarem ladies, also bore testimony to the healthfulness of the climate. The streets are always clean and dry, even in the height of the wet season ; good order is always kept, and the place pretty well supplied with provisions. None but those who have suffered from the difficulty of obtaining the necessaries of life at any price in most of the interior settlements of South

America, can appreciate the advantages of Santarem in this respect. Everything, however, except meat, was dear, and becoming every year more so. Sugar, coffee, and rice, which ought to be produced in surplus in the neighbourhood, are imported from other provinces, and are high in price; sugar indeed, is a little dearer here than in England. There were two or three butchers' shops, where excellent beef could he had daily at twopence or twopence-halfpenny per pound. The cattle have not to be brought from a long distance as at Pará, being bred on the campos, which border the Lago Grande, only one or two days' journey from the town. Fresh fish could be bought in the port on most evenings, but, as the supply did not equal the demand, there was always a race amongst purchasers to the water-side when the canoe of a fisherman hove in sight. Very good bread was hawked around the town every morning, with milk, and a great variety of fruits and vegetables. Amongst the fruits, there was a kind called atta, which I did not see in any other part of the country. It belongs to the Anonaceous order, and the tree which produces it grows apparently wild in the neighbourhood of Santarem. It is a little larger than a good-sized orange, and the rind, which encloses a mass of rich custardy pulp, is scaled like the pine-apple, but green when ripe, and encrusted on the inside with sugar. To finish this account of the advantages of Santarem, the delicious bathing in the clear waters of the Tapajos may be mentioned. There is here no fear of alligators; when the east wind blows, a long swell rolls in on the clean sandy beach, and the bath is most exhilarating.

There is one great drawback to the merits of Santarem. This is the prevalence here of the terrible leprosy. It seems, however, confined to certain families, and I did not hear of a well-authenticated case of a European being attacked by it. I once visited many of the lepers in company of an American physician. They do not live apart; family ties are so strong, that all attempts to induce people to separate from their leprous relatives have failed; but many believe that the malady is not contagious. The disease commences with glandular swellings in different parts of the body, which are succeeded by livid patches on the skin, and at the tips of the fingers and toes. These spread, and the parts embraced by them lose their sensibility, and decay. In course of time, as the frightful atrophy extends to the internal organs, some vital part is affected, and the sufferer dies. Some of the best families in the place are tainted with leprosy; but it falls on all races alike; white, Indian, and negro. I saw some patients who had been ill of it for ten and a dozen years; they were hideously disfigured, but bore up cheerfully; in fact, a hopeful spirit, and free, generous living had been the means of retarding in them the progress of the disorder; none were ever known to be cured of it. One man tried a voyage to Europe, and was healed whilst there, but the malady broke out again on his return. I do not know whether the dry and hot soil of Santarem has anything to do with the prevalence of this disease; it is not confined to this place, many cases having occurred at Pará, and in other provinces, but it is nowhere so rife as here; the evil fame of the settlement indeed has spread to Portugal, where Santarem is known as the "Cidade dos Lazaros," or City of Lepers.

When the Portuguese first ascended the Amazons towards the middle of the 17th century, they found the banks of the Tapajos in the neighbourhood of Santarem, peopled by a warlike tribe of Indians, called the Tapajócos. From these, the river and the settlement (Santarem in the Indian language is called Tapajós), derive their name. The Tapajos, however, amongst the Brazilian settlers in this part, is most generally known by the Portuguese name of Rio Preto, or the Black River. According to Acunna, the historian of the Teixeira expedition (in 1637-9), the Tapajócos were very numerous, one village alone having contained more than 500 families. Their weapons were poisoned darts. Notwithstanding their numbers and courage, they quickly gave way before the encroaching Portuguese settlers, who are said to have treated them with great barbarity. The name of the tribe is no longer known in the neighbourhood, but it is probable their descendants still linger on the banks of the Lower Tapajos, a traditional hatred towards the Portuguese having been preserved amongst the semi-civilised inhabitants to the present day. The fact of the Urarí poison having been in use amongst the Tapajócos is curious, inasmuch as it shows there was at that time communication between distant tribes along the course of the main Amazons. The Indians now living on the banks of the Tapajos are ignorant of the Urarí, the drug being prepared only by tribes which live on the rivers flowing into the Upper Amazons from the north, 1200 miles distant from the Tapajos.

The city of Santarem suffered greatly during the disorders of 1835-6. According to the accounts I received, it must have been just before that time a much more flourishing place than it is now. There were many more large proprietors, rich in slaves and cattle; the produce of cacao was greater; and a much larger trade was done with the miners of Matto Grosso, who descended the Tapajos with their gold and diamonds, to exchange for salt, hardware, and other heavy European goods. An old Scotch gentleman, Captain Hislop, who had lived here for about thirty-five years, told me that Santarem was then a most delightful place to live in. Provisions were abundant and cheap; labour was easily obtained; and the greatest order, friendliness, and contentment prevailed. The political squabble amongst the whites, which began the troubles, ended, in this part of the country, in a revolt of the Indians. At the beginning of the disorders two parties were formed, one tolerant of the "Bicudos" (long-snouts), as the Portuguese were nicknamed, and supporters of the legal Brazilian Government; the other in favour of revolution, expulsion of the Portuguese, and native rule. The latter co-operated with a large body of rebels who had collected at a place on the banks of the river, not far distant; and on a certain day, according to agreement, the town was invaded by the horde of scoundrels and mistaken patriots. All the Portuguese and those who befriended them, that these infuriated people could lay their hands on, were brutally massacred. A space filled with mounds, amongst the myrtle bushes in the woods behind Santarem, now marks the spot where these poor fellows were confusedly buried. I could give a long account of the horrors of this time as they were related to me; but I think the details would not serve any useful purpose. It must not be thought, however, that the Amazonian people are habitually a blood-thirsty race; on the contrary, the peaceableness and gentleness of character of the inhabitants of this province, in quiet times, are proverbial throughout Brazil. The rarity or absence of deeds of violence from year to year is always commented upon by the President in his annual report to the Central Government.

When the Cabanas or rebels entered the town, the friends of lawful government retired to a large block of buildings near the water-side, which they held for many days, to cover the embarcation of their families and moveables. The negro slaves generally remained faithful to their masters. Whilst the embarcation was going on many daring feats were performed, chiefly by coloured people: one brave fellow, a mameluco, named Paca, made a bold dash one day, with a few young men of the same stamp, and secured five or six of the rebel leaders, who were carried, gagged and handcuffed, on board a schooner in the port. But the legal party were greatly outnumbered and deficient in arms and ammunition, and they were obliged, soon after Paca's feat, entirely to evacuate the town; retiring to the village of Prayinha, about 150 miles down the river. Those citizens of Santarem who sympathised with the rebels were obliged to follow soon after, as the revolt took the shape of a war between Indians and whites. The red skins, however, made an exception in favour of the few English and French residents. Captain Hislop remained in the town during its occupation by the Cabanas, and told me that he was treated very well by the Indians and rebel chiefs.

After Santarem was recaptured, about nine months subsequent to these events, by a small sea and land force sent from Rio Janeiro, aided by the townspeople who were picked up at Prayinha, it was again attacked by a large force of Indians. This affair showed the blind fearlessness and obstinacy of the Indian character in a striking manner. An attack was expected, as the rebels were known to be concealed in great numbers in the neighbouring woods; so the Commandante of the garrison (Captain Leaõ) had the whites' quarter strongly stockaded, and every man slept under arms. The Indians acted as though inspired by a diabolical fanaticism; they had no arms, except wooden spears, clubs, and bows and arrows; for their powder and lead had been exhausted long before. With these rude weapons they came through forest and campo to the storming of the now fortified town. The attack was made at sunrise; the sentinels were killed or driven in, and the swarms of red skins climbed the stockade and thronged down the principal street. They were soon met by a strong and well-armed force, well posted in houses or behind walls, and the reckless savages were shot down by hundreds. It was not until the street was encumbered by the heaps of slain that the rest turned their

c 2

backs and fled. Their numbers were estimated at 2000 men; the remnant of the force escaped across the campos to the village of Altar do Chao, twenty miles distant, whence they scattered themselves along the shores of the Tapajos, and gave great trouble to the Brazilians for many years afterwards. Several expeditions were sent from Santarem to reduce them, a task in which the Government was aided by the friendly Mundurucús of the Upper Tapajos, a large body of whom, under the leadership of their Tushaúa Joaquim, made war on the hostile Indians on the lower parts both of the Madeira and the Tapajos, until they were nearly exterminated.

The country around Santarem is not clothed with dense and lofty forest, like the rest of the great humid river plain of the Amazons. It is a campo region; a slightly elevated and undulating tract of land, wooded only in patches, or with single scattered trees. A good deal of the country on the borders of the Tapajos, which flows from the great campo area of Interior Brazil, is of this description. On this account I consider the eastern side of the river, towards its mouth, to be a northern prolongation of the continental land, and not a portion of the alluvial flats of the Amazons. The soil is a coarse gritty sand; the substratum, which is visible in some places, consisting of sandstone conglomerate probably of the same formation as that which underlies the Tabatinga clay in other parts of the river valley. The surface is carpeted with slender hairy grasses, unfit for pasture, growing to a uniform height of about a foot. The patches of wood look like copses in the middle of green meadows; they are called by the natives "ilhas de mato," or islands of jungle; the name being, no doubt, suggested by their compactness of outline, neatly demarcated in insular form from the smooth carpet of grass around them. They are composed of a great variety of trees, loaded with succulent parasites, and lashed together by woody climbers, like the forest in other parts. A narrow belt of dense wood, similar in character to these ilhas, and like them sharply limited along its borders, runs everywhere parallel and close to the river. In crossing the campo, the path from the town ascends a little for a mile or two, passing through this marginal strip of wood; the grassy land then slopes gradually to a broad valley, watered by rivulets, whose banks are clothed with lofty and luxuriant forest. Beyond this, a range of hills extends as far as the eye can reach towards the yet untrodden interior. Some of these hills are long ridges, wooded or bare; others are isolated conical peaks, rising abruptly from the valley. The highest are probably not more than a thousand feet above the level of the river. One remarkable hill, the Serra de Muruarú, about fifteen miles from Santarem, which terminates the prospect to the south, is of the same truncated pyramidal form as the range of hills near Almeyrim. Complete solitude reigns over the whole of this stretch of beautiful country. The inhabitants of Santarem know nothing of the interior, and seem to feel little curiosity concerning it. A few tracks from the town across the campo lead to some small clearings four or five miles off, belonging to the poorer inhabitants of the place; but, excepting these, there are no roads, or signs of the proximity of a civilised settlement.

The sandy soil and scanty clothing of trees are probably the causes of the great dryness of the climate. In some years no rain falls from August to February; whilst in other parts of the Amazons plains, both above and below this middle part of the river, heavy showers are frequent throughout the dry season. I have often watched the rain-clouds in November and December, when the shrubby vegetation is parched up by the glowing sun of the preceding three months, rise as they approached the hot air over the campos, or diverge from it to discharge their contents on the low forest-clad islands of the opposite shore. The trade-wind, however, blows with great force during the dry months; the hotter the weather the stronger is the breeze, until towards the end of the season it amounts to a gale, stopping the progress of downward-bound vessels.

Some of the trees which grow singly on the campos are very curious. The caju is very abundant; indeed, some parts of the district might be called orchards of this tree, which seems to prefer sandy or gravelly soils. There appear to be several distinct species of it growing in company, to judge by the differences in the colour, flavour, and size of the fruit. This, when ripe, has the colour and figure of a codlin apple, but it has a singular appearance owing to the large kidney-shaped kernel growing outside the pulpy portion of the fruit. It ripens in January, and the poorer classes of Santarem then resort to the campos and gather immense quantities, to make a drink or "wine" as it is called, which is considered a remedy in certain cutaneous disorders. The kernels are roasted and eaten. Another wild fruit-tree is the Murishí (Byrsomina), which yields an abundance of small yellow acid berries. A decoction of its bark dyes cloth a maroon colour. It is employed for this purpose chiefly by the Indians, and coarse cotton shirts tinted with it were the distinctive badges of the native party during the revolution. A very common tree in the Ilhas do Mato is the Breio branco, which secretes from the inner bark a white resin, resembling camphor in smell and appearance. The fruit is a small black berry, and the whole tree, fruit, leaf, and stem, has the same aromatic fragrance. By loosening the bark and allowing the resin to flow freely, I collected a large quantity, and found it of great service in preserving my insect collections from the attacks of ants and mites. Another tree, much rarer than the Breio branco, namely the Umirí (Humirium floribundum), growing in the same localities, distils in a similar way an oil of the most *recherché* fragrance. The yield, however, is very small. The native women esteem it highly as a scent. To obtain a supply of the precious liquid, large strips of bark are loosened and pieces of cotton left in soak underneath. By visiting the tree daily, and pressing the oil from the cotton, a small phial containing about an ounce may be filled in the course of a month. One of the most singular of the vegetable productions of the campos is the Súcu-úba tree (Plumieria phagedænica). It grows in the greatest luxuriance in the driest parts, and with its long, glossy, dark-green leaves, fresh and succulent even in the most arid seasons, and white jasmine-like flowers, forms the greatest decoration of these solitary places. The bark, leaves, and leaf-stalks, yield a copious supply of milky sap, which the natives use very generally as plaister in local inflammations, laying the liquid on the skin with a brush, and covering the place with cotton. I have known it to work a cure in many cases; but, perhaps, the good effect is attributable to the animal heat drawn to the place by the pad of cotton. The milk flows most freely after the occasional heavy rains in the intervals between the dry and wet seasons; it then spurts out with great force from any part of the tree if hacked with a knife in passing.

The appearance of the campos changes very much according to the season. There is not that grand uniformity of aspect throughout the year which is observed in the virgin forest, and which makes a deeper impression on the naturalist the longer he remains in this country. The seasons in this part of the Amazons region are sharply contrasted, but the difference is not so great as in some tropical countries, where, during the dry monsoon, insects and reptiles æstivate, and the trees simultaneously shed their leaves. As the dry season advances (August, September), the grass on the campos withers, and the shrubby vegetation near the town becomes a mass of parched yellow stubble. The period, however, is not one of general torpidity or repose for animal or vegetable life. Birds certainly are not so numerous as in the wet season, but some kinds remain and lay their eggs at this time—for instance, the ground doves (Chamæpelia). The trees retain their verdure throughout, and many of them flower in the dry months. Lizards do not become torpid, and insects are seen both in the larva and the perfect states, showing that the aridity of the climate has not a general influence on the development of the species. Some kinds of butterflies, especially the little hair-streaks (Theclæ), whose larvæ feed on the trees, make their appearance only when the dry season is at its height. The land molluscs of the district, are the only animals which æstivate; they are found in clusters, Bulimi and Helices, concealed in hollow trees, the mouths of their shells closed by a film of mucus. The fine weather breaks up often with great suddenness about the beginning of February. Violent squalls from the west or the opposite direction to the trade-wind then occur. They give very little warning, and the first generally catches the people unprepared. They fall in the night, and blowing directly into the harbour, with the first gust sweep all vessels from their anchorage; in a few minutes, a mass of canoes, large and small, including schooners of fifty tons burthen, are clashing together, pell mell, on the beach. I have reason to remember these storms, for I was once caught in one myself, whilst crossing the river in an undecked boat, about a day's journey from Santarem. They are accompanied with terrific electric explosions, the sharp claps of thunder falling almost simultaneously with the blinding flashes of lightning. Torrents of rain follow the first outbreak; the wind then gradually abates, and the rain subsides into a steady drizzle, which continues often for the greater part of the succeeding day. After a week or two of showery weather the aspect of the country is completely changed. The parched ground in the neighbourhood of Santarem breaks out, so to speak, in a rash of greenery; the dusty, languishing trees gain, without having shed their old leaves, a new clothing of tender green foliage; a wonderful variety of quick-growing leguminous plants springs up, and leafy creepers overrun the ground, the bushes, and the trunks of trees. One is reminded of the sudden advent of spring after a few warm showers in northern climates; I was the more struck by it as nothing similar is witnessed in the virgin forests amongst which I had passed the four years previous to my stay in this part. The grass on the campos is renewed, and many of the campo trees, especially the myrtles, which grow abundantly in one portion of the district, begin to flower, attracting by the fragrance of their blossoms a great number and variety of insects, more particularly Coleoptera. Many kinds of birds; parrots, toucans, and barbets, which live habitually in the forest, then visit the open places. A few weeks of comparatively dry weather generally intervene in March, after a month or two of rain. The heaviest rains fall in April, May, and June; they come in a succession of showers, with sunny gleamy weather in the intervals. June and July are the months when the leafy luxuriance of the campos, and the activity of life, are at their highest. Most birds have then completed their moulting, which extends over the period from February to May. The flowering shrubs are then mostly in bloom, and numberless kinds of Dipterous and Hymenopterous insects appear simultaneously with the flowers. This season might be considered the equivalent of summer in temperate climates, as the bursting forth of the foliage in February represents the spring; but under the equator there is not that simultaneous march in the annual life of animals and plants, which we see in high latitudes; some species, it is true, are dependent upon others in their periodical acts of life, and go hand-in-hand with them, but they are not all simultaneously and similarly affected by the physical changes of the seasons.

I will now give an account of some of my favourite collecting places in the neighbourhood of Santarem, incorporating with the description a few of the more interesting observations made on the Natural History of the localities. To the west of the town there was a pleasant path along the beach to a little bay, called Mapirí, about five miles within the mouth of the Tapajos. The road was practicable only in the dry season. The river at Santarem rises on the average about thirty feet, varying in different years about ten feet; so that in the four months, from April to July, the water comes up to the edge of the marginal belt of wood already spoken of. This Mapirí excursion was most pleasant and profitable in the months from January to March, before the rains become too continuous. The sandy beach beyond the town is very irregular; in some places forming long spits on which, when the east wind is blowing, the waves break in a line of foam; at others receding to shape out quiet little bays and pools. On the outskirts of the town a few scattered huts of Indians and coloured people are passed, prettily situated on the margin of the white beach, with a background of glorious foliage; the cabin of the pure-blood Indian being distinguished from the mud hovels of the free negroes and mulattoes by its light construction, half of it being an open shed where the dusky tenants are seen at all hours of the day lounging in their open-meshed grass hammocks. About two miles on the road we come to a series of shallow pools, called the Laguinhos, which are connected with the river in the wet season, but separated from it by a high bank of sand topped with bushes at other times. There is a break here in the fringe of wood, and a glimpse is obtained of the grassy campo. When the waters have risen to the level of the pools this place is frequented by many kinds of wading birds. Snow-white egrets of two species stand about the margins of the water, and dusky-striped herons may be seen half hidden under the shade of the bushes. The pools are covered with a small kind of water-lily, and surrounded by a dense thicket. Amongst the birds which inhabit this spot is the rosy-breasted Troupial (Trupialis Guianensis), a bird resembling our starling in size and habits, and not unlike it in colour, with the exception of the rich rosy vest. The water at this time of the year overflows a large level tract of campo bordering the pools, and the Troupials come to feed on the larvæ of insects which then abound in the moist soil.

Beyond the Laguinhos there succeeds a tract of level beach covered with trees which form a beautiful grove.

About the month of April, when the water rises to this level, the trees are covered with blossom, and a handsome orchid, an Epidendron with large white flowers, which clothes thickly the trunks, is profusely in bloom. Several kinds of kingfisher resort to the place: four species may be seen within a small space: the largest as big as a crow, of a mottled-grey hue, and with an enormous beak; the smallest not larger than a sparrow. The large one makes its nest in clay cliffs, three or four miles distant from this place. None of the kingfishers are so brilliant in colour as our English species. The blossoms on the trees attract two or three species of humming-birds, the most conspicuous of which is a large swallow-tailed kind (Eupetomena macroura), with a brilliant livery of emerald green and steel blue. I noticed that it did not remain so long poised in the air before the flowers as the other smaller species; it perched more frequently, and sometimes darted after small insects on the wing. Emerging from the grove there is a long stretch of sandy beach; the land is high and rocky, and the belt of wood which skirts the river banks is much broader than it is elsewhere. At length, after rounding a projecting bluff, the bay of Mapirí is reached. The river view is characteristic of the Tapajos: the shores are wooded, and on the opposite side is a line of clay cliffs, with hills in the background clothed with rolling forest. A long spit of sand extends into mid-river, beyond which is an immense expanse of dark water, the further shore of the Tapajos being barely visible as a thin grey line of trees on the horizon. The transparency of air and water in the dry season when the brisk east wind is blowing, and the sharpness of outline of hills, woods, and sandy beaches, give a great charm to this spot.

The little pools along the beach were tenanted by several species of fresh-water mollusks. The most abundant was a long turret-shaped Melania, which swarmed in them in the same way as Limnææ do in ponds at home. I found no Limnæa, nor indeed any European genus of fresh-water mollusk, in the Amazons region. After the first storms of February the coast is strewn with large apple-shells (Ampullaria). They are not inhabitants of the pools on this side of the river, but are involuntary visitors, being driven across by the wind and waves with masses of marsh plants from the low land of the opposite shore. A great many are dead shells, and more or less worn. In showery weather I seldom came this way without seeing one or more water snakes of the genus Helicops. They were generally concealed under the heaps of thick aquatic grasses cast ashore by storms; and when exposed, always made off straight for the water. They glided along with such agility that I rarely succeeded in capturing one, and on reaching the river they sought at once the bottom in the deepest parts. I believe these snakes are swept from the marshy land of the western shore with the patches of grass and the Ampullariæ just mentioned. Other reptiles and a great number of insects are blown or floated over in the same way by the violent squalls which occur in January or February. None of the species take root on the Santarem side of the river. Sometimes myriads of Coleopterous insects, belonging to about half a dozen kinds, are blown across, and become perfect pests to the town's people for two or three nights, swarming about the lights in every chamber. They get under one's clothing, or down one's back, and pass from the oil-lamp on to the furniture, books, and papers, smearing everything they touch. The open shops facing the beach become filled with them, and customers have to make a dash in and out through the showers that fall about the large brass lamps over the counter, when they want to make a purchase. The species are certainly not indigenous to the eastern side of the river; the hosts soon disappear; those which cannot get back must perish helplessly, for the soil, vegetation, and climate of the Santarem side are ill suited to the inhabitants of the opposite shore.

The pools I have mentioned were tenanted by a considerable variety of insects.* I found also a very large number, chiefly of carnivorous land-beetles under the pebbles and rejectamenta along the edge of the water during my many rambles. I was much struck with the similarity of the Dragon-flies (whose early states are passed in the water) to those of Britain. A species of Libellula with pointed tail, which darted about over the bushes near the ponds, is very closely

* The water-beetles found in the pools belonged to seventeen genera, thirteen of which are European. Those European genera which form the greater part of the pond population in Coleoptera in northern latitudes, are quite absent in the Amazons region: these are, Haliplus, Cnemidotus, Pelobius, Noterus, Ilybius, Agabus, Colymbetes, Dyticus, and Acilius: Hydropori, also, are very rare. The most common species belong to the genera Hydracanthus, Copelatus, Cybister, Tropisternus, and Berosus, three of which are unknown in Europe.

allied to our English L. quadrimaculata. But the resemblance was greater in the small, slender-bodied and slow-flying species, the Agrions, which every lover of rural walks must have noticed in England by river sides. There was one pretty kind with a pale blue ring at the tip of the body which resembled to a remarkable degree a common British species. Although very near akin, neither this nor any of the other kinds, were perfectly identical with European ones. The strikingly peculiar dragon-flies from Tropical America which are seen in our collections are denizens of the forest, being bred in the shady brooks and creeks in their recesses, and not in the weedy ponds of open places. Some of these forest species are strange creatures with slender bodies measuring seven inches in length; their elegant lace-work wings tipped with white or yellow. They fly slowly amongst the trees, preying on small Diptera, and in their flight look like animated spindles; the wings, placed at the fore extremity of the long, horizontally-extended body, moving rapidly and creating the impression of rotary motion.

Whilst resting in the shade during the great heat of the early hours of afternoon, I used to find amusement in watching the proceedings of the sand-wasps. A small pale green kind of Bembex (Bembex ciliata), was plentiful near the bay of Mapirí. When they are at work, a number of little jets of sand are seen shooting over the surface of the sloping bank. The little miners excavate with their fore feet, which are strongly built and furnished with a fringe of stiff bristles; they work with wonderful rapidity, and the sand thrown out beneath their bodies issues in continuous streams. They are solitary wasps, each female working on her own account. After making a gallery two or three inches in length in a slanting direction from the surface, the owner backs out and takes a few turns round the orifice apparently to see whether it is well made, but in reality, I believe, to take note of the locality, that she may find it again. This done, the busy workwoman flies away ; but returns, after an absence varying in different cases from a few minutes to an hour or more, with a fly in her grasp, with which she re-enters her mine. On again emerging, the entrance is carefully closed with sand. During this interval she has laid an egg on the body of the fly which she had previously benumbed with her sting, and which is to serve as food for the soft, footless grub soon to be hatched from the egg. From what I could make out, the Bembex makes a fresh excavation for every egg to be deposited ; at least in two or three of the galleries which I opened there was only one fly enclosed.

I have said that the Bembex on leaving her mine took note of the locality : this seemed to be the explanation of the short delay previous to her taking flight ; on rising in the air also the insects generally flew round over the place before making straight off. Another nearly allied but much larger species, the Monedula signata, whose habits I observed on the banks of the Upper Amazons, sometimes excavates its mine solitarily on sand-banks recently laid bare in the middle of the river, and closes the orifice before going in search of prey. In these cases the insect has to make a journey of at least half a mile to procure the kind of fly, the Motúca (Hadaüs lepidotus), with which it provisions its cell. I often noticed it to take a few turns in the air round the place before starting; on its return it made without hesitation straight for the closed moüth of the mine. I was convinced that the insects noted the bearings of their nests and the direction they took in flying from them. The proceeding in this and similar cases (I have read of something analogous having been noticed in hive bees) seems to be a mental act of the same nature as that which takes place in ourselves when recognising a locality. The senses, however, must be immeasurably more keen and the mental operation much more certain in them than it is in man; for to my eye there was absolutely no land-mark on the even surface of sand which could serve as guide, and the borders of the forest were not nearer than half a mile. The action of the wasp would be said to be instinctive; but it seems plain that the instinct is no mysterious and unintelligible agent, but a mental process in each individual, differing from the same in man only by its unerring certainty. The mind of the insect appears to be so constituted that the impression of external objects or the want felt, causes it to act with a precision which seems to us like that of a machine constructed to move in a certain given way. I have noticed in Indian boys a sense of locality almost as keen as that possessed by the sand-wasp. An old Portuguese and myself, accompanied by a young lad about ten years of age, were once lost in the forest in a most solitary place on the banks of the main river. Our case seemed hopeless, and it did not, for some time occur to us to consult our little companion, who had been playing with his bow and arrow all the way whilst we were hunting, apparently taking no note of the route. When asked, however, he pointed out, in a moment, the right direction of our canoe. He could not explain how he knew; I believe he had noted the course we had taken almost unconsciously: the sense of locality in his case seemed instinctive.

The Monedula signata is a good friend to travellers in those parts of the Amazons which are infested with the blood-thirsty Motúca. I first noticed its habit of preying on this fly one day when we landed to make our fire and dine on the borders of the forest adjoining a sand-bank. The insect is as large as a hornet, and has a most waspish appearance. I was rather startled when one out of the flock which was hovering about us flew straight at my face: it had espied a Motúca on my neck and was thus pouncing upon it. It seizes the fly not with its mandibles but with its fore and middle feet, and carries it off tightly held to its breast. Wherever the traveller lands on the Upper Amazons in the neighbourhood of a sand-bank he is sure to be attended by one or more of these useful vermin-killers.

The bay of Mapirí was the limit of my day excursions by the river-side to the west of Santarem. A person may travel, however, on foot, as Indians frequently do, in the dry season for fifty or sixty miles along the broad clean sandy beaches of the Tapajos. The only obstacles are the rivulets, most of which are

D 2

fordable when the waters are low. To the east my rambles extended to the banks of the Mahicá inlet. This enters the Amazons about three miles below Santarem, where the clear stream of the Tapajos begins to be discoloured by the turbid waters of the main river. The broad, placid channel of the Mahicá separates the Tapajos mainland from the alluvial low lands of the great river plain. It communicates in the interior with other inlets, and the whole forms a system of inland water-paths navigable by small vessels from Santarem to the river Curuá, forty miles distant. The Mahicá has a broad margin of rich, level pasture, limited on each side by the straight, tall hedge of forest. On the Santarem side it is skirted by high wooded ridges. A landscape of this description always produced in me an impression of sadness and loneliness which the riant virgin forests that closely hedge in most of the by-waters of the Amazons never created. The pastures are destitute of flowers, and also of animal life, with the exception of a few small plain-coloured birds and solitary Caracára eagles whining from the topmost branches of dead trees on the forest borders. A few settlers have built their palm-thatched and mud-walled huts on the banks of the Mahicá, and occupy themselves chiefly in tending small herds of cattle. They seemed to be all wretchedly poor. The oxen however, though small, were sleek and fat, and the district most promising for agricultural and pastoral employments. In the wet season the waters gradually rise and cover the meadows, but there is plenty of room for the removal of the cattle to higher ground. The lazy and ignorant people seem totally unable to profit by these advantages. The houses have no gardens or plantations near them. I was told it was useless to plant anything, because the cattle devoured the young shoots. In this country, grazing and planting are very rarely carried on together; for the people seem to have no notion of enclosing patches of ground for cultivation. They say it is too much trouble to make enclosures. The construction of a durable fence is certainly a difficult matter, for it is only two or three kinds of tree which will serve the purpose in being free from the attacks of insects, and these are scattered far and wide through the woods.

In one place, where there was a pretty bit of pasture surrounded by woods, I found a grazier established, who supplied Santarem daily with milk. He was a strong, wiry half-breed, a man endowed with a little more energy than his neighbours, and really a hard-working fellow. The land was his own, and the dozen or so well-conditioned cows which grazed upon it. It was melancholy, however, to see the miserable way in which the man lived. His house, a mere barn, scarcely protecting its owner from the sun and rain, was not much better built or furnished than an Indian's hut. He complained that it was impossible to induce any of the needy free people to work for wages. The poor fellow led a dull, solitary life; he had no family, and his wife had left him for some cause or other. He was up every morning by four o'clock, milked his cows with the help of a neighbour, and carried the day's yield to the town in stone bottles packed in leather bags on horseback by sunrise. His wretched little farm produced nothing else. The house stood in the middle of the bare pasture, without garden or any sort of plantation; a group of stately palms stood close by, to the trunks of which he secured the cows whilst milking. Butter-making is unknown in this country; the milk, I was told, is too poor; it is very rare indeed to see even the thinnest coating of cream on it, and the yield for each cow is very small. Our dairyman had to bring from Santarem every morning the meat, bread, and vegetables for the day's consumption. The other residents of Mahicá were not even so well off as this man. I always had to bring my own provisions when I came this way, for a perennial famine seemed to reign in the place. I could not help picturing to myself the very different aspect this fertile tract of country would wear if it were peopled by a few families of agricultural settlers from Northern Europe.

Although the meadows were unproductive ground to a Naturalist, the woods on their borders teemed with life: the number and variety of curious insects of all orders which occurred here was quite wonderful. The belt of forest was intersected by numerous pathways leading from one settler's house to another. The ground was moist, but the trees were not so lofty or their crowns so densely packed together as in other parts; the sun's light and heat therefore had freer access to the soil, and the underwood was much more diversified than in the virgin forest. I never saw so many kinds of dwarf palms together as here; pretty miniature species; some not more than five feet high, and bearing little clusters of round fruit not larger than a good bunch of currants. A few of the forest trees had the size and strongly-branched figures of our oaks, and a similar bark. One noble palm grew here in great abundance, and gave a distinctive character to the district. This was the Œnocarpus distichus, one of the kinds called Bacába by the natives. It grows to a height of forty to fifty feet. The crown is of a lustrous dark-green colour, and of a singularly flattened or compressed shape; the leaves being arranged on each side in nearly the same plane. When I first saw this tree on the campos, where the east wind blows with great force night and day for several months, I thought the shape of the crown was due to the leaves being prevented from radiating equally by the constant action of the breezes. But the plane of growth is not always in the direction of the wind, and the crown has the same shape when the tree grows in the sheltered woods. The fruit of this fine palm ripens towards the end of the year, and is much esteemed by the natives, who manufacture a pleasant drink from it similar to the assai described in a former chapter, by rubbing off the coat of pulp from the nuts, and mixing it with water. A bunch of fruit weighs thirty or forty pounds. The beverage has a milky appearance, and an agreeable nutty flavour. The tree is very difficult to climb, on account of the smoothness of its stem; consequently the natives, whenever they want a bunch of fruit for a bowl of Bacába, cut down and thus destroy a tree which has taken a score or two of years to grow, in order to get at it.

In the lower part of the Mahicá woods, towards the river, there is a bed of stiff white clay, which supplies the people of Santarem with material for the manufacture of coarse pottery and cooking utensils: all the kettles, saucepans, mandioca ovens, coffee-pots, washing-vessels, and so forth, of the poorer classes throughout the country, are made of this same plastic clay, which occurs at short intervals over the whole surface of the Amazons valley, from the neighbourhood of Pará to within the Peruvian borders, and forms part of the great Tabatinga marl deposit. To enable the vessels to stand the fire, the bark of a certain tree, called Caraipé, is burnt and mixed with the clay, which gives tenacity to the ware. Caraipé is an article of commerce, being sold, packed in baskets, at the shops in most of the towns. The shallow pits, excavated in the marly soil at Mahicá, were very attractive to many kinds of mason bees and wasps, who make use of the clay to build their nests with. I spent many an hour, watching their proceedings: a short account of the habits of some of these busy creatures may be interesting.

The most conspicuous was a large yellow and black wasp, with a remarkably long and narrow waist, the Pelopæus fistularis. It collected the clay in little round pellets, which it carried off, after rolling them into a convenient shape in its mandibles. It came straight to the pit with a loud hum, and, on alighting, lost not a moment in beginning to work; finishing the kneading of its little load in two or three minutes. The nest of this species is shaped like a pouch, two inches in length, and is attached to a branch or other projecting object. One of these restless artificers once began to build on the handle of a chest in the cabin of my canoe, when we were stationary at a place for several days. It was so intent on its work that it allowed me to inspect the movements of its mouth with a lens whilst it was laying on the mortar. Every fresh pellet was brought in with a triumphant song, which changed to a cheerful busy hum when it alighted and began to work. The little ball of moist clay was laid on the edge of the cell, and then spread out around the circular rim by means of the lower lip guided by the mandibles. The insect placed itself astride over the rim to work, and, on finishing each addition to the structure, took a turn round, patting the sides with its feet inside and out before flying off to gather a fresh pellet. It worked only in sunny weather, and the previous layer was sometimes not quite dry when the new coating was added. The whole structure takes about a week to complete. I left the place before the gay little builder had quite finished her task: she did not accompany the canoe, although we moved along the bank of the river very slowly. On opening closed nests of this species, which are common in the neighbourhood of Mahicá, I always found them to be stocked

Pelopæus Wasp building nest.

with small spiders of the genus Gastracantha, in the usual half-dead state to which the mother wasps reduce the insects which are to serve as food for their progeny.

Besides the Pelopæus there were three or four kinds of Trypoxylon, a genus also found in Europe, and which some Naturalists have supposed to be parasitic, because the legs are not furnished with the usual row of strong bristles for digging, characteristic of the family to which it belongs. The species of Trypoxylon, however, are all building wasps; two of them which I observed (T. albitarse and an undescribed species) provision their nests with spiders, a third (T. aurifrons) with small caterpillars. Their habits are similar to those of the Pelopæus; namely, they carry off the clay in their mandibles, and have a different song when they hasten away with the burthen, to that which they sing whilst at work. Trypoxylon albitarse, which is a large black kind, three-quarters of an inch in length, makes a tremendous fuss whilst building its cell. It often chooses the walls or doors of chambers for this purpose, and when two or three are at work in the same place their loud humming keeps the house in an uproar. The cell is a tubular structure about three inches in length. T. aurifrons, a much smaller species, makes a neat little nest shaped like a carafe; building rows of them together in the corners of verandahs.

Cells of Trypoxylon aurifrons.

But the most numerous and interesting of the clay artificers are the workers of a species of social bee, the Melipona fasciculata. The Meliponæ in tropical America take the place of the true Apides, to which the European hive-bee belongs, and which are here unknown; they are generally much smaller insects than the hive-bees and have no sting. The M. fasciculata is about a third shorter than the Apis mellifica: its colonies are composed of an immense number of individuals; the workers are generally seen collecting pollen in the same way as other bees, but great numbers are employed gathering clay. The rapidity and precision of their movements whilst thus engaged are wonderful. They first scrape the clay with their man-

Melipona Bees gathering clay.

dibles; the small portions gathered are then cleared by the anterior paws and passed to the second pair of feet, which, in their turn, convey them to the large foliated expansions of the hind shanks which are adapted normally in bees, as every one knows, for the collection of pollen. The middle feet pat the growing pellets of mortar on the hind legs to keep them in a compact shape as the particles are successively added. The little hodsmen soon have as much as they can carry, and they then fly off. I was for some time puzzled to know what the bees did with the clay; but I had afterwards plenty of opportunity for ascertaining. They construct their combs in any suitable crevice in trunks of trees or perpendicular banks, and the clay is required to build up a wall so as to close the gap, with the exception of a small orifice for their own entrance and exit. Most kinds of Meliponæ are in this way masons as well as workers in wax and pollen-gatherers. One little species (undescribed) not more than two lines long, builds a neat tubular gallery of clay, kneaded with some viscid substance outside the entrance to its hive, besides blocking up the crevice in the tree within which it is situated. The mouth of the tube is trumpet-shaped, and at the entrance a number of the pigmy bees are always stationed apparently acting as sentinels.

It is remarkable that none of the American bees have attained that high degree of architectural skill in the construction of their comb which is shown by the European hive bee. The wax cells of the Meliponæ are generally oblong, showing only an approximation to the hexagonal shape in places where several of them are built in contact. It would appear that the Old World has produced in bees, as well as in other families of animals, far more advanced forms than the tropics of the New World.

A hive of the Melipona fasciculata, which I saw opened, contained about two quarts of pleasantly-tasted liquid honey. The bees, as already remarked, have no sting, but they bite furiously when their colonies are disturbed. The Indian who plundered the hive was completely covered by them; they took a particular fancy to the hair of his head, and fastened on it by hundreds. I found forty-five species of these bees in different parts of the country; the largest was half an inch in length; the smallest were extremely minute, some kinds being not more than one-twelfth of an inch in size. These tiny fellows are often very troublesome in the woods, on account of their familiarity; they settle on one's face and hands; and, in crawling about, get into the eyes and mouth, or up the nostrils.

The broad expansion of the hind shanks of bees is applied in some species to other uses besides the conveyance of clay and pollen. The female of the handsome golden and black Euglossa Surinamensis has this palette of very large size. This species builds its solitary nest also in crevices of walls or trees; but it closes up the chink with fragments of dried leaves and sticks cemented together, instead of clay. It visits the cajú trees, and gathers with its hind legs a small quantity of the gum which exudes from their trunks. To this it adds the other materials required from the neighbouring bushes, and when laden flies off to its nest.

Whilst on the subject of bees, I may mention that the neighbourhoods of Santarem and Villa Nova yielded me about 140 species. The genera are for the most part different from those inhabiting Europe. A very large number make their cells in hollow twigs and branches. As in our own country, the industrious nest-building kinds are attended by other species which do not work or store up food for their progeny, but deposit their ova in the cells of their comrades. Some of these, it is well known, counterfeit the dress and general figure of their victims. To all appearance this similarity of shape and colours between the parasite and its victim is given for the purpose of deceiving the poor hard-working bee, which would otherwise revenge itself by slaying its plunderers. Some parasitic bees, however, have no resemblance to the species they impose upon; probably they live together on more friendly terms, or have some other means of disarming suspicion. Many Dipterous insects are also parasitic on bees, and wear the same dress as the species they live upon. That the dress of the victimisers is arranged with especial reference to their prey, I think is proved by what I observed at Santarem. The genera of the parasites here are not the same as in Europe; and when they counterfeit working bees, it is the peculiarly-coloured species of their own country that are imitated, and not those of any other region. The European genus Apathus, which mimics European Humble-bees, is not found in South America; but the common Bombus of Santarem, which is remarkable in being wholly of a sooty-black colour, is attended by a sooty black parasite of a widely-different genus, the Eurytis funereus. Many of the little Meliponæ have their counterfeits in small Diptera of the family Syrphidæ; and the brilliant green or blue bees of the country (Euglossa) have their imitators in parasitic bees of equally bright colours, belonging to genera unknown out of the countries where the Euglossæ are found.*

* These are Melissa, Mesocheira, Thalestria, &c.

To the south my rambles never extended further than the banks of the Irurá, a stream which rises amongst the hills already spoken of, and running through a broad valley, wooded along the margins of the water-courses, falls into the Tapajos, at the head of the bay of Mapirí. All beyond, as before remarked, is terra incognita to the inhabitants of Santarem. The Brazilian settlers on the banks of the Amazons seem to have no taste for explorations by land, and I could find no person willing to accompany me on an excursion further towards the interior. Such a journey would be exceedingly difficult in this country, even if men could be obtained willing to undertake it. Besides, there were reports of a settlement of fierce runaway negroes on the Serra de Mururarú, and it was considered unsafe to go far in that direction, except with a large armed party. I visited the banks of the Irurá and the rich woods accompanying it, and two other streams in the same neighbourhood, one called the Paněma, and the other the Urumarí, once or twice a week during the whole time of my residence in Santarem, and made large collections of their natural productions. These forest brooks, with their clear cold waters brawling over their sandy or pebbly beds through wild tropical glens, always had a great charm for me. The beauty of the moist, cool, and luxuriant glades was heightened by the contrast they afforded to the sterile country around them. The bare or scantily wooded hills which surround the valley are parched by the rays of the vertical sun. One of them, the Pico do Irurá, forms a nearly perfect cone, rising from a small grassy plain to a height of 500 or 600 feet, and its ascent is excessively fatiguing after the long walk from Santarem over the campos. I tried it one day, but did not reach the summit. A dense growth of coarse grasses clothed the steep sides of the hill, with here and there a stunted tree of kinds found in the plain beneath. In bared places, a red crumbly soil is exposed; and in one part a mass of rock, which appeared to me, from its compact texture and the absence of stratification, to be porphyritic; but I am not Geologist sufficient to pronounce on such questions. Mr. Wallace states that he found fragments of scoriæ, and believes the hill to be a volcanic cone. To the south and east of this isolated peak, the elongated ridges or table-topped hills attain a somewhat greater elevation.

The forest in the valley is limited to a tract a few hundred yards in width on each side the different streams: in places where these run along the bases of the hills the hill-sides facing the water are also richly wooded, although their opposite declivities are bare or nearly so. The trees are lofty and of great variety; amongst them are colossal examples of the Brazil nut tree (Bertholletia excelsa), and the Pikiá. This latter bears a large eatable fruit, curious in having a hollow chamber between the pulp and the kernel, beset with hard spines which produce serious wounds if they enter the skin. The eatable part appeared to me not much more palatable than a raw potato; but the inhabitants of Santarem are very fond of it, and undertake the most toilsome journeys on foot to gather a basketful. The tree which yields the tonka bean

(Dipteryx odorata), used in Europe for scenting snuff, is also of frequent occurrence here. It grows to an immense height, and the fruit, which, although a legume, is of a rounded shape, and has but one seed, can be gathered only when it falls to the ground. A considerable quantity (from 1000 to 3000 pounds) is exported annually from Santarem, the produce of the whole region of the Tapajos. An endless diversity of trees and shrubs, some beautiful in flower and foliage, others bearing curious fruits, grow in this matted wilderness. It would be tedious to enumerate many of them. I was much struck with the variety of trees with large and diversely-shaped fruits growing out of the trunk and branches, some within a few inches of the ground, like the cacao. Most of them are called by the natives Cupú, and the trees are of inconsiderable height. One of them called Cupú-aï bears a fruit of elliptical shape and of a dingy earthen colour six or seven inches long, the shell of which is woody and thin, and contains a small number of seeds loosely enveloped in a juicy pulp of very pleasant flavour. The fruits hang like clayey ants'-nests from the branches. Another kind more nearly resembles the cacao; this is shaped something like the cucumber, and has a green ribbed husk. It bears the name of Cacao de macaco, or monkey's chocolate, but the seeds are smaller than those of the common cacao. I tried once or twice to make chocolate from them. They contain plenty of oil of similar fragrance to that of the ordinary cacao-nut, and make up very well into paste; but the beverage has a repulsive clayey colour and an inferior flavour.

My excursions to the Irurá had always a picnic character. A few rude huts are scattered through the valley, but they are tenanted only for a few days in the year, when their owners come to gather and roast the mandioca of their small clearings. We used generally to take with us two boys—one negro, the other Indian—to carry our provisions for the day; a few pounds of beef or fried fish, farinha and bananas, with plates, and a kettle for cooking. José carried the guns, ammunition and game-bags, and I the apparatus for entomologizing—the insect net, a large leathern bag with compartments for corked boxes, phials, glass tubes, and so forth. It was our custom to start soon after sunrise, when the walk over the campos was cool and pleasant, the sky without a cloud, and the grass wet with dew. The paths are mere faint tracks; in our early excursions it was difficult to avoid missing our way. We were once completely lost, and wandered about for several hours over the scorching soil without recovering the road. A fine view is obtained of the country from the rising ground about half way across the waste. Thence to the bottom of the valley is a long, gentle, grassy slope, bare of trees. The strangely-shaped hills; the forest at their feet, richly varied with palms; the bay of Mapirí on the right, with the dark waters of the Tapajos and its white glistening shores, are all spread out before one as if depicted on canvas. The extreme transparency of the atmosphere gives to all parts of the landscape such clearness of outline that the idea of distance is destroyed, and one fancies the whole to be almost within reach of the hand.

Descending into the valley, a small brook has to be crossed, and then half a mile of sandy plain, whose vegetation wears a peculiar aspect, owing to the predominance of a stemless palm, the Curuá (Attalea spectabilis), whose large, beautifully pinnated, rigid leaves rise directly from the soil. The fruit of this species is similar to the coco-nut, containing milk in the interior of the kernel, but it is much inferior to it in size. Here, and indeed all along the road, we saw, on most days in the wet season, tracks of the Jaguar. We never, however, met with the animal, although we sometimes heard his loud "hough" in the night whilst lying in our hammocks at home, in Santarem, and knew he must be lurking somewhere near us.

My best hunting ground was a part of the valley sheltered on one side by a steep hill whose declivity, like the swampy valley beneath, was clothed with magnificent forest. We used to make our halt in a small cleared place, tolerably free from ants and close to the water. Here we assembled after our toilsome morning's hunt in different directions through the woods, took our well-earned meal on the ground—two broad leaves of the wild banana serving us for a tablecloth—and rested for a couple of hours during the great heat of the afternoon. The diversity of animal productions was as wonderful as that of the vegetable forms in this rich locality. I find by my register that it was not unusual to meet with thirty or forty new species of conspicuous insects during a day's search, even after I had made a great number of trips to the same spot. It was pleasant to lie down during the

E 2

hottest part of the day, when my people lay asleep, and watch the movements of animals. Sometimes a troop of Anús (Crotophaga), a glossy black-plumaged bird, which lives in small societies in grassy places, would come in from the campos, one by one, calling to each other as they moved from tree to tree. Or a Toucan (Rhamphastos ariel) silently hopped or ran along and up the branches, peeping into chinks and crevices. Notes of solitary birds resounded from a

The Jacuarú (Teius teguexim).

distance through the wilderness. Occasionally a sulky Trogon would be seen, with its brilliant green back and rose-coloured breast, perched for an hour without moving on a low branch. A number of large, fat lizards two feet long, of a kind called by the natives Jacuarú (Teius teguexim) were always observed in the still hours of mid-day scampering with great clatter over the dead leaves, apparently in chase of each other. The fat of this bulky lizard is much prized by the natives, who apply it as a poultice to draw palm spines or even grains of shot from the flesh. Other lizards of repulsive aspect, about three feet in length when full grown, splashed about and swam in the water; sometimes emerging to crawl into hollow trees on the banks of the stream, where I once found a female and a nest of eggs. The lazy flapping flight of large blue and black morpho butterflies high in the air, the hum of insects, and many inanimate sounds, contributed their share to the total impression this strange solitude produced. Heavy fruits from the crowns of trees which were mingled together at a giddy height overhead, fell now and then with a startling "plop" into the water. The breeze, not felt below, stirred in the topmost branches, setting the twisted and looped sipós in motion, which creaked and groaned in a great variety of notes. To these noises were added the monotonous ripple of the brook, which had its little cascade at every score or two yards of its course.

We frequently fell in with an old Indian woman, named Cecilia, who had a small clearing in the woods. She had the reputation of being a witch (feiticeira), and I found, on talking with her, that she prided herself on her knowledge of the black art. Her slightly curled hair showed that she was not a pure-blood Indian: I was told her father was a dark mulatto. She was always very civil to our party; showing us the best paths, explaining the virtues and uses of different plants, and so forth. I was much amused at the accounts she gave of the place. Her solitary life and the gloom of the woods seemed to have filled her with superstitious fancies. She said gold was contained in the bed of the brook, and that the murmur of the water over the little cascades was the voice of the "water-mother" revealing the hidden treasure. A narrow pass between two hill sides was the portaõ or gate, and all within, along the wooded banks of the stream, was enchanted ground. The hill underneath which we were encamped was the enchanter's abode, and she gravely told us she often had long conversations with him. These myths were of her own invention, and in the same way an endless number of other similar ones have originated in the childish imaginations of the poor Indian and half-breed inhabitants of different parts of the country. It is to be remarked, however, that the Indian men all become sceptics after a little intercourse with the whites. The witchcraft of poor Cecilia was of a very weak quality. It consisted in throwing pinches of powdered bark of a certain tree and other substances into the fire whilst muttering a spell—a prayer repeated backwards—and adding the name of the person on whom she wished the incantation to operate. Some of the feiticeiras, however, play more dangerous tricks than this harmless mummery. They are acquainted with many poisonous plants, and although they seldom have the courage to administer a fatal dose, sometimes contrive to convey to their victim sufficient to cause serious illness. The motive by which they are actuated is usually jealousy of other women in love matters. Whilst I resided in Santarem a case of what was called witchcraft was tried by the sub-delegado, in which a highly respectable white lady was the complainant. It appeared that some feiticeira had sprinkled a quantity of the acrid juice of a large arum on her linen as it was hanging out to dry, and it was thought this had caused a serious eruption under which the lady suffered.

I seldom met with any of the larger animals in these excursions. We never saw a mammal of any kind on the campos; but tracks of three species were seen occasionally besides those of the Jaguar: these belonged to a small tiger cat, a deer, and an opossum; all of which animals must have been very rare, and probably nocturnal in their habits, with the exception of the deer. I saw in the woods, on one occasion, a small flock of monkeys, and once had an opportunity of watching the movements of a sloth. The monkeys belonged to a very pretty and rare species, a kind of marmoset, I think the Hapale humeralifer described by Geoffroy St. Hilaire. I did not succeed in obtaining a specimen, but saw a living example afterwards in the possession of a shopkeeper at Santarem. It seems to occur nowhere else except in the dry woods bordering the campos in the interior parts of Brazil. The colours of its fur are beautifully varied; the fore part of the body is white, with the hands gray; the hind part black, with the rump and underside reddish-tawny; the tail is banded with gray and black. Its face is partly naked and flesh-coloured, and the ears are fringed with long white hairs. The specimen was not more than eight inches in length, exclusive of the tail. Altogether I thought it the prettiest species of its family I had yet seen. One would mistake it, at first sight, for a kitten, from its small size, varied colours, and the softness of its fur. It was a most timid creature, screaming and biting when any one attempted to handle it; it became familiar, however, with the people of the house a few days after it came into their possession. When hungry or uneasy it uttered a weak querulous cry, a shrill note, which was sometimes prolonged so as to resemble the stridulation of a grasshopper. The sloth was of the kind called by Cuvier Bradypus tridactylus, which is clothed with shaggy gray hair. The natives call it, in the Tupí language, Aï ybyreté (in Portuguese, Preguiça da terra firme), or sloth of the mainland, to distinguish it from the Bradypus infuscatus, which has a long, black and tawny stripe between the shoulders, and is called Aï Ygapó (Preguiça das vargens), or sloth of the flooded lands. Some travellers in South America have described the sloth as very nimble in its native woods, and have disputed the justness of the name which has been bestowed on it. The inhabitants of the Amazons region, however, both Indians and descendants of the Portuguese, hold to the common opinion, and consider the sloth as the type of laziness. It is very common for one native to call another, in reproaching him for idleness, "bicho do Embaüba" (beast of the Cecropia tree); the leaves of the Cecropia being the food of the sloth. It is a strange sight to watch the uncouth creature, fit production of these silent shades, lazily moving from branch to branch. Every movement betrays, not indolence exactly, but extreme caution. He never looses his hold from one branch without first securing himself to the next, and when he does not immediately find a bough to grasp with the rigid hooks into which his paws are so curiously transformed, he raises his body, supported on his hind legs, and claws around in search of a fresh foothold. After watching the animal for about half an hour I gave him a charge of shot; he fell with a terrific crash, but caught a bough, in his descent, with his powerful claws, and remained suspended. Our Indian lad tried to climb the tree, but was driven back by swarms of stinging ants; the poor little fellow slid down in a sad predicament, and plunged into the brook to free himself. Two days afterwards I found the body of the sloth on the ground: the animal having dropped on the relaxation of the muscles a few hours after death. In one of our voyages, Mr. Wallace and I saw a sloth (B. infuscatus) swimming across a river, at a place where it was probably 300 yards broad. I believe it is not generally known that this animal takes to the water. Our men caught the beast, cooked, and ate him.

In returning from these trips we were sometimes benighted on the campos. We did not care for this on moonlit nights, when there was no danger of losing the path. The great heat felt in the middle hours of the day is much mitigated by four o'clock in the afternoon; a few birds then make their appearance; small flocks of ground doves run about the stony hillocks; parrots pass over and sometimes settle in the ilhas; pretty little finches of several species, especially one kind, streaked with olive-brown and yellow, and somewhat resembling our yellow-hammer, but I believe not belonging to the same genus, hop about the grass, enlivening the place with a few musical notes. The Carashúe (Mimus) also then resumes its mellow, blackbird-like song ; and two or three species of humming-bird, none of which however are peculiar to the district, flit about from tree to tree. On the other hand, the little blue and yellow-striped lizards, which abound amongst the herbage during the scorching heats of midday, retreat towards this hour to their hiding-places ; together with the day-flying insects and the numerous campo butterflies. Some of these latter resemble greatly our English species found in heathy places, namely, a fritillary, Argynnis (Euptoieta) Hegesia, and two smaller kinds, which are deceptively like the little Nemeobius Lucina. After sunset the air becomes delightfully cool and fragrant with fruits and flowers. The nocturnal animals then come forth. A monstrous hairy spider, five inches in expanse (Mygale Blondii), of a brown colour with yellowish lines along its stout legs — which is very common here, inhabiting broad tubular galleries smoothly lined with silken web—may be then caught on the watch at the mouth of its burrow. It is only seen at night, and I think does not wander far from its den ; the gallery is about two inches in diameter, and runs in a slanting direction, about two feet from the surface of the soil. As soon as it is night, swarms of goat-suckers suddenly make their appearance, wheeling about in a noiseless, ghostly manner, in chase of night-flying insects. They sometimes descend and settle on a low branch, or even on the pathway close to where one is walking, and then squatting down on their heels, are difficult to distinguish from the surrounding soil. One kind (Hydropsalis psalidurus?) has a long forked tail. In the daytime they are concealed in the wooded ilhas, where I very often saw them crouched and sleeping on the ground in the dense shade. They make no nest, but lay their eggs on the bare ground. Their breeding time is in the rainy season, and fresh eggs are found from December to June. Birds have not one uniform time for nidification here, as in temperate latitudes. Gulls and plovers lay in September, when the sand-banks are exposed in midriver in the dry season. Later in the evening, the singular notes of the goat-suckers are heard, one species crying Quao, Quao, another Chuck-co-co-cao; and these are repeated at intervals far into the night in the most monotonous manner. A great number of toads are seen on the bare sandy path-ways soon after sunset. One of them was quite a colossus, about seven inches in length and three in height. This big fellow would never move out of the way until we were close to him. If we jerked him out of the path with a stick, he would slowly recover himself, and then turn round to have a good impudent stare. I have counted as many as thirty of these mon-sters within a distance of half a mile.

The surface of the campos is disfigured in all direc-tions by earthy mounds and conical hillocks, the work of many different species of white ants. Some of these structures are five feet high, and formed of particles of earth worked into a material as hard as stone; others are smaller, and constructed in a looser manner. The ground is everywhere streaked with the narrow covered galleries which are built up by the insects of grains of earth different in colour from the surrounding soil, to protect themselves whilst conveying materials wherewith to build their cities—for such the tumuli may be considered—or carrying their young from one hillock to another. The same covered ways are spread over all the dead timber, and about the decaying roots of herbage, which serve as food to the white ants. An examination of these tubular passages or arcades in any part of the district, or a peep into one of the tumuli, reveals always a throng of eager, busy creatures. I became very much interested in these insects while staying at Santarem, where many circumstances favoured the study of their habits, and examined several hundred colonies in endeavouring to clear up obscure points in their natural history. Very little, up to that date, had been recorded of the constitution and economy of their communities, owing doubtless to their not being found in northern and central Europe, and, therefore, not within reach of European observers. I will give a short summary of my observations, and with this we shall have done with Santarem and its neighbourhood.*

White ants are small, pale-coloured, soft-bodied insects, having scarcely anything in common with true

* My original notes on the Termites, comprising all details, were sent to Professor Westwood (Oxford) in 1854 and 1855; they were not printed in England, but have been translated into German, and published by Dr. Hagen, with his monograph of the family, in the Linnæa Entomologica, 12 Band, Stettin, 1858, p. 207, ff.

ants, except their consisting, in each species and family, of several distinct orders of individuals or castes which live together in populous, organized communities. In both there are, besides the males and females, a set of individuals of no fully-developed sex, immensely more numerous than their brothers and sisters, whose task is to work and care for the young brood. In true ants this class of the community consists of undeveloped females, and when it comprises, as is the case in many species, individuals of different structure, the functions of these do not seem to be rigidly defined. The contrary happens in the Termites, and this perhaps shows that the organization of their communities has reached a higher stage, the division of labour being more complete. The neuters in these wonderful insects are always divided into two classes—fighters and workers; both are blind, and each keeps to its own task; the one to build, make covered roads, nurse the young brood from the egg upwards, take care of the king and queen, who are the progenitors of the whole colony, and secure the exit of the males and females, when they acquire wings and fly out to pair and disseminate the race: the other to defend the community against all comers. Ants and termites are also widely different in their mode of growth, or, as it is called, metamorphosis. Ants in their early stage are footless grubs, which, before they reach the adult state, pass through an intermediate quiescent stage (pupa) inclosed in a membrane. Termites, on the contrary, have a similar form when they emerge from the egg to that which they retain throughout life; the chief difference being the gradual acquisition of eyes and wings in the sexual individuals during the later stages of growth. Termites and true ants, in fact, belong to two widely dissimilar orders of insects, and the analogy between them is only a general one of habits. The mode of growth of Termites and the active condition of their younger stages (larva and pupa) make the constitution of their communities much more difficult of comprehension than that of ants; hence how many castes existed, and what sort of individuals they were composed of, if not males and females, have always been puzzles to naturalists in the absence of direct observation.

What a strange spectacle is offered to us in the organisation of these insect communities! Nothing analogous occurs amongst the higher animals. Social instincts exist in many species of mammals and birds, where numerous individuals unite to build common habitations, as we see in the case of weaver-birds and beavers; but the principle of division of labour, the setting apart of classes of individuals for certain employments, occurs only in human societies in an advanced state of civilisation. In all the higher animals there are only two orders of individuals as far as bodily structure is concerned, namely, males and females. The wonderful part in the history of the Termites is, that not only is there a rigid division of labour, but nature has given to each class a structure of body adapting it to the kind of labour it has to perform. The males and females form a class apart; they do no kind of work, but in the course of growth acquire wings to enable them to issue forth and disseminate their kind. The workers and soldiers are wingless, and differ solely in the shape and armature of the head. This member in the labourers is smooth and rounded, the mouth being adapted for the working of the materials in building the hive; in the soldiers the head is of very large size, and is provided in almost every kind with special organs of offence or defence in the form of horny processes resembling pikes, tridents, and so forth. Some species do not possess these extraordinary projections, but have, in compensation, greatly lengthened jaws, which are shaped in some kinds as sickles, in others as sabres and saws.

The course of human events in our day seems, unhappily, to make it more than ever necessary for the citizens of civilised and industrious communities to set apart a numerous armed class for the protection of the rest; in this nations only do what nature has of old done for the Termites. The soldier Termes, however, has not only the fighting instinct and function; he is constructed as a soldier, and carries his weapons not in his hand, but growing out of his body.

Whenever a colony of Termites is disturbed, the workers are at first the only members of the community seen; these quickly disappear through the endless ramified galleries of which a Termitarium is composed, and soldiers make their appearance. The observations of Smeathman on the soldiers of a species inhabiting tropical Africa are often quoted in books on Natural History, and give a very good idea of their habits. I was always amused at the pugnacity displayed, when, in making a hole in the earthy cemented archway of their covered roads, a host of these little fellows mounted the breach to cover the retreat of the workers. The edges of the rupture bristled with their armed heads as the courageous warriors ranged themselves in compact line around them. They at-

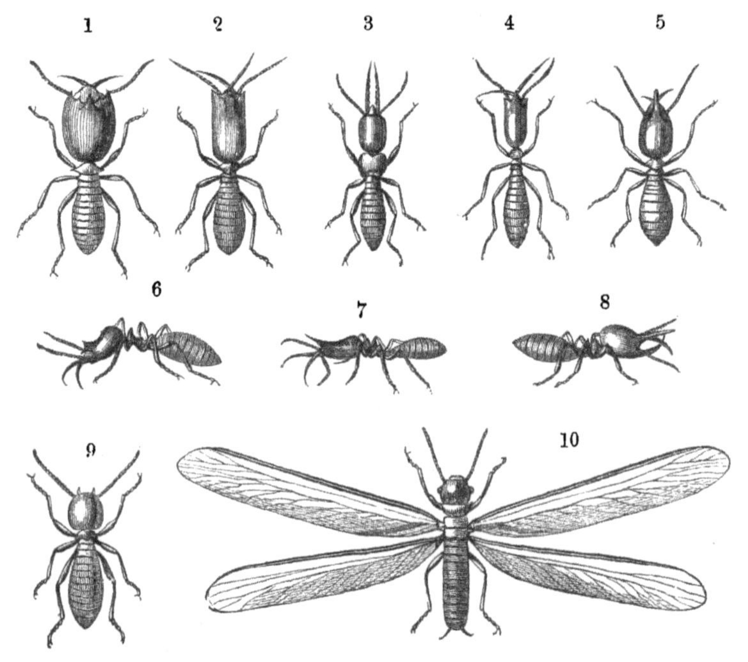

1—8. Soldiers of different species of White Ants.—9. Ordinary shape of worker.—10. Winged class.

tacked fiercely any intruding object, and as fast as their front ranks were destroyed, others filled up their places. When the jaws closed in the flesh, they suffered themselves to be torn in pieces rather than loosen their hold. It might be said that this instinct is rather a cause of their ruin than a protection when a colony is attacked by the well-known enemy of Termites, the ant-bear; but it is the soldiers only which attach themselves to the long worm-like tongue of this animal, and the workers, on whom the prosperity of the young brood immediately depends, are left for the most part unharmed. I always found, on thrusting my finger into a mixed crowd of Termites, that the soldiers only fastened upon it. Thus the fighting caste do in the end serve to protect the species by sacrificing themselves for its good.

A family of Termites consists of workers as the majority, of soldiers, and of the King and Queen. These are the constant occupants of a completed Termitarium. The royal couple are the father and mother of the colony, and are always kept together closely guarded by a detachment of workers in a large chamber in the very heart of the hive, surrounded by much stronger walls than the other cells. They are wingless and both immensely larger than the workers and soldiers. The Queen, when in her chamber, is always found in a gravid condition, her abdomen enormously distended with eggs, which, as fast as they come forth, are conveyed by a relay of workers in their mouths from the royal chamber to the minor cells dispersed throughout the hive. The other members of a Termes family are the winged individuals: these make their appearance only at a certain time of the year, generally in the beginning of the rainy season. It has puzzled naturalists to make out the relationship between the winged Termites and the wingless King and Queen. It has also generally been thought that the soldiers and workers are the larvæ of the others; an excusable mistake, seeing that they much resemble larvæ. I satisfied myself, after studying the habits of these insects daily for several months, that the winged Termites were males and females in about equal numbers, and that some of them, after shedding their wings and pairing, became Kings and Queens of new colonies; also, that the soldiers and workers were individuals which had arrived at their full growth without passing through the same stages as their fertile brothers and sisters.

A Termitarium, although of different shape, size, texture of materials, and built in different situations, according to the species, is always composed of a vast number of chambers and irregular intercommunicating galleries, built up with particles of earth or vegetable matter, cemented together by the saliva of the insects. There is no visible mode of ingress or egress, the entrances being connected with covered roads, which are the sole means of communication with the outer world. The structures are prominent objects in all tropical countries. The very large hillocks at Santarem are the work of many distinct species, each of which uses materials differently compacted, and keeps to its own portion of the tumulus. One kind, Termes arenarius, on which these remarks are chiefly founded, makes little conical hillocks of friable structure, a foot or two in height, and is generally the sole occupier. Another kind (Termes exiguus) builds small dome-shaped papery edifices. Many species live on trees, their earthy nests, of all sizes, looking like ugly excrescences on the trunks and branches. Some are wholly subterranean, and others live under the bark, or in the interior of trees: it is these two latter kinds which get into houses and destroy furniture, books, and clothing. All hives do not contain a queen and her partner. Some are new constructions, and, when taken to pieces, show only a large number of workers occupied in bringing eggs from an old over-stocked Termitarium, with a small detachment of soldiers evidently told off for their protection.

A few weeks before the exodus of the winged males and females a completed Termitarium contains Termites of all castes and in all stages of development. On close examination I found the young of each of the four orders of individuals crowded together, and apparently feeding in the same cells. The full-grown workers showed the greatest attention to the young larvæ, carrying them in their mouths along the galleries from one cell to another, but they took no notice of the full-grown ones. It was not possible to distinguish the larvæ of the four classes when extremely young, but at an advanced stage it was easy to see which were to become males and females, and which workers and soldiers. The workers have the same form throughout, the soldiers showed in their later stages of growth the large head and cephalic processes, but much less developed than in the adult state. The males and females were distinguishable by the possession of rudimentary wings and eyes, which increased in size after three successive changes of skin.

Thus I think I made out that the soldier and worker castes are, like the males and females, distinct from the egg; they are not made so by a difference of food or treatment during their earlier stages, and they never

F 2

become winged insects. The workers and soldiers feed on decayed wood and other vegetable substances; I could not clearly ascertain what the young fed upon, but they are seen of all sizes, larvæ and pupæ, huddled together in the same cells, with their heads converging towards the bottom, and I thought I sometimes detected the workers discharging a liquid from their mouths into the cells. The growth of the young family is very rapid, and seems to be completed within the year: the greatest event of Termite life then takes place, namely, the coming of age of the winged males and females, and their exit from the hive.

It is curious to watch a Termitarium when this exodus is taking place. The workers are set in the greatest activity, as if they were aware that the very existence of their species depended on the successful emigration and marriages of their brothers and sisters. They clear the way for their bulky but fragile bodies, and bite holes through the outer walls for their escape. The exodus is not completed in one day, but continues until all the males and females have emerged from their pupa integuments, and flown away. It takes place on moist, close evenings, or on cloudy mornings: they are much attracted by the lights in houses, and fly by myriads into chambers, filling the air with a loud rustling noise, and often falling in such numbers that they extinguish the lamps. Almost as soon as they touch ground they wriggle off their wings, to aid which operation there is a special provision in the structure of the organs, a seam running across near their roots and dividing the horny nervures. To prove that this singular mutilation was voluntary, on the part of the insects, I repeatedly tried to detach the wings by force, but could never succeed whilst they were fresh, for they always tore out by the roots. Few escape the innumerable enemies which are on the alert at these times to devour them; ants, spiders, lizards, toads, bats, and goat-suckers. The waste of life is astonishing. The few that do survive pair and become the kings and queens of new colonies. I ascertained this by finding single pairs a few days after the exodus, which I always examined and proved to be males and females, established under a leaf, a clod of earth, or wandering about under the edges of new tumuli. The females are then not gravid. I once found a newly-married pair in a fresh cell tended by a few workers.

The office of Termites in these hot countries is to hasten the decomposition of the woody and decaying parts of vegetation. In this they perform what in temperate latitudes is the task of other orders of insects. Many points in their natural history still remain obscure. We have seen that there are males and females, which grow, reach the adult winged state, and propagate their kind like all other insects. Unlike others, however, which are always, each in its sphere, provided with the means of maintaining their own in the battle of life, these are helpless creatures, which, without external aid, would soon perish, entailing the extinction of their kind. The family to which they belong is therefore provided with other members, not males or females, but individuals deprived of the sexual instincts, and so endowed in body and mind that they are adapted and impelled to devote their lives for the good of their species. But I have not explained how these neuter individuals, soldiers and workers, come to be distinct castes. This is still a knotty point, which I could do nothing to solve. Neuter bees and ants are known to be undeveloped females. I thought it a reasonable hypothesis, on account of the total absence of intermediate individuals connecting the two forms, that worker and soldier might be in a similar way female and male whose development had been in some way arrested. A French anatomist, however, M. Lespés,* believes to have found by dissection imperfect males and females in each of the castes. The correctness of his observations is doubted by competent judges;† if his conclusion be true, the biology of Termites is indeed a mystery.

* Recherches sur l'Organization et les Mœurs du Termite Lucifuge, Annales des Sciences Naturelles, 4me serie, tome 5, fasc. 4 et 5. Paris, 1856. M. Lespés states also to have found two distinct forms of pupa in the same species, one only of which he believes to become kings and queens. I observed nothing of the kind in Termes arenarius. Dr. Hagen mentions, in his monograph, cases of beaked workers and winged soldiers. I always found the beaked individuals to be of the fighting caste; with regard to winged soldiers and other curious forms of pupæ which have occurred, they are probably either monstrosities, or belong to species having a peculiar mode of development. I did not meet with such; I found, however, a species whose soldier class did not differ at all, except in the fighting instinct, from the workers.

† Gerstaecker, Bericht über den Leistungen, &c., der Entomologie, 1856. p. 6. Hagen, Linnæa Entomologica, 1858, p. 24.

CHAPTER II.

VOYAGE UP THE TAPAJOS.

Preparations for voyage—First day's sail—Mode of arranging money-matters and remittance of collections in the interior—Loss of boat—Altar do Chaõ—Excursion in forest—Valuable timber—Modes of obtaining fish—Difficulties with crew—Arrival at Aveyros—Excursions in the neighbourhood—White Cebus and habits and dispositions of Cebi monkeys—Tame parrot—Missionary settlement—Enter the River Cuparí—Adventure with Anaconda—Smoke-dried monkey—Boa-constrictor—Village of Mundurucú Indians, and incursion of a wild tribe—Falls of the Cuparí—Hyacinthine macaw—Re-emerge into the broad Tapajos—Descent of river to Santarem.

June, 1852.—I will now proceed to relate the incidents of my principal excursion up the Tapajos, which I began to prepare for, after residing about six months at Santarem.

I was obliged, this time, to travel in a vessel of my own; partly because trading canoes large enough to accommodate a Naturalist very seldom pass between Santarem and the thinly-peopled settlements on the river, and partly because I wished to explore districts at my ease, far out of the ordinary track of traders. I soon found a suitable canoe; a two-masted cuberta, of about six tons' burthen, strongly built of Itaüba or stone-wood, a timber of which all the best vessels in the

Amazons country are constructed, and said to be more durable than teak. This I hired of a merchant at the cheap rate of 500 reis, or about one shilling and twopence per day. I fitted up the cabin, which, as usual in canoes of this class, was a square structure with its floor above the water-line, as my sleeping and working apartment. My chests, filled with store-boxes and trays for specimens, were arranged on each side, and above them were shelves and pegs to hold my little stock of useful books, guns, and game bags, boards and materials for skinning and preserving animals, botanical press and papers, drying cages for insects and birds, and so forth. A rush mat was spread on the floor, and my rolled-up hammock, to be used only when sleeping ashore, served for a pillow. The arched covering over the hold in the fore part of the vessel contained, besides a sleeping place for the crew, my heavy chests, stock of salt provisions and groceries, and an assortment of goods wherewith to pay my way amongst the half-civilised or savage inhabitants of the interior. The goods consisted of cashaça, powder and shot, a few pieces of coarse checked-cotton cloth and prints, fish-hooks, axes, large knives, harpoons, arrow-heads, looking-glasses, beads, and other small wares. José and myself were busy for many days arranging these matters. We had to salt the meat and grind a supply of coffee ourselves. Cooking utensils, crockery, water-jars, a set of useful carpenter's tools, and many other things had to be provided. We put all the groceries and other perishable articles in tin canisters and boxes, having found that this was the only way of preserving them from damp and insects in this climate.

When all was done, our canoe looked like a little floating workshop.

I could get little information about the river, except vague accounts of the difficulty of the navigation, and the famito or hunger which reigned on its banks. As I have before mentioned, it is about a thousand miles in length, and flows from south to north; in magnitude it stands the sixth amongst the tributaries of the Amazons. It is navigable, however, by sailing vessels only for about 160 miles above Santarem. The hiring of men to navigate the vessel was our greatest trouble. José was to be my helmsman, and we thought three other hands would be the fewest with which we could venture. But all our endeavours to procure these were fruitless. Santarem is worse provided with Indian canoemen than any other town on the river. I found, on applying to the tradesmen to whom I had brought letters of introduction and to the Brazilian authorities, that almost any favour would be sooner granted than the loan of hands. A stranger, however, is obliged to depend on them; for it is impossible to find an Indian or half-caste whom some one or other of the head-men do not claim as owing him money or labour. I was afraid at one time I should have been forced to abandon my project on this account. At length, after many rebuffs and disappointments, José contrived to engage one man, a mulatto, named Pinto, a native of the mining country of Interior Brazil, who knew the river well; and with these two I resolved to start, hoping to meet with others at the first village on the road.

We left Santarem on the 8th of June. The waters

were then at their highest point, and my canoe had been anchored close to the back door of our house. The morning was cool and a brisk wind blew, with which we sped rapidly past the white-washed houses and thatched Indian huts of the suburbs. The charming little bay of Mapirí was soon left behind; we then doubled Point Maria Josepha, a headland formed of high cliffs of Tabatinga clay, capped with forest. This forms the limit of the river view from Santarem, and here we had our last glimpse, at a distance of seven or eight miles, of the city, a bright line of tiny white buildings resting on the dark water. A stretch of wild rocky uninhabited coast was before us, and we were fairly within the Tapajos.

Some of my readers may be curious to know how I managed money affairs during these excursions in the interior of the South American continent: it can be explained in a few words. In the first place, I had an agent in London to whom I consigned my collections. During the greater part of the time I drew on him for what sums I wanted, and an English firm at Pará (the only one in the country which traded regularly and directly with England) cashed the drafts. I found no difficulty in the interior of the country, for almost any of the larger Portuguese or Brazilian traders, of whom there are one or two in every village of 600 or 700 inhabitants, would honour my draft on the English house; they having each a correspondent at Pará who deals with the foreign merchants. Sometimes a Portuguese trader would hint at discount, or wish me to take part of the amount in goods, but the Brazilians were generally more liberal. At one period, when I was obliged to wait for remittances from England,* I sometimes ran short of money; but I had only to say a word to one of these generous and considerate men, and the assistance was given without interest to the extent I required. The current money on the Amazons varied much during the eleven years of my stay. At first, nothing but copper coins and Brazilian treasury notes, the smallest representing 1000 reis (2*s.* 3*d.*), were seen; afterwards (1852—1856), with the increase of the India-rubber trade, a large amount of specie was imported,—American gold coins, Spanish and Mexican dollars, and English sovereigns. These were the commonest medium of exchange in Pará and on the Lower Amazons, until India-rubber fell suddenly in price, in 1855, when the gold again quickly disappeared. About the year 1857, new silver coin, issued by the Brazilian Government, was introduced; elegant pieces of money of convenient values, answering nearly to our sixpenny, shilling, and two shilling pieces. Neither gold, silver, nor paper, however, was of much use on a journey like the one I had now undertaken. All travellers on the branch rivers have to carry cloth, cashaça, and small wares, to exchange for produce or food with the Indians; a small quantity of copper money, the only coin whose value is understood amongst the remote settlers, being nevertheless necessary to balance exchanges. When I had to

* I take this opportunity of mentioning my obligations to Mr. George Brocklehurst, of the Pará firm, by whom, during the latter years of my travels in the interior, my wants were attended to in the promptest and kindest manner.

send collections down to Pará to be shipped for England, which happened three or four times a year, I used to arrange with any trader who was dispatching a vessel to the capital with produce; the owners very often charging nothing for the carriage. Sometimes I had to entrust chests full of choice specimens to Indians for a voyage of thirty or forty days: a word to the Pilot recommending him to keep the boxes free from damp was quite sufficient. I never suffered any loss or damage.

Our course lay due west for about twenty miles. The wind increased as we neared Point Cururú, where the river bends from its northern course. A vast expanse of water here stretches to the west and south, and the waves, with a strong breeze, run very high. As we were doubling the Point, the cable which held our montaria in tow astern, parted, and in endeavouring to recover the boat, without which we knew it would be difficult to get ashore on many parts of the coast, we were very near capsizing. We tried to tack down the river; a vain attempt with a strong breeze and no current. Our ropes snapped, the sails flew to rags, and the vessel, which we now found was deficient in ballast, heeled over frightfully. Contrary to José's advice, I ran the cuberta into a little bay, thinking to cast anchor there and wait for the boat coming up with the wind; but the anchor dragged on the smooth sandy bottom, and the vessel went broadside on to the rocky beach. With a little dexterous management, but not until after we had sustained some severe bumps, we managed to get out of this difficulty, clearing the rocky point at a close shave with our jib-sail. Soon after we drifted into the smooth water of a sheltered bay which leads to the charmingly situated village of Altar do Chaõ; and we were obliged to give up our attempt to recover the montaria.

The little settlement, Altar do Chaõ—altar of the ground, or Earth altar—owes its singular name to the existence at the entrance to the harbour of one of those strange flat-topped hills which are so common in this part of the Amazons country, shaped like the high altar in Roman Catholic churches. It is an isolated one and much lower in height than the similarly truncated hills and ridges near Almeyrim, being elevated probably not more than 300 feet above the level of the river. It is bare of trees, but covered in places with a species of fern. At the head of the bay is an inner harbour which communicates by a channel with a series of lakes lying in the valleys between hills and stretching far into the interior of the land. The village is peopled almost entirely by semi-civilised Indians to the number of sixty or seventy families, and the scattered houses are arranged in broad streets on a strip of green sward at the foot of a high, gloriously-wooded ridge.

We stayed here nine days. As soon as we anchored I went ashore and persuaded, by the offer of a handsome reward, two young half-breeds to go in search of my missing boat. The head man of the place, Captain Thomas, a sleepy-looking mameluco, whom I found in his mud-walled cottage in loose shirt and drawers, with a large black rosary round his neck, promised me two Indians to complete my crew, if I would wait a few days until they had finished felling trees for a new plantation. Meantime my men had to make a new sail and repair the rigging, and I explored the rich woods of the vicinity.

Captain Thomas sent his son one day to show me the best paths. A few steps behind the houses we found ourselves in the virgin forest. The soil was sandy, and the broad path sloped gently up towards the high ridge which forms so beautiful a back-ground to the village. From the top of the hill a glimpse of the bay is obtained through the crowns of the trees. The road then descends, and so continues for many miles over hill and dale. There are no habitations, however, in this direction ; the road having been made by people formerly employed in felling timber. The forest at Altar do Chaõ is noted for its riches in choice woods, and its large laurel and Itauba trees, which are used in building river schooners. The beautiful tortoise-shell wood, Moira pinímà, minutely barred and spotted with red and black, which is made into walking-sticks by Brazilian carpenters, and exported as such in some numbers to Portugal, was formerly abundant here ; it is the heart-wood of a tree I believe unknown to science, and is obtainable only in logs a few inches in diameter. The Moira coatiára (striped wood), a most beautiful material for cabinet work, being close-grained and richly streaked with chocolate-brown on a yellow ground, is another of these, and is also the heart-wood of a tree, but obtainable in logs a foot or more in diameter and ten feet in length. A rare wood called Sápu-píra, of excessively hard texture, deep brown in colour, thickly speckled with yellow, is also a product of these forests. Captain Thomas showed me a mortar, four feet high, for pounding coffee, made of it. Many other kinds of ornamental and useful timber are met with, including a kind of box, which I saw made into carpenters' planes; ebony and marupá; the last-mentioned a light whitish wood of the same texture as mahogany. Although the trees have been felled near the village, more of the same kinds are said to exist in the forest, which extends to an unknown distance in the interior. I heard here, also, of the Mururé, a lofty tree which yields a yellow milk of remarkable virtues, on making incisions in the bark. It is called by the Portuguese Mercurio vegetal, or vegetable mercury, from the cures it effects when taken internally in syphilitic rheumatism. It is said to produce terrible pains in the limbs soon after it is taken, but the cure is certain. I was never able to get a sight of this tree. Captain Thomas said that the only specimen he knew of it, had been cut down. Persons in Santarem had attempted to send samples of the milk to Europe for experiment, but they had failed on account of the stone bottles in which it was contained always bursting in transit.

We walked two or three miles along this dark and silent forest road, and then struck off through the thicket to another path running parallel to it, by which we returned to the village. About half way we passed through a tract of wood, densely overgrown with the

Curuá palm tree; the natives call a place of this kind a Pindobál. The rigid, elegantly pinnated leaves, twenty feet in length, grow, as I have before described, directly out of the ground. I had frequently occasion to notice in the virgin forests some one kind of palm, growing abundantly in society in one limited tract although scarce elsewhere, no difference of soil, altitude, or humidity being apparent to account for the phenomenon. The Pindobál covered an area of probably four or five acres, and the whole lay under the shade of the tall forest trees. The last half mile of our road led through a more humid part of the forest near the low shores of the lake. We here saw a Couxio monkey (Pithecia satanas), a large black species which, as I have before mentioned, has a thick cap of hair on the head parted at the crown. He was seated alone on a branch fingering a cluster of flowers that lay within his reach. My companion fired at him, but missed, and he then slowly moved away. The borders of the path were enlivened with troops of small and delicate butterflies. I succeeded in capturing, in about half an hour, no less than eight species of one genus, Mesosemia; a group remarkable for having the wings ornamented with central eye-like spots encircled by fine black and gray concentric lines arranged in different patterns according to the species.

I was so much pleased with the situation of this settlement, and the number of rare birds and insects which tenanted the forest, that I revisited it in the following year, and spent four months making collections. The village itself is a neglected, poverty-stricken place: the governor (Captain of Trabalhadores or Indian workmen) being an old, apathetic half-breed, who had spent all his life here. The priest was a most profligate character; I seldom saw him sober; he was a white, however, and a man of good ability. I may as well mention here, that a moral and zealous priest is a great rarity in this province: the only ministers of religion in the whole country who appeared sincere in their calling, being the Bishop of Pará and the Vicars of Ega on the Upper Amazons and Obydos. The houses in the village swarmed with vermin; bats in the thatch; fire-ants (formiga de fogo) under the floors; cockroaches and spiders on the walls. Very few of them had wooden doors and locks. Altar do Chaõ was originally a settlement of the aborigines, and was called Burarí. The Indians were always hostile to the Portuguese, and during the disorders of 1835-6 joined the rebels in the attack on Santarem. Few of them escaped the subsequent slaughter, and for this reason there is now scarcely an old or middle-aged man in the place. As in all the semi-civilised villages where the original orderly and industrious habits of the Indian have been lost without anything being learnt from the whites to make amends, the inhabitants live in the greatest poverty. The scarcity of fish in the clear waters and rocky bays of the neighbourhood is no doubt partly the cause of the poverty and perennial hunger which reign here. When we arrived in the port our canoe was crowded with the half-naked villagers—men, women, and children, who came to beg each a piece of salt pirarucu "for the love of God." They are not quite so badly off in the dry season. The shallow lakes and bays then contain plenty of fish, and the boys and women go out at night to spear them by torchlight; the torches being made of thin strips of green bark from the leaf-stalks of palms, tied in bundles. Many excellent kinds of fish are thus obtained; amongst them the Pescada, whose white and flaky flesh, when boiled, has the appearance and flavour of cod-fish; and the Tucunaré (Cichla temensis), a handsome species, with a large prettily-coloured, eye-like spot on its tail. Many small Salmonidæ are also met with, and a kind of sole, called Aramassá, which moves along the clear sandy bottom of the bay. At these times a species of sting-ray is common on the sloping beach, and bathers are frequently stung most severely by it. The weapon of this fish is a strong blade with jagged edges, about three inches long, growing from the side of the long fleshy tail. I once saw a woman wounded by it whilst bathing; she shrieked frightfully, and was obliged to be carried to her hammock, where she lay for a week in great pain; I have known strong men to be lamed for many months by the sting.

There was a mode of taking fish here which I had not before seen employed, but found afterwards to be very common on the Tapajos. This is by using a poisonous liana called Timbó (Paullinia pinnata). It will act only in the still waters of creeks and pools. A few rods, a yard in length, are mashed and soaked in the water, which quickly becomes discoloured with the milky deleterious juice of the plant. In about half an hour all the smaller fishes, over a rather wide space around the spot, rise to the surface floating on their sides, and with the gills wide open. The poison acts evidently by suffocating the fishes; it spreads slowly in the water, and a very slight mixture seems sufficient to stupify them. I was surprised, on beating the water in places where no fishes were visible in the clear depths for many yards round, to find, sooner or later, sometimes 24 hours afterwards, a considerable number floating dead on the surface.

The people occupy themselves the greater part of the year with their small plantations of mandioca. All the heavy work, such as felling and burning the timber, planting and weeding, is done in the plantation of each family by a congregation of neighbours, which they call a "pucherum:"—a similar custom to the "bee" in the backwood settlements of North America. They make quite a holiday of each pucherum. When the invitation is issued, the family prepares a great quantity of fermented drink, called in this part Tarobá, from soaked mandioca cakes, and porridge of Manicueira. This latter is a kind of sweet mandioca, very different from the Yuca of the Peruvians and Macasheira of the Brazilians (Manihot Aypi), having oblong juicy roots, which become very sweet a few days after they are gathered. With these simple provisions they regale their helpers. The work is certainly done, but after a very rude fashion; all become soddened with Tarobá, and the day finishes often in a drunken brawl.

The climate is rather more humid than that of Santarem. I suppose this is to be attributed to the neighbouring country being densely wooded, instead of an open campo. In no part of the country did I enjoy more the moonlit nights than here, in the dry season. After the day's work was done I used to go down to the shores of the bay, and lay all my length on the cool sand for two or three hours before bed-time. The soft pale light, resting on the broad sandy beaches and palm-thatched huts, reproduced the effect of a mid-winter scene in the cold north when a coating of snow lies on the landscape. A heavy shower falls about once a week, and the shrubby vegetation never becomes parched up as at Santarem. Between the rains the heat and dryness increase from day to day: the weather on the first day after the rain is gleamy with intervals of melting sunshine and passing clouds; the next day is rather drier, and the east wind begins to blow; then follow days of cloudless sky, with gradually increasing strength of breeze. When this has continued about a week a light mistiness begins to gather about the horizon; clouds are formed; grumbling thunder is heard, and then, generally in the night-time, down falls the refreshing rain. The sudden chill caused by the rains produces colds, which are accompanied by the same symptoms as in our own climate; with this exception the place is very healthy.

June 17*th.*—The two young men returned without meeting with my montaria, and I found it impossible here to buy a new one. Captain Thomás could find me only one hand. This was a blunt-spoken but willing young Indian, named Manoel. He came on board this morning at eight o'clock, and we then got up our anchor and resumed our voyage.

The wind was light and variable all day, and we made only about fifteen miles by seven o'clock in the evening. The coast formed a succession of long, shallow bays with sandy beaches, on which the waves broke in a long line of surf. Ten miles above Altar do Chaõ is a conspicuous headland, called Point Cajetúba. During a lull of the wind, towards midday, we ran the cuberta aground in shallow water and waded ashore, but the woods were scarcely penetrable, and not a bird was to be seen. The only thing observed worthy of note, was the quantity of drowned winged ants along the beach; they were all of one species, the terrible formiga de fogo (Myrmica sævissima); the dead, or half-dead bodies of which were heaped up in a line an inch or two in height and breadth, the line continuing without interruption for miles at the edge of the water. The countless thousands had been doubtless cast into the river whilst flying during a sudden squall the night before, and afterwards cast ashore by the waves. We found ourselves at seven o'clock near the mouth of a creek leading to a small lake, called Aramána-í, and the wind having died away, we anchored, guided by the lights ashore, near the house of a settler, named Jeronymo, whom I knew, and who, soon after, showed us a snug little harbour, where we could remain in safety for the night. The river here cannot be less than ten miles broad; it is quite clear of islands and free from shoals at this season of the year. The opposite coast appeared in the daytime as a long thin line of forest, with dim gray hills in the back ground.

June 18th and 19th.—Senhor Jeronymo promised to sell me a montaria, so I waited for three hours after sunrise the next morning, expecting it to be forthcoming, but in vain. I sent Pinto and afterwards José to enquire about it, but they, instead of performing the errand, joined the easy-natured master of the house in a morning carousal. I was obliged, when my patience was exhausted, to go after them, having to clamber down a projecting bough, in the absence of a boat, to get ashore; and then found my two men, their host, and two or three neighbours, lolling in hammocks, tinkling wire guitars, and drinking cashaça. I mention this as a sample of a very common class of incidents in Brazilian travelling. Master Jeronymo backed out of his promise regarding the montaria. José and Pinto, who seemed to think they had done nothing wrong, sulkily obeyed my order to go on board, and we again got under way. The wind failed us on the 18th towards three p.m. About six miles above Aramána-í we rounded a rocky point, called Acarátingarí, the distance travelled being altogether not more than twelve miles. The greater part of the day was thus lost: we passed the night in a snug little harbour sheltered by trees.

To-day (19th) we had a good wind, which carried us to the mouth of a creek, called Paquiatúba, where the "inspector" of the district lived, Senhor Cypriano, for whom I had brought an order from Captain Thomás to supply me with another hand. We had great difficulty in finding a place to land. The coast in this part was a tract of level, densely-wooded country, through which flowed the winding rivulet, or creek, which gives its name to a small scattered settlement hidden in the wilderness ; the hills here receding two or three miles towards the interior. A large portion of the forest was flooded, the trunks of the very high trees near the mouth of the creek standing 18 feet deep in water. We lost two hours working our way with poles through the inundated woods in search of the port. Every inlet we tried ended in a labyrinth choked up with bushes, but we were at length guided to the right place by the crowing of cocks. On shouting for a montaria an Indian boy made his appearance, guiding one through the gloomy thickets ; but he was so alarmed, I suppose at the apparition of a strange-looking white man in spectacles bawling from the prow of the vessel, that he shot back quickly into the bushes. He returned when Manoel spoke, and we went ashore: the montaria winding along a gloomy overshadowed water-path, made by cutting away the lower branches and underwood. The foot-road to the houses was a narrow, sandy alley, bordered by trees of stupendous height, overrun with creepers, and having an unusual number of long air-roots dangling from the epiphytes on their branches.

After passing one low smoky little hut, half-buried in foliage, the path branched off in various directions, and the boy having left us we took the wrong turn. We were brought to a stand soon after by the barking of dogs ; and on shouting, as is customary on approaching a dwelling, "O da casa!" (Oh of the house!) a dark-skinned native, a Cafuzo, with a most unpleasant expression of countenance, came forth through the tangled maze of bushes, armed with a long knife, with which he pretended to be whittling a stick. He directed us to the house of Cypriano, which was about a mile distant along another forest road. The circumstance of the Cafuzo coming out armed to receive visitors very much astonished my companions, who talked it over at every place we visited for several days afterwards; the freest and most unsuspecting welcome in these retired places being always counted upon by strangers. But, as Manoel remarked, the fellow may have been one of the unpardoned rebel leaders who had settled here after the recapture of Santarem in 1836, and lived in fear of being enquired for by the authorities of Santarem. After all our trouble we found Cypriano absent from home. His house was a large one, and full of people, old and young, women and children, all of whom were Indians or mamelucos. Several smaller huts surrounded the large dwelling, besides extensive open sheds containing mandioca ovens and rude wooden mills for grinding sugar-cane to make molasses. All the buildings were embosomed in trees: it would be scarcely possible to find a more retired nook, and an air of contentment was spread over the whole establishment. Cypriano's wife, a good-looking mameluco girl, was superintending the packing of farinha. Two or three old women, seated on mats, were making baskets with narrow strips of bark from the leaf-stalks of palms, whilst others were occupied lining them with the broad leaves of a species of maranta, and filling them afterwards with farinha, which was previously measured in a rude square vessel. It appeared that Senhor Cypriano was a large producer of the article, selling 300 baskets (sixty pounds' weight each) annually to Santarem traders. I was sorry we were unable to see him, but it was useless waiting, as we were told all the men were at present occupied in "puche-rums," and he would be unable to give me the assistance I required. We returned to the canoe in the evening, and, after moving out into the river, anchored and slept.

June 20th.—We had a light, baffling wind off shore all day on the 20th, and made but fourteen or fifteen miles by six p.m.; when, the wind failing us, we anchored at the mouth of a narrow channel, called Tapaiúna, which runs between a large island and the mainland. About three o'clock we passed in front of Boim, a village on the opposite (western) coast. The breadth of the river is here six or seven miles: a confused patch of white on the high land opposite was all we saw of the village, the separate houses being undistinguishable on account of the distance. The coast along which we sailed to-day is a continuation of the low and flooded land of Paquiatúba.

June 21st.—The next morning we sailed along the Tapaiúna channel, which is from 400 to 600 yards in breadth. We advanced but slowly, as the wind was generally dead against us, and stopped frequently to ramble ashore. Wherever the landing-place was sandy it was impossible to walk about, on account of the swarms of the terrible fire-ant, whose sting is likened by the Brazilians to the puncture of a red-hot needle. There was scarcely a square inch of ground free from them. About three p.m. we glided into a quiet, shady creek, on whose banks an industrious white settler had located himself. I resolved to pass the rest of the day and night here, and endeavour to obtain a fresh supply of provisions, our stock of salt beef being now nearly exhausted. The situation of the house was beautiful; the little harbour being gay with water plants, Pontederiæ, now full of purple blossom, from which flocks of Piosócas started up screaming as we entered. The owner sent a boy with my men to show them the best place for fish up the creek, and in the course of the evening sold me a number of fowls, besides baskets of beans and farinha. The result of the fishing was a good supply of Jandiá, a handsome spotted Siluride fish, and Piránha, a kind of Salmonidæ (Tetragonopterus). Piránhas are of several kinds, many of which abound in the waters of the Tapajos. They are caught with almost any kind of bait, for their taste is indiscriminate and their appetite most ravenous. They often attack the legs of bathers near the shore, inflicting severe wounds with their strong triangular teeth. At Paquiatúba and this place I added about twenty species of small fishes to my collection; caught by hook and line, or with the hand in shallow pools under the shade of the forest.

My men slept ashore, and on their coming aboard in the morning Pinto was drunk and insolent. According to José, who had kept himself sober, and was alarmed at the other's violent conduct, the owner of the house and Pinto had spent the greater part of the night together, drinking aguardente de beijú,—a spirit distilled from the mandioca root. We knew nothing of the antecedents of this man, who was a tall, strong, self-willed fellow, and it began to dawn on us that this was not a very safe travelling companion in a wild country like this. I thought it better now to make the best of our way to the next settlement, Aveyros, and get rid of him. Our course to-day lay along a high, rocky coast, which extended without a break for about eight miles. The height of the perpendicular rocks was from 100 to 150 feet; ferns and flowering shrubs grew in the crevices, and the summit supported a luxuriant growth of forest, like the rest of the river banks. The waves beat with loud roar at the foot of these inhospitable barriers. At two p.m. we passed the mouth of a small picturesque harbour, formed by a gap in the precipitous coast. Several families have here settled; the place is called Itá-puáma, or "standing rock," from a remarkable isolated cliff, which stands erect at the entrance to the little haven. A short distance beyond Itá-puáma we found ourselves opposite to the village of Pinhel, which is perched, like Boim, on high ground, on the western side of the river. The stream is here from six to seven miles wide. A line of low islets extends in front of Pinhel, and a little further to the south is a larger island, called Capitarí, which lies nearly in the middle of the river.

June 23rd.—The wind freshened at ten o'clock in the morning of the 23rd. A thick black cloud then began to spread itself over the sky a long way down the river; the storm which it portended, however, did not reach us, as the dark threatening mass crossed from east to west, and the only effect it had was to impel a column of cold air up river, creating a breeze with which we bounded rapidly forward. The wind in the afternoon strengthened to a gale; we carried on with one foresail only, two of the men holding on to the boom to prevent the whole thing from flying to pieces. The rocky coast continued for about twelve miles above Itá-puáma: then succeeded a tract of low marshy land, which had evidently been once an island whose channel of separation from the mainland had become silted up. The island of Capitarí and another group of islets succeeding it, called Jacaré, on the opposite side, helped also to contract at this point the breadth of the river, which was now not more than about three miles. The little cuberta almost flew along this coast, there being no perceptible current, past extensive swamps, margined with thick floating grasses. At length, on rounding a low point, higher land again appeared on the right bank of the river, and the village of Aveyros hove in sight, in the port of which we cast anchor late in the afternoon.

Aveyros is a small settlement, containing only fourteen or fifteen houses besides the church; but it is the place of residence of the authorities of a large district; the priest, Juiz de Paz, the subdelegado of police, and the Captain of the Trabalhadores. The district includes Pinhel, which we passed about twenty miles lower down on the left bank of the river. Five miles beyond Aveyros, and also on the left bank, is the missionary village of Santa Cruz, comprising thirty or forty families of baptised Mundurucú Indians, who are at present under the management of a Capuchin Friar, and are independent of the Captain of Trabalhadores of Aveyros.

The river view from this point towards the south was very grand; the stream is from two to three miles broad, with green islets resting on its surface, and on each side a chain of hills stretches away in long perspective. I resolved to stay here for a few weeks to make collections. On landing, my first care was to obtain a house or room that I might live ashore. This was soon arranged; the head man of the place, Captain Antonio, having received notice of my coming, so that before night all the chests and apparatus I required were housed and put in order for working.

I here dismissed Pinto, who again got drunk and quarrelsome a few hours after he came ashore. He left the next day to my great relief in a small trading canoe that touched at the place on its way to Santarem. The Indian Manoel took his leave at the same time, having engaged to accompany me only as far as Aveyros; I was then dependent on Captain Antonio for fresh hands. The captains of Trabalhadores are appointed by the Brazilian Government, to embody the scattered Indian labourers and canoe-men of their respective districts, to the end that they may supply passing travellers with men when required. A semi-military organisation is given to the bodies; some of the steadiest amongst the Indians themselves being nominated as sergeants, and all the members mustered at the principal village of their district twice a-year. The captains, however, universally abuse their authority, monopolising the service of the men for their own purposes, so that it is only by favour that the loan of a canoe-hand can be wrung from them. I was treated by Captain Antonio with great consideration, and promised two good Indians when I should be ready to continue my voyage.

Little happened worth narrating, during my forty days' stay at Aveyros. The time was spent in the quiet, regular pursuit of Natural History: every morning I had my long ramble in the forest, which extended to the back-doors of the houses, and the afternoons were occupied in preserving and studying the objects collected. The priest was a lively old man, but rather a bore from being able to talk of scarcely anything except homœopathy, having been smitten with the mania during a recent visit to Santarem. He had a Portuguese Homœopathic Dictionary, and a little leather case containing glass tubes filled with globules, with which he was doctoring the whole village. A bitter enmity seemed to exist between the female members of the priest's family and those of the captain's; the only white women in the settlement. It was amusing to notice how they flaunted past each other, when going to church on Sundays, in their starched muslin dresses. I found an intelligent young man living here, a native of the province of Goyaz, who was exploring the neighbourhood for gold and diamonds. He had made one journey up a branch river, and declared to me, that he had found one diamond, but was unable to continue his researches, because the Indians who accompanied him refused to remain any longer: he was now waiting for Captain Antonio to assist him with fresh men, having offered him in return a share in the results of the enterprise. There appeared to be no doubt, that gold is occasionally found within two or three days' journey of Aveyros; but all lengthened search is made impossible by the scarcity of food and the impatience of the Indians, who see no value in the precious metal, and abhor the tediousness of the gold-searcher's occupation. It is impossible to do without them, as they are required to paddle the canoes.

The weather, during the month of July, was uninterruptedly fine; not a drop of rain fell, and the river sank rapidly. The mornings, for two hours after sunrise, were very cold; we were glad to wrap ourselves in blankets on turning out of our hammocks, and walk about at a quick pace in the early sunshine. But in the afternoons the heat was sickening; for the glowing sun then shone full on the front of the row of whitewashed houses, and there was seldom any wind to moderate its effects. I began now to understand why the branch rivers of the Amazons were so unhealthy, whilst the main stream was pretty nearly free from diseases arising from malaria. The cause lies, without doubt, in the slack currents of the tributaries in the dry season, and the absence of the cooling Amazonian trade-wind, which purifies the air along the banks of the main river. The trade-wind does not deviate from its nearly straight westerly course, so that the branch streams, which run generally at right angles to the Amazons, and have a slack current for a long distance from their mouths, are left to the horrors of nearly stagnant air and water.

Aveyros may be called the head-quarters of the fire-ant, which might be fittingly termed the scourge of this fine river. The Tapajos is nearly free from the insect pests of other parts, mosquitoes, sand-flies, Motúcas and piums; but the formiga de fogo is perhaps a greater plague than all the others put together. It is found only on sandy soils in open places, and seems to thrive most in the neighbourhood of houses and weedy villages, such as Aveyros: it does not occur at all in the shades of the forest. I noticed it in most places on the banks of the Amazons, but the species is not very common on the main river, and its presence is there scarcely noticed, because it does not attack man, and the sting is not so virulent as it is in the same species on the banks of the Tapajos. Aveyros was deserted a few years before my visit on account of this little tormentor, and the inhabitants had only recently returned to their houses, thinking its numbers had decreased. It is a small species, of a shining reddish colour, not greatly differing from the common red stinging ant of our own country (Myrmica rubra), except that the pain and irritation caused by its sting are much greater. The soil of the whole village is undermined by it: the ground is perforated with the entrances to their subterranean galleries, and a little sandy dome occurs here and there, where the insects bring their young to receive warmth near the surface. The houses are overrun with them; they dispute every fragment of food with the inhabitants, and destroy clothing for the sake of the starch. All eatables are obliged to be suspended in baskets from the rafters, and the cords well soaked with copaüba balsam, which is the only means known of preventing them from climbing. They seem to attack persons out of sheer malice: if we stood for a few moments in the street, even at a distance from their nests, we were sure to be overrun and severely punished, for the moment an ant touched the flesh, he secured himself with his jaws, doubled in his tail, and stung with all his might. When we were seated on chairs in the evenings in front of the house to enjoy a chat with our neighbours, we had stools to support our feet, the legs of which as well as those of the chairs, were well anointed with the balsam. The cords of hammocks are obliged to be smeared in the same way to prevent the ants from paying sleepers a visit.

The inhabitants declare that the fire-ant was unknown on the Tapajos, before the disorders of 1835-6, and believe that the hosts sprang up from the blood of the slaughtered Cabanas. They have, doubtless, increased since that time, but the cause lies in the depopulation of the villages and the rank growth of weeds in the previously cleared, well-kept spaces. I have already described the line of sediment formed on the sandy shores lower down the river by the dead bodies of the winged individuals of this species. The exodus from their nests of the males and females takes place at the end of the rainy season (June), when the swarms are blown into the river by squalls of wind, and subsequently cast ashore by the waves. I was told that this wholesale destruction of ant-life takes place annually, and that the same compact heap of dead bodies which I saw only in part, extends along the banks of the river for twelve or fifteen miles.

The forest behind Aveyros yielded me little except insects, but in these it was very rich. It is not too dense, and broad sunny paths skirted by luxuriant beds of Lycopodiums, which form attractive sporting places for insects, extend from the village to a swampy hollow or ygapó, which lies about a mile inland. Of butterflies alone I enumerated fully 300 species, captured or seen in the course of forty days within a half-hour's walk of the village. This is a greater number than is found in the whole of Europe. The only monkey I observed was the Callithrix moloch—one of the kinds called by the Indians Whaiápu-saí. It is a moderately-sized species, clothed with long brown hair, and having hands of a whitish hue. Although nearly allied to the Cebi it has none of their restless vivacity, but is a dull, listless animal. It goes in small flocks of five or six individuals, running along the main boughs of the trees. One of the specimens which I obtained here was caught on a low fruit-tree at the back of our house at sunrise one morning. This was the only instance of a monkey being captured in such a position that I ever heard of. As the tree was isolated it must have descended to the ground from the neighbouring forest and walked some distance to get at it. The species is sometimes kept in a tame state by the natives: it does not make a very amusing pet, and survives captivity only a short time.

I heard that the white Cebus, the Caiarára branca, a kind of monkey I had not yet seen, and wished very much to obtain, inhabited the forests on the opposite side of the river; so one day on an opportunity being afforded by our host going over in a large boat, I crossed to go in search of it. We were about twenty persons in all, and the boat was an old ricketty affair with the gaping seams rudely stuffed with tow and pitch. In addition to the human freight we took three sheep with us, which Captain Antonio had just received from Santarem and was going to add to his new cattle farm on the other side. Ten Indian paddlers carried us quickly across. The breadth of the river could not be less than three miles, and the current was scarcely perceptible. When a boat has to cross the main Amazons, it is obliged to ascend along the banks for half a mile or more to allow for drifting by the current; in this lower part of the Tapajos this is not necessary. When about half-way, the sheep, in moving about, kicked a hole in the bottom of the boat. The passengers took the matter very coolly, although the water spouted up alarmingly, and I thought we should inevitably be swamped. Captain Antonio took off his socks to stop the leak, inviting me and the Juiz de paz, who was one of the party, to do the same, whilst two Indians baled out the water with large cuyas. We thus managed to keep afloat until we reached our destination, when the men patched up the leak for our return journey.

The landing-place lay a short distance within the mouth of a shady inlet, on whose banks, hidden amongst the dense woods, were the houses of a few Indian and mameluco settlers. The path to the cattle farm led first through a tract of swampy forest; it then ascended a slope and emerged on a fine sweep of prairie, varied with patches of timber. The wooded portion occupied the hollows where the soil was of a rich chocolate-

H 2

brown colour, and of a peaty nature. The higher grassy undulating parts of the campo had a lighter and more sandy soil. Leaving our friends, I and José took our guns and dived into the woods in search of the monkeys. As we walked rapidly along I was very near treading on a rattlesnake which lay stretched out nearly in a straight line on the bare sandy pathway. It made no movement to get out of the way, and I escaped the danger by a timely and sudden leap, being unable to check my steps in the hurried walk. We tried to excite the sluggish reptile by throwing handsfull of sand and sticks at it, but the only notice it took was to raise its ugly horny tail and shake its rattle. At length it began to move rather nimbly, when we despatched it by a blow on the head with a pole, not wishing to fire on account of alarming our game.

We saw nothing of the white Caiarára; we met, however, with a flock of the common light-brown allied species (Cebus albifrons?), and killed one as a specimen. A resident on this side of the river told us that the white kind was found further to the south, beyond Santa Cruz. The light-brown Caiarára is pretty generally distributed over the forests of the level country. I saw it very frequently on the banks of the Upper Amazons, where it was always a treat to watch a flock leaping amongst the trees, for it is the most wonderful performer in this line of the whole tribe. The troops consist of thirty or more individuals which travel in single file. When the foremost of the flock reaches the outermost branch of an unusually lofty tree, he springs forth into the air without a moment's hesitation and alights on the dome of yielding foliage belonging to the neighbouring tree, maybe fifty feet beneath; all the rest following the example. They grasp, on falling, with hands and tail, right themselves in a moment, and then away they go along branch and bough to the next tree. The Caiarára owes its name in the Tupí language, macaw or large-headed (Acain, head, and Arára macaw), to the disproportionate size of the head compared with the rest of the body. It is very frequently kept as a pet in houses of natives. I kept one myself for about a year, which accompanied me in my voyages and became very familiar, coming to me always on wet nights to share my blanket. It is a most restless creature, but is not playful like most of the American monkeys; the restlessness of its disposition seeming to arise from great nervous irritability and discontent. The anxious, painful, and changeable expression of its countenance, and the want of purpose in its movements, betray this. Its actions are like those of a wayward child; it does not seem happy even when it has plenty of its favourite food, bananas; but will leave its own meal to snatch the morsels out of the hands of its companions. It differs in these mental traits from its nearest kindred, for another common Cebus, found in the same parts of the forest, the Prego monkey (Cebus cirrhifer?), is a much quieter and better-tempered animal; it is full of tricks, but these are generally of a playful character.

The Caiarára keeps the house in a perpetual uproar where it is kept: when alarmed, or hungry, or excited by envy, it screams piteously; it is always, however, making some noise or other, often screwing up its mouth and uttering a succession of loud notes resembling a whistle. My little pet, when loose, used to run after me, supporting itself for some distance on its hind legs, without, however, having been taught to do it. He offended me greatly one day by killing, in one of his jealous fits, another and much choicer pet—the nocturnal, owl-faced monkey (Nyctipithecus trivirgatus). Some one had given this a fruit, which the other coveted, so the two got to quarrelling. The Nyctipithecus fought only with its paws, clawing out and hissing like a cat; the other soon obtained the mastery, and before I could interfere, finished his rival by cracking its skull with his teeth. Upon this I got rid of him.

After a ramble of four or five hours, during which José shot a beautiful green and black-striped lizard of the Iguana family, from the trunk of a tree, and I filled my insect box with new and rare species (including an extremely beautiful butterfly of the genus Heliconius, H. Hermathena), we rejoined our companions at a hut, in the middle of the campo, where the Indians lived who had charge of the cattle. A tract of land like this, several miles in extent, alternating prairie and woodland, would be a rich possession in a better peopled country. The few oxen seemed to thrive on the nutritious grasses, and to make all complete there was a little lake in the low grounds, surrounded by fan-leaved Caraná palms, where the cattle could be watered all the year round. The farm was at present new, and the men said they had not yet been visited by jaguars. The poor fellows seemed to fare very badly. Captain Antonio treated all his Indians like slaves; paying them no wages and stinting them to scanty rations of salt fish and farinha. There was an air of poverty and misery over the whole establishment, which produced a very disagreeable impression: these are certainly not the people to develope the resources of a fine country like this.

On recrossing the river to Aveyros in the evening, a pretty little parrot fell from a great height headlong into the water near the boat; having dropped from a flock which seemed to be fighting in the air. One of the Indians secured it for me, and I was surprised to find the bird uninjured. There had probably been a quarrel about mates, resulting in our little stranger being temporarily stunned by a blow on the head from the beak of a jealous comrade. The species was the Conurus guianensis, called by the natives Maracaná; the plumage green, with a patch of scarlet under the wings. I wished to keep the bird alive and tame it, but all our efforts to reconcile it to captivity were vain; it refused food, bit every one who went near it, and damaged its plumage in its exertions to free itself. My friends in Aveyros said that this kind of parrot never became domesticated. After trying nearly a week I was recommended to lend the intractable creature to an old Indian woman, living in the village, who was said to be a skilful bird-tamer. In two days she brought it back almost as tame as the familiar love-birds of our aviaries. I kept my little pet for upwards of two years; it learned to talk pretty well, and was considered quite a wonder as being a bird usually so difficult of domestication. I do not know what arts the old woman used: Captain Antonio said she fed it with her saliva. The chief reason why almost all animals become so wonderfully tame in the houses of the natives is, I believe, their being treated with uniform gentleness, and allowed to run at large about the rooms. Our Maracanã used to accompany us sometimes in our rambles, one of the lads carrying it on his head. One day, in the middle of a long forest road, it was missed, having clung probably to an overhanging bough and escaped into the thickets without the boy perceiving it. Three hours afterwards, on our return by the same path, a voice greeted us in a colloquial tone as we passed "Maracanã!" We looked about for some time, but could not see anything until the word was repeated with emphasis "Maracanã-ã!" when we espied the little truant half concealed in the foliage of a tree. He came down and delivered himself up evidently as much rejoiced at the meeting as we were.

After I had obtained the two men promised, stout young Indians, 17 or 18 years of age, one named Ricardo and the other Alberto, I paid a second visit to the western side of the river in my own canoe; being determined, if possible, to obtain specimens of the White Cebus. We crossed over first to the mission village, Santa Cruz. It consists of 30 or 40 wretched-looking mud huts, closely built together in three straight ugly rows on a high gravelly bank. The place was deserted with the exception of two or three old men and women and a few children. The missionary, Fré Isidro, an

Italian monk, was away at another station called Wishitúba, two days' journey farther up the river. Report said of him that he had no zeal for religion or devotion to his calling, but was occupied in trading, using the Indian proselytes to collect salsaparilla and so forth, with a view to making a purse wherewith to retire to his own country. The semi-civilised Indians, who speak the Tupí language, called him Pai tucúra, or Father Grasshopper: his peaked hood having a droll resemblance to the pointed head of the insect. I afterwards became acquainted with Fré Isidoro, and found him a man of superior intelligence and ability. He complained much of the ill treatment the Indians received at the hands of traders and the Brazilian civil authorities, and said that he and his predecessors had incessantly to contend for the rights secured to the aborigines by the laws of the empire. The plan of assembling families in formal, blank-looking settlements, like this of Santa Cruz, seemed to me very ill chosen. The Indians would be happier in their scattered wigwams, embowered in foliage on the banks of shady rivulets where they prefer to settle when left to themselves.

A narrow belt of wood runs behind the village: beyond this is an elevated barren campo, with a clayey and gravelly soil. To the south the coast country is of a similar description; a succession of scantily-wooded hills, bare grassy spaces, and richly-timbered hollows. We traversed forest and campo in various directions during three days without meeting with monkeys, or indeed with anything that repaid us the time and trouble. The soil of the district appeared too dry; at this season of the year I had noticed, in other parts of the country, that mammals and birds resorted to the more humid areas of forest, we therefore proceeded to explore carefully the low and partly swampy tract along the coast to the north of Santa Cruz. We spent two days in this way, landing at many places, and penetrating a good distance in the interior. Although unsuccessful with regard to the White Cebus, the time was not wholly lost, as I added several small birds of species new to my collection. On the second evening we surprised a large flock, composed of about 50 individuals, of a curious eagle with a very long and slender hooked beak, the Rostrhamus hamatus. They were perched on the bushes which surrounded a shallow lagoon separated from the river by a belt of floating grass: my men said they fed on toads and lizards found at the margins of pools. They formed a beautiful sight as they flew up and wheeled about at a great height in the air. We obtained only one specimen.

Before returning to Aveyros, we paid another visit to the Jacaré inlet leading to Captain Antonio's cattle farm, for the sake of securing further specimens of the many rare and handsome insects found there; landing at the port of one of the settlers. The owner of the house was not at home, and the wife, a buxom young woman, a dark mameluca, with clear though dark complexion and fine rosy cheeks, was preparing, in company with another stout-built Amazon, her rod and lines to go out fishing for the day's dinner. It was now the season for Tucunarés, and Senhora Joaquina showed us the fly baits used to take this kind of fish, which she had made with her own hands of parrots' feathers. The rods used are slender bamboos, and the lines made from the fibres of pine-apple leaves. It is not very common for the Indian and half-caste women to provide for themselves in the way these spirited dames were doing, although they are all expert paddlers, and very frequently cross wide rivers in their frail boats without the aid of men. It is possible that parties of Indian women, seen travelling alone in this manner, may have given rise to the fable of a nation of Amazons invented by the first Spanish explorers of the country. Senhora Joaquina invited me and José to a Tucunaré dinner for the afternoon, and then shouldering their paddles and tucking up their skirts, the two dusky fisherwomen marched down to their canoe. We sent the two Indians into the woods to cut palm-leaves to mend the thatch of our cuberta, whilst I and José rambled through the woods which skirted the campo. On our return, we found a most bountiful spread in the house of our hostess. A spotless white cloth was laid on the mat, with a plate for each guest and a pile of fragrant newly-made farinha by the side of it. The boiled Tucunarés were soon taken from the kettles and set before us. I thought the men must be happy husbands who owned such wives as these. The Indian and mameluco women certainly do make excellent managers ; they are more industrious than the men and most of them manufacture farinha for sale on their own account, their credit always standing higher with the traders on the river than that of their male connections. I was quite surprised at the quantity of fish they had taken; there being sufficient for the whole party, including several children, two old men from a neighbouring hut, and my Indians. I made our good-natured entertainers a small present of needles and sewing-cotton, articles very much prized, and soon after we re-embarked, and again crossed the river to Aveyros.

August 2nd.—Left Aveyros; having resolved to ascend a branch river, the Cuparí, which enters the Tapajos about eight miles above this village, instead of going forward along the main stream. I should have liked to visit the settlements of the Mundurucú tribe which lie beyond the first cataract of the Tapajos, if it had been compatible with the other objects I had in view. But to perform this journey a lighter canoe than mine would have been necessary, and six or eight Indian paddlers, which in my case it was utterly impossible to obtain. There would be, however, an opportunity of seeing this fine race of people on the Cuparí, as a horde was located towards the head waters of this stream. The distance from Aveyros to the last civilised settlement on the Tapajos, Itaitúba, is about forty miles. The falls commence a short distance beyond this place. Ten formidable cataracts or rapids then succeed each other at intervals of a few miles: the chief of which are the Coaitá, the Buburé, the Salto Grande about thirty feet high, and the Montanha. The canoes of Cuyabá tradesmen which descend annually to Santarem are obliged to be unloaded at each of these, and the cargoes carried by land on the backs of Indians, whilst the empty vessels are dragged by ropes over the obstructions. The Cuparí was described to me as flowing through a rich moist clayey valley, covered with forests and abounding in game; whilst the banks of the Tapajos beyond Aveyros were barren sandy campos, with ranges of naked or scantily-wooded hills, forming a kind of country which I had always found very unproductive in Natural History objects in the dry season which had now set in.

We entered the mouth of the Cuparí on the evening of the following day (August 3rd). It was not more than 100 yards wide, but very deep: we found no bottom in the middle with a line of eight fathoms. The banks were gloriously wooded; the familiar foliage of the cacao growing abundantly amongst the mass of other trees reminding me of the forests of the main Amazons. We rowed for five or six miles, generally in a south-easterly direction although the river had many abrupt bends, and stopped for the night at a settler's house situated on a high bank and accessible only by a flight of rude wooden steps fixed in the clayey slope. The owners were two brothers, half-breeds, who with their families shared the large roomy dwelling; one of them was a blacksmith, and we found him working with two Indian lads at his forge, in an open shed under the shade of mango trees. They were the sons of a Portuguese immigrant who had settled here forty years previously and married a Mundurucú woman. He must have been a far more industrious man than the majority of his countrymen who emigrate to Brazil now-a-days, for there were signs of former extensive cultivation at the back of the house in groves of orange, lemon, and coffee trees, and a large plantation of cacao occupied the lower grounds.

The next morning one of the brothers brought me a beautiful opossum which had been caught in the fowl-house a little before sunrise. It was not so large as a rat, and had soft brown fur, paler beneath and on the face, with a black stripe on each cheek. This made the third species of marsupial rat I had so far obtained: but the number of these animals is very considerable in Brazil, where they take the place of the shrews of Europe, shrew mice and, indeed, the whole of the insectivorous order of mammals, being entirely absent from Tropical America. One kind of these rat-like opossums is aquatic, and has webbed feet. The terrestrial species are nocturnal in their habits, sleeping during the day in hollow trees, and coming forth at night to prey on birds in their roosting places. It is very difficult to rear poultry in this country on account of these small opossums, scarcely a night passing in some parts in which the fowls are not attacked by them.

August 5th.—The river reminds me of some parts of the Jaburú channel, being hemmed in by two walls of forest rising to the height of at least 100 feet, and the outlines of the trees being concealed throughout by a dense curtain of leafy creepers. The impression of vegetable profusion and overwhelming luxuriance increases at every step. The deep and narrow valley of the Cuparí has a moister climate than the banks of the Tapajos. We have now frequent showers, whereas we left everything parched up by the sun at Aveyros.

After leaving the last sitio we advanced about eight miles and then stopped at the house of Senhor Antonio

Malagueita, a mameluco settler, whom we had been recommended to visit. His house and outbuildings were extensive, the grounds well wooded, and the whole wore an air of comfort and well-being which is very uncommon in this country. A bank of indurated white clay sloped gently up from the tree-shaded port to the house, and beds of kitchen-herbs extended on each side, with (rare sight!) rose and jasmine trees in full bloom. Senhor Antonio, a rather tall middle-aged man with a countenance beaming with good nature, came down to the port as soon as we anchored. I was quite a stranger to him, but he had heard of my coming and seemed to have made preparations. I never met with a heartier welcome. On entering the house, the wife, who had more of the Indian tint and features than her husband, was equally warm and frank in her greeting. Senhor Antonio had spent his younger days at Pará, and had acquired a profound respect for Englishmen. I stayed here two days. My host accompanied me in my excursions; in fact, his attentions, with those of his wife and the host of relatives of all degrees who constituted his household, were quite troublesome, as they left me not a moment's privacy from morning till night.

We had together several long and successful rambles along a narrow pathway which extended several miles into the forest. I here met with a new insect pest, one which the natives may be thankful is not spread more widely over the country: it was a large brown fly of the Tabanidæ family (genus Pangonia), with a proboscis half an inch long and sharper than the finest needle.

It settled on our backs by twos and threes at a time, and pricked us through our thick cotton shirts, making us start and cry out with the sudden pain. I secured a dozen or two as specimens. As an instance of the extremely confined ranges of certain species it may be mentioned that I did not find this insect in any other part of the country except along half a mile or so of this gloomy forest road.

We were amused at the excessive and almost absurd tameness of a fine Mutum or Curassow turkey that ran about the house. It was a large glossy-black species (the Mitu tuberosa) having an orange-coloured beak surmounted by a bean-shaped excrescence of the same hue. It seemed to consider itself as one of the family: attended at all the meals, passing from one person to another round the mat to be fed, and rubbing the sides of its head in a coaxing way against their cheeks or shoulders. At night it went to roost on a chest in a sleeping-room beside the hammock of one of the little girls, to whom it seemed particularly attached, following her wherever she went about the grounds. I found this kind of Curassow bird was very common in the forests of the Cuparí; but it is rare on the Upper Amazons, where an allied species which has a round instead of a bean-shaped waxen excrescence on the beak (Crax globicera) is the prevailing kind. These birds in their natural state never descend from the tops of the loftiest trees, where they live in small flocks and build their nests. The Mitu tuberosa lays two rough-shelled, white eggs; it is fully as large a bird as the common turkey, but the flesh when cooked is drier and not so well flavoured. It is difficult to find the reason why these superb birds have not been reduced to domestication by the Indians, seeing that they so readily become tame. The obstacle offered by their not breeding in confinement, which is probably owing to their arboreal habits, might perhaps be overcome by repeated experiment; but for this the Indians probably had not sufficient patience or intelligence. The reason cannot lie in their insensibility to the value of such birds, for the common turkey, which has been introduced into the country, is much prized by them.

We had an unwelcome visitor whilst at anchor in the port of Joaõ Malagueita. I was awoke a little after midnight as I lay in my little cabin by a heavy blow struck at the sides of the canoe close to my head, which was succeeded by the sound of a weighty body plunging in the water. I got up; but all was again quiet, except the cackle of fowls in our hen-coop, which hung over the sides of the vessel about three feet from the cabin door. I could find no explanation of the circumstance, and, my men being all ashore, I turned in again and slept till morning. I then found my poultry loose about the canoe, and a large rent in the bottom of the hen-coop, which was about two feet from the surface of the water: a couple of fowls were missing. Senhor Antonio said the depredator was a Sucurujú (the Indian name for the Anaconda, or great water serpent—Eunectes murinus), which had for months past been haunting this part of the river, and had carried off many ducks and fowls from the ports of various houses. I was inclined to doubt the fact of a serpent striking at its prey from the water, and thought an alligator more likely to be the culprit, although we had not yet met with alligators in the river. Some days afterwards the young men belonging to the different sitios agreed together to go in search of the serpent. They began in a systematic manner, forming two parties each embarked in three or four canoes, and starting from points several miles apart, whence they gradually approximated, searching all the little inlets on both sides the river. The reptile was found at last sunning itself on a log at the mouth of a muddy rivulet, and despatched with harpoons. I saw it the day after it was killed: it was not a very large specimen, measuring only eighteen feet nine inches in length and sixteen inches in circumference at the widest part of the body. I measured skins of the Anaconda afterwards, twenty-one feet in length and two feet in girth. The reptile has a most hideous appearance, owing to its being very broad in the middle and tapering abruptly at both ends. It is very abundant in some parts of the country; nowhere more so than in the Lago Grande, near Santarem, where it is often seen coiled up in the corners of farm-yards, and detested for its habit of carrying off poultry, young calves, or whatever animal it can get within reach of.

At Ega a large Anaconda was once near making a meal of a young lad about ten years of age belonging to one of my neighbours. The father and his son went one day in their montaria a few miles up the Teffé to gather wild fruit; landing on a sloping sandy shore, where the boy was left to mind the canoe whilst the man entered the forest. The beaches of the Teffé form groves of wild guava and myrtle trees, and during most months of the year are partly overflown by the river. Whilst the boy was playing in the water under the shade of these trees a huge reptile of this species stealthily wound its coils around him, unperceived until it was too late to escape. His cries brought the father quickly to the rescue ; who rushed forward, and seizing the Anaconda boldly by the head, tore his jaws asunder. There appears to be no doubt that this formidable serpent grows to an enormous bulk and lives to a great age, for I heard of specimens having been killed which measured forty-two feet in length, or double the size of the largest I had an opportunity of examining. The natives of the Amazons country universally believe in the existence of a monster water-serpent said to be many score fathoms in length, which appears successively in different parts of the river. They call it the Mai d'agoa—the mother or spirit of the water. This fable, which was doubtless suggested by the occasional appearance of Sucurujús of unusually large size, takes a great variety of forms, and the wild legends form the subject of conversation amongst old and young, over the wood fires in lonely settlements.

August 6th and 7th.—On leaving the sitio of Antonio Malagueita we continued our way along the windings of the river, generally in a south-east and south-south-east direction but sometimes due south, for about fifteen miles, when we stopped at the house of one Paulo Christo, a mameluco whose acquaintance I had made at Aveyros.

I 2

Here we spent the night and part of the next day ; doing in the morning a good five hours' work in the forest, accompanied by the owner of the place. In the afternoon of the 7th we were again under way : the river makes a bend to the east-north-east for a short distance above Paulo Christo's establishment, it then turns abruptly to the south-west, running from that direction about four miles. The hilly country of the interior then commences: the first token of it being a magnificently-wooded bluff rising nearly straight from the water to a height of about 250 feet. The breadth of the stream hereabout was not more than sixty yards, and the forest assumed a new appearance from the abundance of the Uruçurí palm, a species which has a noble crown of broad fronds with symmetrical rigid leaflets.

On the road, we passed a little shady inlet, at the mouth of which a white-haired, wrinkle-faced old man was housed in a temporary shed, washing the soil for gold. He was quite alone : no one knew anything of him in these parts, except that he was a Cuyabano, or native of Cuyabá in the mining districts, and his little boat was moored close to his rude shelter. Whatever success he might have had remained a secret, for he went away, after a three weeks' stay in the place, without communicating with any one.

We reached, in the evening, the house of the last civilised settler on the river, Senhor Joaõ Aracú, a wiry, active fellow and capital hunter, whom I wished to make a friend of and persuade to accompany me to the Mundurucú village and the falls of the Cuparí, some forty miles further up the river.

I stayed at the sitio of Joaõ Aracú until the 19th, and again, in descending, spent fourteen days at the same place. The situation was most favourable for collecting the natural products of the district. The forest was not crowded with underwood, and pathways led through it for many miles and in various directions. I could make no use here of our two men as hunters, so, to keep them employed whilst José and I worked daily in the woods, I set them to make a montaria under Joaõ Aracú's directions. The first day a suitable tree was found for the shell of the boat, of the kind called Itaüba amarello, the yellow variety of the stone-wood. They felled it, and shaped out of the trunk a log nineteen feet in length: this they dragged from the forest, with the help of my host's men, over a road they had previously made with pieces of round wood to act as rollers. The distance was about half a mile, and the ropes used for drawing the heavy load were tough lianas cut from the surrounding trees. This part of the work occupied about a week: the log had then to be hollowed out, which was done with strong chisels through a slit made down the whole length. The heavy portion of the task being then completed, nothing remained but to widen the opening, fit two planks for the sides and the same number of semicircular boards for the ends, make the benches, and caulk the seams.

The expanding of the log thus hollowed out is a critical operation, and not always successful, many a good shell being spoilt by its splitting or expanding irregularly. It is first reared on tressels, with the slit downwards, over a large fire, which is kept up for seven or eight hours, the process requiring unremitting attention to avoid cracks and make the plank bend with the proper dip at the two ends. Wooden straddlers, made by cleaving pieces of tough elastic wood and fixing them with wedges, are inserted into the opening, their compass being altered gradually as the work goes on, but in different degree according to the part of the boat operated upon. Our casca turned out a good one: it took a long time to cool, and was kept in shape whilst it did so by means of wooden cross-pieces. When the boat was finished it was launched with great merriment by the men, who hoisted coloured handkerchiefs for flags, and paddled it up and down the stream to try its capabilities. My people had suffered as much inconvenience from the want of a montaria as myself, so this was a day of rejoicing to all of us.

I was very successful at this place with regard to the objects of my journey. About twenty new species of fishes and a considerable number of small reptiles were added to my collection; but very few birds were met with worth preserving. A great number of the most conspicuous insects of the locality were new to me, and turned out to be species peculiar to this part of the Amazons valley. There is the most striking contrast between the productions of the Cuparí and those of Altar do Chaõ in this department: the majority of the species inhabiting the one district being totally unknown in the other. At the same time a considerable proportion of the Cuparí species were identical with those of Ega on the Upper Amazons, a region eight times further removed than the village just mentioned. The most interesting acquisition at this place was a large and handsome monkey, of a species I had not before met with—the white-whiskered Coaitá, or spider monkey, Ateles marginatus. I saw a pair one day in the forest moving slowly along the branches of a lofty tree, and shot one of them; the next day Joao Aracú brought down another, possibly the companion. The species is of about the same size as the common black kind of which I have given an account in a former chapter, and has a similar lean body with limbs clothed with coarse black hair; but it differs in having the whiskers and a triangular patch on the crown of the head of a white colour. It is never met with in the alluvial plains of the Amazons, nor, I believe, on the northern side of the great river valley, except towards the head waters, near the Andes; where Humboldt discovered it on the banks of the Santiago. I thought the meat the best flavoured I had ever tasted. It resembled beef, but had a richer and sweeter taste. During the time of our stay in this part of the Cuparí, we could get scarcely anything but fish to eat, and as this diet ill agreed with me, three successive days of it reducing me to a state of great weakness, I was obliged to make the most of our Coaitá meat. We smoke-dried the joints instead of salting them; placing them for several hours on a framework of sticks arranged over a fire, a plan adopted by the natives to preserve fish when they have no salt, and which they call "muquiar." Meat putrefies in this climate in less than twenty-four hours, and salting is of no use, unless the pieces are cut in thin slices and dried immediately in the sun. My monkeys lasted me about a fortnight, the last joint being an arm with the clenched fist, which I used with great economy, hanging it in the intervals between my frugal meals on a nail in the cabin. Nothing but the hardest necessity could have driven me so near to cannibalism as this, but we had the greatest difficulty in obtaining here a sufficient supply of animal food. About every three days the work on the montaria had to be suspended and all hands turned out for the day to hunt and fish, in which they were often unsuccessful, for although there was plenty of game in the forest, it was too widely scattered to be available. Ricardo and Alberto occasionally brought in a tortoise or an anteater, which served us for one day's consumption. We made acquaintance here with many strange dishes, amongst them Iguana eggs; these are of oblong form, about an inch in length, and covered with a flexible shell. The lizard lays about two score of them in the hollows of trees. They have an oily taste; the men ate them raw, beaten up with farinha, mixing a pinch of salt in the mess; I could only do with them when mixed with Tucupí sauce, of which we had a large jar full always ready to temper unsavoury morsels.

One day as I was entomologizing alone and unarmed, in a dry Ygapó, where the trees were rather wide apart and the ground coated to the depth of eight or ten inches with dead leaves, I was near coming into collision with a boa constrictor. I had just entered a little thicket to capture an insect, and whilst pinning it was rather startled by a rushing noise in the vicinity. I looked up to the sky, thinking a squall was coming on, but not a breath of wind stirred in the tree-tops. On stepping out of the bushes I met face to face a huge serpent coming down a slope, and making the dry twigs crack and fly with his weight as he moved over them. I had very frequently met with a smaller boa, the Cutim-boia, in a similar way, and knew from the habits of the family that there was no danger, so I stood my ground. On seeing me the reptile suddenly turned, and glided at an accelerated pace down the path. Wishing to take a note of his probable size and the colours and markings of his skin, I set off after him; but he increased his speed, and I was unable to get near enough for the purpose. There was very little of the serpentine movement in his course. The rapidly moving and shining body looked like a stream of brown liquid flowing over the thick bed of fallen leaves, rather than a serpent with skin of varied colours. He descended towards the lower and moister parts of the Ygapó. The huge trunk of an uprooted tree here lay across the road; this he glided over in his undeviating course, and soon after penetrated a dense swampy thicket, where of course I did not choose to follow him.

I suffered terribly from the heat and mosquitoes as the river sank with the increasing dryness of the season, although I made an awning of the sails to work under, and slept at night in the open air with my hammock slung between the masts. But there was no rest in any part; the canoe descended deeper and deeper into the gulley, through which the river flows between high clayey banks, as the water subsided, and with the glowing sun overhead we felt at midday as if in a furnace. I could bear scarcely any clothes in the daytime between eleven in the morning and five in the afternoon, wearing nothing but loose and thin cotton trousers and a light straw hat, and could not be accommodated in Joaõ Aracú's house, as it was a small one and full of noisy children. One night we had a terrific storm. The heat in the afternoon had been greater than ever, and at sunset the sky had a brassy glare: the black patches of cloud which floated in it, being lighted up now and then by flashes of sheet lightning. The mosquitoes at night were more than usually troublesome, and I had just sunk exhausted into a doze towards the early hours of morning when the storm began; a complete deluge of rain with incessant lightning and rattling explosions of thunder. It lasted for eight hours; the grey dawn opening amidst the crash of the tempest. The rain trickled through the seams of the cabin roof on to my collections, the late hot weather having warped the boards, and it gave me immense trouble to secure them in the midst of the confusion. Altogether I had a bad night of it, but what with storms, heat, mosquitoes, hunger, and, towards the last, ill health, I seldom had a good night's rest on the Cuparí.

A small creek traversed the forest behind Joaõ Aracú's house, and entered the river a few yards from our anchoring place. I used to cross it twice a day, on going and returning from my hunting ground. One day early in September, I noticed that the water was two or three inches higher in the afternoon than it had been in the morning. This phenomenon was repeated the next day, and in fact daily, until the creek became dry with the continued subsidence of the Cuparí, the time of rising shifting a little from day to day. I pointed out the circumstance to Joaõ Aracú, who had not noticed it before (it was only his second year of residence in the locality), but agreed with me that it must be the "maré." Yes, the tide! the throb of the great oceanic pulse felt in this remote corner, 530 miles distant from the place where it first strikes the body of fresh water at the mouth of the Amazons. I hesitated at first at this conclusion, but on reflecting that the tide was known to be perceptible at Obydos, more than 400 miles from the sea; that at high water in the dry season a large flood from the Amazons enters the mouth of the Tapajos, and that there is but a very small difference of level between that point and the Cuparí, a fact shown by the absence of current in the dry season; I could have no doubt that this conclusion was a correct one.

The fact of the tide being felt 530 miles up the Amazons, passing from the main stream to one of its affluents 380 miles from its mouth, and thence to a branch in the third degree, is a proof of the extreme flatness of the land which forms the lower part of the Amazonian valley. This uniformity of level is shown also in the broad lake-like expanses of water formed near their mouths by the principal affluents which cross the valley to join the main river.

August 21st.—Joaõ Aracú consented to accompany me to the falls with one of his men, to hunt and fish for me. One of my objects was to obtain specimens of the hyacinthine macaw, whose range commences on all the branch rivers of the Amazons which flow from the south through the interior of Brazil, with the first cataracts. We started on the 19th; our direction on that day being generally south-west. On the 20th our course was southerly and south-easterly. This morning (August 21st) we arrived at the Indian settlement, the first house of which lies about thirty-one miles above the sitio of Joaõ Aracú. The river at this place is from sixty to seventy yards wide, and runs in a zigzag course between steep clayey banks twenty to fifty feet in height. The houses of the Mundurucús to the number of about thirty are scattered along the banks for a distance of six or seven miles. The owners appear to have chosen all the most picturesque sites—tracts of level ground at the foot of wooded heights, or little havens with bits of white sandy beach—as if they had an appreciation of natural beauty. Most of the dwellings are conical huts, with walls of framework filled in with mud and thatched with palm leaves, the broad eaves reaching halfway to the ground. Some are quadrangular, and do not differ in structure from those of the semi-civilised settlers in other parts; others are open sheds or ranchos. They seem generally to contain not more than one or two families each.

At the first house we learnt that all the fighting men had this morning returned from a two days' pursuit of a wandering horde of savages of the Parárauáte tribe, who had strayed this way from the interior lands and robbed the plantations. A little further on we came to the house of the Tushaúa or chief, situated on the top of a high bank, which we had to ascend by wooden steps. There were four other houses in the neighbourhood, all filled with people. A fine old fellow, with face, shoulders, and breast tattooed all over in a cross-bar pattern, was the first strange object that caught my eye. Most of the men lay lounging or sleeping in their hammocks. The women were employed in an adjoining shed making farinha, many of them being quite naked, and rushing off to the huts to slip on their petticoats when they caught sight of us. Our entrance aroused the Tushaúa from a nap; after rubbing his eyes he came forward and bade us welcome with the most formal politeness, and in very good Portuguese. He was a tall, broad-shouldered, well-made man, apparently about thirty years of age, with handsome regular features, not tattooed, and a quiet good-humoured expression of countenance. He had been several times to Santarem and once to Pará, learning the Portuguese language during these journeys. He was dressed in shirt and trousers made of blue-checked cotton cloth and there was not the slightest trace of the savage in his appearance or demeanour. I was told that he had come into the chieftainship by inheritance, and that the Cuparí horde of Mundurucús, over which his fathers had ruled before him, was formerly much more numerous, furnishing 300 bows in time of war. They could now scarcely muster forty; but the horde has no longer a close political connection with the main body of the tribe, which inhabits the banks of the Tapajos, six days' journey from the Cuparí settlement.

I spent the remainder of the day here, sending Aracú and the men to fish, whilst I amused myself with the

Tushaúa and his people. A few words served to explain my errand on the river; he comprehended at once why white men should admire and travel to collect the beautiful birds and animals of his country, and neither he nor his people spoke a single word about trading, or gave us any trouble by coveting the things we had brought. He related to me the events of the preceding three days. The Parárauátes were a tribe of intractable savages with whom the Mundurucús have been always at war. They had no fixed abode, and of course made no plantations, but passed their lives like the wild beasts, roaming through the forest, guided by the sun: wherever they found themselves at night-time there they slept, slinging their bast hammocks, which are carried by the women, to the trees. They ranged over the whole of the interior country, from the head waters of the Itapacurá (a branch of the Tapajos flowing from the east, whose sources lie in about 7° south latitude) to the banks of the Curuá (about 3° south latitude), and from the Mundurucú settlements on the Tapajos (55° west longtitude) to the Pacajaz (50° west longitude). They cross the streams which lie in their course in bark canoes, which they make on reaching the water, and cast away after landing on the opposite side. The tribe is very numerous, but the different hordes obey only their own chieftains. The Mundurucús of the upper Tapajos have an expedition on foot against them at the present time, and the Tushaúa supposed that the horde which had just been chased from his maloca were fugitives from that direction. There were about a hundred of them—including men, women, and children. Before they were discovered the hungry savages had uprooted all the macasheira, sweet potatoes, and sugar cane, which the industrious Mundurucús had planted for the season, on the east side of the river. As soon as they were seen they made off, but the Tushaúa quickly got together all the young men of the settlement, about thirty in number, who armed themselves with guns, bows and arrows, and javelins, and started in pursuit. They tracked them, as before related, for two days through the forest, but lost their traces on the further bank of the Cuparitinga, a branch stream flowing from the north-east. The pursuers thought, at one time, they were close upon them, having found the inextinguished fire of their last encampment. The footmarks of the chief could be distinguished from the rest by their great size and the length of the stride. A small necklace made of scarlet beans was the only trophy of the expedition, and this the Tushaúa gave to me.

I saw very little of the other male Indians, as they were asleep in their huts all the afternoon. There were two other tattooed men lying under an open shed, besides the old man already mentioned. One of them presented a strange appearance, having a semicircular black patch in the middle of his face, covering the bottom of the nose and mouth, crossed lines on his back and breast, and stripes down his arms and legs. It is singular that the graceful curved patterns used by the South Sea Islanders, are quite unknown among the Brazilian red men; they being all tattooed either in simple lines or patches. The nearest approach to elegance of design which I saw, was amongst the Tucúnas of the Upper Amazons, some of whom have a scroll-like mark on each cheek, proceeding from the corner of the mouth. The taste, as far as form is concerned, of the American Indian would seem to be far less refined than that of the Tahitian and New Zealander.

To amuse the Tushaúa, I fetched from the canoe the two volumes of Knight's Pictorial Museum of Animated Nature. The engravings quite took his fancy, and he called his wives, of whom, as I afterwards learnt from Aracú, he had three or four, to look at them; one of them was a handsome girl, decorated with necklace and bracelets of blue beads. In a short time others left their work, and I then had a crowd of women and children around me, who all displayed unusual curiosity for Indians. It was no light task to go through the whole of the illustrations, but they would not allow me to miss a page, making me turn back when I tried to skip. The pictures of the elephant, camels, orang-otangs, and tigers, seemed most to astonish them; but they were interested in almost everything, down even to the shells and insects. They recognised the portraits of the most striking birds and mammals which are found in their own country; the jaguar, howling monkeys, parrots, trogons, and toucans. The elephant was settled to be a large kind of Tapir; but they made but few remarks, and those in the Mundurucú language, of which I understood only two or three words. Their way of expressing surprise was a clicking sound made with the teeth, similar to the one we ourselves use, or a subdued exclamation, Hm ! hm ! Before I finished, from fifty to sixty had assembled ; there was no pushing or rudeness, the grown-up women letting the young girls and children stand before them, and all behaved in the most quiet and orderly manner possible.

The great difference in figure, shape of head, and arrangement of features amongst these people struck me forcibly, and showed how little uniformity there is in these respects amongst the Brazilian Indians, even when belonging to the same tribe. The only points in which they all closely resembled each other were the long, thick, straight, jet-black hair, the warm coppery-brown tint of the skin, and the quiet, rather dull, expression of countenance. I saw no countenance so debased in expression as many seen amongst the Múra tribe, and no head of the Mongolian type—broad, with high cheek bones, and oblique position of the eyes—of which single examples occur amongst the semi-civilised canoemen on the river. Many of them had fine oval faces, with rather long and well-formed features, moderately thin lips, and arched forehead. One little girl, about twelve years of age, had quite a European cast of features, and a remarkably slim figure. They were all clean in their persons ; the petticoats of the women being made of coarse cotton cloth obtained from traders, and their hair secured in a knot behind by combs made of pieces of bamboo. The old men had their heads closely cropped, with the exception of a long fringe which hung down in front over their foreheads.

The Mundurucús are perhaps the most numerous

and formidable tribe of Indians now surviving in the Amazons region. They inhabit the shores of the Tapajos (chiefly the right bank), from 3° to 7° south latitude, and the interior of the country between that part of the river and the Madeira. On the Tapajos alone they can muster, I was told, 2000 fighting men; the total population of the tribe may be about 20,000. They were not heard of until about ninety years ago, when they made war on the Portuguese settlements; their hosts crossing the interior of the country eastward of the Tapajos, and attacking the establishments of the whites in the province of Maranham. The Portuguese made peace with them in the beginning of the present century, the event being brought about by the common cause of quarrel entertained by the two peoples against the hated Múras. They have ever since been firm friends of the whites. It is remarkable how faithfully this friendly feeling has been handed down amongst the Mundurucús, and spread to the remotest of the scattered hordes. Wherever a white man meets a family, or even an individual of the tribe, he is almost sure to be reminded of this alliance. They are the most warlike of the Brazilian tribes, and are considered also the most settled and industrious; they are not, however, superior in this latter respect to the Jurís and Passés on the Upper Amazons, or the Uapés Indians near the head waters of the Rio Negro. They make very large plantations of mandioca, and sell the surplus produce, which amounts on the Tapajos to from 3000 to 5000 baskets (60 lbs. each) annually, to traders who ascend the river from Santarem between the months of August and January. They also gather large quantities of salsaparilla, India-rubber, and Tonka beans, in the forests. The traders, on their arrival at the Campinas (the scantily wooded region inhabited by the main body of Mundurucús beyond the cataracts) have first to distribute their wares—cheap cotton cloths, iron hatchets, cutlery, small wares, and cashaça—amongst the minor chiefs, and then wait three or four months for repayment in produce.

A rapid change is taking place in the habits of these Indians through frequent intercourse with the whites, and those who dwell on the banks of the Tapajos now seldom tattoo their children. The principal Tushaúa of the whole tribe or nation, named Joaquim, was rewarded with a commission in the Brazilian army, in acknowledgment of the assistance he gave to the legal authorities during the rebellion of 1835-6. It would be a misnomer to call the Mundurucús of the Cuparí and many parts of the Tapajos, savages; their regular mode of life, agricultural habits, loyalty to their chiefs, fidelity to treaties, and gentleness of demeanour, give them a right to a better title. Yet they show no aptitude for the civilised life of towns, and, like the rest of the Brazilian tribes, seem incapable of any further advance in culture. In their former wars they exterminated two of the neighbouring peoples, the Júmas and the Jacarés; and make now an annual expedition against the Parárauátes, and one or two other similar wild tribes who inhabit the interior of the land, but are sometimes driven by hunger towards the banks of the great rivers to rob the plantations of the agricul-

K 2

tural Indians. These campaigns begin in July, and last throughout the dry months; the women generally accompanying the warriors to carry their arrows and javelins. They had the diabolical custom, in former days, of cutting off the heads of their slain enemies, and preserving them as trophies around their houses. I believe this, together with other savage practices, has been relinquished in those parts where they have had long intercourse with the Brazilians, for I could neither see nor hear anything of these preserved heads. They used to sever the head with knives made of broad bamboo, and then, after taking out the brain and fleshy parts, soak it in bitter vegetable oil (andiroba), and expose it for several days over the smoke of a fire or in the sun. In the tract of country between the Tapajos and the Madeira, a deadly war has been for many years carried on between the Mundurucús and the Aráras. I was told by a Frenchman at Santarem, who had visited that part, that all the settlements there have a military organization. A separate shed is built outside of each village, where the fighting men sleep at night, sentinels being stationed to give the alarm with blasts of the Turé on the approach of the Aráras, who choose the night for their onslaughts.

Each horde of Mundurucús has its pajé or medicine man, who is the priest and doctor; fixes upon the time most propitious for attacking the enemy; exorcises evil spirits, and professes to cure the sick. All illness whose origin is not very apparent is supposed to be caused by a worm in the part affected. This the pajé pretends to extract; he blows on the seat of pain the smoke from a large cigar, made with an air of great mystery by rolling tobacco in folds of Tauarí, and then sucks the place, drawing from his mouth, when he has finished, what he pretends to be the worm. It is a piece of very clumsy conjuring. One of these pajés was sent for by a woman in Joaõ Malagueita's family, to operate on a child who suffered much from pains in the head. Senhor Joaõ contrived to get possession of the supposed worm after the trick was performed in our presence, and it turned out to be a long white air-root of some plant. The pajé was with difficulty persuaded to operate whilst Senhor Joaõ and I were present. I cannot help thinking that he, as well as all others of the same profession, are conscious impostors, handing down the shallow secret of their divinations and tricks from generation to generation. The institution seems to be common to all tribes of Indians, and to be held to more tenaciously than any other.

The opposite (western) shore of the Tapajos for some distance beyond the falls, and the country thence to the channels behind Villa Nova, are inhabited by the Mauhés tribe, of whom I have spoken in a former chapter. These are also a settled, agricultural people, but speak a totally different language from that of the Mundurucús. I saw at Aveyros several men of this fine tribe, who were descending the river in a trading canoe, and who, on being confronted with a Mundurucú were quite unable to understand him. There are many other points of difference between the two tribes. The Mauhés are much less warlike, and do not practise tattooing. Their villages are composed of a number of small huts, tenanted by single families, whilst the separate hordes of Mundurucús generally live together, each in one large dwelling. The Cuparí horde do not form an exception in this respect, as they also lived together in one of these large huts until very recently. The Mauhés are undistinguishable in physical appearance from their neighbours, being of middle size, with broad muscular chests, and well-shaped limbs and hands. But the individuals of both tribes can be readily distinguished from the Múras; less, however, by the structure and proportions of the body than by the expression of their countenances, which is mild and open instead of brutal, surly and mistrustful, as in those savages. They are invariably friendly to the whites; as I have already mentioned, they use the Paricá snuff, a habit quite unknown to the Mundurucús. They are the only tribe who manufacture Guaraná, a hard substance made of the pounded seeds of a climbing plant (Paullinia sorbilis), which they sell in large quantities to traders, it being used throughout the whole of the interior provinces of Brazil, grated and mixed in water, as a remedy in diarrhœa and intermittent fevers. The Mundurucús have a tradition that they and the Mauhés originally formed one tribe; the two peoples were formerly bitter enemies, but are now, and have been for many years, at peace with each other. Many centuries must have elapsed since the date of their first separation, to have produced the great differences now existing in language and customs between the two tribes. I fancy the so-called tradition is only a myth, but it doubtless conveys the truth. The points of resemblance between all the tribes inhabiting the region of the Amazons are so numerous and striking, that, notwithstanding the equally striking points of difference which some of them exhibit, we must conclude that not only the Mundurucús and Mauhés, but all the various peoples had a common origin—that is, they are derived by immigration from one quarter and one stock, the separate tribes subsequently acquiring their peculiarities by long isolation.

I bought of the Tushaúa two beautiful feather sceptres, with their bamboo cases. These are of cylindrical shape, about three feet in length and three inches in diameter, and are made by gluing with wax the fine white and yellow feathers from the breast of the toucan on stout rods, the tops being ornamented with long plumes from the tails of parrots, trogons, and other birds. The Mundurucús are considered to be the most expert workers in feathers of all the South American tribes. It is very difficult, however, to get them to part with the articles, as they seem to have a sort of superstitious regard for them. They manufacture head-dresses, sashes and tunics, besides sceptres; the feathers being assorted with a good eye to the proper contrast of colours, and the quills worked into strong cotton webs, woven with knitting sticks in the required shape. The dresses are worn only during their festivals, which are celebrated, not at stated times, but whenever the Tushaúa thinks fit. Dancing, singing, sports, and drinking, appear to be the sole objects of these occasional holidays. When a day is fixed upon, the women prepare a great quantity of tarobá, and the monotonous jingle is kept up, with little intermission night and day until the stimulating beverage is finished.

We left the Tushaúa's house early the next morning. The impression made upon me by the glimpse of Indian life in its natural state obtained here, and at another cluster of houses visited higher up, was a pleasant one, notwithstanding the disagreeable incident of the Pará-rauáte visit. The Indians are here seen to the best advantage; having relinquished many of their most barbarous practices, without being corrupted by too close contact with the inferior whites and half-breeds of the civilised settlements. The manners are simpler, the demeanour more gentle, cheerful and frank, than amongst the Indians who live near the towns. I could not help contrasting their well-fed condition, and the signs of orderly, industrious habits, with the poverty and laziness of the semi-civilised people of Altar do Chaõ. I do not think that the introduction of liquors has been the cause of much harm to the Brazilian Indian. He has his drinking bout now and then, like the common working people of other countries. It was his habit in his original state, before Europeans visited his country; but he is always ashamed of it afterwards, and remains sober during the pretty long intervals. The harsh, slave-driving practices of the Portuguese and their descendants have been the greatest curses to the Indians; the Mundurucús of the Cuparí, however, have been now for many years protected against ill-treatment. This is one of the good services rendered by the missionaries, who take care that the Brazilian laws in favour of the aborigines shall be respected by the brutal and unprincipled traders who go amongst them. I think no Indians could be in a happier position than these simple, peaceful and friendly people on the banks of the Cuparí. The members of each family live together, and seem to be much attached to each other; and the authority of the chief is exercised in the mildest manner. Perpetual summer reigns around them; the land is of the highest fertility, and a moderate amount of light work produces them all the necessaries of their simple life. It is difficult to get at their notions on subjects that require a little abstract thought; but the mind of the Indian is in a very primitive condition. I believe he thinks of nothing except the matters that immediately concern his daily material wants. There is an almost total absence of curiosity in his mental disposition, consequently he troubles himself very little concerning the causes of the natural phenomena around him. He has no idea of a Supreme Being; but, at the same time, he is free from revolting superstitions—his religious notions going no farther than the belief in an evil spirit, regarded merely as a kind of hobgoblin, who is at the bottom of all his little failures, troubles in fishing, hunting, and so forth. With so little mental activity, and with feelings and passions slow of excitement, the life of these people is naturally monotonous and dull, and their virtues are, properly speaking, only negative; but the picture of harmless homely contentment they exhibit is very pleasing, compared with the state of savage races in many other parts of the world.

The men awoke me at four o'clock with the sound of

their oars on leaving the port of the Tushaúa. I was surprised to find a dense fog veiling all surrounding objects, and the air quite cold. The lofty wall of forest, with the beautiful crowns of Assai palms standing out from it on their slender, arching stems, looked dim and strange through the misty curtain. The sudden change a little after sunrise had quite a magical effect, for the mist rose up like the gauze veil before the transformation scene at a pantomime, and showed the glorious foliage in the bright glow of morning, glittering with dew-drops. We arrived at the falls about ten o'clock. The river here is not more than forty yards broad, and falls over a low ledge of rock stretching in a nearly straight line across.

We had now arrived at the end of the navigation for large vessels—a distance from the mouth of the river, according to a rough calculation, of a little over seventy miles. I found it the better course now to send José and one of the men forward in the montaria with Joaõ Aracú, and remain myself with the cuberta and our other man, to collect in the neighbouring forest. We stayed here four days; one of the boats returning each evening from the upper river with the produce of the day's chase of my huntsmen. I obtained six good specimens of the hyacinthine macaw, besides a number of smaller birds, a species new to me of Guaríba, or howling monkey, and two large lizards. The Guaríba was an old male, with the hair much worn from his rump and breast, and his body disfigured with large tumours made by the grubs of a gad-fly (Œstrus). The back and tail were of a ruddy-brown colour; the limbs and underside of the body, black. The men ascended to the second falls, which form a cataract several feet in height, about fifteen miles beyond our anchorage. The macaws were found feeding in small flocks on the fruit of the Tucumá palm (Astryocaryum Tucumá), the excessively hard nut of which is crushed into pulp by the powerful beak of the bird. I found the craws of all the specimens filled with the sour paste to which the stone-like fruit had been reduced. Each bird took me three hours to skin, and I was occupied with these and my other specimens every evening until midnight, after my own laborious day's hunt; working on the roof of my cabin by the light of a lamp.

The place where the cuberta was anchored formed a little rocky haven, with a sandy beach sloping to the forest, within which were the ruins of the Indian Maloca, and a large weed-grown plantation. The port swarmed with fishes, whose movements it was amusing to watch in the deep, clear water. The most abundant were the Piránhas. One species, which varied in length, according to age, from two to six inches, but was recognisable by a black spot at the root of the tail, was always the quickest to seize any fragment of meat thrown into the water. When nothing was being given to them, a few only were seen scattered about, their heads all turned one way in an attitude of expectation; but as soon as any offal fell from the canoe, the water was blackened with the shoals that rushed instantaneously to the spot. Those who did not succeed in securing a fragment, fought with those who had been more successful, and many contrived to steal the coveted morsels from their mouths. When a bee or fly passed through the air near the water, they all simultaneously darted towards it as if roused by an electric shock. Sometimes a larger fish approached, and then the host of Piránhas took the alarm and flashed out of sight. The population of the water varied from day to day. Once a small shoal of a handsome black-banded fish, called by the

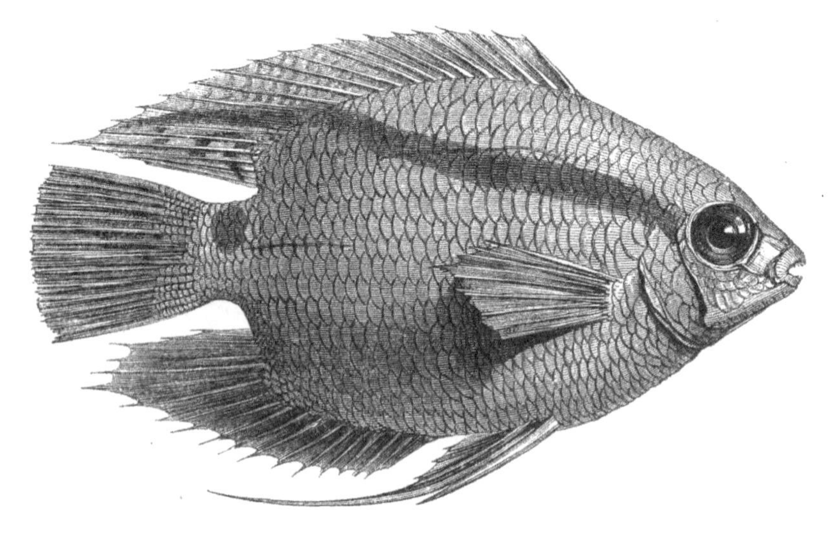

Acará (Mesonauta insignis).

natives Acará bandeira (Mesonauta insignis, of Günther), came gliding through at a slow pace, forming a very pretty sight. At another time, little troops of needle fish, eel-like animals, with excessively long and slender toothed jaws, sailed through the field, scattering before them the hosts of smaller fry; and in the rear of the needle-fishes a strangely-shaped kind called Sarapó came wriggling along, one by one, with a slow movement. We caught with hook and line, baited with pieces of banana, several Curimatá (Anodus Amazonum), a most delicious fish, which, next to the Tucunaré and the Pescada, is most esteemed by the natives. The Curimatá seemed to prefer the middle of the stream, where the waters were agitated beneath the little cascade.

The weather was now settled and dry, and the river sank rapidly—six inches in twenty-four hours. In this remote and solitary spot I can say that I heard for the first and almost the only time the uproar of life at sunset, which Humboldt describes as having witnessed towards the sources of the Orinoco, but which is unknown on the banks of the larger rivers. The noises of animals began just as the sun sank behind the trees after a sweltering afternoon, leaving the sky above of the intensest shade of blue. Two flocks of howling monkeys, one close to our canoe, the other about a furlong distant, filled the echoing forests with their dismal roaring. Troops of parrots, including the hyacinthine macaw we were

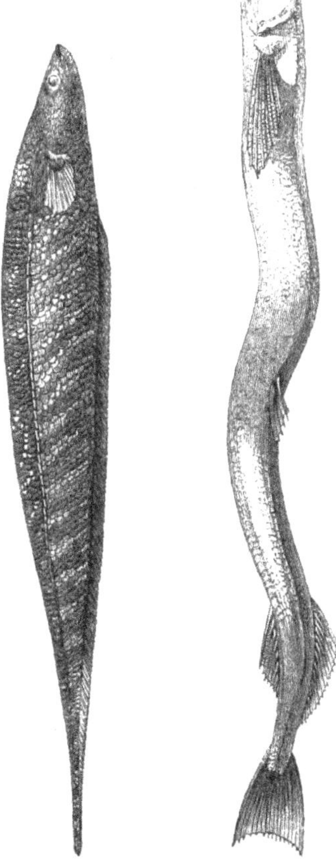

Sarapó (Carapus.) Needle-fish (Hemaramphus).

in search of, began then to pass over; the different styles of cawing and screaming of the various species making a terrible discord. Added to these noises were the songs of strange Cicadas, one large kind perched high on the trees around our little haven setting up a most piercing chirp: it began with the usual harsh jarring tone of its tribe, but this gradually and rapidly became shriller, until it ended in a long and loud note resembling the steam-whistle of a locomotive engine. Half-a-dozen of these wonderful performers made a considerable item in the evening concert. I had heard the same species before at Pará, but it was there very uncommon: we obtained here one of them for my collection by a lucky blow with a stone. The uproar of beasts, birds, and insects lasted but a short time: the sky quickly lost its intense hue, and the night set in. Then began the tree-frogs—quack-quack, drum-drum, hoo-hoo; these, accompanied by a melancholy night-jar, kept up their monotonous cries until very late.

My men encountered on the banks of the stream a Jaguar and a black Tiger, and were very much afraid of falling in with the Parárauátes, so that I could not after their return on the fourth day, induce them to undertake another journey. We began our descent of the river in the evening of the 26th of August. At night forest and river were again enveloped in mist, and the air before sunrise was quite cold. There is a considerable current from the falls to the house of Joaõ Aracú, and we accomplished the distance, with its aid and by rowing, in seventeen hours.

Sept. 21*st.*—At five o'clock in the afternoon we emerged from the confined and stifling gully through which the Cuparí flows, into the broad Tapajos, and breathed freely again. How I enjoyed the extensive view after being so long pent up: the mountainous coasts, the gray distance, the dark waters tossed by a refreshing breeze! Heat, mosquitoes, insufficient and bad food, hard work and anxiety, had brought me to a very low state of health; and I was now anxious to make all speed back to Santarem.

We touched at Aveyros, to embark some chests I had left there and to settle accounts with Captain Antonio: finding nearly all the people sick with fever and vomit, against which the Padre's homœopathic globules were of no avail. The Tapajos had been pretty free from epidemics for some years past, although it was formerly a very unhealthy river. A sickly time appeared to be now returning: in fact, the year following my visit (1853) was the most fatal one ever experienced in this part of the country. A kind of putrid fever broke out, which attacked people of all races alike. The accounts we received at Santarem were most distressing: my Cuparí friends especially suffered very severely. Joaõ Aracú and his family all fell victims, with the exception of his wife: my kind friend Joaõ Malagueita also died, and a great number of people in the Mundurucú village.

The descent of the Tapajos in the height of the dry season, which was now close at hand, is very hazardous on account of the strong winds, absence of current, and shoaly water far away from the coasts. The river towards the end of September is about thirty feet shallower than in June; and in many places, ledges of rock are laid bare, or covered with only a small depth of water. I had been warned of these circumstances by my Cuparí friends, but did not form an adequate idea of what we should have to undergo. Canoes, in descending, only travel at night, when the terral, or light land-breeze, blows off the eastern shore. In the day-time a strong wind rages from down river, against which it is impossible to contend, as there is no current, and the swell raised by its sweeping over scores of miles of shallow water is dangerous to small vessels. The coast for the greater part of the distance affords no shelter: there are, however, a number of little harbours, called *esperas*, which the canoe-men calculate upon, carefully arranging each night-voyage so as to reach one of them before the wind begins the next morning.

We left Aveyros in the evening of the 21st, and sailed gently down with the soft land-breeze, keeping about a mile from the eastern shore. It was a brilliant moonlit night, and the men worked cheerfully at the oars, when the wind was slack; the terral wafting from the forest a pleasant perfume like that of mignonette. At midnight we made a fire and got a cup of coffee, and at three o'clock in the morning reached the sitio of Ricardo's father, an Indian named André, where we anchored and slept.

Sept. 22nd.—Old André with his squaw came aboard this morning. They brought three Tracajás, a turtle, and a basketful of Tracajá eggs, to exchange with me for cotton cloth and cashaça. Ricardo, who had been for some time very discontented, having now satisfied his longing to see his parents cheerfully agreed to accompany me to Santarem. The loss of a man at this juncture would have been very annoying, with Captain Antonio ill at Aveyros, and not a hand to be had anywhere in the neighbourhood; but if we had not called at André's sitio, we should not have been able to have kept Ricardo from running away at the first landing-place. He was a lively, restless lad, and although impudent and troublesome at first, had made a very good servant; his companion, Alberto, was of quite a different disposition, being extremely taciturn, and going through all his duties with the quietest regularity.

We left at 11 a.m., and progressed a little before the wind began to blow from down river, when we were obliged again to cast anchor. The terral began at six o'clock in the evening, and we sailed with it past the long line of rock-bound coast near Itapuáma. At ten o'clock a furious blast of wind came from a cleft between the hills, catching us with the sails close-hauled, and throwing the canoe nearly on its beam-ends, when we were about a mile from the shore. José had the presence of mind to slacken the sheet of the mainsail, whilst I leapt forward and lowered the sprit of the foresail; the two Indians standing stupified in the prow. It was what the canoemen call a *trovoada secca* or white squall. The river in a few minutes became a sheet of foam; the wind ceased in about half an hour, but the terral was over for the night, so we pulled towards the shore to find an anchoring place.

We reached Tapaiuna by midnight on the 23rd, and on the morning of the 24th arrived at the Retiro, where we met a shrewd Santarem trader, whom I knew, Senhor Chico Honorio, who had a larger and much better provided canoe than our own. The wind was strong from below all day, so we remained at this place in his company. He had his wife with him, and a number of Indians, male and female. We slung our hammocks under the trees, and breakfasted and dined together, our cloth being spread on the sandy beach in the shade; after killing a large quantity of fish with *timbó*, of which we had obtained a supply at Itapuáma. At night we were again under way with the land breeze. The water was shoaly to a great distance off the coast, and our canoe having the lighter draught went ahead, our leadsman crying out the soundings to our companion: the depth was only one fathom, half a mile from the coast. We spent the next day (25th) at the mouth of a creek called Piní, which is exactly opposite the village of Boim, and on the following night advanced about twelve miles. Every point of land had a long spit of sand stretching one or two miles towards the middle of the river, which it was necessary to double by a wide circuit. The terral failed us at midnight when we were near an *espera*, called Maraï, the mouth of a shallow creek.

Sept. 26th.—I did not like the prospect of spending the whole dreary day at Maraï, where it was impossible to ramble ashore, the forest being utterly impervious, and the land still partly under water. Besides, we had used up our last stick of firewood to boil our coffee at sunrise, and could not get a fresh supply at this place. So there being a dead calm on the river in the morning,

I gave orders at ten o'clock to move out of the harbour, and try with the oars to reach Paquiatúba, which was only five miles distant. We had doubled the shoaly point which stretches from the mouth of the creek, and were making way merrily across the bay, at the head of which was the port of the little settlement, when we beheld to our dismay, a few miles down the river, the signs of the violent day breeze coming down upon us—a long, rapidly advancing line of foam with the darkened water behind it. Our men strove in vain to gain the harbour; the wind overtook us, and we cast anchor in three fathoms, with two miles of shoaly water between us and the land on our lee. It came with the force of a squall: the heavy billows washing over the vessel and drenching us with the spray. I did not expect that our anchor would hold; I gave out, however, plenty of cable and watched the result at the prow; José placing himself at the helm, and the men standing by the jib and foresail, so as to be ready, if we dragged, to attempt the passage of the Maraï spit, which was now almost dead to leeward. Our little bit of iron, however, held its place; the bottom being fortunately not so sandy as in most other parts of the coast; but our weak cable then began to cause us anxiety. We remained in this position all day without food, for everything was tossing about in the hold; provision-chests, baskets, kettles, and crockery. The breeze increased in strength towards the evening, when the sun set fiery red behind the misty hills on the western shore, and the gloom of the scene was heightened by the strange contrasts of colour; the inky water and the lurid gleam of the sky.

L 2

Heavy seas beat now and then against the prow of our vessel with a force that made her shiver. If we had gone ashore in this place, all my precious collections would have been inevitably lost; but we ourselves could have scrambled easily to land, and re-embarked with Senhor Honorio, who had remained behind in the Piní, and would pass in the course of two or three days. When night came I lay down exhausted with watching and fatigue, and fell asleep, as my men had done some time before. About nine o'clock, I was awoke by the montaria bumping against the sides of the vessel, which had veered suddenly round, and the full moon, previously astern, then shone full in the cabin. The wind had abruptly ceased, giving place to light puffs from the eastern shore, and leaving a long swell rolling into the shoaly bay.

After this I resolved not to move a step beyond Paquiatúba without an additional man, and one who understood the navigation of the river at this season. We reached the landing-place at ten o'clock, and anchored within the mouth of the creek. In the morning I walked through the beautiful shady alleys of the forest, which were water-paths in June when we touched here in ascending the river, to the house of Inspector Cypriano. After an infinite deal of trouble I succeeded in persuading him to furnish me with another Indian. There are about thirty families established in this place, but the able-bodied men had been nearly all drafted off within the last few weeks by the Government, to accompany a military expedition against runaway negroes, settled in villages in the interior.

Senhor Cypriano was a pleasant-looking and extremely civil young Mameluco. He accompanied us, on the night of the 28th, five miles down the river to Point Jaguararí, where the man lived whom he intended to send with me. I was glad to find my new hand a steady, middle-aged, and married Indian; his name was of very good promise, Angelo Custodio (Guardian Angel).

After the 26th of September the north-west day-breeze came every morning with the same strength, beginning at ten or eleven o'clock, and ending suddenly at seven or eight in the evening. The moon was in her third quarter, and we had many successive days and nights of clear, cloudless sky. I believe this wind to be closely connected with the easterly trade-wind of the main Amazons; indeed, to be the same, reflected from the west after the land-surface in that quarter has been cooled by it to a much lower point than the sun-heated surface of the stagnant Tapajos. The wind always arose in the morning after the air in the direction of the north-west had been further cooled by radiation of heat during the night; and it ceased in the evening, when the equilibrium of temperature between the Tapajos and the Amazons had become restored. The light land breeze from the east which always began to blow soon after the strong north-wester ceased, is attributable in like manner to the wooded surface of the land being then cooler than the air on the river. The terral lasted generally from 7 until 11 p.m., but after midnight it usually veered gradually to the north-east, and blew rather freshly from that quarter towards sunrise.

Point Jaguararí forms at this season of the year a high sandbank, which is prolonged as a narrow spit, stretching about three miles towards the middle of the river. We rounded this with great difficulty in the night of the 29th; reaching before daylight a good shelter behind a similar sandbank at Point Acarátingarí, a headland situated not more than five miles in a straight line from our last anchoring place. We remained here all day; the men beating *timbó* in a quiet pool between the sandbank and the mainland, and obtaining a great quantity of fish, from which I selected six species new to my collection. We made rather better progress the two following nights, but the terral now always blew strongly from the north-north-east after midnight, and thus limited the hours during which we could navigate, forcing us to seek the nearest shelter to avoid being driven back faster than we came.

On the 2nd of October we reached Point Cajetúba and had a pleasant day ashore. The river scenery in this neighbourhood is of the greatest beauty. A few houses of settlers are seen at the bottom of the broad bay of Aramána-í at the foot of a range of richly-timbered hills, the high beach of snow-white sand stretching in a bold curve from point to point. The opposite shores of the river are ten or eleven miles distant, but towards the north is a clear horizon of water and sky. The country near Point Cajetúba is similar to the neighbourhood of Santarem: namely, campos with scattered trees. We gathered a large quantity of wild fruit: Cajú, Umirí, and Aápiránga. The Umirí berry (Humirium floribundum) is a black drupe similar in appearance to the damascene plum, and not greatly unlike it in taste. The Aápiránga is a bright vermilion-coloured berry, with a hard skin and a sweet viscid pulp enclosing the seeds. Between the point and Altar do Chaō was a long stretch of sandy beach with moderately deep water; our men, therefore, took a rope ashore and towed the cuberta at merry speed until we reached the village. A long, deeply-laden canoe with miners from the interior provinces here passed us. It was manned by ten Indians, who propelled the boat by poles; the men, five on each side, trotting one after the other along a plank arranged for the purpose from stem to stern.

It took us two nights to double Point Cururú, where, as already mentioned, the river bends from its northerly course beyond Altar do Chaō. A confused pile of rocks, on which many a vessel heavily laden with farinha has been wrecked, extends at the season of low water from the foot of a high bluff far into the stream. We were driven back on the first night (October 3rd) by a squall. The light terral was carrying us pleasantly round the spit, when a small black cloud which lay near the rising moon suddenly spread over the sky to the northward; the land-breeze then ceased, and furious blasts began to blow across the river. We regained, with great difficulty, the shelter of the point. It blew almost a hurricane for two hours, during the whole of which time the sky over our heads was beautifully clear and starlit. Our shelter at first was not very secure, for the wind blew away the lashings of our sails, and caused our anchor to drag. Angelo Custodio, however, seized a rope which was attached to the foremast and leapt ashore; had he not done so, we should probably have been driven many miles backwards up the storm-tossed river. After the cloud had passed, the regular east wind began to blow, and our further progress was effectually stopped for the night. The next day we all went ashore, after securing well the canoe, and slept from eleven o'clock till five under the shade of trees.

The distance between Point Cururú and Santarem was accomplished in three days, against the same difficulties of contrary and furious winds, shoaly water, and rocky coasts. I was thankful at length to be safely housed, with the whole of my collections, made under so many privations and perils, landed without the loss or damage of a specimen. The men, after unloading the canoe and delivering it to its owner, came to receive their payment. They took part in goods and part in money, and after a good supper, on the night of the 7th October, shouldered their bundles and set off to walk by land some eighty miles to their homes. I was rather surprised at the good feeling exhibited by these poor Indians at parting. Angelo Custodio said that whenever I should wish to make another voyage up the Tapajos, he would be always ready to serve me as pilot. Alberto was undemonstrative as usual; but Ricardo, with whom I had had many sharp quarrels, actually shed tears when he shook hands and bid me the final "adeos."

CHAPTER III.

THE UPPER AMAZONS—VOYAGE TO EGA.

Departure from Barra—First day and night on the Upper Amazons—Desolate appearance of river in the flood season—Cucáma Indians—Mental condition of Indians—Squalls—Manatee—Forest—Floating pumice-stones from the Andes—Falling banks—Ega and its inhabitants—Daily life of a Naturalist at Ega—Customs, trade, &c.—The four seasons of the Upper Amazons.

I MUST now take the reader from the picturesque, hilly country of the Tapajos, and its dark, streamless waters, to the boundless, wooded plains and yellow, turbid current of the Upper Amazons or Solimoens. I will resume the narrative of my first voyage up the river, which was interrupted at the Barra of the Rio Negro in the seventh chapter to make way for the description of Santarem and its neighbourhood.

I embarked at Barra on the 26th of March, 1850, three years before steamers were introduced on the upper river, in a cuberta which was returning to Ega, the first and only town of any importance in the vast solitudes of the Solimoens, from Santarem, whither it had been sent with a cargo of turtle oil in earthenware jars. The owner, an old white-haired Portuguese trader of Ega named Daniel Cardozo, was then at Barra, attending the assizes as juryman, a public duty performed without remuneration, which took him six weeks away from his business. He was about to leave Barra himself, in a small boat, and recommended me to send forward my heavy baggage in the cuberta and make the journey with him. He would reach Ega, 370 miles distant from Barra, in twelve or fourteen days; whilst the large vessel would be thirty or forty days on the road. I preferred, however, to go in company with my luggage, looking forward to the many opportunities I should have of landing and making collections on the banks of the river.

I shipped the collections made between Pará and the Rio Negro in a large cutter which was about descending to the capital, and after a heavy day's work got all my chests aboard the Ega canoe by eight o'clock at night. The Indians were then all embarked, one of them being brought dead drunk by his companions, and laid to sober himself all night on the wet boards of the tombadilha. The cabo, a spirited young white, named Estulano Alves Carneiro, who has since risen to be a distinguished citizen of the new province of the Upper Amazons, soon after gave orders to get up the anchor. The men took to the oars, and in a few hours we crossed the broad mouth of the Rio Negro; the night being clear, calm, and starlit, and the surface of the inky waters smooth as a lake.

When I awoke the next morning, we were progressing by espia along the left bank of the Solimoens. The rainy season had now set in over the region through which the great river flows; the sand-banks and all the lower lands were already under water, and the tearing current, two or three miles in breadth, bore along a continuous line of uprooted trees and islets of floating plants. The prospect was most melancholy; no sound was heard but the dull murmur of the waters; the coast along which we travelled all day was encumbered every step of the way with fallen trees, some of which quivered in the currents which set around projecting points of land. Our old pest, the Motúca, began to torment us as soon as the sun gained power in the morning. White egrets were plentiful at the edge of the water, and humming-birds, in some places, were whirring about the flowers overhead. The desolate appearance of the landscape increased after sunset, when the moon rose in mist.

This upper river, the Alto-Amazonas or Solimoens, is always spoken of by the Brazilians as a distinct stream. This is partly owing, as before remarked, to the direction it seems to take at the fork of the Rio Negro; the inhabitants of the country, from their partial knowledge, not being able to comprehend the whole river system in one view. It has, however, many peculiarities to distinguish it from the lower course of the river. The trade-wind or sea-breeze, which reaches, in the height of the dry season, as far as the mouth of the Rio Negro, 900 or 1000 miles from the Atlantic, never blows on the upper river. The atmosphere is therefore more stagnant and sultry, and the winds that do prevail are of irregular direction and short duration. A great part of the land on the borders of the Lower Amazons is hilly; there are extensive campos or open plains, and long stretches of sandy soil clothed with thinner forests. The climate, in consequence, is comparatively dry, many months in succession during the fine season passing without rain. All this is changed on the Solimoens. A fortnight of clear, sunny weather is a rarity: the whole region through which the river and its affluents flow, after leaving the easternmost ridges of the Andes, which Pöppig describes as rising like a wall from the level country 240 miles from the Pacific, is a vast plain, about 1000 miles in length, and 500 or 600 in breadth, covered with one uniform, lofty, impervious, and humid forest. The soil is nowhere sandy, but always either a stiff clay, alluvium, or vegetable mould, which latter, in many places, is seen in water-worn sections of the river banks to be twenty or thirty feet in depth. With such a soil and climate, the luxuriance of vegetation, and the abundance and beauty of animal forms which are already so great in the region nearer the Atlantic, increase on the upper river. The fruits, both wild and cultivated, common to the two sections of the country, reach a progressively larger size in advancing westward, and some trees which blossom only once a year at Pará and Santarem, yield flower and fruit all the year round at Ega. The climate is healthy, although one lives here as in a permanent vapour bath. I must not, however, give here a lengthy description of the region whilst we are yet on its threshold. I resided and travelled on the Solimoens altogether for four years and a half. The country on its borders is a magnificent wilderness where civilized man, as yet, has scarcely obtained a footing; the cultivated ground from the Rio Negro to the Andes amounting only to a few score acres. Man, indeed, in any condition, from his small numbers, makes but an insignificant figure in these vast solitudes. It may be mentioned that the Solimoens is 2130 miles in length, if we reckon from the source of what is usually considered the main stream (Lake Lauricocha, near Lima); but 2500 miles by the route of the Ucayali, the most considerable and practicable fork of the upper part of the river. It is navigable at all seasons by large steamers for upwards of 1400 miles from the mouth of the Rio Negro.

On the 28th we passed the mouth of Ariauü, a narrow inlet which communicates with the Rio Negro, emerging in front of Barra. Our vessel was nearly drawn into this by the violent current which set from the Solimoens. The towing-cable was lashed to a strong tree about thirty yards ahead, and it took the whole strength of crew and passengers to pull across. We passed the Guariba, a second channel connecting the two rivers, on the 30th, and on the 31st sailed past a straggling settlement called Manacápurú, situated on a high, rocky bank. Many citizens of Barra have *sitios*, or country-houses, in this place, although it is eighty miles distant from the town by the nearest road. They come here for a few weeks in the fine season to economise, and pass the time in planting on a small scale, fishing, and trading. The custom of having two places of residence is very general throughout the country, and exists amongst the aborigines, at least the more advanced tribes. Some of the establishments at Manacápurú are large and of old date, shown by the number and size of the mangos and other introduced fruit-trees. The houses, though spacious, were now in a neglected and ruinous condition. Estulano and I landed at one of them, and dined off roasted wild hog with the owner, an uncommonly lively little old man, named Feyres. The place looked dirty and desolate; the stucco and whitewash had peeled off in great pieces from the walls; the doors and window-shutters were broken and off their hinges; the dingy mud-floors were covered with litter, and the cultivated grounds around the house choked with weeds. The high bank, and with it the settlement, terminates at the mouth of a narrow channel which leads to a large interior lake abounding in fish, manatee, and turtle.

Beyond Manacápurú all traces of high land cease; both shores of the river, henceforward for many hundred miles, are flat, except in places where the Tabatinga formation appears in clayey elevations of from twenty to forty feet above the line of highest water. The country is so completely destitute of rocky or gravelly beds that not a pebble is seen during many weeks' journey. Our voyage was now very monotonous. After leaving the last house at Manacápurú we travelled nineteen days without seeing a human habitation, the few settlers being located on the banks of inlets or lakes some distance from the shores of the main river. We met only one vessel during the whole of the time, and this did not come within hail, as it was drifting down in the middle of the current in a broad part of the river two miles from the bank along which we were laboriously warping our course upwards.

After the first two or three days we fell into a regular way of life aboard. Our crew was composed of ten Indians of the Cucáma nation, whose native country is a portion of the borders of the upper river in the neighbourhood of Nauta, in Peru. The Cucámas speak the Tupí language, using, however, a harsher accent than is common amongst the semi-civilized Indians from Ega downwards. They are a shrewd, hard-working people, and are the only Indians who willingly and in a body engage themselves to navigate the canoes of traders. The pilot, a steady and faithful fellow named Vicente, told me that he and his companions had now been fifteen months absent from their wives and families, and that on arriving at Ega they intended to take the first chance of a passage to Nauta. There was nothing in the appearance of these men to distinguish them from canoemen in general. Some were tall and well built, others had squat figures with broad shoulders and excessively thick arms and legs. No two of them were at all similar in the shape of the head: Vicente had an oval visage with fine regular features, whilst a little dumpy fellow, the wag of the party, was quite a Mongolian in breadth and prominence of cheek, spread of nostrils, and obliquity of eyes; these two formed the extremes as to face and figure. None of them were tattooed or disfigured in any way; they were all quite destitute of beard. The Cucámas are notorious on the river for their provident habits. The desire of acquiring property is so rare a trait in Indians that the habits of these people are remarked on with surprise by the Brazilians. The first possession which they strive to acquire on descending the river into Brazil, which all the Peruvian Indians look upon as a richer country than their own, is a wooden trunk with lock and key ; in this they stow away carefully all their earnings converted into clothing, hatchets, knives, harpoon heads, needles and thread, and so forth. Their wages are only fourpence or sixpence a day, which are often paid in goods charged a hundred per cent. above Pará prices, so that it takes them a long time to fill their chest.

It would be difficult to find a better-behaved set of men in a voyage than these poor Indians. During our thirty-five days' journey they lived and worked together in the most perfect good fellowship. I never heard an angry word pass amongst them. Senhor Estulano let them navigate the vessel in their own way, exerting his authority only now and then when they were inclined to be lazy. Vicente regulated the working hours. These depended on the darkness of the nights. In the first and second quarters of the moon they kept it up with *espia*, or oars, until towards midnight; in the third and fourth quarters they were allowed to go to sleep soon after sunset, and aroused at three or four o'clock in the morning to resume their work. On cool, rainy days we all bore a hand at the *espia*, trotting with bare feet on the sloppy deck in Indian file to the tune of some wild boatman's chorus. We had a favourable wind for two days only out of the thirty-five, by which we made about forty miles ; the rest of our long journey was accomplished literally by pulling our way from tree to tree. When we encountered a *remanso* near the shore we got along very pleasantly for a few miles by rowing; but this was a rare occurrence. During leisure hours the Indians employed themselves in sewing. Vicente was a good hand at cutting out shirts and trousers, and acted as master tailor to the whole party. Each had a thick steel thimble and a stock of needles and thread of his own. Vicente made for me a set of blue-check cotton shirts during the passage.

The goodness of these Indians, like that of most others amongst whom I lived, consisted perhaps more in the absence of active bad qualities, than in the possession of good ones; in other words, it was negative rather than positive. Their phlegmatic, apathetic temperament; coldness of desire and deadness of feeling; want of curiosity and slowness of intellect, make the Amazonian Indians very uninteresting companions anywhere. Their imagination is of a dull, gloomy quality, and they seem never to be stirred by the emotions:—love, pity, admiration, fear, wonder, joy, enthusiasm. These are characteristics of the whole race. The good fellowship of our Cucámas seemed to arise, not from warm sympathy, but simply from the absence of eager selfishness in small matters. On the morning when the favourable wind sprung up, one of the crew, a lad of about seventeen years of age, was absent ashore at the time of starting, having gone alone in one of the montarias to gather wild fruit. The sails were spread and we travelled for several hours at great speed, leaving the poor fellow to paddle after us against the strong current. Vicente, who might have waited a few minutes at starting, and the others, only laughed when the hardship of their companion was alluded to. He overtook us at night, having worked his way with frightful labour the whole day without a morsel of food. He grinned when he came on board, and not a dozen words were said on either side.

Their want of curiosity is extreme. One day we had an unusually sharp thunder-shower. The crew were lying about the deck, and after each explosion all set up a loud laugh; the wag of the party exclaiming "There's my old uncle hunting again!" an expression showing the utter emptiness of mind of the spokesman. I asked Vicente what he thought was the cause of lightning and thunder? He said, "Timaá ichoquá,"—I don't know. He had never given the subject a moment's thought! It was the same with other things. I asked him who made the sun, the stars, the trees? He didn't know, and had never heard the subject mentioned amongst his tribe. The Tupí language, at least as taught by the old Jesuits, has a word—Tupána—signifying God. Vicente sometimes used this word, but he showed by his expressions that he did not attach the idea of a Creator to it. He seemed to think it meant some deity or visible image which the whites worshipped in the churches he had seen in the villages. None of the Indian tribes on the Upper Amazons have an idea of a Supreme Being, and consequently have no word to express it in their own languages. Vicente thought the river on which we were travelling encircled the whole earth, and that the land was an island like those seen in the stream, but larger. Here a gleam of curiosity and imagination in the Indian mind is revealed: the necessity of a theory of the earth and water has been felt, and a theory has been suggested. In all other matters not concerning the common wants of life the mind of Vicente was a blank, and such I always found to be the case with the Indian in his natural state. Would a community of any race of men be otherwise, were they isolated for centuries in a wilderness like the Amazonian Indians, associated in small numbers wholly occupied in procuring a mere subsistence, and without a written language, or a leisured class to hand down acquired knowledge from generation to generation?

One day a smart squall gave us a good lift onward; it came with a cold, fine, driving rain, which enveloped the desolate landscape as with a mist: the forest swayed and roared with the force of the gale, and flocks of birds were driven about in alarm over the tree-tops. On another occasion a similar squall came from an unfavourable quarter: it fell upon us quite unawares when we had all our sails out to dry, and blew us broadside foremost on the shore. The vessel was fairly lifted on to the tall bushes which lined the banks, but we sustained no injury beyond the entanglement of our rigging in the branches. The days and nights usually passed in a dead calm, or with light intermittent winds from up river and consequently full against us. We landed twice a day to give ourselves and the Indians a little

M 2

rest and change, and to cook our two meals—breakfast and dinner. There was another passenger beside myself—a cautious, middle-aged Portuguese, who was going to settle at Ega, where he had a brother long since established. He was accommodated in the fore-cabin, or arched covering over the hold. I shared the cabin-proper with Senhores Estulano and Manoel, the latter a young half-caste, son-in-law to the owner of the vessel, under whose tuition I made good progress in learning the Tupí language during the voyage.

Our men took it in turns, two at a time, to go out fishing; for which purpose we carried a spare montaria. The master had brought from Barra, as provisions, nothing but stale, salt pirarucú—half-rotten fish, in large, thin, rusty slabs—farinha, coffee, and treacle. In these voyages passengers are expected to provide for themselves, as no charge is made except for freight of the heavy luggage or cargo they take with them. The Portuguese and myself had brought a few luxuries, such as beans, sugar, biscuits, tea, and so forth; but we found ourselves almost obliged to share them with our two companions and the pilot, so that before the voyage was one-third finished, the small stock of most of these articles was exhausted. In return, we shared in whatever the men brought. Sometimes they were quite unsuccessful, for fish is extremely difficult to procure in the season of high water, on account of the lower lands lying between the inlets and infinite chain of pools and lakes being flooded from the main river, thus increasing tenfold the area over which the finny population has to range. On most days, however, they brought two or three fine fish, and once they harpooned a manatee, or Vacca marina. On this last-mentioned occasion we made quite a holiday; the canoe was stopped for six or seven hours, and all turned out into the forest to help to skin and cook the animal. The meat was cut into cubical slabs, and each person skewered a dozen or so of these on a long stick. Fires were made, and the spits stuck in the ground and slanted over the flames to roast. A drizzling rain fell all the time, and the ground around the fires swarmed with stinging ants, attracted by the entrails and slime which were scattered about. The meat has somewhat the taste of very coarse pork; but the fat, which lies in thick layers between the lean parts, is of a greenish colour, and of a disagreeable, fishy flavour. The animal was a large one, measuring nearly ten feet in length, and nine in girth at the broadest part. The manatee is one of the few objects which excite the dull wonder and curiosity of the Indians, notwithstanding its commonness. The fact of its suckling its young at the breast, although an aquatic animal resembling a fish, seems to strike them as something very strange. The animal, as it lay on its back, with its broad rounded head and muzzle, tapering body, and smooth, thick, lead-coloured skin, reminded me of those Egyptian tombs which are made of dark, smooth stone, and shaped to the human figure.

It rarely happened that we caught anything near the canoe; but one day, as we were slowly progressing along a *remanso* past a thick bed of floating grasses, the men caught sight of a large Pirarucú: the fish which, salted, forms the staple food of all classes in most parts of the Lower Amazons country. It darted past with great speed close to the surface of the water, exhibiting its ornamental coat of mail, the extremely large, broad scales being margined with bright red. One of the Indians seized a harpoon and, jumping into the montaria, was after it in a moment. He killed it at the distance of a few yards, as it was plunging amongst the entangled beds of grass. The fish was a nearly full-grown one, measuring eight feet in length and five in girth, and supplied us all with two plentiful meals. The best parts only were cooked, the rest being thrown most improvidently to the vultures. The Indian name Pirarucú, or Anatto fish (from Pira, fish ; and urucú, anatto or red), is in allusion to the red colour of the borders of its scales, and is a sample of the figurative style of nomenclature of the Tupí nation.

Notwithstanding the hard fare, the confinement of the canoe, the trying weather,—frequent and drenching rains with gleams of fiery sunshine,—and the woful desolation of the river scenery, I enjoyed the voyage on the whole. We were not much troubled by mosquitoes, and therefore passed the nights very pleasantly, sleeping on deck wrapped in blankets or old sails. When the rains drove us below we were less comfortable, as there was only just room in the small cabin for three of us to lie close together, and the confined air was stifling. I became inured to the Piums in the course of the first week ; all the exposed parts of my body, by that time, being so closely covered with black punctures that the little bloodsuckers could not very easily find an unoccupied place to operate upon. Poor Miguel, the Portuguese, suffered horribly from these pests, his ancles and wrists being so much inflamed that he was confined to his hammock, slung in the hold, for weeks. At every landing-place I had a ramble in the forest whilst the red skins made the fire and cooked the meal. The result was a large daily addition to my collection of insects, reptiles, and shells. Sometimes the neighbourhood of our gipsy-like encampment was a tract of dry and spacious forest pleasant to ramble in; but more frequently it was a rank wilderness, into which it was impossible to penetrate many yards, on account of uprooted trees, entangled webs of monstrous woody climbers, thickets of spiny bamboos, swamps, or obstacles of one kind or other. The drier lands were sometimes beautified to the highest degree by groves of the Urucurí palm (Attalea excelsa), which grew by thousands under the crowns of the lofty, ordinary forest trees; their smooth columnar stems being all of nearly equal height (forty or fifty feet), and their broad, finely-pinnated leaves interlocking above to form arches and woven canopies of elegant and diversified shapes. The fruit of this palm ripens on the upper river in April, and during our voyage I saw immense quantities of it strewn about under the trees in places where we encamped. It is similar in size and shape to the date, and has a pleasantly-flavoured juicy pulp. The Indians would not eat it; I was surprised at this, as they greedily devoured many other kinds of palm fruit whose sour and fibrous pulp was much less palatable.

Vicente shook his head when he saw me one day eating a quantity of the Urucurí plums. I am not sure they were not the cause of a severe indigestion under which I suffered for many days afterwards.

In passing slowly along the interminable wooded banks week after week, I observed that there were three tolerably distinct kinds of coast and corresponding forest constantly recurring on this upper river. First, there were the low and most recent alluvial deposits,—a mixture of sand and mud, covered with tall, broad-leaved grasses, or with the arrow-grass before described, whose feathery-topped flower-stem rises to a height of fourteen or fifteen feet. The only large trees which grow in these places are the Cecropiæ. Many of the smaller and newer islands were of this description. Secondly, there were the moderately high banks, which are only partially overflowed when the flood season is at its height; these are wooded with a magnificent, varied forest, in which a great variety of palms and broad-leaved Marantaceæ form a very large proportion of the vegetation. The general foliage is of a vivid light-green hue; the water frontage is sometimes covered with a diversified mass of greenery; but where the current sets strongly against the friable, earthy banks, which at low water are twenty-five to thirty feet high, these are cut away, and expose a section of forest where the trunks of trees loaded with epiphytes appear in massy colonnades. One might safely say that three-fourths of the land bordering the Upper Amazons, for a thousand miles, belong to this second class. The third description of coast is the higher, undulating, clayey land, which appears only at long intervals, but extends sometimes for many miles along the borders of the river. The coast at these places is sloping, and composed of red or variegated clay. The forest is of a different character from that of the lower tracts: it is rounder in outline, more uniform in its general aspect; palms are much less numerous and of peculiar species — the strange bulging-stemmed species, Iriartea ventricosa, and the slender, glossy-leaved Bacâba-í (Œnocarpus minor), being especially characteristic; and, in short, animal life, which imparts some cheerfulness to the other parts of the river, is seldom apparent. This "terra firme," as it is called, and a large portion of the fertile lower land, seemed well adapted for settlement; some parts were originally peopled by the aborigines, but these have long since become extinct or amalgamated with the white immigrants. I

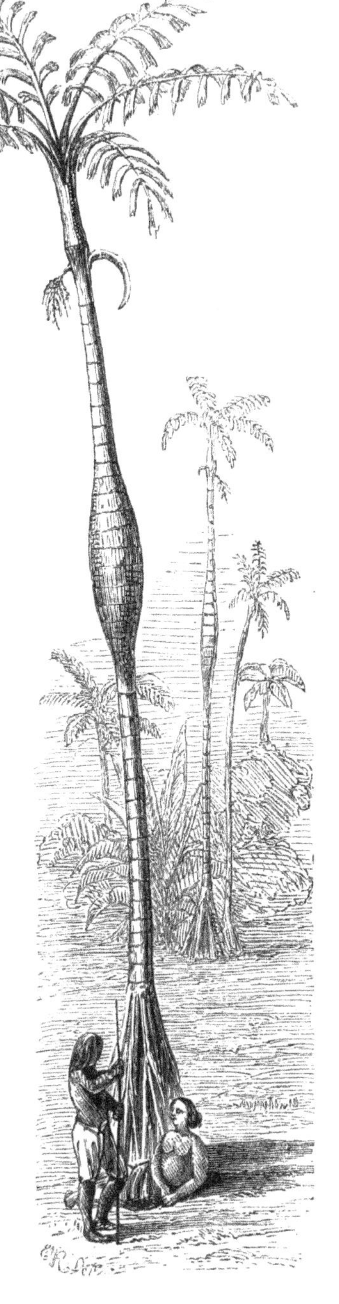

Bulging-stemmed Palm: Pashiúba barrigudo (Iriartea ventricosa).

afterwards learnt that there were not more than eighteen or twenty families settled throughout the whole country from Manacapurú to Quarý, a distance of 240 miles; and these, as before observed, do not live on the banks of the main stream, but on the shores of inlets and lakes.

The fishermen twice brought me small rounded pieces of very porous pumice-stone, which they had picked up floating on the surface of the main current of the river. They were to me objects of great curiosity as being messengers from the distant volcanoes of the Andes: Cotopaxi, Llanganete, or Sangay, which rear their peaks amongst the rivulets that feed some of the early tributaries of the Amazons, such as the Macas, the Pastaza, and the Napo. The stones must have already travelled a distance of 1200 miles. I afterwards found them rather common: the Brazilians use them for cleaning rust from their guns, and firmly believe them to be solidified river foam. A friend once brought me, when I lived at Santarem, a large piece which had been found in the middle of the stream below Monte Alegre, about 900 miles further down the river: having reached this distance, pumice-stones would be pretty sure of being carried out to sea, and floated thence with the north-westerly Atlantic current to shores many thousand miles distant from the volcanoes which ejected them. They are sometimes found stranded on the banks in different parts of the river. Reflecting on this circumstance since I arrived in England, the probability of these porous fragments serving as vehicles for the transportation of seeds of plants, eggs of insects, spawn of fresh-water fish, and so forth, has suggested itself to me. Their rounded, water-worn appearance showed that they must have been rolled about for a long time in the shallow streams near the sources of the rivers at the feet of the volcanoes, before they leapt the waterfalls and embarked on the currents which lead direct for the Amazons. They may have been originally cast on the land and afterwards carried to the rivers by freshets ; in which case the eggs and seeds of land insects and plants might be accidentally introduced and safely enclosed with particles of earth in their cavities. As the speed of the current in the rainy season has been observed to be from three to five miles an hour, they might travel an immense distance before the eggs or seeds were destroyed. I am ashamed to say that I neglected the opportunity, whilst on the spot, of ascertaining whether this was actually the case. The attention of Naturalists has only lately been turned to the important subject of occasional means of wide dissemination of species of animals and plants. Unless such be shown to exist, it is impossible to solve some of the most difficult problems connected with the distribution of plants and animals. Some species, with most limited powers of locomotion, are found in opposite parts of the earth, without existing in the intermediate regions; unless it can be shown that these may have migrated or been accidentally transported from one point to the other, we shall have to come to the strange conclusion that the same species had been created in two separate districts.

Canoemen on the Upper Amazons live in constant

dread of the "terras cahidas," or landslips, which occasionally take place along the steep, earthy banks; especially when the waters are rising. Large vessels are sometimes overwhelmed by these avalanches of earth and trees. I should have thought the accounts of them exaggerated if I had not had an opportunity during this voyage of seeing one on a large scale. One morning I was awoke before sunrise by an unusual sound resembling the roar of artillery. I was lying alone on the top of the cabin; it was very dark, and all my companions were asleep, so I lay listening. The sounds came from a considerable distance, and the crash which had aroused me was succeeded by others much less formidable. The first explanation which occurred to me was that it was an earthquake; for, although the night was breathlessly calm, the broad river was much agitated and the vessel rolled heavily. Soon after, another loud explosion took place, apparently much nearer than the former one; then followed others. The thundering peal rolled backwards and forwards, now seeming close at hand, now far off; the sudden crashes being often succeeded by a pause or a long-continued dull rumbling. At the second explosion, Vicente, who lay snoring by the helm, awoke and told me it was a "terra cahida;" but I could scarcely believe him. The day dawned after the uproar had lasted about an hour, and we then saw the work of destruction going forward on the other side of the river, about three miles off. Large masses of forest, including trees of colossal size, probably 200 feet in height, were rocking to and fro, and falling headlong one after the other into the water. After each avalanche the wave which it caused returned on the crumbly bank with tremendous force, and caused the fall of other masses by undermining them. The line of coast over which the landslip extended was a mile or two in length; the end of it, however, was hid from our view by an intervening island. It was a grand sight: each downfall created a cloud of spray; the concussion in one place causing other masses to give way a long distance from it, and thus the crashes continued, swaying to and fro, with little prospect of a termination. When we glided out of sight, two hours after sunrise, the destruction was still going on.

On the 9th of April we passed the mouth of a narrow channel which leads to an extensive lake called Anurí; it lies at the bottom of a long *enseada* or bay, on the north or left side of the river, around which sets the whole force of the current. The steamboat company have since established a station near this for supplying their vessels with firewood. A few miles beyond, on the opposite side, we saw the principal mouth of the Purús, a very large stream, whose sources are still unknown. Salsaparilla and Copaüba collectors, the only travellers on its waters, have ascended it in small boats a distance of two months' journey without meeting with any obstruction to navigation. This shows that its course lies to a very great extent within the level plain of the Upper Amazons. The mouth is not more than a quarter of a mile broad, and the water is of an olive-green colour.

We passed Cudajá on the 12th. This is a channel

which communicates with an extensive system of back-waters and lakes, lying between this part of the river and the Japurá, 250 miles further west. The inhabitants of the Solimoens give the name of Cupiyó to this little-known interior water-system. A Portuguese, whom I knew very well, once navigated it throughout its whole length. He described the country in glowing terms. The waters are clear ; some of the lakes are of vast extent, and the land everywhere is level and luxuriantly wooded. It is a more complete solitude than the banks of the main river, for the whole region is peopled only by a few families of Múra savages. The inhabitants of Ega, who are employed in the summer season in salting pirarucú, sometimes make their fishing stations on the sandy shores of one or other of these lakes. The largest of them, whose opposite or northern shore is said to be scarcely visible from the south side, is called Lake Múra, and is very seldom visited.

A number of long, straggling islands occur in mid-river beyond Cudajá. We passed the mouth of the Mamiyá, a black-water stream, on the 18th, and on the 19th arrived at the entrance to Lake Quarý. This is not, strictly speaking, a lake, but the expansion of the united beds of several affluents of the Solimoens, caused by the slowly-moving waters of the tributaries originally spreading out over the flat alluvial valley, into which they descend from the higher country of the early part of their course, instead of flowing directly into the full and swift current of the main river. Henceforward most of the branch rivers exhibit these lake-like expansions of their beds. The same phenomenon takes a great variety of forms, and is shown, as already observed in the Tapajos and other tributaries of the Lower Amazons. The mouth of the Quarý, or the channel which connects the lake with the Solimoens, is only 200 or 300 yards broad, and has but a very feeble current. It is about half a mile long, and opens on a broad sheet of water which is not of imposing magnitude, as it is only a small portion of the lake, this having a rather sharp bend in its lower part, so that the whole extent is not visible at one view. There is a small village on the shores of the inner water, distant twelve hours' journey by boat from the entrance. We anchored within the mouth, and visited in the montaria two or three settlers, whose houses are built in picturesque situations on the banks of the lower lake not far inwards. Several small but navigable streams or inlets here fall into the Quarý ; the land appeared to be of the highest fertility ; we crossed a neck of land on foot, from one inlet to another, passing through extensive groves of coffee, planted in a loose manner amongst the forest trees. One of the settlers was a Gibraltar Jew, established here many years, and thoroughly reconciled to the ways of life of the semi-civilised inhabitants. We found him barefoot, with trousers turned up to the knee, busily employed with a number of Indians—men, women, and children—shelling and drying cacao, which grows wild in immense profusion in the neighbourhood. He seemed a lively and sensible fellow ; was a great admirer of the country, the climate, and the people, and had no desire to return to Europe. This was the only Jew I met with on the upper river; there are several settled at Santarem, Cametá, and Pará, where, on account of their dealings being fairer than those of Portuguese traders, they do a good trade, and live on friendly terms with the Brazilians.

Our object here was to purchase a supply of fresh farinha and anything else we could find in the way of provisions, as our farinha had become rotten and unfit to eat, and we had been on short rations for several days. We got all we wanted except sugar; not a pound of this article of luxury was to be had, and we were obliged henceforward to sweeten our coffee with treacle, as is the general custom in this part of Brazil.

We left Quarý before sunrise on the 20th. On the 22nd we threaded the Paraná-mirím of Arauána-í, one of the numerous narrow by-waters which lie conveniently for canoes away from the main river, and often save a considerable circuit round a promontory or island. We rowed for half a mile through a magnificent bed of Victoria water-lilies; the flower-buds of which were just beginning to expand. Beyond the mouth of the Catuá, a channel leading to another great lake which we passed on the 25th, the river appeared greatly increased in breadth. We travelled for three days along a broad reach which both up and down river presented a blank horizon of water and sky: this clear view was owing to the absence of islands, but it renewed one's impressions of the magnitude of the stream, which here, 1200 miles from its mouth, showed so little diminution of width. Further westward a series of large islands commences, which divides the river into two and sometimes three channels, each about a mile in breadth. We kept to the southernmost of these, travelling all day on the 30th April along a high and rather sloping bank.

In the evening we arrived at a narrow opening, which would be taken by a stranger navigating the main channel for the outlet of some insignificant stream : it was the mouth of the Teffé, on whose banks Ega is situated, the termination of our voyage. After having struggled for thirty-five days with the muddy currents and insect pests of the Solimoens, it was unspeakably refreshing to find one's-self again in a dark-water river, smooth as a lake and free from Pium and Motúca. The rounded outline, small foliage, and sombre green of the woods, which seemed to rest on the glassy waters, made a pleasant contrast to the tumultuous piles of rank, glaring, light-green vegetation, and torn, timber-strewn banks to which we had been so long accustomed on the main river. The men rowed lazily until night-fall, when, having done a laborious day's work, they discontinued and went to sleep, intending to make for Ega in the morning. It was not thought worth while to secure the vessel to the trees or cast anchor, as there was no current. I sat up for two or three hours after my companions had gone to rest, enjoying the solemn calm of the night. Not a breath of air stirred; the sky was of a deep blue, and the stars seemed to stand forth in sharp relief; there was no sound of life in the woods, except the occasional melancholy note of some nocturnal bird. I reflected on my own wandering life : I had now reached the end of the third stage of my journey, and was now more than half way across the continent. It was necessary for me, on many accounts, to find a rich locality for Natural History explorations, and settle myself in it for some months or years. Would the neighbourhood of Ega turn out to be suitable, and should I, a solitary stranger on a strange errand, find a welcome amongst its people?

Our Indians resumed their oars at sunrise the next morning (May 1st), and after an hour's rowing along the narrow channel, which varies in breadth from 100 to 500 yards, we doubled a low wooded point, and emerged suddenly on the so-called Lake of Ega; a magnificent sheet of water, five miles broad—the expanded portion of the Teffé. It is quite clear of islands, and curves away to the west and south, so that its full extent is not visible from this side. To the left, on a gentle grassy slope at the point of junction of a broad tributary with the Teffé, lay the little settlement: a cluster of a hundred or so of palm-thatched cottages and whitewashed red-tiled houses, each with its neatly-enclosed orchard of orange, lemon, banana, and guava trees. Groups of palms, with their tall slender shafts and feathery crowns, overtopped the buildings and lower trees. A broad grass-carpeted street led from the narrow strip of white sandy beach to the rudely-built barn-like church with its wooden crucifix on the green before it, in the centre of the town. Cattle were grazing before the houses, and a number of dark-skinned natives were taking their morning bath amongst the canoes of various sizes which were anchored or moored to stakes in the port. We let off rockets and fired salutes, according to custom, in token of our safe arrival, and shortly afterwards went ashore.

A few days' experience of the people and the forests of the vicinity showed me that I might lay myself out for a long, pleasant, and busy residence at this place. An idea of the kind of people I had fallen amongst may be conveyed by an account of my earliest acquaintances in the place. On landing, the owner of the canoe killed an ox in honour of our arrival, and the next day took me round the town to introduce me to the principal residents. We first went to the Delegado of police, Senhor Antonio Cardozo, of whom I shall have to make frequent mention by-and-by. He was a stout, broad-featured man, ranking as a white, but having a tinge of negro blood; his complexion, however, was ruddy, and scarcely betrayed the mixture. He received us in a very cordial, winning manner: I had afterwards occasion to be astonished at the boundless good nature of this excellent fellow, whose greatest pleasure seemed to be to make sacrifices for his friends. He was a Paraense, and came to Ega orignally as a trader; but not succeeding in this, he turned planter on a small scale, and collector of the natural commodities of the country, employing half-a-dozen Indians in the business. We then visited the military commandant, an officer in the Brazilian army, named Praia. He was breakfasting with the vicar, and we found the two in dishabille (morning-gown loose round the neck, and slippers), seated at a rude wooden table in an open mud-floored verandah, at the back of the house. Commander Praia was a little curly-headed

N 2

man (also somewhat of a mulatto), always merry and fond of practical jokes. His wife, Donna Anna, a dressy dame from Santarem, was the leader of fashion in the settlement. The vicar, Father Luiz Gonsalvo Gomez, was a nearly pure-blood Indian, a native of one of the neighbouring villages, but educated in Maranham, a city on the Atlantic seaboard. I afterwards saw a good deal of him, as he was an agreeable, sociable fellow, fond of reading and hearing about foreign countries, and quite free from the prejudices which might be expected in a man of his profession. I found him, moreover, a thoroughly upright, sincere, and virtuous man. He supported his aged mother and unmarried sisters in a very creditable way out of his small salary and emoluments. It is a pleasure to be able to speak in these terms of a Brazilian priest, for the opportunity occurs rarely enough.

Leaving these agreeable new acquaintances to finish their breakfast, we next called on the Director of the Indians of the Japurá, Senhor José Chrysostomo Monteiro, a thin wiry Mameluco, the most enterprising person in the settlement. Each of the neighbouring rivers with its numerous wild tribes is under the control of a Director, who is nominated by the Imperial Government. There are now no missions in the region of the Upper Amazons: the "gentios" (heathens, or unbaptized Indians) being considered under the management and protection of these despots, who, like the captains of Trabalhadores, before mentioned, use the natives for their own private ends; Senhor Chrysostomo had, at this time, 200 of the Japúra Indians in his employ. He was half Indian himself, but was a far worse master to the red-skins than the whites usually are. We finished our rounds by paying our respects to a venerable native merchant, Senor Romaõ de Oliveira, a tall, corpulent, fine-looking old man, who received us with a naïve courtesy quite original in its way. He had been an industrious, enterprising man in his younger days, and had built a substantial range of houses and warehouses. The shrewd and able old gentleman knew nothing of the world beyond the wilderness of the Solimoens and its few thousands of isolated inhabitants; yet he could converse well and sensibly, making observations on men and things as sagaciously as though he had drawn them from long experience of life in a European capital. The semi-civilised Indians respected old Romaõ, and he had, consequently, a great number in his employ in different parts of the river: his vessels were always filled quicker with produce than those of his neighbours. On our leaving, he placed his house and store at my disposal. This was not a piece of empty politeness, for some time afterwards, when I wished to settle for the goods I had had of him, he refused to take any payment.

I made Ega my head-quarters during the whole of the time I remained on the Upper Amazons (four years and a half). My excursions into the neighbouring region extended sometimes as far as 300 and 400 miles from the place. An account of these excursions will be given in subsequent chapters; in the intervals between them I led a quiet, uneventful life in the settlement;

following my pursuit in the same peaceful, regular way as a Naturalist might do in a European village. For many weeks in succession my journal records little more than the notes made on my daily captures. I had a dry and spacious cottage, the principal room of which was made a workshop and study; here a large table was placed, and my little library of reference arranged on shelves in rough wooden boxes. Cages for drying specimens were suspended from the rafters by cords well anointed, to prevent ants from descending, with a bitter vegetable oil: rats and mice were kept from them by inverted *cuyas*, placed half-way down the cords. I always kept on hand a large portion of my private collection, which contained a pair of each species and variety, for the sake of comparing the old with the new acquisitions. My cottage was whitewashed inside and out about once a year by the proprietor, a native trader; the floor was of earth; the ventilation was perfect, for the outside air, and sometimes the rain as well, entered freely through gaps at the top of the walls under the eaves and through wide crevices in the doorways. Rude as the dwelling was, I look back with pleasure on the many happy months I spent in it. I rose generally with the sun, when the grassy streets were wet with dew, and walked down to the river to bathe: five or six hours of every morning were spent in collecting in the forest, whose borders lay only five minutes' walk from my house: the hot hours of the afternoon, between three and six o'clock, and the rainy days, were occupied in preparing and ticketing the specimens, making notes, dissecting, and drawing. I frequently had short rambles by water in a small montaria, with an Indian lad to paddle. The neighbourhood yielded me, up to the last day of my residence, an uninterrupted succession of new and curious forms in the different classes of the animal kingdom, but especially insects.

I lived, as may already have been seen, on the best of terms with the inhabitants of Ega. Refined society, of course, there was none; but the score or so of decent, quiet families which constituted the upper class of the place were very sociable; their manners offered a curious mixture of naïve rusticity and formal politeness; the great desire to be thought civilised leads the most ignorant of these people (and they are all very ignorant, although of quick intelligence) to be civil and kind to strangers from Europe. I was never troubled with that impertinent curiosity on the part of the people in these interior places which some travellers complain of in other countries. The Indians and lower half-castes—at least such of them who gave any thought to the subject—seemed to think it natural that strangers should collect and send abroad the beautiful birds and insects of their country. The butterflies they universally concluded to be wanted as patterns for bright-coloured calico-prints. As to the better sort of people, I had no difficulty in making them understand that each European capital had a public museum, in which were sought to be stored specimens of all natural productions in the mineral, animal, and vegetable kingdoms. They could not comprehend how a man could study science for its own sake; but I told them I was collecting for the "Museo de Londres," and was paid for it; *that* was very intelligible. One day, soon after my arrival, when I was explaining these things to a listening circle seated on benches in the grassy street, one of the audience, a considerable tradesman, a Mameluco native of Ega, got suddenly quite enthusiastic, and exclaimed "How rich are these great nations of Europe! We half-civilised creatures know nothing. Let us treat this stranger well, that he may stay amongst us and teach our children." We very frequently had social parties, with dancing and so forth; of these relaxations I shall have more to say presently. The manners of the Indian population also gave me some amusement for a long time. During the latter part of my residence, three wandering Frenchmen, and two Italians, some of them men of good education, on their road one after the other from the Andes down the Amazons, became enamoured of this delightfully-situated and tranquil spot, and made up their minds to settle here for the remainder of their lives. Three of them ended by marrying native women. I found the society of these friends a very agreeable change.

There were, of course, many drawbacks to the amenities of the place as a residence for a European; but these were not of the nature that my readers would perhaps imagine. There was scarcely any danger from wild animals: it seems almost ridiculous to refute the idea of danger from the natives in a country where even incivility to an unoffending stranger is a rarity. A Jaguar, however, paid us a visit one night. It was considered an extraordinary event, and so much uproar was made by the men who turned out with guns and bows and arrows that the animal scampered off and was heard of no more. Alligators were rather troublesome in the dry season. During these months there was almost always one or two lying in wait near the bathing-place for anything that might turn up at the edge of the water; dog, sheep, pig, child, or drunken Indian. When this visitor was about, every one took extra care whilst bathing. I used to imitate the natives in not advancing far from the bank and in keeping my eye fixed on that of the monster, which stares with a disgusting leer along the surface of the water; the body being submerged to the level of the eyes, and the top of the head, with part of the dorsal crest, the only portions visible. When a little motion was perceived in the water behind the reptile's tail, bathers were obliged to beat a quick retreat. I was never threatened myself, but I often saw the crowds of women and children scared whilst bathing by the beast making a movement towards them; a general scamper to the shore and peals of laughter were always the result in these cases. The men can always destroy these alligators when they like to take the trouble to set out with montarias and harpoons for the purpose, but they never do it unless one of the monsters, bolder than usual, puts some one's life in danger. This arouses them, and they then track the enemy with the greatest pertinacity; when half killed they drag it ashore and despatch it amid loud execrations. Another, however, is sure to appear some days or weeks afterwards, and take the vacant place on the station. Besides alligators, the only animals to be feared are the poisonous serpents. These are certainly common enough in the forest, but no accident happened during the whole time of my residence.

I suffered most inconvenience from the difficulty of getting news from the civilised world down river, from the irregularity of receipt of letters, parcels of books and periodicals, and towards the latter part of my residence from ill health arising from bad and insufficient food. The want of intellectual society, and of the varied excitement of European life, was also felt most acutely, and this, instead of becoming deadened by time, increased until it became almost insupportable. I was obliged, at last, to come to the conclusion that the contemplation of Nature alone is not sufficient to fill the human heart and mind. I got on pretty well when I received a parcel from England by the steamer once in two or four months. I used to be very economical with my stock of reading lest it should be finished before the next arrival and leave me utterly destitute. I went over the periodicals, the "Athenæum," for instance, with great deliberation, going through every number three times; the first time devouring the more interesting articles, the second, the whole of the remainder; and the third, reading all the advertisements from beginning to end. If four months (two steamers) passed without a fresh parcel, I felt discouraged in the extreme. I was worst off in the first year, 1850, when twelve months elapsed without letters or remittances. Towards the end of this time my clothes had worn to rags; I was barefoot, a great inconvenience in tropical forests, notwithstanding statements to the contrary that have been published by travellers; my servant ran away, and I was robbed of nearly all my copper money. I was obliged then to descend to Pará, but returned, after finishing the examination of the middle part of the Lower Amazons and the Tapajos, in 1855, with my Santarem assistant and better provided for making collections on the upper river. This second visit was in pursuit of the plan before mentioned, of exploring in detail the whole valley of the Amazons, which I formed in Pará in the year 1851.

During so long a residence I witnessed, of course, many changes in the place. Some of the good friends who made me welcome on my first arrival, died, and I followed their remains to their last resting-place in the little rustic cemetery on the borders of the surrounding forest. I lived there long enough, from first to last, to see the young people grow up, attended their weddings and the christenings of their children, and, before I left, saw them old married folks with numerous families. In 1850 Ega was only a village, dependent on Pará 1400 miles distant, as the capital of the then undivided province. In 1852, with the creation of the new province of the Amazons, it became a city; returned its members to the provincial parliament at Barra; had its assizes, its resident judges, and rose to be the chief town of a *comarca* or county. A year after this, namely, in 1853, steamers were introduced on the Solimoens, and from 1855, one ran regularly every two months between the Rio Negro and Nauta in Peru, touching at all the villages, and accomplishing the distance in ascending, about 1200 miles, in eighteen days. The trade and population, however, did not increase with these changes. The people became more "civilised," that is, they began to dress according to the latest Parisian fashions, instead of going about in stockingless feet, wooden clogs and shirt sleeves; acquired a taste for money getting and office holding; became divided into parties, and lost part of their former simplicity of manners. But the place remained, when I left it in 1859, pretty nearly what it was when I first arrived in 1850—a semi-Indian village, with much in the ways and notions of its people, more like those of a small country town in Northern Europe than a South American settlement. The place is healthy, and almost free from insect pests; perpetual verdure surrounds it; the soil is of marvellous fertility, even for Brazil; the endless rivers and labyrinths of channels teem with fish and turtle; a fleet of steamers might anchor at any season of the year in the lake, which has uninterrupted water communication straight to the Atlantic. What a future is in store for the sleepy little tropical village!

After speaking of Ega as a city, it will have a ludicrous effect to mention that the total number of its inhabitants is only about 1200. It contains just 107 houses, about half of which are miserably built mud-walled cottages, thatched with palm-leaves. A fourth of the population are almost always absent, trading or collecting produce on the rivers. The neighbourhood within a radius of thirty miles, and including two other small villages, contains probably 2000 more people. The settlement is one of the oldest in the country, having been founded in 1688 by Father Samuel Fritz, a Bohemian Jesuit, who induced several of the docile tribes of Indians, then scattered over the neighbouring region, to settle on the site. From 100 to 200 acres of sloping ground around the place, were afterwards cleared of timber; but such is the encroaching vigour of vegetation in this country, that the site would quickly relapse into jungle if the inhabitants neglected to pull up the young shoots as they arose. There is a stringent municipal law which compels each resident to weed a given space around his dwelling. Every month, whilst I resided here, an inspector came round with his wand of authority, and fined every one who had not complied with the regulation. The Indians of the surrounding country have never been hostile to the European settlers. The rebels of Pará and the Lower Amazons, in 1835-6, did not succeed in rousing the natives of the Solimoens against the whites. A party of forty of them ascended the river for that purpose, but on arriving at Ega, instead of meeting with sympathisers as in other places, they were surrounded by a small body of armed residents, and shot down without mercy. The military commandant at the time, who was the prime mover in this orderly resistance to anarchy, was a courageous and loyal negro, named José Patricio, an officer known throughout the Upper Amazons for his unflinching honesty and love of order, whose acquaintance I had the pleasure of making at St. Paulo in 1858. Ega was the head-quarters of the great scientific commission, which met in the years from 1781 to 1791, to settle the boundaries between the Spanish and Portuguese territories in South America. The chief commissioner for Spain, Don Francisco Requena, lived some time in the village with his family. I found only one person at Ega, my old friend Romaõ de Oliveira, who recollected, or had any knowledge of this important time, when a numerous staff of astronomers, surveyors, and draughtsmen, explored much of the surrounding country, with large bodies of soldiers and natives.

More than half the inhabitants of Ega are mamelucos; there are not more than forty or fifty pure whites; the number of negroes and mulattos is probably a little less, and the rest of the population consists of pure blood Indians. Every householder, including Indians and free negroes, is entitled to a vote in the elections, municipal, provincial, and imperial, and is liable to be called on juries, and to serve in the national guard. These privileges and duties of citizenship do not seem at present to be appreciated by the more ignorant coloured people. There is, however, a gradual improvement taking place in this respect. Before I left there was a rather sharp contest for the Presidency of the Municipal Chamber, and most of the voters took a lively interest in it. There was also an election of members to represent the province in the Imperial Parliament at Rio Janeiro, in which each party strove hard to return its candidate. On this occasion, an unscrupulous lawyer was sent by the government party from the capital to overawe the opposition to its nominee; many of the half-castes, headed by my old friend John da Cunha, who was then settled at Ega, fought hard, but with perfect legality and good humour, against this powerful interest. They did not succeed; and although the government agent committed many tyrannical and illegal acts, the losing party submitted quietly to their defeat. In a larger town, I believe, the government would not have dared to attempt thus to control the elections. I think I saw enough to warrant the conclusion that the machinery of constitutional government would, with a little longer trial, work well amongst the mixed Indian, white, and negro population, even of this remote part of the Brazilian empire. I attended, also, before I left, several assize meetings at Ega, and witnessed the novel sight of negro, white, half-caste, and Indian, sitting gravely side by side on the jury bench.

The way in which the coloured races act under the conditions of free citizenship, is a very interesting subject. Brazilian statesmen seem to have abandoned the idea, if they ever entertained it, of making this tropical empire a nation of whites, with a slave labouring class. The greatest difficulty on the Amazons is with the Indians. The general inflexibility of character of the race, and their abhorrence of the restraints of civilised life, make them very intractable subjects. Some of them, however, who have learned to read and write, and whose dislike to live in towns has been overcome by some cause acting early in life, make very good citizens. I have already mentioned the priest, who is a good example of what early training can do. There can be no doubt that if the docile Amazonian Indians were kindly treated by their white fellow-citizens, and educated, they would not be so quick as they have hitherto shown themselves to be to leave the towns and return into their half wild condition on the advancing civilisation of the places. The inflexibility of character, although probably organic, is seen to be sometimes overcome. The principal blacksmith of Ega, Senhor Macedo, was also an Indian, and a very sensible fellow. He sometimes filled minor offices in the government of the place. He used to come very frequently to my house to chat, and was always striving to acquire solid information about things. When Donati's comet appeared, he took a great interest in it. We saw it at its best from the 3rd to the 10th of October (1858), between which dates it was visible near the western horizon, just after sunset; the tail extending in a broad curve towards the north, and forming a sublime object. Macedo consulted all the old almanacs in the place to ascertain whether it was the same comet as that of 1811, which he said he well remembered. Before the Indians can be reclaimed in large numbers, it is most likely they will become extinct as a race. There is less difficulty with regard to the mamelucos, who, even when the proportion of white blood is small, sometimes become enterprising and versatile people. The Indian element in the blood and character seems to be quite lost, or dominated in the offspring of white and mameluco, that is in the fruits of the second cross. I saw a striking example of this in the family of a French blacksmith, who had lived for many years on the banks of the Solimoens, and had married a mameluco woman. His children might have all passed as natives of Northern Europe, a little tanned by foreign travel. One of them, a charming young girl named Isabel, was quite a blonde, having gray eyes, light brown hair, and fair complexion ; yet her grandmother was a tattooed Indian of the Tucúna tribe.

Many of the Ega Indians, including all the domestic servants, are savages who have been brought from the neighbouring rivers ; the Japurá, the Issá, and the Solimoens. I saw here individuals of at least sixteen different tribes ; most of whom had been bought, when children, of the native chiefs. This species of slave dealing, although forbidden by the laws of Brazil, is winked at by the authorities, because, without it, there would be no means of obtaining servants. They all become their own masters when they grow up, and never show the slightest inclination to return to utter savage life. But the boys generally run away and embark on the canoes of traders; and the girls are often badly treated by their mistresses, the jealous, passionate, and ill-educated Brazilian women. Nearly all the enmities which arise amongst residents at Ega and other places, are caused by disputes about Indian servants. No one who has lived only in old settled countries, where service can be readily bought, can imagine the difficulties and annoyances of a land where the servant class are ignorant of the value of money, and hands cannot be obtained except by coaxing them from the employ of other masters.

Great mortality takes place amongst the poor captive children on their arrival at Ega. It is a singular circumstance, that the Indians residing on the Japurá and other tributaries always fall ill on descending to the Solimoens, whilst the reverse takes place with the inhabitants of the banks of the main river, who never fail of taking intermittent fever when they first ascend these branch rivers, and of getting well when they return. The finest tribes of savages who inhabit the country near Ega are the Jurís and Passés: these are now, however, nearly extinct, a few families only remaining on the banks of the retired creeks connected with the Teffé, and on other branch rivers between the Teffé and the Jutahí. They are a peaceable, gentle, and industrious people, devoted to agriculture and fishing, and have always been friendly to the whites. I shall have occasion to speak again of the Passés, who are a slenderly-built and superior race of Indians, distinguished by a large, square tattooed patch in the middle of their faces. The principal cause of their decay in numbers seems to be a disease which always appears amongst them when a village is visited by people from the civilised settlements—a slow fever, accompanied by the symptoms of a common cold, "defluxo," as the Brazilians term it, ending probably in consumption. The disorder has been known to break out when the visitors were entirely free from it; the simple contact of civilised men, in some mysterious way being sufficient to create it. It is generally fatal to the Jurís and Passés: the first question the poor, patient Indians now put to an advancing canoe is, "Do you bring defluxo?"

My assistant, José, in the last year of my residence at Ega, "resgatou" (ransomed, the euphemism in use for purchased) two Indian children, a boy and a girl, through a Japurá trader. The boy was about twelve years of age, and of an unusually dark colour of skin : he had, in fact, the tint of a Cafuzo, the offspring of Indian and negro. It was thought he had belonged to some perfectly wild and houseless tribe, similar to the Parárauátes of the Tapajos, of which there are several in different parts of the interior of South America. His face was of regular, oval shape, but his glistening black eyes had a wary, distrustful expression, like that of a wild animal; and his hands and feet were small and delicately formed. Soon after his arrival, finding that none of the Indian boys and girls in the houses of our neighbours understood his language, he became sulky and reserved; not a word could be got from him until many weeks afterwards, when he suddenly broke out with complete phrases of Portuguese. He was ill of swollen liver and spleen, the result of intermittent fever, for a long time after coming into our hands. We found it difficult to cure him, owing to his almost invincible habit of eating earth, baked clay, pitch, wax, and other similar substances. Very many children on the upper parts of the Amazons have this strange habit; not only Indians, but negroes and whites. It is not, therefore, peculiar to the famous Otomacs of the Orinoco, described by Humboldt, or to Indians at all, and seems to originate in a morbid craving, the result of a meagre diet of fish, wild-fruits, and mandioca-meal. We gave our little savage the name of Sebastian. The use of these Indian children is to fill water-jars from the river, gather fire-wood in the forest, cook, assist in paddling the montaria in excursions, and

o 2

so forth. Sebastian was often my companion in the woods, where he was very useful in finding the small birds I shot, which sometimes fell in the thickets amongst confused masses of fallen branches and dead leaves. He was wonderfully expert at catching lizards with his hands, and at climbing. The smoothest stems of palm-trees offered little difficulty to him: he would gather a few lengths of tough, flexible lianas; tie them in a short, endless band to support his feet with in embracing the slippery shaft, and then mount upwards by a succession of slight jerks. It was very amusing, during the first few weeks, to witness the glee and pride with which he would bring to me the bunches of fruit he had gathered from almost inaccessible trees. He avoided the company of boys of his own race, and was evidently proud of being the servant of a real white man. We brought him down with us to Pará: but he showed no emotion at any of the strange sights of the capital; the steam-vessels, large ships and houses, horses and carriages, the pomp of church ceremonies, and so forth. In this he exhibited the usual dulness of feeling and poverty of thought of the Indian; he had, nevertheless, very keen perceptions, and was quick at learning any mechanical art. José, who had resumed, some time before I left the country, his old trade of goldsmith, made him his apprentice, and he made very rapid progress; for after about three months' teaching he came to me one day with radiant countenance and showed me a gold ring of his own making.

The fate of the little girl, who came with a second batch of children all ill of intermittent fever, a month or two after Sebastian, was very different. She was brought to our house, after landing, one night in the wet season, when the rain was pouring in torrents, thin and haggard, drenched with wet and shivering with ague. An old Indian who brought her to the door, said briefly, "ecui encommenda" (here's your little parcel or order), and went away. There was very little of the savage in her appearance, and she was of a much lighter colour than the boy. We found she was of the Miránha tribe, all of whom are distinguished by a slit, cut in the middle of each wing of the nose, in which they wear on their holiday occasions a large button made of pearly river-shell. We took the greatest care of our little patient; had the best nurses in the town, fomented her daily, gave her quinine and the most nourishing food; but it was all of no avail: she sank rapidly; her liver was enormously swollen and almost as hard to the touch as stone. There was something uncommonly pleasing in her ways, and quite unlike anything I had yet seen in Indians. Instead of being dull and taciturn, she was always smiling and full of talk. We had an old woman of the same tribe to attend her, who explained what she said to us. She often begged to be taken to the river to bathe; asked for fruit, or coveted articles she saw in the room for play-things. Her native name was Oria. The last week or two she could not rise from the bed we had made for her in a dry corner of the room: when she wanted lifting, which was very often, she would allow no one to help her but me, calling me by the name of "Caríwa" (white man), the only word of Tupí she seemed to know. It was inexpressibly touching to hear her as she lay, repeating by the hour the verses which she had been taught to recite with her companions in her native village: a few sentences repeated over and over again with a rhythmic accent, and relating to objects and incidents connected with the wild life of her tribe. We had her baptized before she died, and when this latter event happened, in opposition to the wishes of the big people of Ega, I insisted on burying her with the same honours as a child of the whites; that is, as an "anjinho" (little angel), according to the pretty Roman Catholic custom of the country. We had the corpse clothed in a robe of fine calico, crossed her hands on her breast over a "palma" of flowers, and made also a crown of flowers for her head. Scores of helpless children like our poor Oria die at Ega, or on the road; but generally not the slightest care is taken of them during their illness. They are the captives made during the merciless raids of one section of the Miránha tribe on the territories of another, and sold to the Ega traders. The villages of the attacked hordes are surprised, and the men and women killed or driven into the thickets without having time to save their children. There appears to be no doubt that the Miránhas are cannibals, and, therefore, the purchase of these captives probably saves them from a worse fate. The demand for them at Ega operates, however, as a direct cause of the supply, stimulating the unscrupulous chiefs, who receive all the profits to undertake these murderous expeditions.

It is remarkable how quickly the savages of the various nations, which each have their own, to all appearance, widely different language, learn Tupí on their arrival at Ega, where it is the common idiom. This perhaps may be attributed chiefly to the grammatical forms of all the Indian tongues being the same, although the words are different. As far as I could learn, the feature is common to all, of placing the preposition *after* the noun, making it, in fact, a *post*-position, thus: "he is come the village *from*;" "go him *with*, the plantation *to*," and so forth. The ideas to be expressed in their limited sphere of life and thought are few; consequently the stock of words is extremely small; besides, all Indians have the same way of thinking, and the same objects to talk about; these circumstances also contribute to the ease with which they learn each other's language. Hordes of the same tribe living on the same branch rivers, speak mutually unintelligible languages; this happens with the Mirânhas on the Japurá, and with the Collínas on the Jurúa; whilst Tupí is spoken with little corruption along the banks of the main Amazons for a distance of 2500 miles. The purity of Tupí is kept up by frequent communication amongst the natives, from one end to the other of the main river; how complete and long-continued must be the isolation in which the small groups of savages have lived in other parts, to have caused so complete a segregation of dialects! It is probable that the strange inflexibility of the Indian organisation, both bodily and mental, is owing to the isolation in which each small tribe has lived, and to the narrow round of life and thought, and close intermarriages for countless generations, which are the necessary results. Their fecundity is of a low degree, for it is very rare to find an Indian family having so many as four children, and we have seen how great is their liability to sickness and death on removal from place to place.

I have already remarked on the different way in which the climate of this equatorial region affects Indians and negroes. No one could live long amongst the Indians of the Upper Amazons, without being struck with their constitutional dislike to the heat. Europeans certainly withstand the high temperature better than the original inhabitants of the country: I always found I could myself bear exposure to the sun or unusually hot weather, quite as well as the Indians, although not well-fitted by nature for a hot climate. Their skin is always hot to the touch, and they perspire little. No Indian resident of Ega can be induced to stay in the village (where the heat is felt more than in the forest or on the river), for many days together. They bathe many times a day, but do not plunge in the water, taking merely a *sitz-bath,* as dogs may be seen doing in hot climates, to cool the lower parts of the body. The women and children, who often remain at home, whilst the men are out for many days together fishing, generally find some excuse for trooping off to the shades of the forest in the hot hours of the afternoons. They are restless and discontented in fine dry weather, but cheerful in cool days, when the rain is pouring down on their naked backs. When suffering under fever, nothing but strict watching can prevent them going down to bathe in the river, or from eating immoderate quantities of juicy fruits, although these indulgences are frequently the cause of death. They are very subject to disorders of the liver, dysentery, and other diseases of hot climates, and when any epidemic is about, they fall ill quicker, and suffer more than negroes or even whites. How different all this is with the negro, the true child of tropical climes! The impression gradually forced itself on my mind that the red Indian lives as a stranger, or immigrant in these hot regions, and that his constitution was not originally adapted, and has not since become perfectly adapted to the climate. It is a case of want of fitness; other races of men living on the earth would have been better fitted to enjoy and make use of the rich unappropriated domain. Unlike the lands peopled by Negro and Caucasian, Tropical America had no indigenous man thoroughly suited to its conditions, and was therefore peopled by an ill-suited race from another continent.

The Indian element is very prominent in the amusements of the Ega people. All the Roman Catholic holidays are kept up with great spirit; rude Indian sports being mingled with the ceremonies introduced by the Portuguese. Besides these, the aborigines celebrate their own ruder festivals: the people of different tribes combining; for, in most of their features, the merry-makings were originally alike in all the tribes. The Indian idea of a holiday is bonfires, processions, masquerading, especially the mimicry of different kinds of animals, plenty of confused drumming and fifing, monotonous dancing, kept up hour after hour without intermission, and the most important point of all, getting gradually and completely drunk. But he attaches a kind of superstitious significance to these acts, and thinks that the amusements appended to the Roman Catholic holidays as celebrated by the descendants of the Portuguese, are also an essential part of the religious ceremonies. But in this respect, the uneducated whites and half-breeds are not a bit more enlightened than the poor dull-souled Indian. All look upon a religious holiday as an amusement, in which the priest takes the part of director or chief actor.

Almost every unusual event, independent of saints' days, is made the occasion of a holiday by the sociable, easy-going people of the white and mameluco classes; funerals, christenings, weddings, the arrival of strangers, and so forth. The custom of "waking" the dead is also kept up. A few days after I arrived, I was awoke in the middle of a dark moist night by Cardozo, to sit up with a neighbour whose wife had just died. I found the body laid out on a table, with crucifix and lighted wax-candles at the head, and the room full of women and girls squatted on stools or on their haunches. The men were seated round the open door, smoking, drinking coffee, and telling stories; the bereaved husband exerting himself much to keep the people merry during the remainder of the night. The Ega people seem to like an excuse for turning night into day; it is so cool and pleasant, and they can sit about during these hours in the open air, clad as usual in simple shirt and trowsers, without streaming with perspiration.

The patron saint is Santa Theresa; the festival at whose anniversary lasts, like most of the others, ten days. It begins very quietly with evening litanies sung in the church, which are attended by the greater part of the population, all clean and gaily dressed in calicos and muslins; the girls wearing jasmines and other natural flowers in their hair, no other head-dress being worn by females of any class. The evenings pass pleasantly; the church is lighted up with wax candles, and illuminated on the outside by a great number of little oil lamps —rude clay cups, or halves of the thick rind of the bitter orange, which are fixed all over the front. The congregation seem very attentive, and the responses to the litany of Our Lady, sung by a couple of hundred fresh female voices, ring agreeably through the still village. Towards the end of the festival the fun commences. The managers of the feast keep open houses, and dancing, drumming, tinkling of wire guitars, and unbridled drinking by both sexes, old and young, are kept up for a couple of days and a night with little intermission. The ways of the people at these merry-makings, of which there are many in the course of the year, always struck me as being not greatly different from those seen at an old-fashioned village wake in retired parts of England. The old folks look on and get very talkative over their cups; the children are allowed a little extra indulgence in sitting up; the dull, reserved fellows become loquacious, shake one another by the hand or slap each other on the back, discovering, all at once, what capital friends they are. The cantankerous individual gets quarrelsome, and the amorous unusually loving. The Indian, ordinarily so taciturn, finds the use of his tongue, and gives the minutest details of some little dispute which he had with his master years ago, and which every one else had forgotten; just as I have known lumpish labouring men in England do, when half-fuddled. One cannot help reflecting, when witnessing these traits of manners, on the similarity of human nature everywhere, when classes are compared whose state of culture and conditions of life are pretty nearly the same.

The Indians play a conspicuous part in the amusements at St. John's eve, and at one or two other holidays which happen about that time of the year—the end of June. In some of the sports the Portuguese element is visible, in others the Indian; but it must be recollected that masquerading, recitative singing, and so forth, are common originally to both peoples. A large number of men and boys disguise themselves to represent different grotesque figures, animals, or persons. Two or three dress themselves up as giants, with the help of a tall framework. One enacts the part of the Caypór, a kind of sylvan deity similar to the Curupíra which I have before mentioned. The belief in this being seems to be common to all the tribes of the Tupí stock. According to the figure they dressed up at Ega, he is a bulky, misshapen monster, with red skin and long shaggy red hair hanging half way down his back. They believe that he has subterranean campos and hunting grounds in the forest, well stocked with pacas and deer. He is not at all an object of worship nor of fear, except to children, being considered merely as a kind of hobgoblin. Most of the masquers make themselves up as animals—bulls, deer, magoary storks, jaguars, and so forth, with the aid of light frameworks covered with old cloth dyed or painted and shaped according to the object represented. Some of the imitations which I saw were capital. One ingenious fellow arranged an old piece of canvas in the form of a tapir, placed himself under it, and crawled about on all fours. He constructed an elastic nose to resemble that of the tapir, and made, before the doors of the principal residents, such a good imitation of the beast grazing, that peals of laughter greeted him wherever he went. Another man walked about solitarily, masked as a jabirú crane (a large animal standing about four feet high), and mimicked the gait and habits of the bird uncommonly well. One year an Indian lad imitated me, to the infinite amusement of the townsfolk. He came the previous day to borrow of me an old blouse and straw hat. I felt rather taken in when I saw him, on the night of the performance, rigged out as an entomologist, with an insect net, hunting bag, and pincushion. To make the imitation complete, he had borrowed the frame of an old pair of spectacles, and went about with it straddled over his nose. The jaguar now and then made a raid amongst the crowd of boys who were dressed as deer, goats, and so forth. The masquers kept generally together, moving from house to house, and the performances were directed by an old musician, who sang the orders and explained to the spectators what was going forward in a kind of recitative, accompanying himself on a wire guitar. The mixture of Portuguese and Indian customs is partly owing to the European immigrants in these parts having been uneducated men, who, instead of introducing European civilisation, have descended almost to the level of the Indians, and adopted some of their practices. The performances take place in the evening, and occupy five or six hours; bonfires are lighted along the grassy streets, and the families of the better class are seated at their doors, enjoying the wild but good-humoured fun.

A purely Indian festival is celebrated the first week in February, which is called the Feast of Fruits: several kinds of wild fruit becoming ripe at that time, more particularly the Umarí and the Wishí, two sorts which are a favourite food of the people of this province, although of a bitter taste and unpalatable to Europeans. It takes place at the houses of a few families of the Jurí tribe, hidden in the depths of the forest on the banks of a creek about three miles from Ega. I saw a little of it one year, when hunting in the neighbourhood with an Indian attendant. There were about 150 people assembled, nearly all red-skins, and signs of the orgy having been very rampant the previous night were apparent in the litter and confusion all around, and in the number of drunken men lying asleep under the trees and sheds. The women had manufactured a great quantity of spirits in rude clay stills, from mandioca, bananas, and pine-apples. I doubt whether there was ever much symbolic meaning attached by the aborigines to festivals of this kind. The harvest-time of the Umirí and Wishí is one of their seasons of abundance, and they naturally made it the occasion of one of their mad, drunken holidays. They learnt the art of distilling spirits from the early Portuguese; it is only, however, one or two of the superior tribes, such as the Jurís and Passés, who practise it. The Indians of the Upper Amazons, like those of the Lower river, mostly use fermented drinks (called here Caysúma), made from mandioca cakes and different kinds of fruit.

I did not see much fruit about. A few old women in one of the sheds were preparing and cooking porridge of bananas in large earthenware kettles. It was now near midday, the time when a little rest is taken before resuming the orgy in the evening; but a small party of young men and women were keeping up the dance to the accompaniment of drums made of hollow logs and beaten with the hands. The men formed a curved line on the outside, and the women a similar line on the inside facing their partners. Each man had in his right hand a long reed representing a javelin, and rested his left on the shoulders of his neighbour. They all moved, first to the right and then to the left, with a slow step, singing a drawling monotonous verse, in a language which I did not understand. The same figure was repeated in the dreariest possible way for at least half an hour, and in fact constituted the whole of the dance. The assembled crowd included individuals of most of the tribes living in the region around Ega; but the majority were Miránhas and Jurís. They had no common chief, an active middle-aged Jurí, named Alexandro, in the employ of Senhor Chrysostomo of Ega, seeming to have the principal management. This festival of fruits was the only occasion in which the Indians of the neighbourhood assembled together or exhibited any traces of joint action. It declined in importance every year, and will no doubt soon be discontinued altogether.

The trade of Ega, like that of all places on the Upper Amazons, consists in the collecting of the produce of the forests and waters, and exchanging it for European and North American goods. About a dozen large vessels, schooners and cubertas, owned by the merchants of the place, are employed in the traffic. Only one voyage a year is made to Pará, which occupies from four to five months, and is arranged so that the vessels shall return before the height of the dry season, when they are sent with assortments of goods; cloth, hardware, salt, and a few luxuries, such as biscuits, wine, &c., to the fishing stations, to buy up produce for the next trip to the capital. Although large profits are apparently made both ways, the retail prices of European wares being from 40 to 80 per cent. higher, and the net prices of produce to the same degree lower, than those of Pará, the traders do not get rich very rapidly. An old Portuguese who had traded with success at Ega for thirty years was reputed rich when he died: his savings then amounting to nine contos of reis, or about a thousand pounds sterling. The value of produce fluctuates much, and losses are often sustained in consequence. Excessively long credit is given: the system being to trust the collectors of produce with goods a twelvemonth in advance; and if anything happens in the meantime to a customer, the debt is lost altogether.

The articles of export from the upper river are cacao, salsaparilla, Brazil nuts, bast for caulking vessels (the inner bark of various species of Lecythideæ or Brazil-nut trees), copaüba balsam, India-rubber, salt-fish (pirarucú), turtle-oil, mishíra (potted vacca marina), and grass hammocks. The total value of the produce annually exported from Ega, I calculated at from seven to eight thousand pounds sterling. Most of the articles are collected in the forest by the Ega people, who take their families and live in the woods for months at a time, during the proper seasons. Some of the productions, such as salsaparilla and balsam of copaüba, have been long ago exhausted in the neighbourhood of towns, at least near the banks of the rivers, the only parts that have yet been explored, and are now got only by more adventurous traders during long voyages up the branch streams. The search for India-rubber has commenced but very lately; the tree appears to grow plentifully on some of the rivers, but only an insignificant fraction of the immense forest has yet been examined. Grass hammocks are manufactured by the wild tribes, and purchased of them in considerable quantities by the salsaparilla collectors. They are knitted with simple rods, except the larger kinds, which are woven in clumsy wooden looms. The fibre of which they are made is not grass, but the young leaflets of certain kinds of palm trees (Astryocaryum). These are split, and the strips twisted into two or three-strand cord, by rolling them with the fingers on the naked thigh. Salt-fish and mishíra are prepared by the half-breeds and civilized Indians, who establish fishing stations (feitorias) on the great sandbanks laid bare by the retreating waters, in places where fish, turtle, and manatee abound, and spend the whole of the dry season in this occupation. Turtle oil is made from the eggs of the large river turtle, and is one of the principal productions of the district ; the mode of collecting the eggs and extracting the oil will be described in the next chapter.

I know several men who have been able, with ordinary sobriety and industry, to bring up their families very respectably, and save money at Ega, as collectors of the spontaneous productions of the neighbourhood. Each family, however, besides this trade, has its little plantation of mandioca, coffee, beans, water melons, tobacco, and so forth, which is managed almost solely by the women. Some do not take the trouble to clear a piece of forest for this purpose, but make use of the sloping, bare, earthy banks of the Solimoens, which remain uncovered by water during eight or nine months of the year, and consequently long enough to give time for the ripening of the crops of mandioca, beans, and so forth. The process with regard to mandioca, the bread of the country, is very simple. A party of women take a few bundles of maníva (mandioca shoots) some fine day in July or August, when the river has sunk some few feet, and plant them in the rich alluvial soil, reckoning with the utmost certainty on finding a plentiful crop when they return in January or February. The regular plantations are all situated some distance from Ega, and across the water, nothing being safe on the mainland near the town on account of the cattle, some hundred head of which are kept grazing in the streets by the townsfolk. Every morning, soon after daybreak, the women are seen paddling off in montarias to their daily labours in these *roças* or clearings ; the mistresses of households with their groups of Indian servant girls. The term agriculture cannot be applied to this business; in this primitive country plough, spade, and hoe are unknown even by name. The people idle away most part of the time at their *roças,* and have no system when they do work, so that a family rarely produces more than is required for its own consumption.

The half-caste and Indian women, after middle age, are nearly all addicted to the use of Ypadú, the powdered leaves of a plant (Erythroxylon coca) which is well known as a product of the eastern parts of Peru, and is to the natives of these regions what opium is to the Turks and betel to the Malays. Persons who indulge in Ypadú at Ega are held in such abhorrence, that they keep the matter as secret as possible; so it is said, and no doubt with truth, that the slender result of the women's daily visits to their *roças,* is owing to their excessive use of this drug. They plant their little plots of the tree in retired nooks in the forest, and keep their stores of the powder in hiding-places near the huts which are built on each plantation. Taken in moderation, Ypadú has a stimulating and not injurious effect, but in excess it is very weakening, destroying the appetite, and producing in time great nervous exhaustion. I once had an opportunity of seeing it made at the house of a Marauá Indian on the banks of the Jutahí. The leaves were dried on a mandioca oven, and afterwards pounded in a very long and narrow wooden mortar. When about half pulverised, a number of the large leaves of the Cecropia palmata (candelabrum tree) were

P 2

burnt on the floor, and the ashes dirtily gathered up and mixed with the powder. The Ypadú-eaters say that this prevents the ill-effects which would arise from the use of the pure leaf, but I should think the mixture of so much indigestible filth would be more likely to have the opposite result.

We lived at Ega, during most part of the year, on turtle. The great fresh-water turtle of the Amazons grows on the upper river to an immense size, a full-grown one measuring nearly three feet in length by two in breadth, and is a load for the strongest Indian. Every house has a little pond, called a curral (pen), in the back-yard to hold a stock of the animals through the season of dearth—the wet months; those who have a number of Indians in their employ sending them out for a month when the waters are low, to collect a stock, and those who have not, purchasing their supply; with some difficulty, however, as they are rarely offered for sale. The price of turtles, like that of all other articles of food, has risen greatly with the introduction of steam-vessels. When I arrived in 1850 a middle-sized one could be bought pretty readily for ninepence, but when I left in 1859, they were with difficulty obtained at eight and nine shillings each. The abundance of turtles, or rather the facility with which they can be found and caught, varies with the amount of annual subsidence of the waters. When the river sinks less than the average, they are scarce; but when more, they can be caught in plenty, the bays and shallow lagoons in the forest having then only a small depth of water. The flesh is very tender, palatable, and wholesome; but it is very cloying: every one ends, sooner or later, by becoming thoroughly surfeited. I became so sick of turtle in the course of two years that I could not bear the smell of it, although at the same time nothing else was to be had, and I was suffering actual hunger. The native women cook it in various ways. The entrails are chopped up and made into a delicious soup called *sarapatel,* which is generally boiled in the concave upper shell of the animal used as a kettle. The tender flesh of the breast is partially minced with farinha, and the breast shell then roasted over the fire, making a very pleasant dish. Steaks cut from the breast and cooked with the fat form another palatable dish. Large sausages are made of the thick-coated stomach, which is filled with minced meat and boiled. The quarters cooked in a kettle of Tucupí sauce form another variety of food. When surfeited with turtle in all other shapes, pieces of the lean part roasted on a spit and moistened only with vinegar make an agreeable change. The smaller kind of turtle, the tracajá, which makes its appearance in the main river, and lays its eggs a month earlier than the large species, is of less utility to the inhabitants although its flesh is superior, on account of the difficulty of keeping it alive; it survives captivity but a very few days, although placed in the same ponds in which the large turtle keeps well for two or three years.

Those who cannot hunt and fish for themselves, and whose stomachs refuse turtle, are in a poor way at Ega. Fish, including many kinds of large and delicious salmonidæ, is abundant in the fine season; but each family fishes only for itself, and has no surplus for sale. An Indian fisherman remains out just long enough to draw what he thinks sufficient for a couple of days' consumption. Vacca marina is a great resource in the wet season; it is caught by harpooning, which requires much skill, or by strong nets made of very thick hammock twine, and placed across narrow inlets. Very few Europeans are able to eat the meat of this animal. Although there is a large quantity of cattle in the neighbourhood of the town, and pasture is abundant all the year round, beef can be had only when a beast is killed by accident. The most frequent cause of death is poisoning by drinking raw Tucupí, the juice of the mandioca root. Bowls of this are placed on the ground in the sheds where the women prepare farinha; it is generally done carelessly, but sometimes intentionally through spite when stray oxen devastate the plantations of the poorer people. The juice is almost certain to be drunk if cattle stray near the place, and death is the certain result. The owners kill a beast which shows symptoms of having been poisoned, and retail the beef in the town. Although every one knows it cannot be wholesome, such is the scarcity of meat and the uncontrollable desire to eat beef, that it is eagerly bought, at least by those residents who come from other provinces where beef is the staple article of food. Game of all kinds is scarce in the forest near the town, except in the months of June and July, when immense numbers of a large and handsome bird, Cuvier's toucan (Ramphastos Cuvieri) make their appearance. They come in well-fed condition, and are shot in such quantities that every family has the strange treat of stewed and roasted toucans daily for many weeks. Curassow birds are plentiful on the banks of the Solimoens, but to get a brace or two requires the sacrifice of several days for the trip. A tapir, of which the meat is most delicious and nourishing, is sometimes killed by a fortunate hunter. I have still a lively recollection of the pleasant effects which I once experienced from a diet of fresh tapir meat for a few days, after having been brought to a painful state of bodily and mental depression by a month's scanty rations of fish and farinha.

We sometimes had fresh bread at Ega made from American flour brought from Pará, but it was sold at ninepence a pound. I was once two years without tasting wheaten bread, and attribute partly to this the gradual deterioration of health which I suffered on the Upper Amazons. Mandioca meal is a poor, weak substitute for bread; it is deficient in gluten, and consequently cannot be formed into a leavened mass or loaf, but is obliged to be roasted in hard grains in order to keep any length of time. Cakes are made of the half-roasted meal, but they become sour in a very few hours. A superior kind of meal is manufactured at Ega of the sweet mandioca (Manihot Aypi); it is generally made with a mixture of the starch of the root, and is therefore a much more wholesome article of food than the ordinary sort which, on the Amazons, is made of the pulp after the starch has been extracted by soaking in water. When we could get neither bread nor biscuit, I found tapioca soaked in coffee the best native substitute. We were seldom without butter, as every canoe brought one or two casks on each return voyage from Pará, where it is imported in considerable quantity from Liverpool. We obtained tea in the same way; it being served as a fashionable luxury at wedding and christening parties; the people were at first strangers to this article, for they used to stew it in a saucepan, mixing it up with coarse raw sugar, and stirring it with a spoon. Sometimes we had milk, but this was only when a cow calved; the yield from each cow was very small, and lasted only for a few weeks in each case, although the pasture is good, and the animals are sleek and fat.

Fruit of the ordinary tropical sorts could generally be had. I was quite surprised at the variety of the wild kinds, and of the delicious flavour of some of them. Many of these are utterly unknown in the regions nearer the Atlantic; being the peculiar productions of this highly-favoured, and little known, interior country. Some have been planted by the natives in their clearings. The best was the *Jabutí-púhe,* or tortoise-foot; a scaled fruit probably of the Anonaceous order. It is about the size of an ordinary apple; when ripe the rind is moderately thin, and encloses, with the seeds, a quantity of custardy pulp of a very rich flavour. Next to this stands the Cumá (Collophora sp.) of which there are two species, not unlike, in appearance, small round pears; but the rind is rather hard, and contains a gummy milk, and the pulpy part is almost as delicious as that of the Jabuti-púhe. The Cumá tree is of moderate height, and grows rather plentifully in the more elevated and drier situations. A third kind is the Pamá, which is a stone-fruit, similar in colour and appearance to the cherry, but of oblong shape. The tree is one of the loftiest in the forest, and has never, I believe, been selected for cultivation. To get at the fruit the natives are obliged to climb to the height of about a hundred feet, and cut off the heavily laden branches. I have already mentioned the Umarí and the Wishí: both these are now cultivated. The fatty, bitter pulp which surrounds the large stony seeds of these fruits is eaten mixed with farinha, and is very nourishing. Another cultivated fruit is the Purumá (Puruma cecropiæfolia, Martius), a round juicy berry, growing in large bunches and resembling grapes in taste. The tree is deceptively like a Cecropia in the shape of its foliage. Another smaller kind, called Purumá-i, grows wild in the forest close to Ega, and has not yet been planted. The most singular of all these fruits is the Uikí, which is of oblong shape, and grows apparently crosswise on the end of its stalk. When ripe the thick green rind opens by a natural cleft across the middle, and discloses an oval seed the size of a damascene plum, but of a vivid crimson colour. This bright hue belongs to a thin coating of pulp which, when the seeds are mixed in a plate of stewed bananas, gives to the mess a pleasant rosy tint, and a rich creamy taste and consistence. *Mingau* (porridge) of bananas

Uikí Fruit.

flavoured and coloured with Uikí is a favourite dish at Ega. The fruit, like most of the others here mentioned, ripens in January. Many smaller fruits such as Wajurú (probably a species of Achras), the size of a gooseberry, which grows singly and contains a sweet gelatinous pulp enclosing two large, shining black seeds; Cashipári-arapaá, an oblong scarlet berry; two kinds of Bacurí, the Bacurí-siúma and the B. curúa, sour fruits of a bright lemon colour when ripe, and a great number of others, are of less importance as articles of food.

Pupunha Palm.

The celebrated "Peach palm," *Pupunha* of the Tupí nations (Guilielma speciosa), is a common tree at Ega. The name, I suppose, is in allusion to the colour of the fruit, and not to its flavour, for it is dry and mealy, and in taste may be compared to a mixture of chestnuts and cheese. Vultures devour it eagerly, and come in quarrelsome flocks to the trees when it is ripe. Dogs will also eat it: I do not recollect seeing cats do the same, although they go voluntarily to the woods to eat Tucumá, another kind of palm fruit. The tree, as it grows in clusters beside the palm-thatched huts, is a noble ornament, being, when full grown, from fifty to sixty feet in height and often as straight as a scaffold-pole. A bunch of fruit when ripe is a load for a strong man, and each tree bears several of them. The Pupunha grows wild nowhere on the Amazons. It is one of those few vegetable productions (including three kinds of mandioca and the American species of Banana) which the Indians have cultivated from time immemorial, and brought with them in their original migration to Brazil. It is only, however, the more advanced tribes who have kept up the cultivation. The superiority of the fruit on the Solimoens to that grown on the Lower Amazons and in the neighbourhood of Pará is very striking. At Ega it is generally as large as a full-sized peach, and when boiled almost as mealy as a potatoe; whilst at Pará it is no bigger than a walnut, and the pulp is fibrous. Bunches of sterile or seedless fruits sometimes occur in both districts. It is one of the principal articles of food at Ega when in season, and is boiled and eaten with treacle or salt. A dozen of the seedless fruits makes a good nourishing meal for a grown-up person. It is the general belief that there is more nutriment in Pupunha than in fish or Vacca marina.

The seasons in the Upper Amazons region offer some points of difference from those of the lower river and the district of Pará, which two sections of the country we have already seen also differ considerably. The year at Ega is divided according to the rises and falls of the river, with which coincide the wet and dry periods. All the principal transactions of life of the inhabitants are regulated by these yearly recurring phenomena. The peculiarity of this upper region consists in there being two rises and two falls within the year. The great annual rise commences about the end of February, and continues to the middle of June, during which the rivers and lakes, confined during the dry periods to their ordinary beds, gradually swell and overflow all the lower lands. The inundation progresses gently inch by inch, and is felt everywhere, even in the interior of the forests of the higher lands, miles away from the river; as these are traversed by numerous gullies, forming in the fine season dry, spacious dells, which become gradually transformed by the pressure of the flood into broad creeks navigable by small boats under the shade of trees. All the countless swarms of turtle of various species then leave the main river for the inland pools: the sand-banks go under water, and the flocks of wading birds migrate northerly to the upper waters of the tributaries which flow from that direction, or to the Orinoco; which streams during the wet period of the Amazons are enjoying the cloudless skies of their dry season. The families of fishermen who have been employed, during the previous four or five months, in harpooning and salting pirarucú and shooting turtle in the great lakes, now return to the towns and villages; their temporarily constructed fishing establishments becoming gradually submerged, with the sand islets or beaches on which they were situated. This is the season, however, in which the Brazil nut and wild cacao ripen, and many persons go out to gather these harvests, remaining absent generally throughout the months of March and April. The rains during this time are not continuous; they fall very heavily at times, but rarely last so long at a stretch as twenty-four hours, and many days intervene of pleasant, sunny weather. The sky, however, is generally overcast and gloomy, and sometimes a drizzling rain falls.

About the first week in June the flood is at its highest; the water being then about forty-five feet above its lowest point; but it varies in different years to the extent of about fifteen feet. The "enchente," or flow, as it is called by the natives, who believe this great annual movement of the waters to be of the same nature as the tide towards the mouth of the Amazons, is then completed, and all begin to look forward to the "vasante," or ebb. The provision made for the dearth of the wet season is by this time pretty nearly exhausted; fish is difficult to procure, and many of the less provident inhabitants have become reduced to a diet of fruits and farinha porridge.

The fine season begins with a few days of brilliant weather—furious, hot sun, with passing clouds. Idle men and women, tired of the dulness and confinement of the flood season, begin to report, on returning from their morning bath, the cessation of the flow: *as agoas estaō paradas*, "the waters have stopped." The muddy streets, in a few days, dry up: groups of young fellows are now seen seated on the shady sides of the cottages making arrows and knitting fishing-nets with tucúm twine; others are busy patching up and caulking their canoes, large and small: in fact, preparations are made on all sides for the much longed-for "veraō," or summer, and the "migration," as it is called, of fish and turtle; that is, their descent from the inaccessible pools in the forest to the main river. Towards the middle of July the sand-banks begin to reappear above the surface of the waters, and with this change come flocks of sandpipers and gulls, which latter make known the advent of the fine season, as the cuckoo does of the European spring; uttering almost incessantly their plaintive cries as they fly about over the shallow waters of sandy shores. Most of the gaily-plumaged birds have now finished moulting, and begin to be more active in the forest.

The fall continues to the middle of October, with the interruption of a partial rise called "repiquet," of a few inches in the midst of very dry weather in September, caused by the swollen contribution of some large affluent higher up the river. The amount of subsidence also varies considerably, but it is never so great as to interrupt navigation by large vessels. The greater it is the more abundant is the season. Every one is prosperous when the waters are low; the shallow bays and pools being then crowded with the concentrated population of fish and turtle. All the people, men, women, and children, leave the villages and spend the few weeks of glorious weather rambling over the vast undulating expanses of sand in the middle of the Solimoens, fishing, hunting, collecting eggs of turtle and plovers, and thoroughly enjoying themselves. The inhabitants pray always for a "vasante grande," or great ebb.

From the middle of October to the beginning of January, the second wet season prevails. The rise is sometimes not more than about fifteen feet, but it is, in some years, much more considerable, laying the large sand islands under water before the turtle eggs are hatched. In one year, whilst I resided at Ega, this second annual inundation reached to within ten feet of the highest water point as marked by the stains on the trunks of trees by the river side.

The second dry season comes on in January, and lasts throughout February. The river sinks sometimes to the extent of a few feet only, but one year (1856) I saw it ebb to within about five feet of its lowest point in September. This is called the summer of the Umarí, "Veraõ do Umarí," after the fruit of this name already described, which ripens at this season. When the fall is great, this is the best time to catch turtles. In the year above mentioned, nearly all the residents who had a canoe, and could work a paddle, went out after them in the month of February, and about 2000 were caught in the course of a few days. It appears that they had been arrested in their migration towards the interior pools of the forest by the sudden drying up of the water-courses, and so had become easy prey.

Thus the Ega year is divided into four seasons; two of dry weather and falling waters, and two of the reverse. Besides this variety, there is, in the month of May, a short season of very cold weather, a most surprising circumstance in this otherwise uniformly sweltering climate. This is caused by the continuance of a cold wind, which blows from the south over the humid forests that extend without interruption from north of the equator to the eighteenth parallel of latitude in Bolivia. I had, unfortunately, no thermometer with me at Ega; the only one I brought with me from England having been lost at Pará. The temperature is so much lowered, that fishes die in the river Teffé, and are cast in considerable quantities on its shores. One year I saw and examined numbers of these benumbed and dead fishes. They were all small fry of different species of Characini. The wind is not strong; but it brings cloudy weather, and lasts from three to five or six days in each year. The inhabitants all suffer much from the cold, many of them wrapping themselves up with the warmest clothing they can get (blankets are here unknown), and shutting themselves in-doors with a charcoal fire lighted. I found, myself, the change of temperature most delightful, and did not require extra clothing. It was a bad time, however, for my pursuit, as birds and insects all betook themselves to places of concealment, and remained inactive. The period during which this wind prevails is called the "tempo da friagem," or the season of coldness. The phenomenon, I presume, is to be accounted for by the fact that in May it is winter in the southern temperate zone, and that the cool currents of air travelling thence northwards towards the equator, become only moderately heated in their course, owing to the intermediate country being a vast, partially-flooded plain, covered with humid forests.

CHAPTER IV.

EXCURSIONS IN THE NEIGHBOURHOOD OF EGA.

The river Teffé—Rambles through groves on the beach—Excursion to the house of a Passé chieftain—Character and customs of the Passé tribe—First excursion to the sand islands of the Solimoens—Habits of great river-turtle—Second excursion—Turtle-fishing in the inland pools—Third excursion—Hunting-rambles with natives in the forest—Return to Ega.

I WILL now proceed to give some account of the more interesting of my shorter excursions in the neighbourhood of Ega. The incidents of the longer voyages, which occupied each several months, will be narrated in a separate chapter.

The settlement, as before described, is built on a small tract of cleared land at the lower or eastern end of the lake, six or seven miles from the main Amazons, with which the lake communicates by a narrow channel. On the opposite shore of the broad expanse stands a small village, called Nogueira, the houses of which are not visible from Ega, except on very clear days; the coast on the Nogueira side is high, and stretches away into the grey distance towards the south-west. The upper part of the river Teffé is not visited by the Ega people, on account of its extreme unhealthiness, and its barrenness in salsaparilla and other wares. To Europeans it will seem a most surprising thing that the people of a civilised settlement, 170 years old, should still be ignorant of the course of the river on whose banks their native place, for which they proudly claim the title of city, is situated. It would be very difficult for a private individual to explore it, as the necessary number of Indian paddlers could not be obtained. I knew only one person who had ascended the Teffé to any considerable distance, and he was not able to give me a distinct account of the river. The only tribe known to live on its banks are the Catauishís, a people who perforate their lips all round, and wear rows of slender sticks in the holes: their territory lies between the Purús and the Juruá, embracing both shores of the Teffé. A very considerable stream, the Bararuá, enters the lake from the west, about thirty miles above Ega; the breadth of the lake is much contracted a little below the mouth of this tributary, but it again expands further south, and terminates abruptly where the Teffé proper, a narrow river with a strong current, forms its head water.

The whole of the country for hundreds of miles is covered with picturesque but pathless forests, and there are only two roads along which excursions can be made by land from Ega. One is a narrow hunter's track, about two miles in length, which traverses the forest in the rear of the settlement. The other is an extremely pleasant path along the beach to the west of the town. This is practicable only in the dry season, when a flat strip of white sandy beach is exposed at the foot of the high wooded banks of the lake, covered with trees, which, as there is no underwood, form a spacious shady grove. I rambled daily, during many weeks of each successive dry season, along this delightful road. The trees, many of which are myrtles (Eugenia Egaensis of Martius) and wild Guavas (Psidium), with smooth yellow stems, were in flower at this time; and the rippling waters of the lake, under the cool shade, everywhere bordered the path. The place was the resort of kingfishers, green and blue tree-creepers, purple-headed tanagers, and humming-birds. Birds generally, however, were not numerous. Every tree was tenanted by Cicadas, the reedy notes of which produced that loud, jarring, insect music which is the general accompaniment of a woodland ramble in a hot climate. One species was very handsome, having wings adorned with patches of bright green and scarlet. It was very common; sometimes three or four tenanting a single tree, clinging as usual to the branches. On approaching a tree thus peopled, a number of little jets of a clear liquid would be seen squirted from aloft. I have often received the well-directed discharge full on my face; but the liquid is harmless, having a sweetish taste, and is ejected by the insect from the anus, probably in self-defence, or from fear. The number and variety of gaily-tinted butterflies, sporting about in this grove on sunny days, were so great that the bright moving flakes of colour gave quite a character to the physiognomy of the place. It was impossible to walk far without disturbing flocks of them from the damp sand at the edge of the water, where they congregated to imbibe the moisture. They

Q 2

were of almost all colours, sizes, and shapes: I noticed here altogether eighty species, belonging to twenty-two different genera. It is a singular fact that, with very few exceptions, all the individuals of these various species thus sporting in sunny places were of the male sex; their partners, which are much more soberly dressed and immensely less numerous than the males, being confined to the shades of the woods. Every afternoon, as the sun was getting low, I used to notice these gaudy sunshine-loving swains trooping off to the forest, where I suppose they would find their sweethearts and wives. The most abundant, next to the very common sulphur-yellow and orange-coloured kinds (Callidryas, seven species), were about a dozen species of Cybdelis, which are of large size, and are conspicuous from their liveries of glossy dark-blue and purple. A superbly-adorned creature, the Callithea Markii, having wings of a thick texture, coloured sapphire-blue and orange, was only an occasional visitor. On certain days, when the weather was very calm, two small gilded-green species (Symmachia Trochilus and Colubris) literally swarmed on the sands, their glittering wings lying wide open on the flat surface. The beach terminates, eight miles beyond Ega, at the mouth of a rivulet; the character of the coast then changes, the river banks being masked by a line of low islets amid a labyrinth of channels.

In all other directions my very numerous excursions were by water; the most interesting of those made in the immediate neighbourhood were to the houses of Indians on the banks of retired creeks; an account of one of these trips will suffice.

On the 23rd of May, 1850, I visited, in company with Antonio Cardozo, the Delegado, a family of the Passé tribe, who live near the head waters of the igarapé, which flows from the south into the Teffé, entering it at Ega. The creek is more than a quarter of a mile broad near the town, but a few miles inland it gradually contracts, until it becomes a mere rivulet flowing through a broad dell in the forest. When the river rises it fills this dell; the trunks of the lofty trees then stand many feet deep in the water, and small canoes are able to travel the distance of a day's journey under the shade, regular paths or alleys being cut through the branches and lower trees. This is the general character of the country of the Upper Amazons; a land of small elevation and abruptly undulated, the hollows forming narrow valleys in the dry months, and deep navigable creeks in the wet months. In retired nooks on the margins of these shady rivulets, a few families or small hordes of aborigines still linger in nearly their primitive state, the relicts of their once numerous tribes. The family we intended to visit on this trip was that of Pedro-uassú (Peter the Great, or Tall Peter), an old chieftain or Tushaúa of the Passés.

We set out at sunrise, in a small igarité, manned by six young Indian paddlers. After travelling about three miles along the broad portion of the creek—which, being surrounded by woods, had the appearance of a large pool—we came to a part where our course seemed to be stopped by an impenetrable hedge of trees and bushes. We were some time before finding the entrance, but when fairly within the shades, a remarkable scene presented itself. It was my first introduction to these singular water-paths. A narrow and tolerably straight alley stretched away for a long distance before us; on each side were the tops of bushes and young trees, forming a kind of border to the path, and the trunks of the tall forest trees rose at irregular intervals from the water, their crowns interlocking far over our heads, and forming a thick shade. Slender air roots hung down in clusters, and looping sipós dangled from the lower branches; bunches of grass, tillandsiæ, and ferns, sat in the forks of the larger boughs, and the trunks of trees near the water had adhering to them round dried masses of freshwater sponges. There was no current perceptible, and the water was stained of a dark olive-brown hue, but the submerged stems could be seen through it to a great depth. We travelled at good speed for three hours along this shady road; the distance of Pedro's house from Ega being about twenty miles. When the paddlers rested for a time, the stillness and gloom of the place became almost painful: our voices waked dull echoes as we conversed, and the noise made by fishes occasionally whipping the surface of the water was quite startling. A cool, moist, clammy air pervaded the sunless shade.

The breadth of the wooded valley, at the commencement, is probably more than half a mile, and there is a tolerably clear view for a considerable distance on each side of the water-path through the irregular colonnade of trees: other paths also, in this part, branch off right and left from the principal road, leading to the scattered houses of Indians on the mainland. The dell contracts gradually towards the head of the rivulet, and the forest then becomes denser; the water-path also diminishes in width, and becomes more winding, on account of the closer growth of the trees. The boughs of some are stretched forth at no great height over one's head, and are seen to be loaded with epiphytes; one orchid I noticed particularly, on account of its bright yellow flowers growing at the end of flower-stems several feet long. Some of the trunks, especially those of palms, close beneath their crowns, were clothed with a thick mass of glossy shield-shaped Pothos plants, mingled with ferns. Arrived at this part we were, in fact, in the heart of the virgin forest. We heard no noises of animals in the trees, and saw only one bird, the sky-blue chatterer, sitting alone on a high branch. For some distance the lower vegetation was so dense that the road runs under an arcade of foliage, the branches having been cut away only sufficiently to admit of the passage of a small canoe. These thickets are formed chiefly of Bamboos, whose slender foliage and curving stems arrange themselves in elegant, feathery bowers: but other social plants,—slender green climbers with tendrils so eager in aspiring to grasp the higher boughs that they seem to be endowed almost with animal energy, and certain low trees having large elegantly-veined leaves, contribute also to the jungly masses. Occasionally we came upon an uprooted tree lying across the path, its voluminous crown still held up by thick cables of sipó, connecting it with standing trees: a wide circuit had to be made in these cases, and it was sometimes difficult to find the right path again.

At length we arrived at our journey's end. We were then in a very dense and gloomy part of the forest: we could see, however, the dry land on both sides of the creek, and to our right a small sunny opening appeared, the landing-place to the native dwellings. The water was deep close to the bank, and a clean pathway ascended from the shady port to the buildings, which were about a furlong distant. My friend Cardozo was godfather to a grandchild of Pedro-uassú, whose daughter had married an Indian settled in Ega. He had sent word to the old man that he intended to visit him: we were therefore expected.

As we landed, Pedro-uassú himself came down to the port to receive us; our arrival having been announced by the barking of dogs. He was a tall and thin old man, with a serious, but benignant expression of countenance, and a manner much freer from shyness and distrust than is usual with Indians. He was clad in a shirt of coarse cotton cloth, dyed with murishí, and trowsers of the same material turned up to the knee. His features were sharply delineated—more so than in any Indian face I had yet seen; the lips thin and the nose rather high and compressed. A large, square, blue-black tattooed patch occupied the middle of his face, which, as well as the other exposed parts of his body, was of a light reddish-tan colour, instead of the usual coppery-brown hue. He walked with an upright, slow gait, and on reaching us saluted Cardozo with the air of a man who wished it to be understood that he was dealing with an equal. My friend introduced me, and I was welcomed in the same grave, ceremonious manner.

He seemed to have many questions to ask: but they were chiefly about Senhora Felippa, Cardozo's Indian housekeeper at Ega, and were purely complimentary. This studied politeness is quite natural to Indians of the advanced agricultural tribes. The language used was Tupí: I heard no other spoken all the day. It must be borne in mind that Pedro-uassú had never had much intercourse with whites: he was, although baptised, a primitive Indian, who had always lived in retirement; the ceremony of baptism having been gone through, as it generally is by the aborigines, simply from a wish to stand well with the whites.

Arrived at the house, we were welcomed by Pedro's wife: a thin, wrinkled, active old squaw, tattooed in precisely the same way as her husband. She had also sharp features, but her manner was more cordial and quicker than that of her husband: she talked much, and with great inflection of voice; whilst the tones of the old man were rather drawling and querulous. Her clothing was a long petticoat of thick cotton cloth, and a very short chemise, not reaching to her waist. I was rather surprised to find the grounds around the establishment in neater order than in any sitio, even of civilised people, I had yet seen on the Upper Amazons: the stock of utensils and household goods of all sorts was larger, and the evidences of regular industry and plenty more numerous than one usually perceives in the farms of civilised Indians and whites. The buildings were of the same construction as those of the humbler settlers in all other parts of the country. The family lived in a large, oblong, open shed built under the shade of trees. Two smaller buildings, detached from the shed and having mud-walls with low doorways, contained apparently the sleeping apartment of different members of the large household. A small mill for grinding sugar-cane, having two cylinders of hard notched wood; wooden troughs, and kettles for boiling the *guarápa* (cane juice), to make treacle, stood under a separate shed, and near it was a large enclosed mud-house for poultry. There was another hut and shed a short distance off, inhabited by a family dependent on Pedro, and a narrow pathway through the luxuriant woods led to more dwellings of the same kind. There was an abundance of fruit trees around the place, including the never-failing banana, with its long, broad, soft green leaf-blades, and groups of full-grown Pupúnhas, or peach palms. There was also a large number of cotton and coffee trees. Amongst the utensils I noticed baskets of different shapes, made of flattened maranta stalks, and dyed various colours. The making of these is an original art of the Passés, but I believe it is also practised by other tribes, for I saw several in the houses of semi-civilised Indians on the Tapajos.

There were only three persons in the house besides the old couple, the rest of the people being absent; several came in, however, in the course of the day. One was a daughter of Pedro's, who had an oval tattooed spot over her mouth; the second was a young grandson; and the third the son-in-law from Ega, Cardozo's *compadre.* The old woman was occupied, when we entered, in distilling spirits from cará, an eatable root similar to the potato, by means of a clay still, which had been manufactured by herself. The liquor had a reddish tint, but not a very agreeable flavour. A cup of it warm from the still, however, was welcome after our long journey. Cardozo liked it, emptied his cup, and replenished it in a very short time. The old lady was very talkative, and almost fussy in her desire to please her visitors. We sat in tucúm hammocks, suspended between the upright posts of the shed. The young woman with the blue mouth—who, although married, was as shy as any young maiden of her race—soon became employed in scalding and plucking fowls for the dinner, near the fire on the ground at the other end of the dwelling. The son-in-law, Pedro-uassú, and Cardozo now began a long conversation on the subject of their deceased wife, daughter, and *comadre.** It appeared she had died of consumption—"tisica," as they called it, a word adopted by the Indians from the Portuguese. The widower repeated over and over again, in nearly the same words, his account of her illness, Pedro chiming in like a chorus, and Cardozo moralising and condoling. I thought the *cauím* (grog) had a good deal to do with the flow of talk and warmth of feeling of all three: the widower drank and wailed until he became maundering, and finally fell asleep.

I left them talking, and went a long ramble into the forest, Pedro sending his grandson, a smiling well-behaved lad of about fourteen years of age, to show me the paths, my companion taking with him his *Zaraba-*

* Co-mother; the term expressing the relationship of a mother to the godfather of her child.

tana, or blowpipe. This instrument is used by all the Indian tribes on the Upper Amazons. It is generally nine or ten feet long, and is made of two separate lengths of wood, each scooped out so as to form one half of the tube. To do this with the necessary accuracy requires an enormous amount of patient labour, and considerable mechanical ability, the tools used being simply the incisor teeth of the Páca and Cutía. The two half tubes, when finished, are secured together by a very close and tight spirally-wound strapping, consisting of long flat strips of Jacitára, or the wood of the climbing palm-tree; and the whole is smeared afterwards with

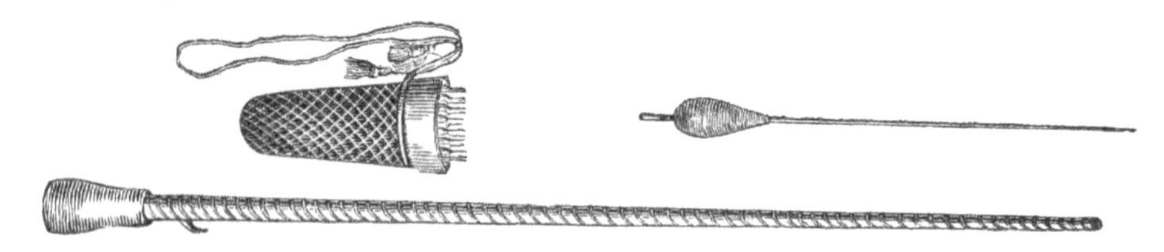

Blow-pipe, quiver, and arrow.

black wax, the production of a Melipona bee. The pipe tapers towards the muzzle, and a cup-shaped mouthpiece, made of wood, is fitted in the broad end. A full-sized *Zarabatana* is heavy, and can only be used by an adult Indian who has had great practice. The young lads learn to shoot with smaller and lighter tubes. When Mr. Wallace and I had lessons at Barra in the use of the blowpipe, of Julio, a Jurí Indian, then in the employ of Mr. Hauxwell, an English bird-collector, we found it very difficult to hold steadily the long tubes. The arrows are made from the hard rind of the leaf-stalks of certain palms, thin strips being cut, and rendered as sharp as needles by scraping the ends with a knife or the tooth of an animal. They are winged with a little oval mass of samaüma silk (from the seed-vessels of the silk-cotton tree, Eriodendron samaüma), cotton being too heavy. The ball of samaüma should fit to a nicety the bore of the blowpipe; when it does so, the arrow can be propelled with such force by the breath that it makes a noise almost as loud as a pop-gun on flying from the muzzle. My little companion was armed with a quiver full of these little missiles, a small number of which, sufficient for the day's sport, were tipped with the fatal Urarí poison. The quiver was an ornamental affair, the broad rim being made of highly-polished wood of a rich cherry-red colour (the Moira-piránga, or red-wood of the Japurá). The body was formed of neatly-plaited strips of Maranta stalks, and the belt by which it was suspended from the shoulder was decorated with cotton fringes and tassels.

We walked about two miles along a well-trodden pathway, through high caäpoeira (second-growth forest). A large proportion of the trees were Melastomas, which bore a hairy yellow fruit, nearly as large and as well flavoured as our gooseberry. The season, however, was nearly over for them. The road was bordered every inch of the way by a thick bed of elegant Lycopodiums. An artificial arrangement of trees and bushes could scarcely have been made to wear so finished an appearance as this naturally decorated avenue. The path at length terminated at a plantation of mandioca, the largest I had yet seen since I left the neighbourhood of Pará. There were probably ten acres of cleared land, and part of the ground was planted with Indian corn, water-melons, and sugar-cane. Beyond this field there was only a faint hunter's track, leading towards the untrodden interior. My companion told me he had never heard of there being any inhabitants in that direction (the south). We crossed the forest from this place to another smaller clearing, and then walked, on our road home, through about two miles of caäpoeira of various ages, the sites of old plantations. The only fruits of our ramble were a few rare insects and a Japú (Cassicus cristatus), a handsome bird with chestnut and saffron-coloured plumage, which wanders through the tree-tops in large flocks. My little companion brought this down from a height which I calculated at thirty yards. The blowpipe, however, in the hands of an expert adult Indian, can be made to propel arrows so as to kill at a distance of fifty and sixty yards. The aim is most certain when the tube is held vertically, or nearly so. It is a far more useful weapon in the forest than a gun, for the report of a firearm alarms the whole flock of birds or monkeys feeding on a tree, whilst the silent poisoned dart brings the animals down one by one until the sportsman has a heap of slain by his side. None but the stealthy Indian can use it effectively. The poison, which must be fresh to kill speedily, is obtained only of the Indians who live beyond the cataracts of the rivers flowing from the north, especially the Rio Negro and the Japurá. Its principal ingredient is the wood of the Strychnos toxifera, a tree which does not grow in the humid forests of the river plains. A most graphic account of the Urarí, and of an expedition undertaken in search of the tree in Guiana, has been given by Sir Robert Schomburgk.*

When we returned to the house after mid-day, Cardozo was still sipping cauím, and now looked exceedingly merry. It was fearfully hot: the good fellow sat in his hammock with a cuya full of grog in his hands; his broad honest face all of a glow, and the perspiration streaming down his uncovered breast, the unbuttoned shirt having slipped half-way over his broad shoulders. Pedro-uassú had not drunk much; he was noted, as I afterwards learnt, for his temperance. But he was standing up as I had left him two hours previous, talking to Cardozo in the same monotonous tones, the conversation apparently not having flagged all the time. I had never heard so much talking amongst Indians. The widower was asleep: the stirring, managing old lady with her daughter were preparing dinner. This, which was ready soon after I entered, consisted of boiled fowls and rice, seasoned with large green peppers and lemon juice, and piles of new, fragrant farinha and raw bananas. It was served on plates of English manufacture on a tupé, or large plaited rush mat, such as is made by the natives pretty generally on the Amazons. Three or four other Indians, men and women of middle age, now made their appearance, and joined in the meal. We all sat round on the floor: the women, according to custom, not eating until after the men had done. Before sitting down, our host apologised in his usual quiet, courteous manner for not having knives and forks; Cardozo and I ate by the aid of wooden

* Annals and Magazine of Natural History, vol. vii. p. 411.

spoons, the Indians using their fingers. The old man waited until we were all served before he himself commenced. At the end of the meal, one of the women brought us water in a painted clay basin of Indian manufacture, and a clean but coarse cotton napkin, that we might wash our hands.

The horde of Passés of which Pedro-uassú was Tushaúa or chieftain, was at this time reduced to a very small number of individuals. The disease mentioned in the last chapter had for several generations made great havoc amongst them; many, also, had entered the service of whites at Ega, and, of late years, intermarriages with whites, half-castes, and civilised Indians had been frequent. The old man bewailed the fate of his race to Cardozo with tears in his eyes. "The people of my nation," he said, "have always been good friends to the Caríwas (whites), but before my grandchildren are old like me the name of Passé will be forgotten." In so far as the Passés have amalgamated with European immigrants or their descendants, and become civilised Brazilian citizens, there can scarcely be ground for lamenting their extinction as a nation; but it fills one with regret to learn how many die prematurely of a disease which seems to arise on their simply breathing the same air as the whites. The original territory of the tribe must have been of large extent, for Passés are said to have been found by the early Portuguese colonists on the Rio Negro; an ancient settlement on that river, Barcellos, having been peopled by them when it was first established; and they formed also part of the original population of Fonte-boa on the Solimoens. Their hordes were therefore spread over a region 400 miles in length from east to west. It is probable, however, that they have been confounded by the colonists with other neighbouring tribes who tattoo their faces in a similar manner; such as the Jurís, Uáinumás, Shumánas, Araúas, and Tucúnas. The extinct tribe of Yurimaúas, or Sorimóas, from which the river Solimoens derives its name, according to traditions extant at Ega, resembled the Passés in their slender figures and friendly disposition. These tribes (with others lying between them) peopled the banks of the main river and its by-streams from the mouth of the Rio Negro to Peru. True Passés existed in their primitive state on the banks of the Issá, 240 miles to the west of Ega, within the memory of living persons. The only large body of them now extant are located on the Japurá, at a place distant about 150 miles from Ega: the population of this horde, however, does not exceed, from what I could learn, 300 or 400 persons. I think it probable that the lower part of the Japurá and its extensive delta lands formed the original home of this gentle tribe of Indians.

The Passés are always spoken of in this country as the most advanced of all the Indian nations in the Amazons region. I saw altogether about thirty individuals of the tribe, and found them generally distinguishable from other Indians by their lighter colour, sharper features, and more open address. But these points of distinction were not invariable, for I saw individuals of the Jurí and Miránha tribes from the Upper Japurá; of the Catoquínos, who inhabit the banks of the Jurúa, 300 miles from its mouth; and of the Tucúnas of St. Paulo, who were scarcely distinguishable from Passés in all the features mentioned. It is remarkable that a small tribe, the Caishánas, who live in the very midst of all these superior tribes, are almost as debased physically and mentally as the Múras, the lowest of all the Indian tribes on the Amazons. Yet were they seen separately, many Caishánas could not be distinguished from Miránhas or Jurís, although none have such slender figures or are so frank in their ways as to be mistaken for Passés. I make these remarks to show that the differences between the nations or tribes of Indians are not absolute, and therefore that there is no ground for supposing any of them to have had an origin entirely different from the rest. Under what influences certain tribes, such as the Passés, have become so strongly modified in mental, social, and bodily features, it is hard to divine. The industrious habits, fidelity, and mildness of disposition of the Passés, their docility and, it may be added, their personal beauty, especially of the children and women, made them from the first very attractive to the Portuguese colonists. They were, consequently, enticed in great number from their villages and brought to Barra and other settlements of the whites. The wives of governors and military officers from Europe were always eager to obtain children for domestic servants: the girls being taught to sew, cook, weave hammocks, manufacture pillow-lace, and so forth. They have been generally treated with kindness, especially by the educated families in the settlements. It is pleasant to have to record that I never heard of a deed of violence perpetrated, on the one side or the other, in the dealings between European settlers and this noble tribe of savages.

Very little is known of the original customs of the Passés. The mode of life of our host Pedro-uassú did not differ much from that of the civilised Mamelucos; but he and his people showed a greater industry, and were more open, cheerful, and generous in their dealings than many half-castes. The authority of Pedro, like that of the Tushaúas generally, was exercised in a mild manner. These chieftains appear able to command the services of their subjects, since they furnish men to the Brazilian authorities when requested; but none of them, even those of the most advanced tribes, appear to make use of this authority for the accumulation of property; the service being exacted chiefly in time of war. Had the ambition of the chiefs of some of these industrious tribes been turned to the acquisition of wealth, probably we should have seen indigenous civilised nations in the heart of South America similar to those found on the Andes of Peru and Mexico. It is very probable that the Passés adopted from the first to some extent the manners of the whites. Ribeiro, a Portuguese official who travelled in these regions in 1774-5, and wrote an account of his journey, relates that they buried their dead in large earthenware vessels (a custom still observed amongst other tribes on the Upper Amazons), and that, as to their marriages, the young men earned their brides by valiant deeds in war. He also states that they possessed a cosmogony, in which the belief that the sun was a fixed body with the earth revolving around it, was a prominent feature. He says, moreover, that

R 2

they believed in a Creator of all things; a future state of rewards and punishments, and so forth. These notions are so far in advance of the ideas of all other tribes of Indians, and so little likely to have been conceived and perfected by a people having no written language or leisured class, that we must suppose them to have been derived by the docile Passés from some early missionary or traveller. I never found that the Passés had more curiosity or activity of intellect than other Indians. No trace of a belief in a future state exists amongst Indians who have not had much intercourse with the civilised settlers, and even amongst those who have it is only a few of the more gifted individuals who show any curiosity on the subject. Their sluggish minds seem unable to conceive or feel the want of a theory of the soul, and of the relations of man to the Creator or the rest of Nature. But is it not so with totally uneducated and isolated people even in the most highly civilised parts of the world? The good qualities of the Passés belong to the moral part of the character: they lead a contented, unambitious, and friendly life, a quiet, domestic, orderly existence, varied by occasional drinking bouts and summer excursions. They are not so shrewd, energetic, and masterful as the Mundurucús, but they are more easily taught, because their disposition is more yielding than that of the Mundurucús or any other tribe.

We started on our return to Ega at half-past four o'clock in the afternoon. Our generous entertainers loaded us with presents. There was scarcely room for us to sit in the canoe, as they had sent down ten large bundles of sugar-cane, four baskets of farinha, three cedar planks, a small hamper of coffee, and two heavy bunches of bananas. After we were embarked the old lady came with a parting gift for me—a huge bowl of smoking-hot banana porridge. I was to eat it on the road "to keep my stomach warm." Both stood on the bank as we pushed off, and gave us their adeos, "Ikuána Tupána eirúm" (Go with God): a form of salutation taught by the old Jesuit missionaries. We had a most uncomfortable passage, for Cardozo was quite tipsy and had not attended to the loading of the boat. The cargo had been placed too far forward, and to make matters worse my heavy friend obstinately insisted on sitting astride on the top of the pile, instead of taking his place near the stern; singing from his perch a most indecent love-song, and disregarding the inconvenience of having to bend down almost every minute to pass under the boughs and hanging sipós as we sped rapidly along. The canoe leaked, but not, at first, alarmingly. Long before sunset, darkness began to close in under these gloomy shades, and our steersman could not avoid now and then running the boat into the thicket. The first time this happened a piece was broken off the square prow (rodella); the second time we got squeezed between two trees. A short time after this latter accident, being seated near the stern with my feet on the bottom of the boat, I felt rather suddenly the cold water above my ankles. A few minutes more and we should have sunk, for a seam had been opened forward under the pile of sugar-cane. Two of us began to bale, and by the most strenuous efforts managed to keep afloat without throwing overboard our cargo. The Indians were obliged to paddle with extreme slowness to avoid shipping water, as the edge of our prow was nearly level with the surface; but Cardozo was now persuaded to change his seat. The sun set, the quick twilight passed, and the moon soon after began to glimmer through the thick canopy of foliage. The prospect of being swamped in this hideous solitude was by no means pleasant, although I calculated on the chance of swimming to a tree and finding a nice snug place in the fork of some large bough wherein to pass the night. At length, after four hours' tedious progress, we suddenly emerged on the open stream where the moonlight glittered in broad sheets on the gently rippling waters. A little extra care was now required in paddling. The Indians plied their strokes with the greatest nicety; the lights of Ega (the oil lamps in the houses) soon appeared beyond the black wall of forest, and in a short time we leapt safely ashore.

A few months after the excursion just narrated, I accompanied Cardozo in many wanderings on the Solimoens, during which we visited the praias (sand-islands), the turtle pools in the forests, and the by-streams and lakes of the great desert river. His object was mainly to superintend the business of digging up turtle eggs on the sand-banks, having been elected commandante for the year, by the municipal council of Ega, of the "praia real" (royal sand-island) of Shimuní, the one lying nearest to Ega. There are four of these royal praias within the Ega district, (a distance of 150 miles from the town), all of which are visited annually by the Ega people for the purpose of collecting eggs and extracting oil from their yolks. Each has its commander, whose business is to make arrangements for securing to every inhabitant an equal chance in the egg harvest by placing sentinels to protect the turtles whilst laying, and so forth. The pregnant turtles descend from the interior pools to the main river in July and August, before the outlets dry up, and then seek in countless swarms their favourite sand-islands; for it is only a few praias that are selected by them out of the great number existing. The young animals remain in the pools throughout the dry season. These breeding places of turtles then lie twenty to thirty or more feet above the level of the river, and are accessible only by cutting roads through the dense forest.

We left Ega on our first trip, to visit the sentinels whilst the turtles were yet laying, on the 26th of September. Our canoe was a stoutly-built igarité, arranged for ten paddlers, and having a large arched toldo at the stern, under which three persons could sleep pretty comfortably. In passing down the narrow channel to the mouth of the Teffé, I noticed that the yellow waters of the Solimoens were flowing slowly inwards towards the lake, showing how much fuller and stronger, at this season, was the current of the main river than that of its tributary. On reaching the broad stream, we descended rapidly on the swift current to the south-eastern or lower end of the large wooded island of Bariá, which here divides the river into two great channels. The distance was about twelve miles: the island of Shimuní lies in the middle of the north-easterly channel, and is reached by passing round the end of Bariá. Two miles further down the broad, wild, and turbid river, lies the small island of Curubarú, skirted like the others by a large praia ; this is not, however, frequented by turtles, on account of the coarse, gritty nature of the deposit. The sand-banks appear to be formed only where there is a remanso or still water, and the wooded islands to which they are generally attached probably first originated in accumulations of sand.

We landed on Curubarú ; Cardozo wishing to try the poços (wells, or deep pools) which lie here as in other praias between the sand-bank and its island, for fish and tracajás. The sun was now nearly vertical, and the coarse, heated sand burnt our feet as we trod. We walked or rather trotted nearly a mile before reaching the pools : there was not a breath of wind nor a cloud to moderate the heat of mid-day, and the Indians who carried the fishing-net suffered greatly. On arriving at the ponds we found the water was quite warm ; the net brought up only two or three small fishes, and we thus had our toilsome journey for nothing.

Re-embarking, we paddled across to Shimuní, reaching the commencement of the praia an hour before sunset. The island-proper is about three miles long and half a mile broad : the forest with which it is covered rises to an immense and uniform height, and presents all round a compact, impervious front. Here and there a singular tree, called Pao mulatto (mulatto wood), with polished dark-green trunk, rose conspicuously amongst the mass of vegetation. The sand-bank, which lies at the upper end of the island extends several miles, and presents an irregular, and in some parts, strongly waved surface, with deep hollows and ridges. When upon it, one feels as though treading an almost boundless field of sand: for towards the south-east, where no forest-line terminates the view, the white, rolling plain stretches away to the horizon. The north-easterly channel of the river lying between the sands and the further shore of the river is at least two miles in breadth; the middle one, between the two islands, Shimuní and Bariá, is not much less than a mile.

We found the two sentinels lodged in a corner of the praia, where it commences at the foot of the towering forest-wall of the island; having built for themselves a little rancho with poles and palm-leaves. Great precautions are obliged to be taken to avoid disturbing the sensitive turtles, who, previous to crawling ashore to lay, assemble in great shoals off the sand-bank. The men, during this time, take care not to show themselves and warn off any fisherman who wishes to pass near the place. Their fires are made in a deep hollow near the borders of the forest, so that the smoke may not be visible. The passage of a boat through the shallow waters where the animals are congregated, or the sight of a man or a fire on the sand-bank, would prevent the turtles from leaving the water that night to lay their eggs, and if the causes of alarm were repeated once or twice, they would forsake the praia for some other quieter place. Soon after we arrived, our men were sent with the net to catch a supply of fish for supper. In half an hour, four or five large basketsful of Acarí were brought in. The sun set soon after our meal was cooked; we were then obliged to extinguish the fire and remove our supper materials to the sleeping ground, a spit of sand about a mile off; this course being necessary on account of the mosquitoes which swarm at night on the borders of the forest.

One of the sentinels was a taciturn, morose-looking, but sober and honest Indian, named Daniel; the other was a noted character of Ega, a little wiry mameluco, named Carepíra (Fish-hawk); known for his waggery, propensity for strong drink, and indebtedness to Ega traders. Both were intrepid canoemen and huntsmen, and both perfectly at home anywhere in these fearful wastes of forest and water. Carepíra had his son with him, a quiet little lad of about nine years of age. These men in a few minutes constructed a small shed with four upright poles and leaves of the arrow-grass, under which I and Cardozo slung our hammocks. We did not go to sleep, however, until after midnight: for when supper was over we lay about on the sand with a flask of rum in our midst, and whiled away the still hours in listening to Carepíra's stories.

I rose from my hammock by daylight, shivering with cold; a praia, on account of the great radiation of heat in the night from the sand, being towards the dawn the coldest place that can be found in this climate. Cardozo and the men were already up watching the turtles. The sentinels had erected for this purpose a stage about fifty feet high, on a tall tree near their station, the ascent to which was by a roughly-made ladder of woody lianas. They are enabled, by observing the turtles from this watch-tower, to ascertain the date of successive deposits of eggs, and thus guide the commandante in fixing the time for the general invitation to the Ega people. The turtles lay their eggs by night, leaving the water when nothing disturbs them, in vast crowds, and crawling to the central and highest part of the praia. These places are, of course, the last to go under water when, in unusually wet seasons, the river rises before the eggs are hatched by the heat of the sand. One could almost believe, from this, that the animals used forethought in choosing a place; but it is simply one of those many instances in animals where unconscious habit has the same result as conscious prevision. The hours between midnight and dawn are the busiest. The turtles excavate with their broad, webbed paws deep holes in the fine sand: the first comer, in each case, making a pit about three feet deep, laying its eggs (about 120 in number) and covering them with sand; the next making its deposit at the top of that of its predecessor, and so on until every pit is full. The whole body of turtles frequenting a praia does not finish laying in less than fourteen or fifteen days, even when there is no interruption. When all have done, the area (called by the Brazilians *taboleiro*) over which they have excavated, is distinguishable from the rest of the praia only by signs of the sand having been a little disturbed.

On rising I went to join my friends. Few recollections of my Amazonian rambles are more vivid and agreeable than that of my walk over the white sea of sand on this cool morning. The sky was cloudless ; the just-risen sun was hidden behind the dark mass of woods on Shimuní, but the long line of forest to the west, on Bariá, with its plumy decorations of palms, was lighted up with his yellow, horizontal rays. A faint chorus of singing birds reached the ears from across the water, and flocks of gulls and plovers were crying plaintively over the swelling banks of the praia, where their eggs lay in nests made in little hollows of the sand. Tracks of stray turtles were visible on the smooth white surface of the praia. The animals which thus wander from the main body are lawful prizes of the sentinels ; they had caught in this way two before sunrise, one of which we had for dinner. In my walk I disturbed several pairs of the chocolate and drab-coloured wild goose (Anser jubatus) which set off to run along the edge of the water. The enjoyment one feels in rambling over these free, open spaces, is no doubt enhanced by the novelty of the scene, the change being very great from the monotonous landscape of forest which everywhere else presents itself.

On arriving at the edge of the forest I mounted the sentinel's stage, just in time to see the turtles retreating to the water on the opposite side of the sand-bank, after having laid their eggs. The sight was well worth the trouble of ascending the shaky ladder. They were about a mile off, but the surface of the sands was blackened with the multitudes which were waddling towards the river ; the margin of the praia was rather steep, and they all seemed to tumble head first down the declivity into the water.

I spent the morning of the 27th collecting insects in the woods of Shimuní; assisting my friend in the afternoon to beat a large pool for Tracajás, Cardozo wishing to obtain a supply for his table at home. The pool was nearly a mile long, and lay on one side of the island between the forest and the sand-bank. The sands are heaped up very curiously around the margins of these isolated sheets of water; in the present case they formed a steeply-inclined bank, from five to eight feet in height. What may be the cause of this formation I cannot imagine. The pools always contain a quantity of imprisoned fish, turtles, tracajás, and Aiyussás.* The turtles and Aiyussás crawl out voluntarily in the course of a few days, and escape to the main river, but the Tracajás remain and become an easy prey to the natives. The ordinary mode of obtaining them is to whip the water in every part with rods for several hours during the day; this treatment having the effect of driving the animals out. They wait, however, until the night following the beating before making their exit. Our Indians were occupied for many hours in this work, and when night came they and the sentinels were placed at intervals along the edge of the water to be ready to capture the runaways. Cardozo and I, after supper, went and took our station at one end of the pool.

We did not succeed, after all our trouble, in getting many Tracajás. This was partly owing to the intense darkness of the night, and partly, doubtless, to the

* Specimens of this species of turtle are named in the British Museum collection, Podocnemis expansa.

sentinels having already nearly exhausted the pool, notwithstanding their declarations to the contrary. In waiting for the animals it was necessary to keep silence: not a pleasant way of passing the night; speaking only in whispers, and being without fire in a place liable to be visited by a prowling jaguar. Cardozo and I sat on a sandy slope with our loaded guns by our side, but it was so dark we could scarcely see each other. Towards midnight a storm began to gather around us. The faint wind which had breathed from over the water since the sun went down, ceased; thick clouds piled themselves up, until every star was obscured, and gleams of watery lightning began to play in the midst of the black masses. I hinted to Cardozo that I thought we had now had enough of watching, and suggested a cigarette. Just then a quick pattering movement was heard on the sands, and grasping our guns, we both started to our feet. Whatever it might have been it seemed to pass by, and a few moments afterwards a dark body appeared to be moving in another direction on the opposite slope of the sandy ravine where we lay. We prepared to fire, but luckily took the precaution of first shouting "Quem va lá?" (Who goes there?). It turned out to be the taciturn sentinel, Daniel, who asked us mildly whether we had heard a "raposa" pass our way. The raposa is a kind of wild dog, with very long, tapering muzzle, and black and white speckled hair.*

* I had once only an opportunity of examining a specimen of this animal. It is probably new to science, at least I have not been able to find a published description that suits the species. The one mentioned was taken from a burrow in the earth in the forests bordering the Teffé, near Ega.

Daniel could distinguish all kinds of animals in the dark by their footsteps. It now began to thunder, and our position was getting very uncomfortable. Daniel had not seen anything of the other Indians, and thought it was useless waiting any longer for Tracajás; we therefore sent him to call in the whole party, and made off, ourselves, as quickly as we could for the canoe. The rest of the night was passed most miserably; as indeed were very many of my nights on the Solimoens. A furious squall burst upon us; the wind blew away the cloths and mats we had fixed up at the ends of the arched awning of the canoe to shelter ourselves, and the rain beat right through our sleeping-place. There we lay, Cardozo and I, huddled together and wet through, waiting for the morning.

A cup of strong and hot coffee put us to rights at sunrise; but the rain was still coming down, having changed to a steady drizzle. Our men were all returned from the pool, having taken only four Tracajás. The business which had brought Cardozo hither being now finished, we set out to return to Ega, leaving the sentinels once more to their solitude on the sands. Our return route was by the rarely frequented north-easterly channel of the Solimoens, through which flows part of the waters of its great tributary stream, the Japurá. We travelled for five hours along the desolate, broken, timber-strewn shore of Bariá. The channel is of immense breadth, the opposite coast being visible only as a long, low line of forest. At three o'clock in the afternoon we doubled the upper end of the island, and then crossed towards the mouth of the Teffé by a broad transverse channel running between Bariá and another island called Quanarú. There is a small sand-bank at the north-westerly point of Bariá, called Jacaré; we stayed here to dine and afterwards fished with the net. A fine rain was still falling, and we had capital sport, in three hauls taking more fish than our canoe would conveniently hold. They were of two kinds only, the

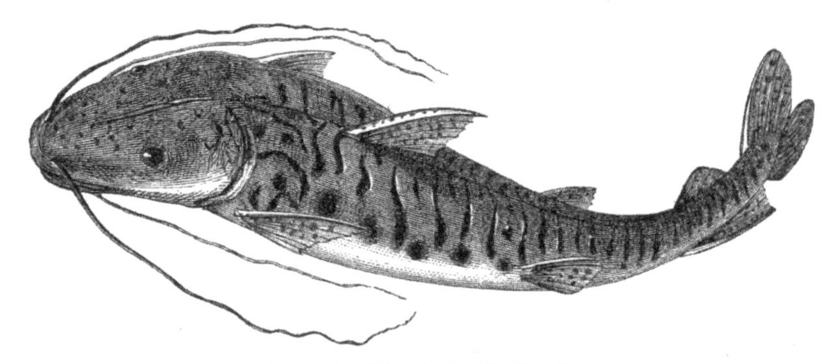

Surubim (Pimelodus tigrinus).

Surubim and the Piraepiéüa (species of Pimelodus), very handsome fishes four feet in length, with flat spoon-shaped heads, and prettily-spotted and striped skins.

On our way from Jacaré to the mouth of the Teffé we had a little adventure with a black tiger or jaguar. We were paddling rapidly past a long beach of dried mud, when the Indians became suddenly excited, shouting "Ecuí Jauareté; Jauarí-pixúna!" (Behold the jaguar, the black jaguar!). Looking ahead we saw the animal quietly drinking at the water's edge. Cardozo ordered the steersman at once to put us ashore. By the time we were landed the tiger had seen us, and was retracing his steps towards the forest. On the spur of the moment and without thinking of what we were doing, we took our guns (mine was a double-barrel, with one charge of B B and one of dust-shot) and gave chase. The animal increased his speed, and reaching the forest border dived into the dense mass of broad-leaved grass which formed its frontage. We peeped through the gap he had made, but, our courage being by this time cooled, did not think it wise to go into the thicket after him. The black tiger appears to be more abundant than the spotted form of jaguar in the neighbourhood of Ega. The most certain method of finding it is to hunt, assisted by a string of Indians shouting and driving the game before them, in the narrow *restingas* or strips of dry land in the forest, which are isolated by the flooding of their neighbourhood in the wet season. We reached Ega by eight o'clock at night.

On the 6th of October we left Ega on a second excursion; the principal object of Cardozo being, this time, to search certain pools in the forest for young turtles. The exact situation of these hidden sheets of water is known only to a few practised huntsmen; we took one of these men with us from Ega, a mameluco named Pedro, and on our way called at Shimuní for Daniel to serve as an additional guide. We started from the praia at sunrise on the 7th, in two canoes containing twenty-three persons, nineteen of whom were Indians. The morning was cloudy and cool, and a fresh wind blew from down river, against which we had to struggle with all the force of our paddles, aided by the current; the boats were tossed about most disagreeably, and shipped a great deal of water. On passing the lower end of Shimuní, a long reach of the river was before us, undivided by islands; a magnificent expanse of water stretching away to the south-east. The country on the left bank is not, however, terra firma, but a portion of the alluvial land which forms the extensive and complex delta region of the Japurá. It is flooded every year at the time of high water, and is traversed by many narrow and deep channels which serve as outlets to the Japurá, or, at least, are connected with that river by means of the interior water-system of the Cupiyó. This inhospitable tract of country extends for several hundred miles, and contains in its midst an endless number of pools and lakes tenanted by multitudes of turtles, fishes, alligators, and water serpents. Our destination was a point on this coast situated about twenty miles below Shimuní, and a short distance from the mouth of the Ananá, one of the channels just alluded to as connected with the Japurá. After travelling for three hours in mid-stream we steered for the land and brought to under a steeply-inclined bank of crumbly earth, shaped into a succession of steps or terraces, marking the various halts which the waters of the river make in the course of subsidence. The coast line was nearly straight for many miles, and the bank averaged about thirty feet in height above the present level of the river: at the top rose the unbroken hedge of forest. No one could have divined that pools of water existed on that elevated land. A narrow level space extended at the foot of the bank. On landing the first business was to get breakfast. Whilst a couple of Indian lads were employed in making the fire, roasting the fish, and boiling the coffee, the rest of the party mounted the bank, and with their long hunting-knives commenced cutting a path through the forest ; the pool, called the Aningal, being about half a mile distant. After breakfast a great number of short poles were cut and laid crosswise on the path, and then three light montarias which we had brought with us were dragged up the bank by lianas, and rolled away to be embarked on the pool. A large net, seventy yards in length, was then disembarked and carried to the place. The work was done very speedily, and when Cardozo and I went to the spot at eleven o'clock we found some of the older Indians, including Pedro and Daniel, had begun their sport. They were mounted on little stages called moutás, made of poles and cross-pieces of wood secured with lianas, and were shooting the turtles, as they came near the surface, with bows and arrows. The Indians seemed to think that netting the animals, as Cardozo proposed doing, was not lawful sport, and wished first to have an hour or two's old-fashioned practice with their weapons.

The pool covered an area of about four or five acres, and was closely hemmed in by the forest, which in picturesque variety and grouping of trees and foliage exceeded almost everything I had yet witnessed. The margins for some distance were swampy, and covered with large tufts of a fine grass called Matupá. These tufts in many places were overrun with ferns, and exterior to them a crowded row of arborescent arums, growing to a height of fifteen or twenty feet, formed a

s 2

green palisade. Around the whole stood the taller forest trees; palmate-leaved Cecropiæ; slender Assai palms, thirty feet high, with their thin feathery heads crowning the gently-curving, smooth stems; small fan-leaved palms; and as a back-ground to all these airy shapes, lay the voluminous masses of ordinary forest trees, with garlands, festoons, and streamers of leafy climbers hanging from their branches. The pool was nowhere more than five feet deep, one foot of which was not water, but extremely fine and soft mud.

Cardozo and I spent an hour paddling about. I was astonished at the skill which the Indians display in shooting turtles. They did not wait for their coming to the surface to breathe, but watched for the slight movements in the water, which revealed their presence underneath. These little tracks on the water are called the Sirirí; the instant one was perceived an arrow flew from the bow of the nearest man, and never failed to pierce the shell of the submerged animal. When the turtle was very distant, of course the aim had to be taken at a considerable elevation, but the marksmen preferred a longish range, because the arrow then fell more perpendicularly on the shell, and entered it more deeply.

The arrow used in turtle shooting has a strong lancet-shaped steel point, fitted into a peg which enters the tip of the shaft. The peg is secured to the shaft by twine made of the fibres of pine-apple leaves, the twine being some thirty or forty yards in length, and neatly wound round the body of the arrow. When the missile enters the shell the peg drops out, and the pierced animal descends with it towards the bottom, leaving the shaft floating on the surface. This being done the sportsman paddles in his montaria to the place, and gently draws the animal by the twine, humouring it by giving it the rein when it plunges, until it is brought again near the surface, when he strikes it with a second arrow. With the increased hold given by the two cords he has then no difficulty in landing his game.

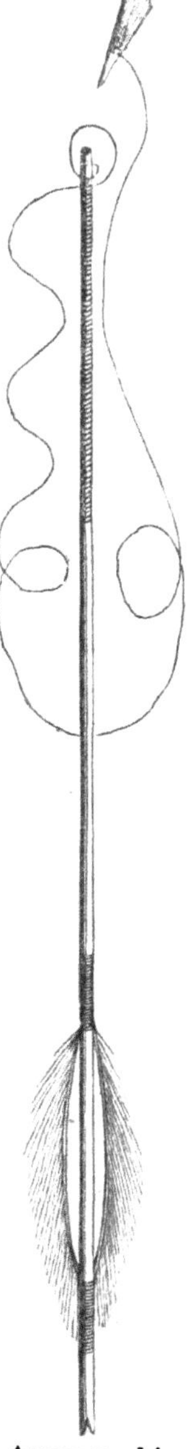

Arrow used in turtle shooting.

By mid-day the men had shot about a score of nearly full-grown turtles. Cardozo then gave orders to spread the net. The spongy, swampy nature of the banks made it impossible to work the net so as to draw the booty ashore; another method was therefore adopted. The net was taken by two Indians and extended in a curve at one extremity of the oval-shaped pool, holding it when they had done so by the perpendicular rods fixed at each end; its breadth was about equal to the depth of the water, its shotted side therefore rested on the bottom, whilst the floats buoyed it up on the surface, so that the whole, when the ends were brought together, would form a complete trap. The rest of the party then spread themselves around the swamp at the opposite end of the pool and began to beat, with stout poles, the thick tufts of Matupá, in order to drive the turtles towards the middle. This was continued for an hour or more, the beaters gradually drawing nearer to each other, and driving the host of animals before them; the number of little snouts constantly popping above the surface of the water showing that all was going on well. When they neared the net the men moved more quickly, shouting and beating with great vigour. The ends of the net were then seized by several strong hands and dragged suddenly forwards, bringing them at the same time together, so as to enclose all the booty in a circle. Every man now leapt into the enclosure, the boats were brought up, and the turtles easily captured by the hand and tossed into them. I jumped in along with the rest, although I had just before made the discovery that the pool abounded in ugly, red, four-angled leeches, having seen several of these delectable animals, which sometimes fasten on the legs of fishermen, although they did not, on this day, trouble us, working their way through cracks in the bottom of our montaria. Cardozo, who remained with the boats, could not turn the animals on their backs fast enough, so that a great many clambered out and got free again. However, three boatloads, or about eighty, were secured in about twenty minutes. They were then taken ashore, and each one secured by the men tying the legs with thongs of bast.

When the canoes had been twice filled, we desisted, after a very hard day's work. Nearly all the animals were young ones, chiefly, according to the statement of Pedro, from three to ten years of age; they varied from six to eighteen inches in length, and were very fat. Cardozo and I lived almost exclusively on them for several months afterwards. Roasted in the shell they form a most appetizing dish. These younger turtles never migrate with their elders on the sinking of the waters, but remain in the tepid pools, fattening on fallen fruits, and, according to the natives, on the fine nutritious mud. We captured a few full-grown mother-turtles, which were known at once by the horny skin of their breast-plates being worn, telling of their having crawled on the sands to lay eggs the previous year. They had evidently made a mistake in not leaving the pool at the proper time, for they were full of eggs, which, we were told, they would, before the season was over, scatter in despair over the swamp. We also found several male turtles, or Capitarís, as they are called by the natives. These are immensely less numerous than the females, and are distinguishable by their much smaller size, more circular shape, and the greater length and thickness of their tails. Their flesh is considered unwholesome, especially to sick people having external signs of inflammation. All diseases in these parts, as well as their remedies and all articles of food, are classed by the inhabitants as "hot" and "cold," and the meat of the Capitarí is settled by unanimous consent as belonging to the "hot" list.

We dined on the banks of the river, a little before sunset. The mosquitoes then began to be troublesome, and finding it would be impossible to sleep here, we all embarked and crossed the river to a sand-bank, about three miles distant, where we passed the night. Cardozo and I slept in our hammocks slung between upright poles, the rest stretching themselves on the sand round a large fire. We lay awake conversing until past midnight. It was a real pleasure to listen to the stories told by one of the older men, they were given with so much spirit. The tales always related to struggles with some intractable animal—jaguar, manatee, or alligator. Many interjections and expressive gestures were used, and at the end came a sudden "Pa! terra!" when the animal was vanquished by a shot or a blow. Many mysterious tales were recounted about the Bouto, as the large Dolphin of the Amazons is called. One of them was to the effect that a Bouto once had the habit of assuming the shape of a beautiful woman, with hair hanging loose to her heels, and walking ashore at night in the streets of Ega, to entice the young men down to the water. If any one was so much smitten as to follow her to the water-side, she grasped her victim round the waist and plunged beneath the waves with a triumphant cry. No animal in the Amazons region is the subject of so many fables as the Bouto; but it is probable these did not originate with the Indians but with the Portuguese colonists. It was several years before I could induce a fisherman to harpoon Dolphins for me as specimens, for no one ever kills these animals voluntarily, although their fat is known to yield an excellent oil for lamps. The superstitious people believe that blindness would result from the use of this oil in lamps. I succeeded at length with Carepira, by offering him a high reward when his finances were at a very low point; but he repented of his deed ever afterwards, declaring that his luck had forsaken him from that day.

TURTLE-FISHING AND ADVENTURE WITH ALLIGATOR.

Vol. II., page 265.

The next day we again beat the pool. Although we had proof of there being a great number of turtles yet remaining, we had very poor success. The old Indians told us it would be so, for the turtles were "ladino" (cunning), and would take no notice of the beating a second day. When the net was formed into a circle, and the men had jumped in, an alligator was found to be inclosed. No one was alarmed, the only fear expressed being that the imprisoned beast would tear the net. First one shouted, "I have touched his head;" then another, "he has scratched my leg;" one of the men, a lanky Mirânha, was thrown off his balance, and then there was no end to the laughter and shouting. At last a youth of about fourteen years of age, on my calling to him, from the bank, to do so, seized the reptile by the tail, and held him tightly until, a little resistance being overcome, he was able to bring it ashore. The net was opened, and the boy slowly dragged the dangerous but cowardly beast to land through the muddy water, a distance of about a hundred yards. Meantime, I had cut a strong pole from a tree, and as soon as the alligator was drawn to solid ground, gave him a smart rap with it on the crown of his head, which killed him instantly. It was a good-sized individual; the jaws being considerably more than a foot long, and fully capable of snapping a man's leg in twain. The species was the large cayman, the Jacaré-uassú of the Amazonian Indians (Jacare nigra).

On the third day we sent our men in the boats to net turtles in another larger pool, about five miles further down the river, and on the fourth returned to Ega.

It will be well to mention here a few circumstances relative to the large Cayman, which, with the incident just narrated, afford illustrations of the cunning, cowardice and ferocity of this reptile.

I have hitherto had but few occasions of mentioning alligators, although they exist by myriads in the waters of the Upper Amazons. Many different species are spoken of by the natives. I saw only three, and of these two only are common: one, the Jacaré-tinga, a small kind (five feet long when full grown) having a long slender muzzle and a black-banded tail; the other, the Jacaré-uassú, to which these remarks more especially relate; and the third the Jacaré-curúa, mentioned in a former chapter. The Jacaré-uassú, or large Cayman, grows to a length of eighteen or twenty feet, and attains an enormous bulk. Like the turtles, the alligator has its annual migrations, for it retreats to the interior pools and flooded forests in the wet season, and descends to the main river in the dry season. During the months of high water, therefore, scarcely a single individual is to be seen in the main river. In the middle part of the Lower Amazons, about Obydos and Villa Nova, where many of the lakes with their channels of communication with the trunk stream, dry up in the fine months, the alligator buries itself in the mud and becomes dormant, sleeping till the rainy season returns. On the Upper Amazons, where the dry season is never excessive, it has not this habit, but is lively all the year round. It is scarcely exaggerating to say that the waters of the Solimoens are as well-stocked with large alligators in the dry season, as a ditch in England is in summer with tadpoles. During a journey of five days which I once made in the Upper Amazons steamer, in November, alligators were seen along the coast almost every step of the way, and the passengers amused themselves, from morning till night, by firing at them with rifle and ball. They were very numerous in the still bays, where the huddled crowds jostled together, to the great rattling of their coats of mail, as the steamer passed.

The natives at once despise and fear the great cayman. I once spent a month at Caiçara, a small village of semi-civilised Indians, about twenty miles to the west of Ega. My entertainer, the only white in the place, and one of my best and most constant friends, Senhor Innocencio Alves Faria, one day proposed a half-day's fishing with net in the lake,—the expanded bed of the small river on which the village is situated. We set out in an open boat with six Indians and two of Innocencio's children. The water had sunk so low that the net had to be taken out into the middle by the Indians, whence at the first draught, two medium-sized alligators were brought to land. They were disengaged from the net and allowed, with the coolest unconcern, to return to the water, although the two children were playing in it, not many yards off. We continued fishing, Innocencio and I lending a helping hand, and each time drew a number of the reptiles of different ages and sizes, some of them Jacaré-tingas; the lake in fact, swarmed with alligators. After taking a very large quantity of fish (I took pains to count the different species, and found there were no less than thirty-five), we prepared to return, and the Indians, at my suggestion, secured one of the alligators with the view of letting it loose amongst the swarms of dogs in the village. An individual was selected about eight feet long: one man holding his head and another his tail, whilst a third took a few lengths of a flexible liana, and deliberately bound the jaws and the legs. Thus secured, the beast was laid across the benches of the boat, on which we sat during the hour and a half's journey to the settlement. We were rather crowded, but our amiable passenger gave us no trouble during the transit. On reaching the village, we took the animal into the middle of the green, in front of the church, where the dogs were congregated, and there gave him his liberty, two of us arming ourselves with long poles to intercept him if he should make for the water, and the others exciting the dogs. The alligator showed great terror, although the dogs could not be made to advance, and made off at the top of its speed for the water, waddling like a duck. We tried to keep him back with the poles, but he became enraged, and seizing the end of the one I held, in his jaws, nearly wrenched it from my grasp. We were obliged, at length, to kill him to prevent his escape.

These little incidents show the timidity or cowardice of the alligator. He never attacks man when his intended victim is on his guard: but he is cunning enough to know when he may do this with impunity: of this we had proof at Caiçara, a few days afterwards. The river had sunk to a very low point, so that the port and bathing-place of the village now lay at the foot of a long sloping bank, and a large cayman made his appearance in the shallow and muddy water. We were all obliged to be very careful in taking our bath ; most of the people simply using a calabash, pouring the water over themselves whilst standing on the brink. A large trading canoe, belonging to a Barra merchant named Soares, arrived at this time, and the Indian crew, as usual, spent the first day or two after their coming in port, in drunkenness and debauchery ashore. One of the men, during the greatest heat of the day when almost every one was enjoying his afternoon's nap, took it into his head whilst in a tipsy state to go down alone to bathe. He was seen only by the Juiz de Paz, a feeble old man who was lying in his hammock, in the open verandah at the rear of his house on the top of the bank, and who shouted to the besotted Indian to beware of the alligator. Before he could repeat his warning, the man stumbled, and a pair of gaping jaws, appearing suddenly above the surface, seized him round the waist and drew him under the water. A cry of agony "Ai Jesús!" was the last sign made by the wretched victim. The village was aroused: the young men with praiseworthy readiness seized their harpoons and hurried down to the bank; but of course it was too late, a winding track of blood on the surface of the water, was all that could be seen. They embarked, however, in montarias, determined on vengeance: the monster was traced, and when, after a short lapse of time, he came up to breathe—one leg of the man sticking out from his jaws—was dispatched with bitter curses.

The last of these minor excursions which I shall

narrate, was made (again in company of Senhor Cardozo, with the addition of his housekeeper Senhora Felippa), in the season when all the population of the villages turns out to dig up turtle eggs, and revel on the praias. Placards were posted on the church doors at Ega, announcing that the excavation on Shimuní would commence on the 17th of October, and on Catuá, sixty miles below Shimuní, on the 25th. We set out on the 16th, and passed on the road, in our well-manned igarité, a large number of people, men, women, and children in canoes of all sizes, wending their way as if to a great holiday gathering. By the morning of the 17th, some 400 persons were assembled on the borders of the sandbank; each family having erected a rude temporary shed of poles and palm leaves to protect themselves from the sun and rain. Large copper kettles to prepare the oil, and hundreds of red earthenware jars, were scattered about on the sand.

The excavation of the taboleiro, collecting the eggs and purifying the oil, occupied four days. All was done on a system established by the old Portuguese governors, probably more than a century ago. The commandante first took down the names of all the masters of households, with the number of persons each intended to employ in digging; he then exacted a payment of 140 reis (about fourpence) a head, towards defraying the expense of sentinels. The whole were then allowed to go to the taboleiro. They ranged themselves round the circle, each person armed with a paddle, to be used as a spade, and then all began simultaneously to dig on a signal being given—the roll of drums—by order of the commandante. It was an animating sight to behold the wide circle of rival diggers throwing up clouds of sand in their energetic labours, and working gradually towards the centre of the ring. A little rest was taken during the great heat of mid-day, and in the evening the eggs were carried to the huts in baskets. By the end of the second day, the taboleiro was exhausted: large mounds of eggs, some of them four to five feet in height, were then seen by the side of each hut, the produce of the labours of the family.

In the hurry of digging some of the deeper nests are passed over; to find these out the people go about provided with a long steel or wooden probe, the presence of the eggs being discoverable by the ease with which the spit enters the sand. When no more eggs are to be found, the mashing process begins. The egg, it may be here mentioned, has a flexible or leathery shell; it is quite round, and somewhat larger than a hen's egg. The whole heap is thrown into an empty canoe and mashed with wooden prongs; but sometimes naked Indians and children jump into the mass and tread it down, besmearing themselves with yolk and making about as filthy a scene as can well be imagined. This being finished, water is poured into the canoe, and the fatty mess then left for a few hours to be heated by the sun, on which the oil separates and rises to the surface. The floating oil is afterwards skimmed off with long spoons, made by tying large mussel-shells to the end of rods, and purified over the fire in copper kettles.

The destruction of turtle eggs every year by these

proceedings is enormous. At least 6000 jars, holding each three gallons of the oil, are exported annually from the Upper Amazons and the Madeira to Pará, where it is used for lighting, frying fish, and other purposes. It may be fairly estimated that 2000 more jars-full are consumed by the inhabitants of the villages on the river. Now, it takes at least twelve basketsfull of eggs, or about 6000, by the wasteful process followed, to make one jar of oil. The total number of eggs annually destroyed amounts, therefore, to 48,000,000. As each turtle lays about 120, it follows that the yearly offspring of 400,000 turtles is thus annihilated. A vast number, nevertheless, remain undetected; and these would probably be sufficient to keep the turtle population of these rivers up to the mark, if the people did not follow the wasteful practice of lying in wait for the newly-hatched young, and collecting them by thousands for eating; their tender flesh and the remains of yolk in their entrails being considered a great delicacy. The chief natural enemies of the turtle are vultures and alligators, which devour the newly-hatched young as they descend in shoals to the water. These must have destroyed an immensely greater number before the European settlers began to appropriate the eggs than they do now. It is almost doubtful if this natural persecution did not act as effectively in checking the increase of the turtle as the artificial destruction now does. If we are to believe the tradition of the Indians, however, it had not this result; for they say that formerly the waters teemed as thickly with turtles as the air now does with mosquitoes. The universal opinion of the settlers on the Upper

Amazons is, that the turtle has very greatly decreased in numbers, and is still annually decreasing.

We left Shimuní on the 20th with quite a flotilla of canoes, and descended the river to Catuá, an eleven hours' journey by paddle and current. Catuá is about six miles long, and almost entirely encircled by its praia. The turtles had selected for their egg-laying a part of the sandbank which was elevated at least twenty feet above the present level of the river; the animals, to reach the place, must have crawled up a slope. As we approached the island, numbers of the animals were seen coming to the surface to breathe, in a small shoaly bay. Those who had light montarias sped forward with bows and arrows to shoot them. Carepíra was foremost: having borrowed a small and very unsteady boat of Cardozo, and embarked in it with his little son. After bagging a couple of turtles, and whilst hauling in a third, he overbalanced himself: the canoe went over, and he with his child had to swim for their lives, in the midst of numerous alligators, about a mile from the land. The old man had to sustain a heavy fire of jokes from his companions for several days after this mishap. Such accidents are only laughed at by this almost amphibious people.

The number of persons congregated on Catuá was much greater than on Shimuní, as the population of the banks of several neighbouring lakes was here added. The line of huts and sheds extended half a mile, and several large sailing vessels were anchored at the place. The commandant was Senhor Macedo, the Indian blacksmith of Ega before mentioned, who maintained excellent order during the fourteen days the process of excavation and oil manufacture lasted. There were also many primitive Indians here from the neighbouring rivers, amongst them a family of Shumánas, good-tempered, harmless people from the Lower Japurá. All of them were tattooed round the mouth, the blueish tint forming a border to the lips, and extending in a line on the cheeks towards the ear on each side. They were not quite so slender in figure as the Passés of Pedro-uassú's family; but their features deviated quite as much as those of the Passés from the ordinary Indian type. This was seen chiefly in the comparatively small mouth, pointed chin, thin lips, and narrow, high nose. One of the daughters, a young girl of about seventeen years of age, was a real beauty. The colour of her skin approached the light tanned shade of the Mameluco women; her figure was almost faultless, and the blue mouth, instead of being a disfigurement, gave quite a captivating finish to her appearance. Her neck, wrists, and ankles were adorned with strings of blue beads. She was, however, extremely bashful, never venturing to look strangers in the face, and never quitting, for many minutes together, the side of her father and mother. The family had been shamefully swindled by some rascally trader on another praia; and, on our arrival, came to lay their case before Senhor Cardozo, as the delegado of police of the district. The mild way in which the old man, without a trace of anger, stated his complaint in imperfect Tupí, quite enlisted our sympathies in his favour. But Cardozo could give him no redress; he invited the family, however, to make their rancho near to ours, and in the end gave them the highest price for the surplus oil which they manufactured.

It was not all work at Catuá; indeed there was rather more play than work going on. The people make a kind of holiday of these occasions. Every fine night parties of the younger people assembled on the sands, and dancing and games were carried on for hours together. But the requisite liveliness for these sports was never got up without a good deal of preliminary rum-drinking. The girls were so coy that the young men could not get sufficient partners for the dances, without first subscribing for a few flagons of the needful cashaça. The coldness of the shy Indian and Mameluco maidens never failed to give way after a little of this strong drink, but it was astonishing what an immense deal they could take of it in the course of an evening. Coyness is not always a sign of innocence in these people, for most of the half-caste women on the Upper Amazons lead a little career of looseness before they marry and settle down for life; and it is rather remarkable that the men do not seem to object much to their brides having had a child or two by various fathers before marriage. The women do not lose reputation unless they become utterly depraved, but in that case they are condemned pretty strongly by public opinion. Depravity is, however, rare, for all require more or less to be wooed before they are won. I did not see (although I mixed pretty freely with the young people) any breach of propriety on the praias. The merry-

T 2

makings were carried on near the ranchos, where the more staid citizens of Ega, husbands with their wives and young daughters, all smoking gravely out of long pipes, sat in their hammocks and enjoyed the fun. Towards midnight we often heard, in the intervals between jokes and laughter, the hoarse roar of jaguars prowling about the jungle in the middle of the praia. There were several guitar-players amongst the young men, and one most persevering fiddler, so there was no lack of music.

The favourite sport was the Pira-purasséya, or fish-dance, one of the original games of the Indians, though now probably a little modified. The young men and women, mingling together, formed a ring, leaving one of their number in the middle, who represented the fish. They then all marched round, Indian file, the musicians mixed up with the rest, singing a monotonous but rather pretty chorus, the words of which were invented (under a certain form) by one of the party who acted as leader. This finished, all joined hands, and questions were put to the one in the middle, asking what kind of fish he or she might be. To these the individual has to reply. The end of it all is that he makes a rush at the ring, and if he succeeds in escaping, the person who allowed him to do so has to take his place; the march and chorus then recommence, and so the game goes on hour after hour. Tupí was the language mostly used, but sometimes Portuguese was sung and spoken. The details of the dance were often varied. Instead of the names of fishes being called over by the person in the middle, the name of some animal, flower, or other object was given to every fresh occupier of the place. There was then good scope for wit in the invention of nick-names, and peals of laughter would often salute some particularly good hit. Thus a very lanky young man was called the Magoary, or the gray stork; a moist gray-eyed man with a profile comically suggestive of a fish was christened Jarakí (a kind of fish), which was considered quite a witty sally; a little Mameluco girl, with light-coloured eyes and brown hair, got the gallant name of Rosa branca, or the white rose; a young fellow who had recently singed his eyebrows by the explosion of fireworks was dubbed Pedro queimado (burnt Peter); in short every one got a nickname, and each time the cognomen was introduced into the chorus as the circle marched round.

It is said by the Portuguese and Brazilian towns-people lower down the river, that much disorder and all kinds of immorality prevail amongst these assemblages of Upper Amazons rustics on the turtle praias. I can only say that nothing of the kind was seen on the occasions when I attended. But it may be added that there were no traders from the "civilised" parts present to set a bad example. Town-bred Indians and half-castes will be disorderly and quarrelsome, like uneducated people everywhere, when they can get their fill of in-toxicating drinks. When low Portuguese traders, who are most certainly the inferiors of these rustics whom they despise, attend the praias, they corrupt the women, and bribe the Indians with cashaça to steal their masters' oil; these proceedings, of course, give rise to disturbances in many ways. There were none of these shining examples of the superior civilisation of Europe in attendance at Catuá. The masters kept their Indians well under control; the young people enjoyed themselves upon the whole innocently, and sociability was pretty general amongst all classes and colours.

Our rancho was a large one, and was erected in a line with the others, near the edge of the sandbank which sloped rather abruptly to the water. During the first week the people were all, more or less, troubled by alligators. Some half-dozen full-grown ones were in attendance off the praia, floating about on the lazily-flowing, muddy water. The dryness of the weather had increased since we had left Shimuní, the currents had slackened, and the heat in the middle part of the day was almost insupportable. But no one could descend to bathe without being advanced upon by one or other of these hungry monsters. There was much offal cast into the river, and this, of course, attracted them to the place. One day I amused myself by taking a basketful of fragments of meat beyond the line of ranchos, and drawing the alligators towards me by feeding them. They behaved pretty much as dogs do when fed; catching the bones I threw them in their huge jaws, and coming nearer and showing increased eagerness after every morsel. The enormous gape of their mouths, with their blood-red lining and long fringes of teeth, and the uncouth shapes of their bodies, made a picture of unsurpassable ugliness. I once or twice fired a heavy charge of shot at them, aiming at the vulnerable part of their bodies, which is a small space situated behind the eyes, but this had no other effect than to make them

NIGHT ADVENTURE WITH ALLIGATOR.

Vol. II., page 279.

give a hoarse grunt and shake themselves; they immediately afterwards turned to receive another bone which I threw to them.

Every day these visitors became bolder; at length they reached a pitch of impudence that was quite intolerable. Cardozo had a poodle dog named Carlito, which some grateful traveller whom he had befriended had sent him from Rio Janeiro. He took great pride in this dog, keeping it well sheared, and preserving his coat as white as soap and water could make it. We slept in our rancho in hammocks slung between the outer posts; a large wood fire (fed with a kind of wood abundant on the banks of the river, which keeps alight all night) being made in the middle, by the side of which slept Carlito on a little mat. Well, one night I was awoke by a great uproar. It was caused by Cardozo hurling burning firewood with loud curses at a huge cayman which had crawled up the bank and passed beneath my hammock (being nearest the water) towards the place where Carlito lay. The dog had raised the alarm in time; the reptile backed out and tumbled down the bank to the water, the sparks from the brands hurled at him flying from his bony hide. To our great surprise the animal (we supposed it to be the same individual) repeated his visit the very next night, this time passing round to the other side of our shed. Cardozo was awake, and threw a harpoon at him, but without doing him any harm. After this it was thought necessary to make an effort to check the alligators; a number of men were therefore persuaded to sally forth in their montarias and devote a day to killing them.

The young men made several hunting excursions during the fourteen days of our stay on Catuá, and I, being associated with them in all their pleasures, made generally one of the party. These were, besides, the sole occasions on which I could add to my collections, whilst on these barren sands. Only two of these trips afforded incidents worth relating.

The first, which was made to the interior of the wooded island of Catuá, was not a very successful one. We were twelve in number, all armed with guns and long hunting-knives. Long before sunrise, my friends woke me up from my hammock, where I lay, as usual, in the clothes worn during the day; and after taking each a cup-full of cashaça and ginger (a very general practice in early morning on the sandbanks), we commenced our walk. The waning moon still lingered in the clear sky, and a profound stillness pervaded sleeping camp, forest, and stream. Along the line of ranchos glimmered the fires made by each party to dry turtle-eggs for food, the eggs being spread on little wooden stages over the smoke. The distance to the forest from our place of starting was about two miles, being nearly the whole length of the sandbank, which was also a very broad one; the highest part, where it was covered with a thicket of dwarf willows, mimosas, and arrow grass, lying near the ranchos. We loitered much on the way, and the day dawned whilst we were yet on the road: the sand at this early hour feeling quite cold to the naked feet. As soon as we were able to distinguish things, the surface of the praia was seen to be dotted with small black objects. These were newly-hatched

Aiyussá turtles, which were making their way in an undeviating line to the water, at least a mile distant. The young animal of this species is distinguishable from that of the large turtle and the Tracajá, by the edges of the breast-plate being raised on each side, so that in crawling it scores two parallel lines on the sand. The mouths of these little creatures were full of sand, a circumstance arising from their having to bite their way through many inches of superincumbent sand to reach the surface on emerging from the buried eggs. It was amusing to observe how constantly they turned again in the direction of the distant river, after being handled and set down on the sand with their heads facing the opposite quarter. We saw also several skeletons of the large cayman (some with the horny and bony hide of the animal nearly perfect) embedded in the sand : they reminded me of the remains of Ichthyosauri fossilized in beds of lias, with the difference of being buried in fine sand instead of in blue mud. I marked the place of one which had a well-preserved skull, and the next day returned to secure it. The specimen is now in the British Museum collection. There were also many foot-marks of Jaguars on the sand.

We entered the forest, as the sun peeped over the tree-tops far away down river. The party soon after divided ; I keeping with a section which was led by Bento, the Ega carpenter, a capital woodsman. After a short walk we struck the banks of a beautiful little lake, having grassy margins and clear dark water, on the surface of which floated thick beds of water-lilies. We then crossed a muddy creek or watercourse that entered the lake, and then found ourselves on a *restinga*, or tongue of land between two waters. By keeping in sight of one or the other of these there was no danger of our losing our way: all other precautions were therefore unnecessary. The forest was tolerably clear of underwood, and consequently easy to walk through. We had not gone far before a soft, long-drawn whistle was heard aloft in the trees, betraying the presence of Mutums (Curassow birds). The crowns of the trees, a hundred feet or more over our heads, were so closely interwoven, that it was difficult to distinguish the birds: the practised eye of Bento, however, made them out, and a fine male was shot from the flock; the rest flying away and alighting at no great distance: the species was the one of which the male has a round red ball on its beak (Crax globicera). The pursuit of the others led us a great distance, straight towards the interior of the island, in which direction we marched for three hours, having the lake always on our right.

Arriving at length at the head of the lake, Bento struck off to the left across the restinga, and we then soon came upon a treeless space choked up with tall grass, which appeared to be the dried-up bed of another lake. Our leader was obliged to climb a tree to ascertain our position, and found that the clear space was part of the creek, whose mouth we had crossed lower down. The banks were clothed with low trees, nearly all of one species, a kind of araça (Psidium), and the ground was carpeted with a slender delicate grass, now in flower. A great number of crimson and vermilion-coloured butterflies (Catagramma

Peristera, male and female) were settled on the smooth, white trunks of these trees. I had also here the great pleasure of seeing for the first time, the rare and curious

Umbrella Bird.

Umbrella Bird (Cephalopterus ornatus), a species which resembles in size, colour, and appearance our common crow, but is decorated with a crest of long, curved, hairy feathers having long bare quills, which, when raised, spread themselves out in the form of a fringed sun-shade over the head. A strange ornament, like a pelerine, is also suspended from the neck, formed by a thick pad of glossy steel-blue feathers, which grow on a long fleshy lobe or excrescence. This lobe is connected (as I found on skinning specimens) with an unusual development of the trachea and vocal organs, to which the bird doubtless owes its singularly deep, loud, and long-sustained fluty note. The Indian name of this strange creature is Uirá-mimbéu, or fife-bird,* in allusion to the tone of its voice. We had the good luck, after remaining quiet a short time, to hear its performance. It drew itself up on its perch, spread widely the umbrella-formed crest, dilated and waved its glossy breast-lappet, and then, in giving vent to its loud piping note, bowed its head slowly forwards. We obtained a pair, male and female: the female has only the rudiments of the crest and lappet, and is duller-coloured altogether than the male. The range of this bird appears to be quite confined to the plains of the Upper Amazons (especially the Ygapó forests), not having been found to the east of the Rio Negro.

Bento and our other friends being disappointed in finding no more Curassows, or indeed any other species of game, now resolved to turn back. On reaching the edge of the forest we sat down and ate our dinners under the shade; each man having brought a little bag containing a few handsfull of farinha, and a piece of fried fish or roast turtle. We expected our companions of the other division to join us at mid-day, but after waiting till past one o'clock without seeing anything of them (in fact, they had returned to the huts an hour or two

* Mimbéu is the Indian name for a rude kind of pan-pipes used by the Caishánas and other tribes.

previously), we struck off across the praia towards the encampment. An obstacle here presented itself on which we had not counted. The sun had shone all day through a cloudless sky untempered by a breath of wind, and the sands had become heated by it to a degree that rendered walking over them with our bare feet impossible. The most hardened footsoles of the party could not endure the burning soil. We made several attempts ; we tried running : wrapped the cool leaves of Heliconiæ round our feet, but in no way could we step forward many yards. There was no means of getting back to our friends before night, except going round the praia, a circuit of about four miles, and walking through the water or on the moist sand. To get to the waterside from the place where we then stood was not difficult, as a thick bed of a flowering shrub, called tintarána, an infusion of the leaves of which is used to dye black, lay on that side of the sand-bank. Footsore and wearied, burthened with our guns, and walking for miles through the tepid shallow water under the brain-scorching vertical sun, we had, as may be imagined, anything but a pleasant time of it. I did not, however, feel any inconvenience afterwards. Every one enjoys the most lusty health whilst living this free and wild life on the rivers.

The other hunting trip which I have alluded to was undertaken in company with three friendly young half-castes. Two of them were brothers, namely, Joaõ (John) and Zephyrino Jabutí : Jabutí, or tortoise, being a nickname which their father had earned for his slow gait, and which, as is usual in this country, had descended as the surname of the family. The other was José Frazaõ, a nephew of Senhor Chrysostomo, of Ega, an active, clever, and manly young fellow whom I much esteemed. He was almost a white, his father being a Portuguese and his mother a Mameluco. We were accompanied by an Indian named Lino, and a Mulatto boy, whose office was to carry our game.

Our proposed hunting-ground on this occasion lay across the water, about fifteen miles distant. We set out in a small montaria, at four o'clock in the morning, again leaving the encampment asleep, and travelled at a good pace up the northern channel of the Solimoens, or that lying between the island Catuá and the left bank of the river. The northern shore of the island had a broad sandy beach reaching to its western extremity. We reached our destination a little after day-break ; this was the banks of the Carapanatúba,* a channel some 150 yards in width, which, like the Ananá already mentioned, communicates with the Cupiyó. To reach this we had to cross the river, here nearly two miles wide. Just as day dawned we saw a Cayman seize a large fish, a Tambakí, near the surface ; the reptile seemed to have a difficulty in securing its prey, for it reared itself above the water, tossing the fish in its jaws and making a tremendous commotion. I was much struck also by the singular appearance presented by certain diving birds having very long and snaky necks (the Plotus Anhinga). Occasionally a long serpentine form would suddenly wriggle itself to a height of a

* Meaning in Tupí, the river of many mosquitoes : from carapaná, mosquito, and itúba, many.

foot and a half above the glassy surface of the water, producing such a deceptive imitation of a snake that at first I had some difficulty in believing it to be the neck of a bird; it did not remain long in view, but soon plunged again beneath the stream.

We ran ashore in a most lonely and gloomy place, on a low sandbank covered with bushes, secured the montaria to a tree, and then, after making a very sparing breakfast on fried fish and mandioca meal, rolled up our trousers and plunged into the thick forest, which here, as everywhere else, rose like a lofty wall of foliage from the narrow strip of beach. We made straight for the heart of the land, John Jabutí leading, and breaking off at every few steps a branch of the lower trees, so that we might recognise the path on our return. The district was quite new to all my companions, and being on a coast almost totally uninhabited by human beings for some 300 miles, to lose our way would have been to perish helplessly. I did not think at the time of the risk we ran of having our canoe stolen by passing Indians; unguarded montarias being never safe even in the ports of the villages, Indians apparently considering them common property, and stealing them without any compunction. No misgivings clouded the lightness of heart with which we trod forwards in warm anticipation of a good day's sport.

The tract of forest through which we passed was Ygapó, but the higher parts of the land formed areas which went only a very few inches under water in the flood season. It consisted of a most bewildering diversity of grand and beautiful trees, draped, festooned, corded, matted, and ribboned with climbing plants, woody and succulent, in endless variety. The most prevalent palm was the tall Astryocaryum Jauarí, whose fallen spines made it necessary to pick our way carefully over the ground, as we were all barefoot. There was not much green underwood, except in places where Bamboos grew; these formed impenetrable thickets of plumy foliage and thorny, jointed stems, which always compelled us to make a circuit to avoid them. The earth elsewhere was encumbered with rotting fruits, gigantic bean-pods, leaves, limbs, and trunks of trees, fixing the impression of its being the cemetery as well as the birthplace of the great world of vegetation overhead. Some of the trees were of prodigious height. We passed many specimens of the Moratinga, whose cylindrical trunks, I dare not say how many feet in circumference, towered up and were lost amidst the crowns of the lower trees, their lower branches, in some cases, being hidden from our view. Another very large and remarkable tree was the Assacú (Sapium aucuparium). A traveller on the Amazons, mingling with the people, is sure to hear much of the poisonous qualities of the juices of this tree. Its bark exudes, when hacked with a knife, a milky sap, which is not only a fatal poison when taken internally, but is said to cause incurable sores if simply sprinkled on the skin. My companions always gave the Assacú a wide berth when we passed one. The tree looks ugly enough to merit a bad name, for the bark is of a dingy olive colour, and is studded with short and sharp, venomous-looking spines.

After walking about half a mile we came upon a dry

water-course, where we observed, first, the old footmarks of a tapir, and, soon after, on the margins of a curious circular hole full of muddy water, the fresh tracks of a Jaguar. This latter discovery was hardly made, when a rush was heard amidst the bushes on the top of a sloping bank on the opposite side of the dried creek. We bounded forward; it was, however, too late, for the animal had sped in a few moments far out of our reach. It was clear we had disturbed, on our approach, the Jaguar, whilst quenching his thirst at the water-hole. A few steps further on we saw the mangled remains of an alligator (the Jacarétinga). The head, fore-quarters, and bony shell were the only parts which remained; but the meat was quite fresh, and there were many footmarks of the Jaguar around the carcase; so that there was no doubt this had formed the solid part of the animal's breakfast. My companions now began to search for the alligator's nest, the presence of the reptile so far from the river being accountable for on no other ground than its maternal solicitude for its eggs. We found, in fact, the nest at the distance of a few yards from the place. It was a conical pile of dead leaves, in the middle of which twenty eggs were buried. These were of elliptical shape, considerably larger than those of a duck, and having a hard shell of the texture of porcelain, but very rough on the outside. They make a loud sound when rubbed together, and it is said that it is easy to find a mother alligator in the Ygapó forests, by rubbing together two eggs in this way, she being never far off, and attracted by the sounds.

I put half-a-dozen of the alligator's eggs in my game-bag for specimens, and we then continued on our way. Lino, who was now first, presently made a start backwards, calling out "Jararáca!" This is the name of a poisonous snake (genus Craspedocephalus), which is far more dreaded by the natives than Jaguar or Alligator. The individual seen by Lino lay coiled up at the foot of a tree, and was scarcely distinguishable, on account of the colours of its body being assimilated to those of the fallen leaves. Its hideous, flat triangular head, connected with the body by a thin neck, was reared and turned towards us: Frazaõ killed it with a charge of shot, shattering it completely, and destroying, to my regret, its value as a specimen. In conversing on the subject of Jararácas as we walked onwards, every one of the party was ready to swear that this snake attacks man without provocation, leaping towards him from a considerable distance when he approaches. I met, in the course of my daily rambles in the woods, many Jararácas, and once or twice very narrowly escaped treading on them, but never saw them attempt to spring. On some subjects the testimony of the natives of a wild country is utterly worthless. The bite of the Jararácas is generally fatal. I knew of four or five instances of death from it, and only of one clear case of recovery after being bitten; but in that case the person was lamed for life.

We walked over moderately elevated and dry ground for about a mile, and then descended (three or four feet only) to the dry bed of another creek. This was pierced in the same way as the former water-course, with round holes full of muddy water. They occurred at intervals of a few yards, and had the appearance of having been made by the hand of man. The smallest were about two feet, the largest seven or eight feet in diameter. As we approached the most considerable of the larger ones, I was startled at seeing a number of large serpent-like heads bobbing above the surface. They proved to be those of electric eels, and it now occurred to me that these round holes were made by these animals working constantly round and round in the moist muddy soil. Their depth (some of them were at least eight feet deep) was doubtless due also to the movements of the eels in the soft soil, and accounted for their not drying up, in the fine season, with the rest of the creek. Thus, whilst alligators and turtles in this great inundated forest region retire to the larger pools during the dry season, the electric eels make for themselves little ponds in which to pass the season of drought.

My companions now cut each a stout pole, and proceeded to eject the eels in order to get at the other fishes, with which they had discovered the ponds to abound. I amused them all very much by showing how the electric shock from the eels could pass from one person to another. We joined hands in a line whilst I touched the biggest and freshest of the animals on the head with the point of my hunting-knife. We found that this experiment did not succeed more than three times with the same eel when out of the water: for, the fourth time, the shock was scarcely perceptible. All the fishes found in the holes (besides the eels) belonged to one species, a small kind of Acarí, or Loricaria, a group whose members have a complete bony integument. Lino and the boy

U 2

strung them together through the gills with slender sipós, and hung them on the trees to await our return later in the day.

Leaving the bed of the creek, we marched onwards, always towards the centre of the land ; guided by the sun, which now glimmered through the thick foliage overhead. About eleven o'clock we saw a break in the forest before us, and presently emerged on the banks of a considerable sheet of water. This was one of the interior pools of which there are so many in this district. The margins were elevated some few feet, and sloped down to the water, the ground being hard and dry to the water's edge, and covered with shrubby vegetation. We passed completely round this pool, finding the crowns of the trees on its borders tenanted by curassow birds, whose presence was betrayed as usual by the peculiar note which they emit. My companions shot two of them. At the farther end of the lake lay a deep watercourse, which we traced for about half a mile, and found to communicate with another and smaller pool. This second one evidently swarmed with turtles, as we saw the snouts of many peering above the surface of the water: the same had not been seen in the larger lake, probably because we had made too much noise in hailing our discovery, on approaching its banks. My friends made an arrangement on the spot for returning to this pool, after the termination of the egg harvest on Catuá.

In recrossing the space between the two pools, we heard the crash of monkeys in the crowns of trees overhead. The chace of these occupied us a considerable time. José fired at length at one of the laggards of the troop, and wounded him. He climbed pretty nimbly towards a denser part of the tree, and a second and third discharge failed to bring him down. The poor maimed creature then trailed his limbs to one of the topmost branches, where we descried him soon after, seated and picking the entrails from a wound in his abdomen; a most heart-rending sight. The height from the ground to the bough on which he was perched could not have been less than 150 feet, and we could get a glimpse of him only by standing directly underneath, and straining our eyes upwards. We killed him at last by loading our best gun with a careful charge, and resting the barrel against the tree-trunk to steady the aim. A few shots entered his chin, and he then fell heels over head screaming to the ground. Although it was I who gave the final shot, this animal did not fall to my lot in dividing the spoils at the end of the day. I regret now not having preserved the skin, as it belonged to a very large species of Cebus, and one which I never met with afterwards.

It was about one o'clock in the afternoon when we again reached the spot where we had first struck the banks of the larger pool. We had hitherto had but poor sport, so after dining on the remains of our fried fish and farinha, and smoking our cigarettes, the apparatus for making which, including bamboo tinder-box and steel and flint for striking a light, being carried by every one always on these expeditions, we made off in another (westerly) direction through the forest to try to find better hunting-ground. We quenched our thirst with water from the pool, which I was rather surprised to find quite pure. These pools are, of course, sometimes fouled for a time by the movements of alligators and other tenants in the fine mud which settles at the bottom, but I never observed a scum of confervæ or traces of oil revealing animal decomposition on the surface of these waters, nor was there ever any foul smell perceptible. The whole of this level land, instead of being covered with unwholesome swamps emitting malaria, forms in the dry season (and in the wet also) a most healthy country. How elaborate must be the natural processes of self-purification in these teeming waters!

On our fresh route we were obliged to cut our way through a long belt of bamboo underwood, and not being so careful of my steps as my companions, I trod repeatedly on the flinty thorns which had fallen from the bushes, finishing by becoming completely lame, one thorn having entered deeply into the sole of my foot. I was obliged to be left behind; Lino, the Indian, remaining with me. The careful fellow cleaned my wounds with his saliva, placed pieces of isca (the felt-like substance manufactured by ants) on them to staunch the blood, and bound my feet with tough bast to serve as shoes, which he cut from the bark of a Mongúba tree. He went about his work in a very gentle way and with much skill, but was so sparing of speech that I could scarcely get answers to the questions I put to him. When he had done, I was able to limp about pretty nimbly. An Indian when he performs a service of this kind never thinks of a reward. I did not find so much disinterestedness in negro slaves or half-castes.

We had to wait two hours for the return of our companions; during part of this time I was left quite alone, Lino having started off into the jungle after a peccary (a kind of wild hog) which had come near to where we sat, but on seeing us had given a grunt and bounded off into the thickets. At length our friends hove in sight, loaded with game; having shot twelve curassows and two cujubíms (Penelope Pipile), a handsome black fowl with a white head, which is arboreal in its habits like the rest of this group of Gallinaceous birds inhabiting the South American forests. They had discovered a third pool containing plenty of turtles. Lino rejoined us at the same time, having missed the peccary, but in compensation shot a Quandú, or porcupine. The mulatto boy had caught alive in the pool a most charming little water-fowl, a species of grebe. It was somewhat smaller than a pigeon, and had a pointed beak; its feet were furnished with many intricate folds or frills of skin instead of webs, and resembled very much those of the gecko lizards. The bird was kept as a pet in Jabutí's house at Ega for a long time afterwards, where it became accustomed to swim about in a common hand-basin full of water, and was a great favourite with everybody.

We now retraced our steps towards the water-side, a weary walk of five or six miles, reaching our canoe by half-past five o'clock, or a little before sunset. It was considered by every one at Catuá that we had had an unusually good day's sport. I never knew any small party to take so much game in one day in these forests, over which animals are everywhere so widely and sparingly scattered. My companions were greatly elated, and on approaching the encampment at Catuá made a great commotion with their paddles to announce their successful return, singing in their loudest key one of the wild choruses of the Amazonian boatmen.

The excavation of eggs and preparation of the oil being finished, we left Catuá on the 3rd of November. Carepíra, who was now attached to Cardozo's party, had discovered another lake rich in turtles, about twelve miles distant, in one of his fishing rambles, and my friend resolved, before returning to Ega, to go there with his nets and drag it as we had formerly done the Aningal. Several mameluco families of Ega begged to accompany us to share the labours and booty; the Shumána family also joined the party; we therefore formed a large body, numbering in all eight canoes and fifty persons.

The summer season was now breaking up; the river was rising; the sky was almost constantly clouded, and we had frequent rains. The mosquitoes also, which we had not felt whilst encamped on the sand-banks, now became troublesome. We paddled up the north-westerly channel, and arrived at a point near the upper end of Catuá at ten o'clock p.m. There was here a very broad beach of untrodden white sand, which extended quite into the forest, where it formed rounded hills and hollows like sand dunes, covered with a peculiar vegetation: harsh, reedy grasses, and low trees matted together with lianas, and varied with dwarf spiny palms of the genus Bactris. We encamped for the night on the sands, finding the place luckily free from mosquitoes. The different portions of the party made arched coverings with the toldos or maranta-leaf awnings of their canoes to sleep under, fixing the edges in the sand. No one, however, seemed inclined to go to sleep, so after supper we all sat or lay around the large fires and amused ourselves. We had the fiddler with us, and in the intervals between the wretched tunes which he played, the usual amusement of story-telling beguiled the time: tales of hair-breadth escapes from jaguar, alligator, and so forth. There were amongst us a father and son who had been the actors, the previous year, in an alligator adventure on the edge of the praia we had just left. The son, whilst bathing, was seized by the thigh and carried under water: a cry was raised, and the father, rushing down the bank, plunged after the rapacious beast which was diving away with his victim. It seems almost incredible that a man could overtake and master the large cayman in his own element; but such was the case in this instance, for the animal was reached and forced to release his booty by the man's thrusting his thumb into his eye. The lad showed us the marks of the alligator's teeth in his thighs. We sat up until past midnight listening to these stories and assisting the flow of talk by frequent potations of burnt rum. A large shallow dish was filled with the liquor and fired: when it had burnt for a few minutes the flame was extinguished and each one helped himself by dipping a tea-cup into the vessel.

One by one the people dropped asleep, and then the quiet murmur of talk of the few who remained awake was interrupted by the roar of jaguars in the jungle about a furlong distant. There was not one only, but several of the animals. The older men showed considerable alarm, and proceeded to light fresh fires around the outside of our encampment. I had read in books of travel of tigers coming to warm themselves by the fires of a bivouac, and thought my strong wish to witness the same sight would have been gratified to-night. I had not, however, such good fortune, although I was the last to go to sleep, and my bed was the bare sand under a little arched covering open at both ends. The jaguars, nevertheless, must have come very near during the night, for their fresh footmarks were numerous within a score yards of the place where we slept. In the morning I had a ramble along the borders of the jungle, and found the tracks very numerous and close together on the sandy soil.

We remained in this neighbourhood four days, and succeeded in obtaining many hundred turtles, but we were obliged to sleep two nights within the Carapanatúba channel. The first night passed rather pleasantly, for the weather was fine and we encamped in the forest, making large fires and slinging our hammocks between the trees. The second was one of the most miserable nights I ever spent. The air was close, and a drizzling rain began to fall about midnight, lasting until morning. We tried at first to brave it out under the trees. Several very large fires were made, lighting up with ruddy gleams the magnificent foliage in the black shades around our encampment. The heat and smoke had the desired effect of keeping off pretty well the mosquitoes, but the rain continued until at length everything was soaked, and we had no help for it but to bundle off to the canoes with drenched hammocks and garments. There was not nearly room enough in the flotilla to accommodate so large a number of persons lying at full length ; moreover the night was pitch dark, and it was quite impossible in the gloom and confusion to get at a change of clothing. So there we lay, huddled together in the best way we could arrange ourselves, exhausted with fatigue and irritated beyond all conception by clouds of mosquitoes. I slept on a bench with a sail over me, my wet clothes clinging to my body, and to increase my discomfort, close beside me lay an Indian girl, one of Cardozo's domestics, who had a skin disfigured with black diseased patches, and whose thick clothing, not having been washed during the whole time we had been out (eighteen days), gave forth a most vile effluvia.

We spent the night of the 7th of November pleasantly on the smooth sands, where the jaguars again serenaded us, and on the succeeding morning commenced our return voyage to Ega. We first doubled the upper end of the island of Catuá, and then struck off for the right bank of the Solimoens. The river was here of immense width, and the current was so strong in the middle that it required the most strenuous exertions on the part of our paddlers to prevent us from being carried miles away down the stream. At night we reached Juteca, a small river which enters the Solimoens by a channel so narrow that a man might almost jump across it, but a furlong inwards expands into a very pretty lake several miles in circumference. We slept again in the forest, and again were annoyed by rain and mosquitoes: but this time Cardozo and I preferred remaining where we were to mingling with the reeking crowd in the boats. When the grey dawn arose a steady rain was still falling, and the whole sky had a settled leaden appearance, but it was delightfully cool. We took our net into the lake and gleaned a good supply of delicious fish for breakfast. I saw at the upper end of this lake the native rice of this country growing wild.

The weather cleared up at 10 o'clock a.m. At 3 p.m. we arrived at the mouth of the Cayambé, another tributary stream much larger than the Juteca. The channel of exit to the Solimoens was here also very narrow, but the expanded river inside is of vast dimensions: it forms a lake (I may safely venture to say) several score miles in circumference. Although prepared for these surprises, I was quite taken aback in this case. We had been paddling all day along a monotonous shore, with the dreary Solimoens before us, here three to four miles broad, heavily rolling onward its muddy waters. We come to a little gap in the earthy banks, and find a dark, narrow inlet with a wall of forest over-shadowing it on each side: we enter it, and at a distance of two or three hundred yards a glorious sheet of water bursts upon the view. The scenery of Cayambé is very picturesque. The land, on the two sides visible of the lake, is high and clothed with sombre woods, varied here and there with a white-washed house, in the middle of a green patch of clearing, belonging to settlers. In striking contrast to these dark, rolling forests is the vivid, light-green and cheerful foliage of the woods on the numerous islets which rest like water-gardens on the surface of the lake. Flocks of ducks, storks, and snow-white herons inhabit these islets, and a noise of parrots with the tingling chorus of Tamburí-parás was heard from them as we passed. This has a cheering effect after the depressing stillness and absence of life in the woods on the margins of the main river.

Cardozo and I with two Indians took a small canoe and crossed the lake on a visit to Senhor Gaspar José Rodriguez, a well-to-do farmer, and the principal resident of Cayambé. His eldest daughter, a home-loving, industrious girl, had married the Portuguese Miguel, my old travelling companion, a few days before we left Ega on these rambles. We had attended and danced at the wedding, and this present visit was in fulfilment of a promise to call on the family whenever we should be near Cayambé. Senhor Gaspar was one of those numerous half-caste proprietors, a few of whom I have had occasion to mention, who by their industrious, regular habits, good sense, and fair dealing, do credit to the class to which they belong. We have heard so much in England of the worthlessness of the half-caste population of Tropical America that it is a real pleasure to be able to bear witness that they are not wholly bad. It is, however, in retired country districts where I have chiefly mixed with them. Some of them, such as the friend of whom I am speaking, are, considering their defective education, as worthy men as can be found in any country. There is however, it must be confessed, a considerable number of superlatively lazy, tricky, and sensual characters amongst the half-castes, both in rural places and in the towns. I found the establishment of Senhor Gaspar similar to that of Joaõ Trinidade which I have before described, opposite to the mouth of the Madeira. It was situated on a high bank: the dwelling-house was large and airy, but roughly built, and with unplastered mud-walls. There was a considerable number of outhouses, and in the rear, extensive orchards of fruit and coffee trees, with paths through them leading to the mandioca plantations. Senhor Miguel, with his wife, were absent at a new clearing which they had made for themselves in another part of the banks of the lake. The rest of the family were at home.

We were received with frank hospitality by these shrewd and lively people. Senhor Gaspar had seven children, and had himself taught them all to read and write. The boys were very quick; one of them afterwards became clerk to the Municipal Chamber of Ega. There was an air of cheerfulness and abundance about the place that was quite exhilarating.

We dined, seated on a large mat, over which a clean white towel was spread: the meal consisting of fowls and rice (the general entertainment in this country for visitors), with dessert of "laranjas torradas," or toasted oranges; that is, oranges partially dried in the sun. The fruit, grown with a little greater care in Gaspar's orchard than is usually bestowed on it in this country, was very fine in itself, but treated in this form its sweetness and richness of flavour were far superior to anything I had yet tasted. When we were about leaving, our host, having listened to my praises of the fruit, sent down to our canoe a large basketful as a present. The conversation after dinner turned on the difficulty of getting good houses built at Ega; on the backward condition of the province; the disregard of the interests of the agricultural class shown by the Government in taxing all the produce of the interior on its reaching Pará, and so forth. Senhor Gaspar had just finished the erection of a substantial town-house at Ega. He told me that it was cheaper to send down to Pará (2800 miles there and back) for doors and shutters, than to make them at Ega; for, as there were no large saws anywhere on the Solimoens, every plank had to be hewn out of the tree with a hatchet.

On our return to the mouth of the Cayambé, whilst in the middle of the lake, a squall suddenly arose, in the direction towards which we were going, and for a whole hour we were in great danger of being swamped. The wind blew away the awning and mats, and lashed the waters into foam: the waves rising to a great height. Our boat, fortunately, was excellently constructed, rising well towards the prow, so that with good steering we managed to head the billows as they arose and escaped without shipping much water. We reached our igarité at sunset, and then made all speed to Curubarú, fifteen miles distant, to encamp for the night on the sands. We reached the praia at 10 o'clock. The waters were now mounting fast upon the sloping beach, and we found on dragging the net next morning that fish was beginning to be scarce. Cardozo and his friends talked quite gloomily at breakfast time over the departure of the joyous *veraõ*, and the setting in of the dull, hungry winter season.

At 9 o'clock in the morning of the 10th of November a light wind from down river sprang up, and all who had sails hoisted them. It was the first time during our trip that we had had occasion to use our sails: so continual is the calm on this upper river. We bowled along merrily, and soon entered the broad channel lying between Bariá and the mainland on the south bank. The wind carried us right into the mouth of the Teffé, and at 4 o'clock p.m. we cast anchor in the port of Ega.

CHAPTER V.

ANIMALS OF THE NEIGHBOURHOOD OF EGA.

Scarlet-faced Monkeys—Parauacú Monkey—Owl-faced Night-apes—Marmosets—Jupurá—Comparison of Monkeys of the New World with those of the Old—Bats—Birds—Cuvier's Toucan—Curl-crested Toucan—Insects—Pendulous Cocoons—Foraging Ants—Blind Ants.

As may have been gathered from the remarks already made, the neighbourhood of Ega was a fine field for a Natural History collector. With the exception of what could be learnt from the few specimens brought home, after transient visits, by Spix and Martius and the Count de Castelnau, whose acquisitions have been deposited in the public museums of Munich and Paris, very little was known in Europe of the animal tenants of this region ; the collections that I had the opportunity of making and sending home attracted, therefore, considerable attention. Indeed, the name of my favourite village has become quite a household word amongst a numerous class of Naturalists, not only in England but abroad, in consequence of the very large number of new species (upwards of 3000) which they have had to describe, with the locality "Ega" attached to them. The discovery of new species, however, forms but a small item in the interest belonging to the study of the living creation. The structure, habits, instincts, and geographical distribution of some of the oldest-known forms supply inexhaustible materials for reflection. The few remarks I have to make on the animals of Ega will relate to the mammals, birds, and insects, and will sometimes apply to the productions of the whole Upper Amazons region. We will begin with the monkeys, the most interesting, next to man, of all animals.

Scarlet-faced Monkeys.—Early one sunny morning, in the year 1855, I saw in the streets of Ega, a number of Indians carrying on their shoulders down to the port, to be embarked on the Upper Amazons steamer, a large cage made of strong lianas, some twelve feet in length and five in height, containing a dozen monkeys of the most grotesque appearance. Their bodies (about eighteen inches in height, exclusive of limbs) were clothed from neck to tail with very long, straight, and shining whitish hair; their heads were nearly bald, owing to the very short crop of thin gray hairs, and their faces glowed with the most vivid scarlet hue. As a finish to their striking physiognomy, they had bushy whiskers of a sandy colour, meeting under the chin, and reddish-yellow eyes. They sat gravely and silently in a group, and altogether presented a strange spectacle. These red-faced apes belonged to a species called by the Indians Uakarí, which is peculiar to the Ega district, and the cage with its contents was being sent as a present by Senhor Chrysostomo, the Director of Indians of the Japurá, to one of the Government officials at Rio Janeiro, in acknowledgment of having

SCARLET-FACED AND PARAUACÚ MONKEYS.

Vol. II., page 306.

been made colonel of the new national guard. They had been obtained with great difficulty in the forests which cover the low lands, near the principal mouth of the Japurá, about thirty miles from Ega. It was the first time I had seen this most curious of all the South American monkeys, and one that appears to have escaped the notice of Spix and Martius. I afterwards made a journey to the district inhabited by it, but did not then succeed in obtaining specimens; before leaving the country, however, I acquired two individuals, one of which lived in my house for several weeks.

The scarlet-faced monkey belongs, in all essential points of structure, to the same family (Cebidæ) as the rest of the large-sized American species; but it differs from all its relatives in having only the rudiment of a tail, a member which reaches in some allied kinds the highest grade of development known in the order. It was so unusual to see a nearly tailless monkey from America, that naturalists thought, when the first specimens arrived in Europe, that the member had been shortened artificially. Nevertheless, the Uakarí is not quite isolated from its related species of the same family, several other kinds, also found on the Amazons, forming a graduated passage between the extreme forms as regards the tail. The appendage reaches its perfection in those genera (the Howlers, the Lagothrix and the Spider monkeys) in which it presents on its under-surface near the tip a naked palm, which makes it sensitive and useful as a fifth hand in climbing. In the rest of the genera of Cebidæ (seven in number, containing thirty-eight species), the tail is weaker in structure, entirely covered with hair, and of little or no service in climbing, a few species nearly related to our Uakarí having it much shorter than usual. All the Cebidæ, both long-tailed and short-tailed, are equally dwellers in trees. The scarlet-faced monkey lives in forests, which are inundated during great part of the year, and is never known to descend to the ground ; the shortness of its tail is therefore no sign of terrestrial habits, as it is in the Macaques and Baboons of the Old World. It differs a little from the typical Cebidæ in its teeth, the incisors being oblique and, in the upper jaw, converging, so as to leave a gap between the outermost and the canine teeth. Like all the rest of its family, it differs from the monkeys of the old world, and from man, in having an additional grinding-tooth (premolar) in each side of both jaws, making the complete set thirty-six instead of thirty-two in number.

The white Uakarí (Brachyurus calvus), seems to be found in no other part of America than the district just mentioned, namely, the banks of the Japurá, near its principal mouth ; and even there it is confined, as far as I could learn, to the western side of the river. It lives in small troops amongst the crowns of the lofty trees, living on fruits of various kinds. Hunters say it is pretty nimble in its motions, but is not much given to leaping, preferring to run up and down the larger boughs in travelling from tree to tree. The mother, as in other species of the monkey order, carries her young on her back. Individuals are obtained alive by shooting them with the blow-pipe and arrows tipped with diluted Urarí poison. They run a considerable distance after being pierced, and it requires an experienced hunter to track them. He is considered the most expert who can keep pace with a wounded one, and catch it in his arms when it falls exhausted. A pinch of salt, the antidote to the poison, is then put in its mouth, and the creature revives. The species is rare, even in the limited district which it inhabits. Senhor Chrysostomo sent six of his most skilful Indians, who were absent three weeks before they obtained the twelve specimens which formed his unique and princely gift. When an independent hunter obtains one, a very high price (thirty to forty milreis*) is asked, these monkeys being in great demand for presents to persons of influence down the river.

Adult Uakarís, caught in the way just described, very rarely become tame. They are peevish and sulky, resisting all attempts to coax them, and biting anyone who ventures within reach. They have no particular cry, even when in their native woods; in captivity they are quite silent. In the course of a few days or weeks, if not very carefully attended to, they fall into a listless condition, refuse food and die. Many of them succumb to a disease which I supposed from the symptoms to be inflammation of the chest or lungs. The one which I kept as a pet died of this disorder after I had had it about three weeks. It lost its appetite in a very few days, although kept in an airy verandah; its coat, which was originally long, smooth, and glossy, became dingy and ragged like that of the specimens seen in museums, and the bright scarlet colour of its face changed to a duller hue. This colour, in health, is spread over

* Three pounds seven shillings to four pounds thirteen shillings.

the features up to the roots of the hair on the forehead and temples, and down to the neck, including the flabby cheeks which hang down below the jaws. The animal, in this condition, looks at a short distance as though some one had laid a thick coat of red paint on its countenance. The death of my pet was slow; during the last twenty-four hours it lay prostrate, breathing quickly, its chest strongly heaving; the colour of its face became gradually paler, but was still red when it expired. As the hue did not quite disappear until two or three hours after the animal was quite dead, I judged that it was not exclusively due to the blood, but partly to a pigment beneath the skin which would probably retain its colour a short time after the circulation had ceased.

After seeing much of the morose disposition of the Uakarí, I was not a little surprised one day at a friend's house to find an extremely lively and familiar individual of this species. It ran from an inner chamber straight towards me after I had sat down on a chair, climbed my legs and nestled in my lap, turning round and looking up with the usual monkey's grin, after it had made itself comfortable. It was a young animal which had been taken when its mother was shot with a poisoned arrow; its teeth were incomplete, and the face was pale and mottled, the glowing scarlet hue not supervening in these animals before mature age; it had also a few long black hairs on the eyebrows and lips. The frisky little fellow had been reared in the house amongst the children, and allowed to run about freely, and take its meals with the rest of the household. There are few animals which the Brazilians of these villages have not succeeded in taming. I have even seen young jaguars running loose about a house, and treated as pets. The animals that I had, rarely became familiar, however long they might remain in my possession, a circumstance due no doubt to their being kept always tied up.

The Uakarí is one of the many species of animals which are classified by the Brazilians as "mortál," or of delicate constitution, in contradistinction to those which are "duro," or hardy. A large proportion of the specimens sent from Ega die before arriving at Pará, and scarcely one in a dozen succeeds in reaching Rio Janeiro alive. It appears, nevertheless, that an individual has once been brought in a living state to England, for Dr. Gray relates that one was exhibited in the gardens of the Zoological Society in 1849. The difficulty it has of accommodating itself to changed conditions probably has some connection with the very limited range or confined sphere of life of the species in its natural state, its native home being an area of swampy woods, not more than about sixty square miles in extent, although no permanent barrier exists to check its dispersal, except towards the south, over a much wider space. When I descended the river in 1859, we had with us a tame adult Uakarí, which was allowed to ramble about the vessel, a large schooner. When we reached the mouth of the Rio Negro, we had to wait four days whilst the custom-house officials at Barra, ten miles distant, made out the passports for our crew, and during this time the schooner lay close to the shore, with its bowsprit secured to the trees on the bank. Well, one morning, scarlet-face was missing, having made his escape into the forest. Two men were sent in search of him, but returned after several hours' absence without having caught sight of the runaway. We gave up the monkey for lost, until the following day, when he re-appeared on the skirts of the forest, and marched quietly down the bowsprit to his usual place on deck. He had evidently found the forests of the Rio Negro very different from those of the delta lands of the Japurá, and preferred captivity to freedom in a place that was so uncongenial to him.

A most curious fact connected with this monkey is the existence of an allied form, or brother species, in a tract of country lying to the west of its district. This differs in being clothed with red instead of white hair, and has been described by Isidore Geoffroy St. Hilaire (from specimens brought to Paris in 1847 by the Comte de Castlenau) as a distinct species, under the name of Brachyurus rubicundus. It wholly replaces the white form in the western parts of the Japurá delta: that is to say, in a uniform district of country, 150 miles in length, and sixty to eighty in breadth, the eastern half is tenanted exclusively by white Uakarís, and the western half by red ones. The district, it may be mentioned, is crossed by several channels, which at the present time doubtless serve as barriers to the dispersal of monkeys, but cannot have done so for many centuries, as the position of low alluvial lands, and the direction of channels in the Amazons Valley, change considerably in the course of a few years. The red-haired Uakarí appears to be most frequently found in the forests lying opposite to the mouth of the river which leads to Fonteboa, and ranges thence to the banks of the Uatí-paraná, the most westerly channel of the Japurá, situated near Tunantins. Beyond that point to the west there is no trace of either the red or the white form, nor of any other allied species. Neither do they pass to the eastward of the main mouth of the Japurá, or to the south shore of the Solimoens. How far they range northwards along the banks of the Japurá, I could not precisely ascertain; Senhor Chrysostomo, however, assured me that at 180 miles from the mouth of this river, neither white nor red Uakarí is found, but that a third, black-faced and gray-haired species, takes their place. I saw two adult individuals of Brachyurus rubicundus at Ega, and a young one at Fonteboa; but was unable to obtain specimens myself, as the forests were inundated at the time I visited their locality. I was surprised to find the hair of the young animal much paler in colour than that of the adults, it being of a sandy and not of a brownish-red hue, and consequently did not differ very much from that of the white species; the two forms, therefore, are less distinct from each other in their young than in their adult states. The fact of the range of these singular monkeys being so curiously limited as here described, cannot be said to be established until the country lying between the northern shore of the Solimoens and New Granada be well explored, but there can be no doubt of the separation of the two forms in the Delta lands of the Japurá, and this is a most instructive fact in the geographical distribution of animals.

The Parauacú Monkey.—Another Ega monkey, nearly related to the Uakarís, is the Parauacú (Pithecia hirsuta), a timid inoffensive creature, with a long bear-like coat of harsh speckled-gray hair. The long fur hangs over the head, half concealing the pleasing, diminutive face, and clothes also the tail to the tip, which member is well developed, being eighteen inches in length, or longer than the body. The Parauacú is found on the "terra firma" lands of the north shore of the Solimoens from Tunantins to Peru. It exists also on the south side of the river, namely on the banks of the Teffé, but there under a changed form, which differs from its type in colours about as much as the red differs from the white Uakarí. This form has been described by Dr. Gray as a distinct species, under the name of Pithecia albicans. The Parauacú is also a very delicate animal, rarely living many weeks in captivity ; but anyone who succeeds in keeping it alive for a month or two, gains by it a most affectionate pet. One of the specimens of Pithecia albicans now in the British Museum was, when living, the property of a young Frenchman, a neighbour of mine at Ega. It became so tame in the course of a few weeks that it followed him about the streets like a dog. My friend was a tailor, and the little pet used to spend the greater part of the day seated on his shoulder, whilst he was at work on his board. It showed, nevertheless, great dislike to strangers, and was not on good terms with any other member of my friend's household than himself. I saw no monkey that showed so strong a personal attachment as this gentle, timid, silent little creature. The eager and passionate Cebi seem to take the lead of all the South American monkeys in intelligence and docility, and the Coaitá has perhaps the most gentle and impressible disposition ; but the Parauacú, although a dull, cheerless animal, excels all in this quality of capability of attachment to individuals of our own species. It is not wanting, however, in intelligence as well as moral goodness, proof of which was furnished one day by an act of our little pet. My neighbour had quitted his house in the morning without taking Parauacú with him, and the little creature having missed its friend, and concluded, as it seemed, that he would be sure to come to me, both being in the habit of paying me a daily visit together, came straight to my dwelling, taking a short cut over gardens, trees, and thickets, instead of going the roundabout way of the street. It had never done this before, and we knew the route it had taken only from a neighbour having watched its movements. On arriving at my house and not finding its master, it climbed to the top of my table, and sat with an air of quiet resignation waiting for him. Shortly afterwards my friend entered, and the gladdened pet then jumped to its usual perch on his shoulder.

Owl-faced Night Apes.—A third interesting genus of monkeys, found near Ega, are the Nyctipitheci, or night apes, called Ei-á by the Indians. Of these I found two species, closely related to each other but nevertheless quite distinct, as both inhabit the same forests, namely, those of the higher and drier lands, without mingling with each other or intercrossing. They sleep all day long in hollow trees, and come forth to prey on insects and eat fruits only in the night. They are of small size, the body being about a foot long, and the tail fourteen inches, and are thickly clothed with soft grey and brown fur, similar in substance to that of the rabbit. Their physiognomy reminds one of an owl, or tiger-cat: the face is round and encircled by a ruff of whitish fur; the muzzle is not at all prominent; the mouth and chin are small; the ears are very short, scarcely appearing above the hair of the head; and the eyes are large and yellowish in colour, imparting the staring expression of nocturnal animals of prey. The forehead is whitish, and decorated with three black stripes, which in one of the species (Nyctipithecus trivirgatus) continue to the crown, and in the other (N. felinus) meet on the top of the forehead. N. trivirgatus was first described by Humboldt, who discovered it on the banks of the Cassiquiare, near the head waters of the Rio Negro.

One cannot help being struck by this curious modification of the American type of monkeys, for the owl-faced night-apes have evidently sprung from the same stock as the rest of the Cebidæ, as they do not differ much in all essential points from the Whaiápu-sais (Callithrix), and the Sai-miris (Chrysothrix). They have nails of the ordinary form to all their fingers, and semi-opposable thumbs; but the molar teeth (contrary to what is usual in the Cebidæ) are studded with sharp points, showing that their natural food is principally insects.

I kept a pet animal of the N. trivirgatus for many months, a young one having been given to me by an Indian *compadre,* as a present from my newly-baptised godson. These monkeys, although sleeping by day, are aroused by the least noise; so that, when a person passes by a tree in which a number of them are concealed, he is startled by the sudden apparition of a group of little striped faces crowding a hole in the trunk. It was in this way that my compadre discovered the colony from which the one given to me was taken. I was obliged to keep my pet chained up; it therefore never became thoroughly familiar. I once saw, however, an individual of the other species (N. felinus) which was most amusingly tame. It was as lively and nimble as the Cebi, but not so mischievous and far more confiding in its disposition, delighting to be caressed by all persons who came into the house. But its owner, the Municipal Judge of Ega, Dr. Carlos Mariana, had treated it for many weeks with the greatest kindness, allowing it to sleep with him at night in his hammock, and to nestle in his bosom half the day as he lay reading. It was a great favourite with every one, from the cleanliness of its habits and the prettiness of its features and ways. My own pet was kept in a box, in which was placed a broad-mouthed glass jar; into this it would dive, head foremost, when any one entered the room, turning round inside, and thrusting forth its inquisitive face an instant afterwards to stare at the intruder. It was very active at night, venting at frequent intervals a hoarse cry, like the suppressed barking of a dog, and scampering about the room, to the length of its tether, after cockroaches and spiders. In climbing between the box and the wall, it straddled the space, resting its hands on the palms and tips of the outstretched fingers with the knuckles bent at an acute angle, and thus mounted to the top with the greatest facility. Although seeming to prefer insects, it ate all kinds of fruit, but would not touch raw or cooked meat, and was very seldom thirsty. I was told by persons who had kept these monkeys loose about the house, that they cleared the chambers of bats as well as insect vermin. When approached gently, my Ei-á allowed itself to be caressed; but when handled roughly, it always took alarm, biting severely, striking out its little hands, and making a hissing noise like a cat. As already related, my pet was killed by a jealous Caiarára monkey, which was kept in the house at the same time.

I have mentioned the near relationship of the night apes to the Sai-mirís (Chrysothrix), which are amongst the commonest of the ordinary monkeys of the American forests. This near relationship is the more necessary to be borne in mind, as some zoologists have drawn a comparison between the Nyctipitheci and the Microcebi, Nycticebi, and Loris, nocturnal apes of the Lemur family inhabiting Ceylon and Java, and it might be erroneously inferred that our American Ei-ás were related more closely to these Old World forms than they are to the rest of the New World monkeys. The Nycticebus of Java has also large nocturnal eyes, short ears, and a physiognomy similar to that of our Nyctipitheci; resemblances which might seem to be strong proofs of blood-relationship, but these points are fallacious guides in ascertaining the genealogy of these animals; they are simply *resemblances of analogy*, and merely show that a few species belonging to utterly dissimilar families have been made similar by being adapted to similar modes of life. The Loris and their relatives of Tropical Asia have six incisor teeth to the lower jaws, and belong, in all other essential points of structure, to the Lemur family, which has not a single representative in the New World. The Ei-ás have teeth of the same number, and growing in nearly the same position, as their near relatives the Sai-miris. I obtained, moreover, yet stronger proof of this close relationship between the night and day monkeys of America, in finding a species on the Upper Amazons which supplies a link between them. This one had ears nearly as short as those of the night apes, and also a striped forehead; the stripes being, however, two in number, instead of three: the colours of the body were very similar to those of the well-known Chrysothrix sciureus, and the eyes were fitted for day vision.

Barrigudo Monkeys.—Ten other species of monkeys were found, in addition to those already mentioned, in the forests of the Upper Amazons. All were strictly arboreal and diurnal in their habits, and lived in flocks, travelling from tree to tree, the mothers with their children on their backs; leading, in fact, a life similar to that of the Parárauáte Indians, and, like them, occasionally plundering the plantations which lie near their line of march. Some of them were found also on the Lower Amazons, and have been noticed in former chapters of this narrative. Of the remainder, the most remarkable is the Macaco barrigudo, or big-bellied monkey of the Portuguese colonists, a species of Lagothrix. The genus is closely allied to the Coaitás, or spider monkeys, having, like them, exceedingly strong and flexible tails, which are furnished underneath with a naked palm like a hand, for grasping. The Barrigudos, however, are very bulky animals, whilst the spider monkeys are remarkable for the slenderness of their bodies and limbs. I obtained specimens of what have been considered two species, one (L. olivaceus of Spix ?) having the head clothed with gray, the other (L. Humboldtii) with black fur. They both live together in the same places, and are probably only differently-coloured individuals of one and the same species. I sent home a very large male of one of these kinds, which measured twenty-seven inches in length of trunk, the tail being twenty-six inches long ; it was the largest monkey I saw in America, with the exception of a black Howler, whose body was twenty-eight inches in height. The skin of the face in the Barrigudo is black and wrinkled, the forehead is low, with the eye-brows projecting, and, in short, the features altogether resemble in a striking manner those of an old negro. In the forests, the Barrigudo is not a very active animal ; it lives exclusively on fruits, and is much persecuted by the Indians, on account of the excellence of its flesh as food. From information given me by a collector of birds and mammals, whom I employed, and who resided a long time amongst the Tucuna Indians, near Tabatinga, I calculated that one horde of this tribe, 200 in number, destroyed 1200 of these monkeys annually for food. The species is very numerous in the forests of the higher lands, but, owing to long persecution, it is now seldom seen in the neighbourhood of the larger villages. It is not found at all on the Lower Amazons.

Its manners in captivity are grave, and its temper mild and confiding, like that of the Coaitás. Owing to these traits, the Barrigudo is much sought after for pets; but it is not hardy like the Coaitás, and seldom survives a passage down the river to Pará.

Marmosets.—It now only remains to notice the Marmosets, which form the second family of American monkeys. Our old friend Midas ursulus, of Pará and the Lower Amazons, is not found on the Upper river, but in its stead a closely-allied species presents itself, which appears to be the Midas rufoniger of Gervais, whose mouth is bordered with longish white hairs. The habits of this species are the same as those of the M. ursulus, indeed it seems probable that it is a form or race of the same stock, modified to suit the altered local conditions under which it lives. One day, whilst walking along a forest pathway, I saw one of these lively little fellows miss his grasp as he was passing from one tree to another along with his troop. He fell head foremost, from a height of at least fifty feet, but managed cleverly to alight on his legs in the pathway; quickly turning round he gave me a good stare for a few moments, and then bounded off gaily to climb another tree. At Tunantins, I shot a pair of a very handsome species of Marmoset, the M. rufiventer, I believe, of zoologists. Its coat was very glossy and smooth; the back deep brown, and the underside of the body of rich black and reddish hues. A third species (found at Tabatinga, 200 miles further west) is of a deep black colour, with the exception of a patch of white hair around its mouth. The little animal, at a short distance, looks as though it held a ball of snow-white cotton in its teeth. The last I shall mention is the Hapale pygmæus, one of the most diminutive forms of the monkey order. I obtained, near St. Paulo, three full-grown specimens, which measured only seven inches in length of body. The pretty Lilliputian face is furnished with long brown whiskers, which are naturally brushed back over the ears. The general colour of the animal is brownish-tawny, but the tail is elegantly barred with black. I was surprised, on my return to England, to learn that the pigmy marmoset was found also in Mexico, no other Amazonian monkey being known to wander far from the great river plain. Thus the smallest, and apparently the feeblest, species of the whole order, is one which has, by some means, become the most widely dispersed.

The Jupurá.—A curious animal, known to naturalists as the Kinkajou, but called Jupurá by the Indians of the Amazons, and considered by them as a kind of monkey, may be mentioned in this place. It is the Cercoleptes caudivolvus of zoologists, and has been considered by some authors as an intermediate form between the Lemur family of apes and the plantigrade Carnivora, or Bear family. It has decidedly no close relationship to either of the groups of American monkeys, having six cutting teeth to each jaw, and long claws intead of nails, with extremities of the usual shape of paws instead of hands. Its muzzle is conical and pointed, like that of many Lemurs of Madagascar; the expression of its countenance, and its habits and actions, are also very similar to those of Lemurs. Its tail is very flexible towards the tip, and is used to twine round branches in climbing. I did not see or hear anything of this animal whilst residing on the Lower Amazons, but on the banks of the Upper river, from the Teffé to Peru, it appeared to be rather common. It is nocturnal in its habits, like the owl-faced monkeys, although, unlike them, it has a bright, dark eye. I once saw it in considerable numbers, when on an excursion with an Indian companion along the low Ygapó shores of the Teffé, about twenty miles above Ega. We slept one night at the house of a native family living in the thick of the forest, where a festival was going on, and there being no room to hang our hammocks under shelter, on account of the number of visitors, we lay down on a mat in the open air, near a shed which stood in the midst of a grove of fruit-trees and pupunha palms. After midnight, when all became still, after the uproar of holiday-making, as I was listening to the dull, fanning sound made by the wings of impish hosts of vampire bats crowding round the Cajú trees, a rustle commenced from the side of the woods, and a troop of slender, long-tailed animals were seen against the clear moonlit sky, taking flying leaps from branch to branch through the grove. Many of them stopped at the pupunha trees, and the hustling, twittering, and screaming, with sounds of falling fruits, showed how they were employed. I thought, at first, they were Nyctipitheci, but they proved to be Jupurás, for the owner of the house early next morning caught a young one, and gave it to me. I kept this as a pet animal for several weeks, feeding it on bananas and mandioca-meal mixed with treacle. It became tame in a very short time, allowing itself to be caressed, but making a distinction in the degree of confidence it showed between myself and strangers. My pet was unfortunately killed by a neighbour's dog, which entered the room where it was kept. The animal is so difficult to obtain alive, its place of retreat in the day-time not being known to the natives, that I was unable to procure a second living specimen.

As I shall not have occasion again to enter on the subject of monkeys, a few general remarks will be here in place, as a summary of my observations on this important order of animals in the Amazons region. The total number of species of monkeys which I found inhabiting the margins of the Upper and Lower Amazons, was thirty-eight. They belonged to twelve different genera, forming two distinct families, the number of genera and families, here as well as in other orders of animals or plants, expressing roughly the amount of diversity existing with regard to forms. All the New World genera of apes, except one (Eriodes, closely allied to the Coaitás, but having claw-shaped nails to the fingers), are represented in the Amazons region. With these ample materials before us, let us draw a comparison between the monkeys of the new continent, and their kindred of the Old World. It seems highly probable that the larger land areas, both continents and islands, on the surface of our globe, became separated pretty nearly as they now are, soon after the first forms of this group of animals came into existence: it will be interesting, therefore, to see how differently the subsequent creations of species have proceeded in each of the separated areas.

The American monkeys are distinguished, as a body, from all those found in the Old World. Upon this point, there is no difference of opinion amongst modern zoologists. It is not probable, therefore, that species of the one continent have passed over to the other, since these great tracts of land received their present inhabitants of this order. The American productions present a cluster of forms, namely, about eighty-six species, separated into thirteen genera, which although greatly diversified amongst themselves, in no case show signs of near relationship to any of the still more diversified forms of the same order belonging to the eastern hemisphere. One of the two American families (Cebidæ) has thirty-six teeth, whilst the corresponding family (Pithecidæ) of Old World apes has, like man, only thirty-two teeth; the difference arising from the Cebidæ having an additional false molar tooth* to each side of both jaws. This important character is constant throughout all the varied forms of which the Cebidæ family is composed; being equally present in the prehensile-tailed group, with its four genera containing twenty-seven species, differing in form and clothing, shape of claws, mental characteristics, and condition of thumb of the anterior hands; and in the true Cebi and the group of Sagouins, with six genera and twenty-four species, including day apes and night apes, short

* False molars, or premolars, differ from true molars, through being preceded in growth by milk teeth.

furred and long-haired apes, apes with excessively long tails, and apes with rudimentary tails. The second American family, the Marmosets, have thirty-two teeth, like the Old World monkeys and man ; but this identity of number arises from one of the true molars being absent ; the Marmosets have three premolar teeth, like the Cebidæ, and are therefore quite as far removed as the Cebidæ from all the forms of the Old World. They are, moreover, a low type of apes, having a smooth brain, and claws instead of nails, although they are gentle and playful in disposition, and have a visage which presents an open facial angle.

The Old World apes, as just observed, are far more diversified amongst themselves, than are those of the New World. They form, in the first place, two widely distinct groups or sub-orders, Pithecidæ and Lemurs, and comprise about 125 species, divided into twenty-one genera. The Lemur group contains a remarkably great diversity of forms ; this is shown by their being naturally divisible into four families,* and twelve genera, although containing only twenty-five species. Their teeth are very irregular in number and position, but never correspond with those of the Pithecidæ or Cebidæ. These four families, in structure, are more widely separated from each other than are the two American groups of the same denomination. The Lemurs also contain a number of anomalous or isolated forms, which, by their teeth, number of teats, and other features, connect the monkeys with other and lower orders of the mammal class ; namely, the Rodents, the

* True Lemurs, Tarsiens, Aye-Ayes, and Galeopitheci.

Insectivora, and the Bats. All the typical Lemurs, which constitute the great majority of the family, inhabit exclusively the Island of Madagascar.

The Pithecidæ are divisible into three groups, which again are much more distinct from each other than the subordinate groups of Cebidæ. These are the Anthropoid section, to which some zoologists consider man himself belongs, comprising the Gorilla, the Chimpanzee, the Orangs and the Gibbons ; the Guenons (which, in their forms, tempers, and habits, resemble the Cebidæ), and lastly, the Baboons, whose extreme forms—the dog-faced species, with nose extending to the tip of the muzzle—seem like a degradation of the monkey type. There is nothing at all resembling the Anthropoid apes and the Baboons existing on the American continent. The Guenons, too, have only a superficial resemblance to American monkeys; for they have all thirty-two teeth, nostrils opening in a downward direction (instead of on the sides, like the Cebidæ and Marmosets), and are, moreover, linked to the Baboons through intermediate forms (Macacus), and the possession of callosities on the breech, and other signs of blood-relationship.

A few more words on the peculiar way in which these groups of monkeys are distributed over the earth's surface. We may consider, in connection with this subject, the great land masses of the warmer parts of the earth to be four in number. 1. Australia, with New Guinea and its neighbouring islands : 2. Madagascar : 3. America : 4. The Continental mass of the Old World, comprising Europe, Africa, Asia, and the Islands of the Malay Archipelago, which latter are connected with

Asia by a shallow sea, whilst they are separated from New Guinea by a channel of very deep water; the shallow sea pointing to a former, but recent, union of the lands which it connects, the deep channel a complete and enduring severance of the lands which it separates. Now, with regard to monkeys, these four land masses seem to have had these animals allotted to them in the most capricious way possible, if we are to take for granted that the species were arbitrarily created on the lands where they are now found. Australia, with soil and climate as well adapted for Baboons as Africa, where they abound, and New Guinea, with rich humid forests as suitable for Orangs and Gibbons as the very similar island of Borneo, have, neither of them, a single species of native monkey. Madagascar possesses only Lemurs, the most lowly-organised group of apes, although the neighbouring continent of Africa contains numerous species of all families of Old World apes. America, as we have seen, has no Lemurs, and not a single representative of the Old World groups of the order, but is well peopled by genera and species belonging to two distinct groups peculiar to the continent. Lastly, the Old World continental mass, with a few anomalous forms of Lemurs scattered here and there, is the exclusive home of the whole of the Pithecidæ family, which presents a series of forms graduating from the debased Baboon to the Gorilla, which some zoologists consider to approach near to man in his organisation.

What does all this mean? Why are the different forms apportioned in this way to the various lands of the earth? Why is Australia with New Guinea destitute of monkeys, and why should Madagascar have stopped short at Lemurs, whilst America has gone on to prehensile-tailed Cebidæ, and the Old-World continent continued to Gibbons, Orangs, Chimpanzee, and Gorilla? Is it that the greater land masses have seen a larger amount of geological and climatal changes with corresponding changes in the geographical relations of species? Moreover, why should the smaller groups of the order be confined to smaller areas within the greater areas peopled by the families to which they belong? For, it must be added, the true Lemurs are confined to Madagascar, the Gibbons and others to South Eastern Asia, the dog-faced baboons to Africa, and, as we have seen, the scarlet-faced monkeys to a limited area on the Upper Amazons. May we be allowed to explain the absence of these animals from New Guinea with Australia, by the supposition that those lands were separated from South Eastern Asia before the first forms of the order came into existence? If so, it may be concluded that Madagascar became separated from Africa, and America from the continental mass of the old world before the Pithecidæ originated. But, if these explanations, founded on natural causes, be entertained, we commit ourselves, by the fact of entertaining them, to the admission that natural causes are competent to explain the existence or non-existence of forms in a given area, and why may not the exercise of our reason, founded on carefully observed and collated facts, be carried a step farther, namely to the origin of the species of monkeys themselves? I have already shown how singularly species of monkeys vary in different localities, and have given the striking case of the white and red-haired Uakarís. If these two forms, which are considered by the most eminent naturalists as distinct species, have originated, as the facts of their distribution plainly tell us they have, from one and the same stock, why may not the various species of Lemurs, of Baboons, of Gibbons, and so forth, given the necessary amount of time and climatal changes, have originated in the same way? And if we can thus account for the origin of the species of one genus, on what grounds can we deny that the genera of the same family, or the families of the same order, have also proceeded from a common stock? I throw out these suggestions simply for the consideration of thoughtful readers, but must add, that unless the common origin, at least, of the species of a family be admitted, the problem of the distribution of monkeys over the earth's surface must remain an inexplicable mystery, whilst, if admitted, a flood of light illuminates the subject, and promises an early solution to honest and patient investigation. These questions, also, show how interesting and difficult are the problems which Natural History, granted the right and ability of the human mind to deal with them, has to solve.

It is a suggestive fact that all the fossil monkeys which have been found in Europe and America, belong in each case to the types which are still peculiar to the continent which they inhabit. The European fossils are all of the Pithecidæ family, the South American all belong to the Cebidæ and Marmoset families. The separation of the two continental masses (at least of their warm zones) must therefore be of great geological antiquity. It is interesting to trace how the diversification of forms (if the expression may be allowed), since the separation, has gone on in Tropical America. What wide divergence as to size, forms, habits, and mental dispositions, between the silver marmoset so small that it may be inclosed in the two hands, and the strong and savage black Howler, nearly two feet and a half in length of trunk! Yet there has been no direct advance in the organisation of the order towards a higher type, such as is exhibited in the old world. America, for her share, has produced the most perfectly arboreal monkey in the world; but beyond the perfection of the arboreal type she does not go. The retention of arboreal forms throughout long geological ages, may teach geologists that there must always have been extensive land areas covered by forests on the site of the tropical zone of America. It is curious to reflect, in conjunction with the fact of the advance of the American Quadrumana having halted at a low stage, that ethnologists have almost unanimously come to the conclusion that the race of men now inhabiting the American continent are not Autochthones of America, the land of the Cebidæ, but immigrants from the Old World continent, the land of the Anthropoid group of the order Quadrumana.

Bats.—The only other mammals that I shall mention are the bats, which exist in very considerable numbers and variety in the forest, as well as in the buildings of the villages. Many small and curious species living in the woods, conceal themselves by day under the broad leaf-blades of Heliconiæ and other plants which grow in shady places; others cling to the trunks of trees. Whilst walking through the forest in the daytime, especially along gloomy ravines, one is almost sure to startle bats from their sleeping-places; and at night they are often seen in great numbers flitting about the trees on the shady margins of narrow channels. I captured altogether, without giving especial attention to bats, sixteen different species at Ega.

The Vampire Bat.—The little gray bloodsucking Phyllostoma, mentioned in a former chapter as found in my chamber at Caripí, was not uncommon at Ega, where everyone believes it to visit sleepers and bleed them in the night. But the vampire was here by far the most abundant of the family of leaf-nosed bats. It is the largest of all the South American species, measuring twenty-eight inches in expanse of wing. Nothing in animal physiognomy can be more hideous than the countenance of this creature when viewed from the front; the large, leathery ears standing out from the sides and top of the head, the erect spear-shaped appendage on the tip of the nose, the grin and the glistening black eye all combining to make up a figure that reminds one of some mocking imp of fable. No wonder that imaginative people have inferred diabolical instincts on the part of so ugly an animal. The vampire, however, is the most harmless of all bats, and its inoffensive character is well known to residents on the banks of the Amazons. I found two distinct species of it, one having the fur of a blackish colour, the other of a ruddy hue, and ascertained that both feed chiefly on fruits. The church at Ega was the head-quarters of both kinds; I used to see them, as I sat at my door during the short evening twilights, trooping forth by scores from a large open window at the back of the altar, twittering cheerfully as they sped off to the borders of the forest. They sometimes enter houses; the first time I saw one in my chamber, wheeling heavily round and round, I mistook it for a pigeon, thinking that a tame one had escaped from the premises of one of my neighbours. I opened the stomachs of several of these bats, and found them to contain a mass of pulp and seeds of fruits, mingled with a few remains of insects.* The natives say they devour ripe cajús and guavas on trees in the gardens, but on comparing the seeds taken from their stomachs with those of all cultivated trees at Ega, I found they were unlike any of them; it is therefore probable that they generally resort to the forest to feed, coming to the village in the morning to sleep, because they find it more secure from animals of prey than their natural abodes in the woods.

Birds.—I have already had occasion to mention several of the more interesting birds found in the Ega district. The first thing that would strike a new-comer in the forests of the Upper Amazons would be the general scarcity of birds; indeed, it often happened that I did not meet with a single bird during a whole day's ramble in the richest and most varied parts of the woods. Yet

* The remains of insects belonged to species of Scarites (Coleoptera) having blunt maxillary blades, several of which fly abroad in great numbers on warm nights.

the country is tenanted by many hundred species, many of which are, in reality, abundant, and some of them conspicuous from their brilliant plumage. The cause of their apparent rarity is to be sought in the sameness and density of the thousand miles of forest which constitute their dwelling-place. The birds of the country are gregarious, at least during the season when they are most readily found; but the frugivorous kinds are to be met with only when certain wild fruits are ripe, and to know the exact localities of the trees requires months of experience. It would not be supposed that the insectivorous birds are also gregarious; but they are so, numbers of distinct species, belonging to many different families, joining together in the chase or search of food. The proceedings of these associated bands of insect-hunters are not a little curious, and merit a few remarks.

Whilst hunting along the narrow pathways that are made through the forest in the neighbourhood of houses and villages, one may pass several days without seeing many birds; but now and then the surrounding bushes and trees appear suddenly to swarm with them. There are scores, probably hundreds of birds, all moving about with the greatest activity—woodpeckers and Dendrocolaptidæ (from species no larger than a sparrow to others the size of a crow) running up the tree trunks; tanagers,* ant-thrushes, humming-birds, fly-catchers, and barbets flitting about the leaves and lower branches.

* Tachyphonus surinamus and cristatus, Tanagrella elegantissima. I very often found fruit-eating birds, such as Cassicus icteronotus and Capito Amazoninus mingled with these bands.

The bustling crowd loses no time, and although moving in concert, each bird is occupied, on its own account, in searching bark or leaf or twig; the barbets visiting every clayey nest of termites on the trees which lie in the line of march. In a few minutes the host is gone, and the forest path remains deserted and silent as before. I became, in course of time, so accustomed to this habit of birds in the woods near Ega, that I could generally find the flock of associated marauders whenever I wanted it. There appeared to be only one of these flocks in each small district; and, as it traversed chiefly a limited tract of woods of second growth, I used to try different paths until I came up with it.

The Indians have noticed these miscellaneous hunting parties of birds, but appear not to have observed that they are occupied in searching for insects. They have supplied their want of knowledge, in the usual way of half-civilised people, by a theory which has degenerated into a myth, to the effect that the onward moving bands are led by a little grey bird, called the Papá-uirá, which fascinates all the rest, and leads them a weary dance through the thickets. There is certainly some appearance of truth in this explanation; for sometimes stray birds, encountered in the line of march, are seen to be drawn into the throng, and purely frugivorous birds are now and then found mixed up with the rest, as though led away by some will-o'-the-wisp. The native women, even the white and half-caste inhabitants of the towns, attach a superstitious value to the skin and feathers of the Papá-uirá, believing that if they keep them in their clothes' chest, the relics will have the effect of attracting for the happy possessors a train of lovers and followers. These birds are consequently in great demand in some places, the hunters selling them at a high price to the foolish girls, who preserve the bodies by drying flesh and feathers together in the sun. I could never get a sight of this famous little bird in the forest. I once employed Indians to obtain specimens for me; but, after the same man (who was a noted woodsman) brought me, at different times, three distinct species of birds as the Papá-uirá, I gave up the story as a piece of humbug. The simplest explanation appears to be this; that the birds associate in flocks from the instinct of self-preservation, and in order to be a less easy prey to hawks, snakes, and other enemies than they would be if feeding alone.

Toucans.—Cuvier's Toucan.—Of this family of birds, so conspicuous from the great size and light structure of their beaks, and so characteristic of Tropical American forests, five species* inhabit the woods of Ega. The largest of all the Toucans found on the Amazons, namely, the Ramphastos toco, called by the natives Tocáno pacova, from its beak resembling in size and shape a banana or pacova, appears not to reach so far up the river as Ega. It is abundant near Pará, and is found also on the low islands of the Rio Negro, near Barra, but does not seem to range much farther to the west. The commonest species at Ega is Cuvier's

* Ramphastos Cuvieri, Pteroglossus Beauharnaisii, Pt. Langsdorfii, Pt. castanotis, Pt. flavirostris. Further westward, namely, near St. Paulo, a sixth species makes its appearance, the Pteroglossus Humboldtii.

Toucan, a large bird, distinguished from its nearest relatives by the feathers at the bottom of the back being of a saffron hue instead of red. It is found more or less numerously throughout the year, as it breeds in the neighbourhood, laying its eggs in holes of trees, at a great height from the ground. During most months of the year, it is met with in single individuals or small flocks, and the birds are then very wary. Sometimes one of these little bands of four or five is seen perched, for hours together, amongst the topmost branches of high trees, giving vent to their remarkably loud, shrill, yelping cries, one bird, mounted higher than the rest, acting, apparently, as leader of the inharmonious chorus ; but two of them are often heard yelping alternately, and in different notes. These cries have a vague resemblance to the syllables Tocáno, Tocáno, and hence the Indian name of this genus of birds. At these times it is difficult to get a shot at Toucans, for their senses are so sharpened that they descry the hunter before he gets near the tree on which they are perched, although he may be half-concealed amongst the underwood, 150 feet below them. They stretch their necks downwards to look beneath, and on espying the least movement amongst the foliage, fly off to the more inaccessible parts of the forest. Solitary Toucans are sometimes met with at the same season, hopping silently up and down the larger boughs, and peering into crevices of the tree-trunks. They moult in the months from March to June, some individuals earlier, others later. This season of enforced quiet being passed, they make their appearance suddenly in the dry forest, near Ega, in large flocks, probably, assemblages of birds gathered together from the neighbouring Ygapó forests, which are then flooded and cold. The birds have now become exceedingly tame, and the troops travel with heavy laborious flight from bough to bough amongst the lower trees. They thus become an easy prey to hunters, and every one at Ega, who can get a gun of any sort and a few charges of powder and shot, or a blow-pipe, goes daily to the woods to kill a few brace for dinner; for, as already observed, the people of Ega live almost exclusively on stewed and roasted Toucans during the months of June and July. The birds are then very fat, and the meat exceedingly sweet and tender. I did not meet with Cuvier's Toucan on the Lower Amazons; in that region, the sulphur and white-breasted Toucan (Ramphastos Vitellinus) seems to take its place, this latter species, on the other hand, being quite unknown on the Upper Amazons. It is probable they are local modifications of one and the same stock.

No one, on seeing a Toucan, can help asking what is the use of the enormous bill, which, in some species, attains a length of seven inches, and a width of more than two inches. A few remarks on this subject may be here introduced. The early naturalists, having seen only the bill of a Toucan, which was esteemed as a marvellous production by the *virtuosi* of the sixteenth and seventeenth centuries, concluded that the bird must have belonged to the aquatic and web-footed order, as this contains so many species of remarkable development of beak, adapted for seizing fish. Some travellers also related fabulous stories of Toucans resorting to the banks of rivers to feed on fish, and these accounts also encouraged the erroneous views of the habits of the birds, which, for a long time, prevailed. Toucans, however, are now well known to be eminently arboreal birds, and to belong to a group (including trogons, parrots, and barbets*), all of whose members are fruit-eaters. On the Amazons, where these birds are very common, no one pretends ever to have seen a Toucan walking on the ground in its natural state, much less acting the part of a swimming or wading bird. Professor Owen found, on dissection, that the gizzard in Toucans is not so well adapted for the trituration of food as it is in other vegetable feeders, and concluded, therefore, as Broderip had observed the habit of chewing the cud in a tame bird, that the great toothed bill was useful in holding and re-masticating the food. The bill can scarcely be said to be a very good contrivance for seizing and crushing small birds, or taking them from their nests in crevices of trees, habits which have been imputed to Toucans by some writers. The hollow, cellular structure of the interior of the bill, its curved and clumsy shape, and the deficiency of force and precision when it is used to seize objects, suggest a want of fitness, if this be the function of the member. But fruit is undoubtedly the chief food of Toucans, and it is in reference to their mode of obtaining it that the use of their uncouth bills is to be sought.

Flowers and fruits on the crowns of the large trees of South American forests grow, principally, towards the end of slender twigs, which will not bear any con-

* Capitoninæ, G. R. Gray.

siderable weight; all animals, therefore, which feed upon fruit, or on insects contained in flowers, must, of course, have some means of reaching the ends of the stalks from a distance. Monkeys obtain their food by stretching forth their long arms and, in some instances, their tails, to bring the fruit near to their mouths. Humming-birds are endowed with highly-perfected organs of flight, with corresponding muscular development, by which they are enabled to sustain themselves on the wing before blossoms whilst rifling them of their contents. These strong-flying creatures, however, will, whenever they get a chance, remain on their perches whilst probing neighbouring flowers for insects. Trogons have feeble wings, and a dull, inactive temperament. Their mode of obtaining food is to station themselves quietly on low branches in the gloomy shades of the forest, and eye the fruits on the surrounding trees, darting off, as if with an effort, every time they wish to seize a mouthful, and returning to the same perch. Barbets (Capitoninæ) seem to have no especial endowment, either of habits or structure, to enable them to seize fruits; and in this respect they are similar to the Toucans, if we leave the bill out of question, both tribes having heavy bodies, with feeble organs of flight, so that they are disabled from taking their food on the wing. The purpose of the enormous bill here becomes evident. Barbets and Toucans are very closely related; indeed a genus has lately been discovered towards the head waters of the Amazons,*

* Tetragonops. Dr. Sclater has lately given a figure of this bird in the Ibis, vol. iii. p. 182.

which tends to link the two families together; the superior length of the Toucan's bill gives it an advantage over the Barbet, with its small, conical beak; it can reach and devour immense quantities of fruit whilst remaining seated, and thus its heavy body and gluttonous appetite form no obstacles to the prosperity of the species. It is worthy of note, that the young of the Toucan has a very much smaller beak than the full-grown bird. The relation between the extraordinarily lengthened bill of the Toucan and its mode of obtaining food, is precisely similar to that between the long neck and lips of the Giraffe and the mode of browsing of the animal. The bill of the Toucan can scarcely be considered a very perfectly-formed instrument for the end to which it is applied, as here explained; but nature appears not to shape organs at once for the functions to which they are now adapted, but avails herself, here of one already-existing structure or instinct, there of another, according as they are handy when need for their further modification arises.

One day, whilst walking along the principal pathway in the woods near Ega, I saw one of these Toucans seated gravely on a low branch close to the road, and had no difficulty in seizing it with my hand. It turned out to be a runaway pet bird; no one, however, came to own it, although I kept it in my house for several months. The bird was in a half-starved and sickly condition, but after a few days of good living it recovered health and spirits, and became one of the most amusing pets imaginable. Many excellent accounts of the habits of tame Toucans, have been published, and therefore I need not describe them in detail, but I do not recollect to have seen any notice of their intelligence and confiding disposition under domestication, in which qualities my pet seemed to be almost equal to parrots. I allowed Tocáno to go free about the house, contrary to my usual practice with pet animals; he never, however, mounted my working-table after a smart correction which he received the first time he did so. He used to sleep on the top of a box in a corner of the room, in the usual position of these birds, namely, with the long tail laid right over on the back, and the beak thrust underneath the wing. He ate of everything that we eat; beef, turtle, fish, farinha, fruit, and was a constant attendant at our table—a cloth spread on a mat. His appetite was most ravenous, and his powers of digestion quite wonderful. He got to know the meal hours to a nicety, and we found it very difficult, after the first week or two, to keep him away from the dining-room, where he had become very impudent and troublesome. We tried to shut him out by enclosing him in the back-yard, which was separated by a high fence from the street on which our front door opened, but he used to climb the fence and hop round by a long circuit to the dining-room, making his appearance with the greatest punctuality as the meal was placed on the table. He acquired the habit, afterwards, of rambling about the street near our house, and one day he was stolen, so we gave him up for lost. But, two days afterwards, he stepped through the open doorway at dinner hour, with his old gait, and sly, magpie-like expression, having escaped from the house where he had been guarded by the person who had stolen him, and which was situated at the further end of the village.

The Curl-crested Toucan (*Pteroglossus Beauharnaisii*).—Of the four smaller Toucans or Arassarís found near Ega, the Pteroglossus flavirostris is perhaps the most beautiful in colours, its breast being adorned with broad belts of rich crimson and black; but the most curious species, by far, is the Curl-crested, or Beauharnais Toucan. The feathers on the head of this singular

Curl-crested Toucan.

bird are transformed into thin, horny plates, of a lustrous black colour, curled up at the ends, and resembling shavings of steel or ebony wood: the curly crest being arranged on the crown in the form of a wig. Mr. Wallace and I first met with this species, on ascending the Amazons, at the mouth of the Solimoens; from that point it continues as a rather common bird on the terra firma, at least on the south side of the river, as far as Fonte Boa, but I did not hear of its being found further to the west. It appears in large flocks in the forest near Ega in May and June, when it has completed its moult. I did not find these bands congregated at fruit-trees, but always wandering through the forest, hopping from branch to branch amongst the lower trees, and partly concealed amongst the foliage. None of the Arassarís, to my knowledge, make a yelping noise like that uttered by the larger Toucans (Ramphastos); the notes of the curl-crested species are very singular, resembling the croaking of frogs. I had an amusing adventure one day with these birds. I had shot one from a rather high tree in a dark glen in the forest, and leaving my gun leaning against a tree-trunk in the pathway, went into the thicket where the bird had fallen, to secure my booty. It was only wounded, and on my attempting to seize it, it set up a loud scream. In an instant, as if by magic, the shady nook seemed alive with these birds, although there was certainly none visible when I entered the thicket. They descended towards me, hopping from bough to bough, some of them swinging on the loops and cables of woody lianas, and all croaking and fluttering their wings like so many furies. Had I had a long stick in my hand I could have knocked several of them over. After killing the wounded one I rushed out to fetch my gun, but, the screaming of their companion having ceased, they remounted the trees, and before I could reload, every one of them had disappeared.

Insects.—Upwards of 7000 species of insects were found in the neighbourhood of Ega. I must confine myself, in this place, to a few remarks on the order

Lepidoptera, and on the ants, several kinds of which, found chiefly on the Upper Amazons, exhibit the most extraordinary instincts.

I found about 550 distinct species of butterflies at Ega. Those who know a little of Entomology will be able to form some idea of the riches of the place in this department, when I mention that eighteen species of true Papilio (the swallow-tail genus) were found within ten minutes' walk of my house. No fact could speak more plainly for the surpassing exuberance of the vegetation, the varied nature of the land, the perennial warmth and humidity of the climate. But no description can convey an adequate notion of the beauty and diversity in form and colour of this class of insects in the neighbourhood of Ega. I paid especial attention to them, having found that this tribe was better adapted than almost any other group of animals or plants, to furnish facts in illustration of the modifications which all species undergo in nature, under changed local conditions. This accidental superiority is owing partly to the simplicity and distinctness of the specific characters of the insects, and partly to the facility with which very copious series of specimens can be collected and placed side by side for comparison. The distinctness of the specific characters is due probably to the fact that all the superficial signs of change in the organisation are exaggerated, and made unusually plain, by affecting the framework, shape, and colour of the wings, which, as many anatomists believe, are magnified extensions of the skin around the breathing orifices of the thorax of the insects. These expansions are clothed with minute feathers or scales, coloured in regular patterns, which vary in accordance with the slightest change in the conditions to which the species are exposed. It may be said, therefore, that on these expanded membranes Nature writes, as on a tablet, the story of the modifications of species, so truly do all changes of the organisation register themselves thereon. Moreover, the same colour-patterns of the wings generally show, with great regularity, the degrees of blood-relationship of the species. As the laws of Nature must be the same for all beings, the conclusions furnished by this group of insects must be applicable to the whole organic world; therefore, the study of butterflies—creatures selected as the types of airiness and frivolity—instead of being despised, will some day be valued as one of the most important branches of Biological science.

I have mentioned, in a former chapter, the general sultry condition of the atmosphere on the Upper Amazons, where the sea-breezes which blow from Pará to the mouth of the Rio Negro (1000 miles up stream) are unknown. This simple difference of meteorological conditions would hardly be thought to determine what genera of butterflies should inhabit each region, yet it does so in a very decisive manner. The Upper Amazons, from Ega upwards, and the eastern slopes of the Andes, whence so large a number of the most richly-coloured species of this tribe have been received in Europe, owe the most ornamental part of their insect population to the absence of strong and regular winds. Nineteen of the most handsome genera of Ega, containing altogether about 100 species, are either entirely absent or very sparingly represented on the Lower Amazons within reach of the trade winds. The range of these nineteen genera is affected by a curiously complicated set of circumstances. In all the species of which they are composed, the males are more than 100 to one more numerous than the females, and being very richly coloured, whilst the females are of dull hues, they spend their lives in sporting about in the sunlight, imbibing the moisture which constitutes their food, from the mud on the shores of streams, their spouses remaining hid in the shades of the forest. The very existence of these species depends on the facilities which their males have for indulgence in the pleasures of this sunshiny life. The greatest obstacle to this is the prevalence of strong winds, which not only dries rapidly all moisture in open places, but prevents the richly-attired dandies from flying daily to their feeding-places. I noticed this particularly whilst residing at Santarem, where the moist margins of water, localities which on the Upper Amazons swarm with these insects, were nearly destitute of them; and at Villa Nova (where a small number exists) I have watched them buffeting with the strong winds at the commencement of the dry season, and, as the dryness increased, disappearing from the locality. On ascending the Tapajos to the calm and sultry banks of the Cuparí, a great number of these insects re-appeared, most of them being the same as those found on the Upper Amazons, thus showing clearly that their existence in the district depended on the absence of winds.

Before proceeding to describe the ants, a few remarks

may be made on the singular cases and cocoons woven by the caterpillars of certain moths found at Ega. The

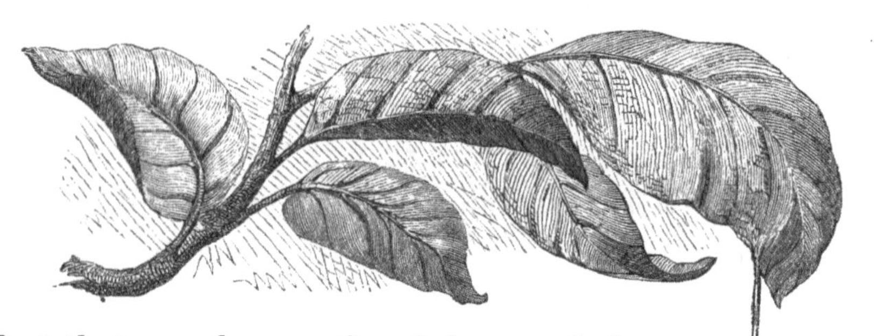

first that may be mentioned, is one of the most beautiful examples of insect workmanship I ever saw. It is a cocoon, about the size of a sparrow's egg, woven by a caterpillar in broad meshes of either buff or rose-coloured silk, and is frequently seen in the narrow alleys of the forest, suspended from the extreme tip of an outstanding leaf by a strong silken thread five or six inches in length. It forms a very conspicuous object, hanging thus in mid-air. The glossy threads with which it is knitted are stout, and the structure is therefore not liable to be torn by the beaks of insectivorous birds, whilst its pendulous position makes it doubly secure against their attacks, the apparatus giving way when they peck at it. There is a small orifice at each end of the egg-shaped bag, to admit of the escape of the moth when it changes from the little chrysalis which sleeps tranquilly in its airy cage. The moth is of a

Suspended cocoon of Moth.

dull slaty colour and belongs to the Lithosiide group of the silk-worm family (Bombycidæ). When the caterpillar begins its work, it lets itself down from the tip of the leaf which it has chosen, by spinning a thread of silk, the thickness of which it slowly increases as it descends. Having given the proper length to the cord, it proceeds to weave its elegant bag, placing itself in the centre and spinning rings of silk at regular intervals, connecting them at the same time by means of cross threads; so that the whole, when finished, forms a loose web, with quadrangular meshes of nearly equal size throughout. The task occupies about four days: when finished, the enclosed caterpillar becomes sluggish, its skin shrivels and cracks, and there then remains a motionless chrysalis of narrow shape, leaning against the sides of its silken cage.

Many other kinds are found at Ega belonging to the same cocoon-weaving family, some of which differ from the rest in their caterpillars possessing the art of fabricating cases with fragments of wood or leaves, in which they live secure from all enemies whilst they are feeding and growing. I saw many species of these; some of them knitted together, with fine silken threads, small bits of stick, and so made tubes similar to those of caddice-worms; others (Saccophora) chose leaves for the same purpose, forming with them an elongated bag open at both ends, and having the inside lined with a thick web. The tubes of full-grown caterpillars of Saccophora are two inches in length, and it is at this stage of growth that I have generally seen them. They feed on the leaves of Melastomæ, and as, in crawling, the weight of so large a dwelling would be greater than the contained caterpillar could sustain, the insect

Sack-bearing Caterpillar (Saccophora).

attaches the case by one or more threads to the leaves or twigs near which it is feeding.

Foraging Ants.—Many confused statements have been published in books of travel, and copied in Natural History works, regarding these ants, which appear to have been confounded with the Saüba, a sketch of whose habits has been given in the first chapter of this work. The Saüba is a vegetable feeder, and does not attack other animals; the accounts that have been published regarding carnivorous ants which hunt in vast armies, exciting terror wherever they go, apply only to the Ecitons, or foraging ants, a totally different group of this tribe of insects. The Ecitons are called Tauóca by the Indians, who are always on the look-out for their armies when they traverse the forest, so as to avoid being attacked. I met with ten distinct species of them, nearly all of which have a different system of marching; eight were new to science when I sent them to England. Some are found commonly in every part of the country, and one is peculiar to the open campos of Santarem; but, as nearly all the species are found together at Ega, where the forest swarmed with their armies, I have left an account of the habits of the whole genus for this part of my narrative. The Ecitons resemble, in their habits, the Driver-ants of Tropical Africa; but they have no close relationship with them in structure, and indeed belong to quite another sub-group of the ant-tribe.

Like many other ants, the communities of Ecitons are composed, besides males and females, of two classes of workers, a large-headed (worker-major) and a small-headed (worker-minor) class; the large-heads have, in some species, greatly lengthened jaws, the small-heads have jaws always of the ordinary shape; but the two classes are not sharply-defined in structure and function, except in two of the species. There is, in all of them a little difference amongst the workers regarding the size of the head; but in some species (E. legionis) this is not sufficient to cause a separation into classes, with division of labour; in others (E. hamata) the jaws are so monstrously lengthened in the worker-majors, that they are incapacitated from taking part in the labours which the worker-minors perform; and again, in others (E. erratica and E. vastator), the difference is so great that the distinction of classes becomes complete, one acting the part of soldiers, and the other that of workers.* The peculiar feature in the habits of the Eciton genus is their hunting for prey in regular bodies, or armies. It is this which chiefly distinguishes them from the genus of common red stinging-ants (Myrmica), several species of which inhabit England, whose habit is to search for food in the usual irregular manner. All the Ecitons hunt in large organised bodies; but almost every species has its own special manner of hunting.

Eciton rapax.—One of the foragers, Eciton rapax, the giant of its genus, whose worker-majors are half-an-inch in length, hunts in single file through the forest. There is no division into classes amongst its workers, although the difference in size is very great, some being scarcely one-half the length of others. The head and jaws, however, are always of the same shape, and a gradation in size is presented from the largest to the

* There is one numerous genus of South American ants in which the two classes of workers are nearly always sharply defined in structure, not only the head, but other parts of the body, being strikingly different. This is the genus Cryptocerus, of which I found fifteen species, but in no case was able to discover the distinctive function of the worker-major class. The contrast between the two classes reaches its acme in C. discocephalus, whose worker-majors have a strange dish-shaped expansion on the crown of the head. All the species inhabit hollow twigs or branches of trees, the monstrous-headed individuals being always found quiescent and mixed with crowds of worker-minors. It cannot be considered wonderful that the function of worker-majors has not been discovered in exotic ants, when Huber, who devoted a life-time to the study of European ants, was unable to detect it in a common species, the Formica rufescens.

smallest, so that all are able to take part in the common labours of the colony. The chief employment of the species seems to be plundering the nests of a large and defenceless ant of another genus (Formica), whose mangled bodies I have often seen in their possession, as they were marching away. The armies of Eciton rapax are never very numerous.

Eciton legionis.—Another species, E. legionis, agrees with E. rapax in having workers not rigidly divisible into two classes ; but it is much smaller in size, not differing greatly, in this respect, from our common English red ant (Myrmica rubra), which it also resembles in colour. The Eciton legionis lives in open places, and was seen only on the sandy campos of Santarem. The movements of its hosts were, therefore, much more easy to observe than those of all other kinds, which inhabit solely the densest thickets; its sting and bite, also, were less formidable than those of other species. The armies of E. legionis consist of many thousands of individuals, and move in rather broad columns. They are just as quick to break line, on being disturbed, and attack hurriedly and furiously any intruding object as the other Ecitons. The species is not a common one, and I seldom had good opportunities of watching its habits. The first time I saw an army, was one evening near sunset. The column consisted of two trains of ants, moving in opposite directions ; one train empty-handed, the other laden with the mangled remains of insects, chiefly larvæ and pupæ of other ants. I had no difficulty in tracing the line to the spot from which they were conveying their booty: this was a low thicket; the Ecitons were moving rapidly about a heap of dead leaves; but as the short tropical twilight was deepening rapidly, and I had no wish to be benighted on the lonely campos, I deferred further examination until the next day.

On the following morning, no trace of ants could be found near the place where I had seen them the preceding day, nor were there signs of insects of any description in the thicket; but at the distance of eighty or one hundred yards, I came upon the same army, engaged, evidently, on a razzia of a similar kind to that of the previous evening; but requiring other resources of their instinct, owing to the nature of the ground. They were eagerly occupied, on the face of an inclined bank of light earth, in excavating mines, whence, from a depth of eight or ten inches, they were extracting the bodies of a bulky species of ant, of the genus Formica. It was curious to see them crowding round the orifices of the mines, some assisting their comrades to lift out the bodies of the Formicæ, and others tearing them in pieces, on account of their weight being too great for a single Eciton; a number of carriers seizing each a fragment, and carrying it off down the slope. On digging into the earth with a small trowel near the entrances of the mines, I found the nests of the Formicæ, with grubs and cocoons, which the Ecitons were thus invading, at a depth of about eight inches from the surface. The eager freebooters rushed in as fast as I excavated, and seized the ants in my fingers as I picked them out, so that I had some difficulty in rescuing a few entire for specimens. In digging the numerous mines to get at their prey, the little Ecitons seemed to be divided into parties, one set excavating, and another set carrying away the grains of earth. When the shafts became rather deep, the mining parties had to climb up the sides each time they wished to cast out a pellet of earth; but their work was lightened for them by comrades, who stationed themselves at the mouth of the shaft, and relieved them of their burthens, carrying the particles, with an appearance of foresight which quite staggered me, a sufficient distance from the edge of the hole to prevent them from rolling in again. All the work seemed thus to be performed by intelligent co-operation amongst the host of eager little creatures; but still there was not a rigid division of labour, for some of them, whose proceedings I watched, acted at one time as carriers of pellets, and at another as miners, and all shortly afterwards assumed the office of conveyors of the spoil.

In about two hours, all the nests of Formicæ were rifled, though not completely, of their contents, and I turned towards the army of Ecitons, which were carrying away the mutilated remains. For some distance there were many separate lines of them moving along the slope of the bank; but a short distance off, these all converged, and then formed one close and broad column, which continued for some sixty or seventy yards, and terminated at one of those large termitariums already described in a former chapter as being constructed of a material as hard as stone. The broad and compact column of ants moved up the steep sides of the hillock in a continued stream; many, which had hitherto

A A 2

trotted along empty-handed, now turned to assist their comrades with their heavy loads, and the whole descended into a spacious gallery or mine, opening on the top of the termitarium. I did not try to reach the nest, which I supposed to lie at the bottom of the broad mine, and therefore in the middle of the base of the stony hillock.

Eciton drepanophora.—The commonest species of foraging ants are the Eciton hamata and E. drepanophora, two kinds which resemble each other so closely that it requires attentive examination to distinguish

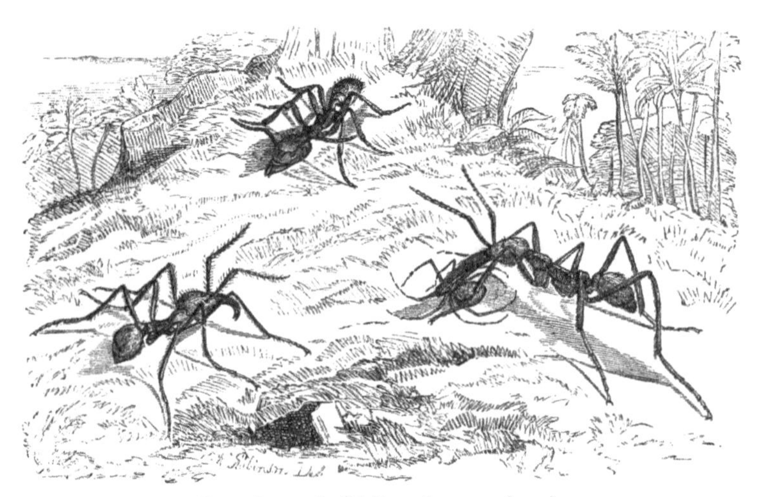

Foraging ants (Eciton drepanophora).

them; yet their armies never intermingle, although moving in the same woods and often crossing each other's tracks. The two classes of workers look, at first sight, quite distinct, on account of the wonderful amount of difference between the largest individuals of the one, and the smallest of the other. There are dwarfs not more than one-fifth of an inch in length, with small heads and jaws, and giants half an inch in length with monstrously enlarged head and jaws, all belonging to the same family. There is not, however, a distinct separation of classes, individuals existing which connect together the two extremes. These Ecitons are seen in the pathways of the forest at all places on the banks of the Amazons, travelling in dense columns of countless thousands. One or other of them is sure to be met with in a woodland ramble, and it is to them probably, that the stories we read in books on South America apply, of ants clearing houses of vermin, although I heard of no instance of their entering houses, their ravages being confined to the thickest parts of the forest.

When the pedestrian falls in with a train of these ants, the first signal given him is a twittering and restless movement of small flocks of plain-coloured birds (ant-thrushes) in the jungle. If this be disregarded until he advances a few steps further, he is sure to fall into trouble, and find himself suddenly attacked by numbers of the ferocious little creatures. They swarm up his legs with incredible rapidity, each one driving its pincer-like jaws into his skin, and with the purchase thus obtained, doubling in its tail, and stinging with all its might. There is no course left but to run for it; if he is accompanied by natives they will be sure to give the alarm, crying "Tauóca!" and scampering at full speed to the other end of the column of ants. The tenacious insects who have secured themselves to his legs then have to be plucked off one by one, a task which is generally not accomplished without pulling them in twain, and leaving heads and jaws sticking in the wounds.

The errand of the vast ant-armies is plunder, as in the case of Eciton legionis; but from their moving always amongst dense thickets, their proceedings are not so easy to observe as in that species. Wherever they move, the whole animal world is set in commotion, and every creature tries to get out of their way. But it is especially the various tribes of wingless insects that have cause for fear, such as heavy-bodied spiders, ants of other species, maggots, caterpillars, larvæ of cockroaches and so forth, all of which live under fallen leaves, or in decaying wood. The Ecitons do not mount very high on trees, and therefore the nestlings of birds are not much incommoded by them. The mode of operation of these armies, which I ascertained only after long-continued observation, is as follows. The main column, from four to six deep, moves forward in a given direction, clearing the ground of all animal matter dead or alive, and throwing off here and there, a thinner column to forage for a short time on the flanks of the main army, and re-enter it again after their task is accomplished. If some very rich place be encountered anywhere near the line of march, for example, a mass of rotten wood abounding in insect larvæ, a delay takes place, and a very strong force of ants is concentrated upon it. The excited creatures search every cranny and tear in pieces all the large grubs they drag to light. It is curious to see them attack wasps' nests, which are sometimes built on low shrubs. They gnaw away the papery covering to get at the larvæ, pupæ, and newly-hatched wasps, and cut everything to tatters, regardless of the infuriated owners which are flying about them. In bearing off their spoil in fragments, the pieces are apportioned to the carriers with some degree of regard to fairness of load: the dwarfs taking the smallest pieces, and the strongest fellows with small heads the heaviest portions. Sometimes two ants join together in carrying one piece, but the worker-majors with their unwieldy and distorted jaws, are incapacitated from taking any part in the labour. The armies never march far on a beaten path, but seem to prefer the entangled thickets where it is seldom possible to follow them. I have traced an army sometimes for half a mile or more, but was never able to find one that had finished its day's course and returned to its hive. Indeed, I never met with a hive; whenever the Ecitons were seen, they were always on the march.

I thought one day, at Villa Nova, that I had come upon a migratory horde of this indefatigable ant. The place was a tract of open ground near the river side, just outside the edge of the forest, and surrounded by rocks and shrubbery. A dense column of Ecitons was seen extending from the rocks on one side of the little haven, traversing the open space, and ascending the opposite declivity. The length of the procession was from sixty to seventy yards, and yet neither van nor rear was visible. All were moving in one and the same direction, except a few individuals on the outside of the column, which were running rearward, trotting along for a short distance, and then turning again to follow the same course as the main body. But these rearward movements were going on continually from one end to the other of the line, and there was every appearance of their being a means of keeping up a common understanding amongst all the members of the army, for the retrograding ants stopped very often for a moment to touch one or other of their ownward-moving comrades with their antennæ; a proceeding which has been noticed in other ants, and supposed to be their mode of conveying intelligence. When I interfered with the column or abstracted an individual from it, news of the disturbance was very quickly communicated to a distance of several yards towards the rear, and the column at that point commenced retreating. All the small-headed workers carried in their jaws a little cluster of white maggots, which I thought, at the time, might be young larvæ of their own colony, but afterwards found reason to conclude were the grubs of some other species whose nests they had been plundering, the procession being most likely not a migration, but a column on a marauding expedition.

The position of the large-headed individuals in the marching column was rather curious. There was one of these extraordinary fellows to about a score of the smaller class; none of them carried anything in their mouths, but all trotted along empty-handed and outside the column, at pretty regular intervals from each other, like subaltern officers in a marching regiment of soldiers. It was easy to be tolerably exact in this observation, for their shining white heads made them very conspicuous amongst the rest, bobbing up and down as the column passed over the inequalities of the road. I did not see them change their position, or take any notice of their small-headed comrades marching in the column, and when I disturbed the line, they did not prance forth or show fight so eagerly as the others. These large-headed members of the community have been considered by some authors as a soldier class, like the similarly-armed caste in Termites; but I found no proof of this, at least in the present species, as they always seemed to be rather less pugnacious than the worker-minors, and their distorted jaws disabled them from fastening on a plane surface like the skin of an attacking animal. I am inclined, however, to think that they may act, in a less direct way, as protectors of the community, namely, as indigestible morsels to the flocks of ant-thrushes which follow the marching columns of these Ecitons, and are the most formidable enemies of the species. It is possible that the hooked and twisted jaws of the large-headed class may be effective weapons of annoyance when in the gizzards or stomachs of these birds, but I unfortunately omitted to ascertain whether this was really the fact.

The life of these Ecitons is not all work, for I frequently saw them very leisurely employed in a way that looked like recreation. When this happened, the place was always a sunny nook in the forest. The main column of the army and the branch columns, at these times, were in their ordinary relative positions; but, instead of pressing forward eagerly, and plundering right and left, they seemed to have been all smitten with a sudden fit of laziness. Some were walking slowly about, others were brushing their antennæ with their fore-feet; but the drollest sight was their cleaning one another. Here and there an ant was seen stretching forth first one leg and then another, to be brushed or washed by one or more of its comrades, who performed the task by passing the limb between the jaws and the tongue, finishing by giving the antennæ a friendly wipe. It was a curious spectacle, and one well calculated to increase one's amazement at the similarity between the instinctive actions of ants and the acts of rational beings, a similarity which must have been brought about by two different processes of development of the primary qualities of mind. The actions of these ants looked like simple indulgence in idle amusement. Have these little creatures, then, an excess of energy beyond what is required for labours absolutely necessary to the welfare of their species, and do they thus expend it in mere sportiveness, like young lambs or kittens, or in idle whims like rational beings? It is probable that these hours of relaxation and cleaning may be indispensable to the effective performance of their harder labours, but whilst looking at them, the conclusion that the ants were engaged merely in play was irresistible.

Eciton prædator.—This is a small dark-reddish species, very similar to the common red stinging-ant of England. It differs from all other Ecitons in its habit of hunting, not in columns, but in dense phalanxes consisting of myriads of individuals, and was first met with at Ega, where it is very common. Nothing in insect movements is more striking than the rapid march of these large and compact bodies. Wherever they pass all the rest of the animal world is thrown into a state of alarm. They stream along the ground and climb to the summits of all the lower trees, searching every leaf to its apex, and whenever they encounter a mass of decaying vegetable matter, where booty is plentiful, they concentrate, like other Ecitons, all their forces upon it, the dense phalanx of shining and quickly-moving bodies, as it spreads over the surface, looking like a flood of dark-red liquid. They soon penetrate every part of the confused heap, and then, gathering together again in marching order, onward they move. All soft-bodied and inactive insects fall an easy prey to them, and, like other Ecitons, they tear their victims in pieces for facility of carriage. A phalanx of this species, when passing over a tract of smooth ground, occupies a space of from four to six square yards; on examining the ants closely they are seen to move, not altogether in one straightforward direction, but in variously-spreading contiguous columns, now separating a little from the general mass, now re-uniting with it. The margins of the phalanx spread out at times like a cloud of skirmishers from the flanks of an army. I was never able to find the hive of this species.

Blind Ecitons.—I will now give a short account of the blind species of Eciton. None of the foregoing kinds have eyes of the facetted or compound structure such as are usual in insects, and which ordinary ants (Formica) are furnished with, but all are provided with organs of vision composed each of a single lens. Connecting them with the utterly blind species of the genus, is a very stout-limbed Eciton, the E. crassicornis, whose eyes are sunk in rather deep sockets. This ant goes on foraging expeditions like the rest of its tribe, and attacks even the nests of other stinging species (Myrmica), but it avoids the light, moving always in concealment under leaves and fallen branches. When its columns have to cross a cleared space, the ants construct a temporary covered way with granules of earth, arched over, and holding together mechanically; under this the procession passes in secret, the indefatigable creatures repairing their arcade as fast as breaches are made in it.

Next in order comes the Eciton vastator, which has no eyes, although the collapsed sockets are plainly

Foraging ants (Eciton erratica) constructing a covered road—Soldiers sallying out on being disturbed.

visible; and, lastly, the Eciton erratica, in which both sockets and eyes have disappeared, leaving only a faint ring to mark the place where they are usually situated. The armies of E. vastator and E. erratica move, as far as I could learn, wholly under covered roads, the ants constructing them gradually but rapidly as they advance. The column of foragers pushes forward step by step, under the protection of these covered passages, through the thickets, and on reaching a rotting log, or other promising hunting-ground, pour into the crevices in search of booty. I have traced their arcades, occasionally, for a distance of one or two hundred yards; the grains of earth are taken from the soil over which the column is passing, and are fitted together without cement. It is this last-mentioned feature that distinguishes them from the similar covered roads made by Termites, who use their glutinous saliva to cement the grains together. The blind Ecitons, working in numbers, build up simultaneously the sides of their convex arcades, and contrive, in a surprising manner, to approximate them and fit in the key-stones without letting the loose uncemented structure fall to pieces. There was a very clear division of labour between the two classes of neuters in these blind species. The large-headed class, although not possessing monstrously-lengthened jaws like the worker-majors in E. hamata and E. drepanophora, are rigidly defined in structure from the small-headed class, and act as soldiers, defending the working community (like soldier Termites) against all comers. Whenever I made a breach in one of their covered ways, all the ants underneath were set in commotion, but the worker-minors remained behind to repair the damage, whilst the large-heads issued forth in a most menacing manner, rearing their heads and snapping their jaws with an expression of the fiercest rage and defiance.

The armies of all Ecitons are accompanied by small swarms of a kind of two-winged fly, the females of which have a very long ovipositor, and which belongs to the genus Stylogaster (family Conopsidæ). These swarms hover with rapidly-vibrating wings, at a height of a foot or less from the soil over which the Ecitons are moving, and occasionally one of the flies darts with great quickness towards the ground. I found they were not occupied in transfixing ants, although they have a long needle-shaped proboscis, which suggests that conclusion, but most probably in depositing their eggs in the soft bodies of insects, which the ants were driving away from their hiding-places. These eggs would hatch after the ants had placed their booty in their hive as food for their young. If this supposition be correct, the Stylogaster would offer a case of parasitism of quite a novel kind. Flies of the genus Tachinus exhibit a similar instinct, for they lie in wait near the entrances to bees' nests, and slip their eggs into the food which the deluded bees are in the act of conveying for their young.

CHAPTER VI.

EXCURSIONS BEYOND EGA.

Steamboat travelling on the Amazons—Passengers—Tunantins—Caishána Indians—The Jutahí—Indian tribes on the Jutahí and the Jurúa—The Sapó—Marauá Indians—Fonte Boa—Journey to St. Paulo—Tucúna Indians—Illness—Descent to Pará—Changes at Pará—Departure for England.

November 7th, 1856.—Embarked on the Upper Amazons steamer, the "Tabatinga," for an excursion to Tunantins, a small semi-Indian settlement, lying 240 miles beyond Ega. The Tabatinga is an iron boat of about 170 tons burthen, built at Rio de Janeiro, and fitted with engines of fifty horse power. The saloon, with berths on each side for twenty passengers, is above deck, and open at both ends to admit a free current of air. The captain, or "commandante," was a lieutenant in the Brazilian navy, a man of polished, sailor-like address, and a rigid disciplinarian; his name, Senhor Nunes Mello Cardozo. I was obliged, as usual, to take with me a stock of all articles of food, except meat and fish, for the time I intended to be absent (three months); and the luggage, including hammocks, cooking utensils, crockery, and so forth, formed fifteen large packages. One volume consisted of a mosquito tent, an article I had not yet had occasion to use on the river, but which was indispensable in all excursions beyond Ega, every person, man woman and child, requiring one, as without it existence would be scarcely possible. My tent was about eight feet long and five feet broad, and was made of coarse calico in an oblong shape, with sleeves at each end through which to pass the cords of a hammock. Under this shelter, which is fixed up every evening before sundown, one can read and write, or swing in one's hammock during the long hours which intervene before bed-time, and feel one's sense of comfort increased by having cheated the thirsty swarms of mosquitoes which fill the chamber.

We were four days on the road. The pilot, a mameluco of Ega, whom I knew very well, exhibited a knowledge of the river and powers of endurance which were quite remarkable. He stood all this time at his post, with the exception of three or four hours in the middle of each day, when he was relieved by a young man who served as apprentice, and he knew the breadth and windings of the channel, and the extent of all the yearly-shifting shoals from the Rio Negro to Loreto, a distance of more than a thousand miles. There was no slackening of speed at night, except during the brief but violent storms which occasionally broke upon us, and then the engines were stopped by the command of Lieutenant Nunes, sometimes against the wish of the pilot. The nights were often so dark that we passengers on the poop deck could not discern the hardy fellow on the bridge, but the steamer drove on at full speed, men being stationed on the look-out at the prow, to watch for floating logs, and one man placed to pass orders to the helmsman; the keel scraped against a sand-bank only once during the passage.

The passengers were chiefly Peruvians, mostly thin, anxious, Yankee-looking men, who were returning home to the cities of Moyobamba and Chachapoyas, on the Andes, after a trading trip to the Brazilian towns on the Atlantic sea-board, whither they had gone six months previously, with cargoes of Panamá hats to exchange for European wares. These hats are made of the young leaflets of a palm-tree, by the Indians and half-caste people who inhabit the eastern parts of Peru. They form almost the only article of export from Peru by way of the Amazons, but the money value is very great compared with the bulk of the goods, as the hats are generally of very fine quality, and cost from twelve shillings to six pounds sterling each; some traders bring down two or three thousand pounds' worth, folded into small compass in their trunks. The return cargoes consist of hardware, crockery, glass, and other bulky or heavy goods, but not of cloth, which, being of light weight, can be carried across the Andes from the ports on the Pacific to the eastern parts of Peru. All kinds of European cloth can be obtained at a much cheaper rate by this route than by the more direct way of the Amazons, the import duties of Peru being, as I was told, lower than those of Brazil, and the difference not being counter-balanced by increased expense of transit, on account of weight, over the passes of the Andes.

There was a great lack of amusement on board. The

table was very well served, professed cooks being employed in these Amazonian steamers, and fresh meat insured by keeping on deck a supply of live bullocks and fowls, which are purchased whenever there is an opportunity on the road. The river scenery was similar to that already described as presented between the Rio Negro and Ega: long reaches of similar aspect, with two long, low lines of forest, varied sometimes with cliffs of red clay, appearing one after the other; an horizon of water and sky on some days limiting the view both up stream and down. We travelled, however, always near the bank, and, for my part, I was never weary of admiring the picturesque grouping and variety of trees, and the varied mantles of creeping plants which clothed the green wall of forest every step of the way. With the exception of a small village called Fonte Boa, retired from the main river, where we stopped to take in firewood, and which I shall have to speak of presently, we saw no human habitation the whole of the distance. The mornings were delightfully cool; coffee was served at sunrise, and a bountiful breakfast at ten o'clock; after that hour the heat rapidly increased until it became almost unbearable; how the engine-drivers and firemen stood it without exhaustion I cannot tell; it diminished after four o'clock in the afternoon, about which time dinner-bell rung, and the evenings were always pleasant.

A few miles below Tunantins, and to the west of the most westerly mouth of the Japurá, on the same side of the Solimoens, I saw, to my surprise, a bed of stratified rock, apparently a fine-grained sandstone, exposed on the banks of the river. It was elevated not more than three or four feet above the present level of the river, which was now, the season having been an unusually wet one, about half full. I had not seen rocks of any kind on the river banks since leaving Manacapurú, 450 miles distant, and this bed seems to have escaped the notice of Spix and Poeppig. The bank, at the foot of which alone the rock was visible, was connected with a tract of land lying higher than the purely alluvial district that extends eastward to a distance of several hundred miles, and was clothed with the rounded, dark-green forest which is distinctive of the *terra firmas* of the Amazons valley. The slightly elevated land continues, with scarcely a break, to the mouth of the Tunantins, which we entered, after making a long circuit to avoid a shoal, on the 11th of November.

November 11th to 30th.—The Tunantins is a sluggish black-water stream, about sixty miles in length, and towards its mouth from 100 to 200 yards in breadth. The vegetation on its banks has a similar aspect to that of the Rio Negro, the trees having small foliage of a sombre hue, and the dark piles of greenery resting on the surface of the inky water. The village is situated on the left bank, about a mile from the mouth of the river, and contains twenty habitations, nearly all of which are merely hovels, built of lath-work and mud. The short streets, after rain, are almost impassable, on account of the many puddles, and are choked up with weeds,—leguminous shrubs, and scarlet-flowered asclepias. The atmosphere in such a place, hedged in as it is by the lofty forest, and surrounded by swamps, is always close, warm, and reeking; and the hum and chirp of insects and birds cause a continual din. The small patch of weedy ground around the village swarms with plovers, sandpipers, striped herons, and scissor-tailed fly-catchers; and alligators are always seen floating lazily on the surface of the river in front of the houses.

On landing, I presented myself to Senhor Paulo Bitancourt, a good-natured half-caste, director of Indians of the neighbouring river Issá, who quickly ordered a small house to be cleared for me. This exhilarating abode contained only one room, the walls of which were disfigured by large and ugly patches of mud, the work of white ants. The floor was the bare earth, dirty and damp; the wretched chamber was darkened by a sheet of calico being stretched over the windows, a plan adopted here to keep out the Pium-flies, which float about in all shady places like thin clouds of smoke, rendering all repose impossible in the daytime wherever they can effect an entrance. My baggage was soon landed, and before the steamer departed I had taken gun, insect-net, and game-bag, to make a preliminary exploration of my new locality.

I remained here nineteen days, and, considering the shortness of the time, made a very good collection of monkeys, birds, and insects. A considerable number of the species (especially of insects) were different from those of the four other stations, which I examined on the south side of the Solimoens, and as many of these were "representative forms" * of others found on the opposite banks of the broad river, I concluded that

* Species or races which take the place of other allied species or races.

there could have been no land connection between the two shores during, at least, the recent geological period. This conclusion is confirmed by the case of the Uakarí monkeys, described in the last chapter. All these strongly modified local races of insects confined to one side of the Solimoens (like the Uakarís), are such as have not been able to cross a wide treeless space such as a river. The acquisition which pleased me most, in this place, was a new species of butterfly (a Catagramma), which has since been named C. excelsior, owing to its surpassing in size and beauty all the previously-known species of its singularly beautiful genus. The upper surface of the wings is of the richest blue, varying in shade with the play of light, and on each side is a broad curved stripe of an orange colour. It is a bold flyer, and is not confined, as I afterwards found, to the northern side of the river, for I once saw a specimen amidst a number of richly-coloured butterflies, flying about the deck of the steamer when we were anchored off Fonte Boa, 200 miles lower down the river.

With the exception of three mameluco families and a stray Portuguese trader, all the inhabitants of the village and neighbourhood are semi-civilised Indians of the Shumána and Passé tribes. The forests of the Tunantins, however, are inhabited by a tribe of wild Indians called Caishánas, who resemble much, in their social condition and manners, the debased Múras of the Lower Amazons, and have, like them, shown no aptitude for civilised life in any shape. Their huts commence at the distance of an hour's walk from the village, along gloomy and narrow forest-paths. The territory of the tribe extends to the Moco, an affluent of the Japurá, with which there is communication by land higher up the Tunantins, the two rivers approximating within about fifteen miles. From what I saw nd heard of the Caishánas, I was led to the conclusion that they had no close genealogical relationship with the Múras, but were more likely a degraded section of the Shumána, or some other neighbouring tribe. Scarcely any of them had the coarse features, the large trunk, broad chest, thick arms, and protuberant abdomen of the Múras, and their features, although presenting a wild, unsteady, and distrustful expression like the Múras, were often as finely shaped as those of the Shumánas and Passés. Senhor Bitancourt told me their "girio," or tribal language, had much resemblance to that of the Shumánas. I have before shown how scattered hordes have segregrated from their original tribes, and by long isolation, themselves become tribes, acquiring totally different languages, habits, and, to a lesser extent, different corporeal structure.

My first and only visit to a Caishána dwelling, was accidental. One day, having extended my walk further than usual, and followed one of the forest-roads until it became a mere *picada*, or hunters' track, I came suddenly upon a well-trodden pathway, bordered on each side with Lycopodia of the most elegant shapes, the tips of the fronds stretching almost like tendrils down the little earthy slopes which formed the edge of the path. The road, though smooth, was narrow and dark, and in many places blocked up by trunks of felled trees, which had been apparently thrown across by the timid Indians on purpose to obstruct the way to their habitations. Half-a-mile of this shady road brought me to a small open space on the banks of a brook or creek, on the skirts of which stood a conical hut with a very low doorway. There was also an open shed, with stages made of split palm-stems, and a number of large wooden troughs. Two or three dark-skinned children, with a man and woman, were in the shed; but, immediately on espying me, all of them ran to the hut, bolting through the little doorway like so many wild animals scared into their burrows. A few moments after, the man put his head out with a look of great distrust; but, on my making the most friendly gestures I could think of, he came forth with the children. They were all smeared with black mud and paint; the only clothing of the elders was a kind of apron made of the inner bark of the sapucaya-tree, and the savage aspect of the man was heightened by his hair hanging over his forehead to the eyes. I stayed about two hours in the neighbourhood, the children gaining sufficient confidence to come and help me to search for insects. The only weapon used by the Caishánas is the blow-pipe, and this is employed only in shooting animals for food. They are not a warlike people, like most of the neighbouring tribes on the Japurá and Issá. Their utensils consist of earthenware cooking-vessels, wooden stools, drinking-cups of gourds, and the usual apparatus for making farinha, of which they produce a considerable quantity, selling the surplus to traders at Tunantins.

The whole tribe of Caishánas does not exceed in number 400 souls. None of them are baptised Indians, and they do not dwell in villages, like the more advanced sections of the Tupí stock; but each family has its own solitary hut. They are quite harmless, do not practise tattooing, or perforate their ears and noses in any way. Their social condition is of a low type, very little removed, indeed, from that of the brutes living in the same forests. They do not appear to obey any common chief, and I could not make out that they had Pajés, or medicine-men, those rudest beginnings of a priest class. Symbolical or masked dances, and ceremonies in honour of the Juruparí, or demon, customs which prevail amongst all the surrounding tribes, are unknown to the Caishánas. There is amongst them a trace of festival-keeping; but the only ceremony used is the drinking of cashirí beer, and fermented liquors made of Indian-corn, bananas, and so forth. These affairs, however, are conducted in a degenerate style, for they do not drink to intoxication, or sustain the orgies for several days and nights in succession, like the Jurís, Passés, and Tucúnas. The men play a musical instrument, made of pieces of stem of the arrow-grass cut in different lengths and arranged like pan-pipes. With this they while away whole hours, lolling in ragged bast hammocks slung in their dark, smoky huts. The Tunantins people say that the Caishánas have persecuted the wild animals and birds to such an extent near their settlements that there is now quite a scarcity of animal food. If they kill a Toucan, it is considered an important event, and the bird is made to serve as a meal for a score or more persons. They boil the meat in earthenware kettles filled with Tucupí sauce, and eat it with beiju, or mandioca-cakes. The women are not allowed to taste of the meat, but forced to content themselves with sopping pieces of cake in the liquor.

I obtained a little information here concerning the inhabitants of the banks of the Issá, a stream 700 miles in length, which, having its sources at the foot of the volcanoes near Pasto, in New Granada, enters the Amazons about twenty miles to the west of Tunantins. I once met a mulatto of Pasto and his wife, who had descended this river from its source to its mouth. They lost all their luggage in passing the cataracts; but found, after the first fifteen days of their journey (about 150 miles), no more obstructions to navigation down to the Solimoens. It is not so unhealthy a river as the Japurá; but the natives are much less friendly to the whites than those inhabiting that river. To the distance of about 400 miles from Tunantins, its banks are now almost destitute of inhabitants. A few half-civilised and peaceable Passés, Juris, and Shumánas, are settled near its mouth; but higher up the Marietés occupy the domain, and towards the frontiers of New Granada, Miránhas are the only Indians met with, whose territory extends overland thence to the Japurá. The Marietés and Miránhas have been for many years constantly at war, and the depopulation of the country is owing partly to this circumstance, and partly to diseases introduced by the whites. These wars are not carried on by the whole of each tribe at once, but in a series of partial hostilities between separate hordes or clans. The hordes of each nation live apart; indeed these tribes have no villages, but are scattered in families over the country, and are connected together by no other ties than a common name and the tradition of general enmity towards the hordes bearing the name of the other nation. Moreover, hordes belonging to the same tribe or nation sometimes quarrel with each other. These petty wars originate in this fashion: a member of a family falls ill, and his or her relations, or the rest of the horde, get hold of the idea that the Pajé of a neighbouring horde has caused the illness by witchcraft; all then assemble for a grand drinking-bout, during which they excite each other by reciting their wrongs. The armed men meet on the following day, and march by intricate paths or circuitous streams, so as to take their enemies by surprise, and then pounce upon them with loud shouts, killing all they can, and burning their huts to the ground.

November 30th.—I left Tunantins in a trading schooner of eighty tons burthen belonging to Senhor Batalha, a tradesman of Ega, which had been out all the summer collecting produce, and was commanded by a friend of mine, a young Paraense, named Francisco Raiol. We arrived, on the 3rd of December, at the mouth of the Jutahí, a considerable stream about half a mile broad, and flowing with a very sluggish current. This is one of a series of six rivers, from 400 to 1000 miles in length, which flow from the south-west through unknown lands lying between Bolivia and the Upper Amazons, and enter this latter river between the Madeira and the Ucayáli. The sources of none of them are known. The longest of the six is the

Purús, the first met with in ascending the Solimoens. I gleaned very little information concerning the Jutahí, which was not visited much by traders, but, as far as I could learn, its banks were peopled by nearly the same wild tribes as those of the next parallel stream, the Juruá, about which I gathered a good deal from my friend John da Cunha, who ascended it as far as it was navigable on a trading expedition. The Juruá flows wholly through a flat country covered with light-green forests, and its waters are tinged ochreous, by the quantity of clayey and earthy matter held in suspension, like those of the Solimoens. At the end of the navigation there is a road by land to the Purús, the two great streams being there only about thirty or forty miles distant from each other. The Jutahí must be a much shorter river than the Juruá, for, as Senhor Cunha told me, the Conibos, an advanced tribe of agricultural Indians living on the banks of the Juruá near its source, have at that point a direct road by land to the Ucayáli, which must pass to the south of the sources both of the Jutahí and Jauarí, the two rivers lying between the Juruá and Ucayáli. Eight distinct tribes of Indians inhabit the banks of the Juruá, all of which, except the most remote (the Conibos) pass overland to the Jutahí.* Each tribe has its peculiar language, and to a great extent, also its peculiar customs. I heard, however, of no new feature in Indian character or customs, except

* The order in which they are met with on ascending the river is as follows :—1. Marauás.—2. Catauishís.—3. Canamarés.—4. Araúas.—5. Collinas (rivers Shiruán and Invíra, affluents of the right bank).—6. Catoquínos (R. Shiruán).—7. Naüas.—8. Coníbos, with their hordes Mauishís, Zaminaüas, and true Coníbos.

that the Coníbos practise the art of knitting cotton cloth, which they fashion into long cloaks. The cloth, of which I saw many specimens, forms a regular, durable, and not inelegant web of tolerably close texture. The Coníbos, like the Indians of Peru, do not grow the poisonous kind of mandioca, but simply the sweet kind, or Macasheira (Manihot Aypi). I estimate the length of the Jutahí at about 400 miles, and that of the Juruá at 600 miles.

We remained at anchor four days within the mouth of the Sapó, a small tributary of the Jutahí flowing from the south-east; Senhor Raiol having to send an igarité to the Cupatána, a large tributary some few miles further up the river, to fetch a cargo of salt fish. During this time we made several excursions in the montaria to various places in the neighbourhood. Our longest trip was to some Indian houses, a distance of fifteen or eighteen miles up the Sapó, a journey made with one Indian paddler, and occupying a whole day. The stream is not more than forty or fifty yards broad; its waters are darker in colour than those of the Jutahí, and flow, as in all these small rivers, partly under shade between two lofty walls of forest. We passed, in ascending, seven habitations, most of them hidden in the luxuriant foliage of the banks; their sites being known only by small openings in the compact wall of forest, and the presence of a canoe or two tied up in little shady ports. The inhabitants are chiefly Indians of the Marauá tribe, whose original territory comprised all the small by-streams lying between the Jutahí and the Juruá, near the mouths of both these great tributaries. They live in separate families or small hordes; have no common chief, and are considered as a tribe little disposed to adopt civilised customs or be friendly with the whites. One of the houses belonged to a Jurí family, and we saw the owner, an erect, noble-looking old fellow, tattooed, as customary with his tribe, in a large patch over the middle of his face, fishing under the shade of a colossal tree in his port with hook and line. He saluted us in the usual grave and courteous manner of the better sort of Indians as we passed by.

We reached the last house, or rather two houses, about ten o'clock, and spent there several hours during the great heat of mid-day. The houses, which stood on a high clayey bank, were of quadrangular shape, partly open like sheds, and partly enclosed with rude mud-walls, forming one or more chambers. The inhabitants, a few families of Marauás, comprising about thirty persons, received us in a frank, smiling manner: a reception which may have been due to Senhor Raiol being an old acquaintance and somewhat of a favourite. None of them were tattooed; but the men had great holes pierced in their ear-lobes, in which they insert plugs of wood, and their lips were drilled with smaller holes. One of the younger men, a fine strapping fellow nearly six feet high, with a large aquiline nose, who seemed to wish to be particularly friendly with me, showed me the use of these lip-holes, by fixing a number of little white sticks in them, and then twisting his mouth about and going through a pantomime to represent defiance in the presence of an enemy. Nearly all the people were disfigured by dark blotches on the skin, the effect of a cutaneous disease very prevalent in this part of the country. The face of one old man was completly blackened, and looked as though it had been smeared with black lead, the blotches having coalesced to form one large patch. Others were simply mottled; the black spots were hard and rough, but not scaly, and were margined with rings of a colour paler than the natural hue of the skin. I had seen many Indians and a few half-castes at Tunantins, and afterwards saw others at Fonte Boa blotched in the same way. The disease would seem to be contagious, for I was told that a Portuguese trader became disfigured with it after cohabiting some years with an Indian woman. It is curious that, although prevalent in many places on the Solimoens, no resident of Ega exhibited signs of the disease: the early explorers of the country, on noticing spotted skins to be very frequent in certain localities, thought they were peculiar to a few tribes of Indians. The younger children in these houses on the Sapó were free from spots; but two or three of them, about ten years of age, showed signs of their commencement in rounded yellowish patches on the skin, and these appeared languid and sickly, although the blotched adults seemed not to be affected in their general health. A middle-aged half-caste at Fonte Boa told me he had cured himself of the disorder by strong doses of salsaparilla; the black patches had caused the hair of his beard and eyebrows to fall off, but it had grown again since his cure.

When my tall friend saw me, after dinner, collecting insects along the paths near the houses, he approached, and, taking me by the arm, led me to a mandioca shed, making signs, as he could speak very little Tupí, that he had something to show. I was not a little surprised when, having mounted the girao, or stage of split palm-stems, and taken down an object transfixed to a post, he exhibited, with an air of great mystery, a large chrysalis suspended from a leaf, which he placed carefully in my hands, saying, "Pána-paná curí" (Tupí: butterfly by-and-by). Thus I found that the metamorphoses of insects were known to these savages; but being unable to talk with my new friend, I could not ascertain what ideas such a phenomenon had given rise to in his mind. The good fellow did not leave my side during the remainder of our stay; but, thinking apparently that I had come here for information, he put himself to considerable trouble to give me all he could. He made a quantity of Hypadú powder, that I might see the process; going about the task with much action and ceremony, as though he were a conjuror performing some wonderful trick.

We left these friendly people about four o'clock in the afternoon, and in descending the umbrageous river, stopped, about half-way down, at another house built in one of the most charming situations I had yet seen in this country. A clean, narrow, sandy pathway led from the shady port to the house, through a tract of forest of indescribable luxuriance. The buildings stood on an eminence in the middle of a level cleared space; the firm sandy soil, smooth as a floor, forming a broad terrace around them. The owner was a semi-civilised Indian, named Manoel; a dull, taciturn fellow, who, together with his wife and children, seemed by no means pleased at being intruded on in their solitude. The family must have been very industrious; for the plantations were very extensive, and included a little of almost all kinds of cultivated tropical productions: fruit trees, vegetables, and even flowers for ornament. The silent old man had surely a fine appreciation of the beauties of nature: for the site he had chosen commanded a view of surprising magnificence over the summits of the forest; and, to give finish to the prospect, he had planted a large quantity of banana trees in the foreground, thus concealing the charred and dead stumps which would otherwise have marred the effect of the rolling sea of greenery. The only information I could get out of Manoel was, that large flocks of richly-coloured birds came down in the fruit season and despoiled his trees. I collected here a great number of insects, including several new species. The sun set over the tree-tops before we left this little Eden, and the remainder of our journey was made slowly and pleasantly, under the chequered shades of the river banks, by the light of the moon.

December 7th.—Arrived at Fonte Boa; a wretched, muddy, and dilapidated village, situated two or three miles within the mouth of a narrow by-stream called the Cayhiar-hy, which runs almost as straight as an artificial canal between the village and the main Amazons. The character of the vegetation and soil here was different from that of all other localities I had hitherto examined; I had planned, therefore, to devote six weeks to the place. Having written beforehand to one of the principal inhabitants, Senhor Venancio, a house was ready for me on landing. The only recommendation of the dwelling was its coolness. It was, in fact, rather damp; the plastered walls bore a crop of green mould, and a slimy moisture oozed through the black, dirty floor; the rooms were large, but lighted by miserable little holes in place of windows. The village is built on a clayey plateau, and the ruinous houses are arranged round a large square, which is so choked up with tangled bushes that it is quite impassable, the lazy inhabitants having allowed the fine open space to relapse into jungle. The stiff clayey eminence is worn into deep gullies which slope towards the river, and the ascent from the port in rainy weather is so slippery that one is obliged to crawl up to the streets on all fours. A large tract of ground behind the place is clear of forest, but this, as well as the streets and gardens, is covered with a dense, tough carpet of shrubs, having the same wiry nature as our common heath. Beneath its deceitful covering the soil is always moist and soft, and in the wet season the whole is converted into a glutinous mud swamp. There is a very pretty church in one corner of the square, but in the rainy months of the year (nine out of the twelve) the place of worship is almost inaccessible to the inhabitants on account of the mud, the only means of getting to it being by hugging closely the walls and palings, and so advancing sideways step by step.

I remained in this delectable place until the 25th of January, 1857. Fonte Boa, in addition to its other amenities, has the reputation throughout the country of being the head-quarters of mosquitoes, and it fully deserves the title. They are more annoying in the houses by day than by night, for they swarm in the dark and damp rooms, keeping, in the daytime, near the floor, and settling by half-dozens together, on the legs. At night the calico tent is a sufficient protection; but this is obliged to be folded every morning, and in letting it down before sunset, great care is required to prevent any of the tormentors from stealing in beneath, their insatiable thirst for blood, and pungent sting, making these enough to spoil all comfort. In the forest the plague is much worse; but the forest-mosquito belongs to a different species from that of the town, being much larger, and having transparent wings; it is a little cloud that one carries about one's person every step on a woodland ramble, and their hum is so loud that it prevents one hearing well the notes of birds. The town-mosquito has opaque speckled wings, a less severe sting, and a silent way of going to work; the inhabitants ought to be thankful the big, noisy fellows never come out of the forest. In compensation for the abundance of mosquitoes, Fonte Boa has no piums; there was, therefore, some comfort outside one's door in the daytime; the comfort, however, was lessened by there being scarcely any room in front of the house to sit down or walk about, for, on our side of the square, the causeway was only two feet broad, and to step over the boundary, formed by a line of slippery stems of palms, was to sink up to the knees in a sticky swamp.

Notwithstanding damp and mosquitoes, I had capital health, and enjoyed myself much at Fonte Boa; swampy and weedy places being generally more healthy than dry ones on the Amazons, probably owing to the absence of great radiation of heat from the ground. The forest was extremely rich and picturesque, although the soil was everywhere clayey and cold, and broad pathways threaded it for many a mile over hill and dale. In every hollow flowed a sparkling brook, with perennial and crystal waters. The margins of these streams were paradises of leafiness and verdure; the most striking feature being the variety of ferns, with immense leaves, some terrestrial, others climbing over trees, and two, at least, arborescent. I saw here some of the largest trees I had yet seen; there was one especially, a cedar, whose colossal trunk towered up for more than a hundred feet, straight as an arrow; I never saw its crown, which was lost to view, from below, beyond the crowd of lesser trees which surrounded it. Birds and monkeys in this glorious forest were very abundant; the bear-like Pithecia hirsuta being the most remarkable of the monkeys, and the Umbrella Chatterer and Curl-crested Toucans amongst the most beautiful of the birds. The Indians and half-castes of the village have made their little plantations, and built huts for summer residence on the banks of the rivulets, and my rambles generally terminated at one or other of these places. The people were always cheerful and friendly, and seemed to be glad when I proposed to join them at their meals, contributing the contents of my provision-bag to the dinner, and squatting down amongst them on the mat.

The village was formerly a place of more importance than it now is, a great number of Indians belonging to the most industrious tribes, Shumánas, Passés, and Cambévas, having settled on the site and adopted civilised habits, their industry being directed by a few whites, who seem to have been men of humane views as well as enterprising traders. One of these old employers, Senhor Guerreiro, a well-educated Paraense, was still trading on the Amazons when I left the country, in 1859: he told me that forty years previously Fonte Boa was a delightful place to live in. The neighbourhood was then well cleared, and almost free from mosquitoes, and the Indians were orderly, industrious, and happy. What led to the ruin of the settlement was the arrival of several Portuguese and Brazilian traders of a low class, who in their eagerness for business taught the easy-going Indians all kinds of trickery and immorality. They enticed the men and women away from their old employers, and thus broke up the large establishments, compelling the principals to take their capital to other places. At the time of my visit there were few pure-blood Indians at Fonte Boa, and no true whites. The inhabitants seemed to be nearly all mamelucos, and were a loose-living, rustic, plain-spoken and ignorant set of people. There was no priest or schoolmaster within 150 miles, and had not been any for many years: the people seemed to be almost without government of any kind, and yet crime and deeds of violence appeared to be of very rare occurrence. The principal man of the village, one Senhor Justo, was a big, coarse, energetic fellow, sub-delegado of police, and the only tradesman who owned a large vessel running directly between Fonte Boa and Pará. He had recently built a large house, in the style of middle-class dwellings of towns, namely, with brick floors and tiled roof, the bricks and tiles having been brought from Pará, 1500 miles distant, the nearest place where they are manufactured in surplus. When Senhor Justo visited me he was much struck with the engravings in a file of "Illustrated London News," which lay on my table. It was impossible to resist his urgent entreaties to let him have some of them "to look at," so one day he carried off a portion of the papers on loan. A fortnight afterwards, on going to request him to return them, I found the engravings had been cut out, and stuck all over the newly whitewashed walls of his chamber, many of them upside down. He thought a room thus decorated with foreign views would increase his importance amongst his neighbours, and when I yielded to his wish to keep them, was boundless in demonstrations of gratitude, ending by shipping a boat-load of turtles for my use at Ega.

These neglected and rude villagers still retained many religious practices which former missionaries or priests had taught them. The ceremony which they observed at Christmas, like that described as practised by negroes in a former chapter, was very pleasing for its simplicity, and for the heartiness with which it was conducted. The church was opened, dried, and swept clean a few days before Christmas-eve, and on the morning all the women and children of the village were busy decorating it with festoons of leaves and wild flowers. Towards midnight it was illuminated inside and out with little oil lamps, made of clay, and the image of the "menino Deus," or Child-God, in its cradle, was placed below the altar, which was lighted up with rows of wax candles, very lean ones, but the best the poor people could afford. All the villagers assembled soon afterwards, dressed in their best, the women with flowers in their hair, and a few simple hymns, totally irrelevant to the occasion, but probably the only ones known by them, were sung kneeling; an old half-caste, with black-spotted face, leading off the tunes. This finished, the congregation rose, and then marched in single file up one side of the church and down the other, singing together a very pretty marching chorus, and each one, on reaching the little image, stooping to kiss the end of a ribbon which was tied round its waist. Considering that the ceremony was got up of their own free-will, and at considerable expense, I thought it spoke well for the good intentions and simplicity of heart of these poor, neglected villagers.

I left Fonte Boa, for Ega, on the 25th of January, making the passage by steamer, down the middle of the current, in sixteen hours. The sight of the clean and neat little town, with its open spaces, close-cropped grass, broad lake, and white sandy shores, had a most exhilarating effect, after my trip into the wilder parts of the country. The district between Ega and Loreto, the first Peruvian village on the river, is, indeed, the most remote, thinly-peopled, and barbarous of the whole line of the Amazons, from ocean to ocean. Beyond Loreto, signs of civilisation, from the side of the Pacific, begin to be numerous, and, from Ega downwards, the improvement is felt from the side of the Atlantic.

September 5th, 1857.—Again embarked on the "Tabatinga," this time for a longer excursion than the last, namely to St. Paulo de Olivença, a village higher up than any I had yet visited, being 260 miles distant, in a straight line, from Ega, or about 400 miles following the bends of the river.

The waters were now nearly at their lowest point; but this made no difference to the rate of travelling, night or day. Several of the Paraná mirims, or by-channels, which the steamer threads in the season of full-water, to save a long circuit, were now dried up, their empty beds looking like deep sandy ravines in the midst of the thick forest. The large sand-islands, and miles of sandy beach, were also uncovered, and these, with the swarms of large aquatic birds, storks, herons, ducks, waders, and spoon-bills, which lined their margins in certain places, made the river view much more varied and animated than it is in the season of the flood. Alligators of large size were common near the shores, lazily floating, and heedless of the passing steamer. The passengers amused themselves by shooting at them from the deck with a double-barrelled rifle we had on board. The sign of a mortal hit was the monster turning suddenly over, and remaining floating, with its white belly upwards. Lieutenant Nunes wished to have one of the dead animals on board, for the purpose of opening the abdomen, and, if a male, extracting a part which is held in great estimation amongst Brazilians as a "remedio," charm or medicine. The steamer was stopped, and a boat sent, with four strong men, to embark the beast; the body, however, was found too heavy to be lifted into the boat; so a rope was passed round it, and the hideous creature towed alongside, and hoisted on deck by means of the crane, which was rigged for the purpose. It had still some sparks of life, and when the knife was applied, lashed its tail, and opened its enormous jaws, sending the crowd of bystanders flying in all directions. A blow with a hatchet on the crown of the head, gave him his quietus at last. The length of the animal was fifteen feet; but this statement can give but an imperfect idea of its immense bulk and weight. The numbers of turtles which were seen swimming in quiet shoaly bays passed on the road, also gave us much amusement. They were seen by dozens ahead, with their snouts peering above the surface of the water; and, on the steamer approaching, turning round to stare, but not losing confidence, till the vessel had nearly passed, when they appeared to be suddenly smitten with distrust, diving like ducks under the stream.

We had on board, amongst our deck-passengers, a middle-aged Indian, of the Jurí tribe; a short, thick-set man, with features resembling much those of the late Daniel O'Connell. His name was Caracára-í (Black Eagle), and his countenance seemed permanently twisted into a grim smile, the effect of which was heightened by the tattooed marks—a blue rim to the mouth, with a diagonal pointed streak from each corner towards the ear. He was dressed in European style—black hat, coat, and trousers—looking very uncomfortable in the dreadful heat which, it is unnecessary to say, exists on board a steamer, under a vertical sun, during mid-day hours. This Indian was a man of steady resolution, ambitious and enterprising; very rare qualities in the race to which he belonged, weakness of resolution being one of the fundamental defects in the Indian character. He was now on his return home to the banks of the Issá from Pará, whither he had been to sell a large quantity of salsaparilla that he had collected, with the help of a number of Indians, whom he induces, or forces, to work for him. One naturally feels inclined to know what ideas such a favourable specimen of the Indian race may have acquired after so much experience amongst civilised scenes. On conversing with our fellow-passenger, I was greatly disappointed in him; he had seen nothing, and thought of nothing, beyond what concerned his little trading speculation, his mind being, evidently, what it had been before, with regard to all higher subjects or general ideas, a blank. The dull, mean, practical way of thinking of the Amazonian Indians, and the absence of curiosity and speculative thought which seems to be organic or confirmed in their character, although they are improveable to a certain extent, make them, like common-place people everywhere, most uninteresting companions. Caracára-í disembarked at Tunantins with his cargo, which consisted of a considerable number of packages of European wares.

The river scenery about the mouth of the Japurá is extremely grand, and was the subject of remark amongst the passengers. Lieutenant Nunes gave it as his opinion, that there was no diminution of width or grandeur in the mighty stream up to this point, a distance of 1500 miles from the Atlantic; and yet we did not here see the two shores of the river on both sides at once; lines of islands, or tracts of alluvial land, having by-channels in their rear, intercepting the view of the northern mainland, and sometimes also of the southern. Beyond the Issá, however, the river becomes evidently narrower, being reduced to an average width of about a mile; there were then no longer those magnificent reaches, with blank horizons, which occur lower down. We had a dark and rainy night after passing Tunantins, and the passengers were all very uneasy on account of the speed at which we were travelling, twelve miles an hour, with every plank vibrating with the force of the engines. Many of them could not sleep, myself amongst the number. At length, a little after midnight, a sudden shout startled us; "back her!" (English terms being used in matters relating to steam-engines). The pilot instantly sprung to the helm, and in a few moments we felt our paddle-box brushing against the wall of forest into which we had nearly driven headlong. Fortunately the water was deep close up to the bank. Early in the morning of the 10th of September we anchored in the port of St. Paulo, after five days' quick travelling from Ega.

St. Paulo is built on a high hill, on the southern bank of the river. The hill is formed of the same Tabatinga clay, which occurs at intervals over the whole valley of the Amazons, but nowhere rises to so great an elevation as here, the height being about 100 feet above the mean level of the river. The ascent from the port is steep and slippery; steps and resting-places have been made to lighten the fatigue of mounting, otherwise the village would be almost inaccessible, especially to porters of luggage and cargo, for there are no means of making a circuitous road of more moderate slope, the hill being steep on all sides, and surrounded by dense forests and swamps. The place contains about 500 inhabitants, chiefly half-castes and Indians of the Tucúna and Collína tribes, who are very little improved from their primitive state. The streets are narrow, and in rainy weather inches deep in mud; many houses are of substantial structure, but in a ruinous condition, and the place altogether presents the appearance, like Fonte Boa, of having seen better days. Signs of commerce, such as meet the eye at Ega, could scarcely be expected in this remote spot, situate 1800 miles, or seven months' round voyage by sailing-vessels, from Pará, the nearest market for produce. A very short experience showed that the inhabitants were utterly debased, the few Portuguese and other immigrants having, instead of promoting industry, adopted the lazy mode of life of the Indians, spiced with the practice of a few strong vices of their own introduction.

The head man of the village, Senhor Antonio Ribeiro, half-white half-Tucúna, prepared a house for me on landing, and introduced me to the principal people. The summit of the hill is grassy table-land, of two or three hundred acres in extent. The soil is not wholly clay, but partly sand and gravel; the village, itself, however, stands chiefly on clay, and the streets therefore, after heavy rains, become filled with muddy puddles. On damp nights, the chorus of frogs and toads which swarm in weedy back-yards, creates such a bewildering uproar, that it is impossible to carry on a conversation in-doors except by shouting. My house was damper even than the one I occupied at Fonte Boa, and this made it extremely difficult to keep my collections from being spoilt by mould. But the general humidity of the atmosphere in this part of the river was evidently much greater than it is lower down ; it appears to increase gradually in ascending from the Atlantic to the Andes. It was impossible at St. Paulo to keep salt for many days in a solid state, which was not the case at Ega, when the baskets in which it is contained were well wrapped in leaves. Six degrees further westward, namely, at the foot of the Andes, the dampness of the climate of the Amazonian forest region appears to reach its acme, for Poeppig found at Chinchao that the most refined sugar, in a few days, dissolved into syrup, and the best gunpowder became liquid, even when enclosed in canisters. At St. Paulo, refined sugar kept pretty well in tin boxes, and I had no difficulty in keeping my gunpowder dry in canisters, although a gun loaded over night could very seldom be fired off in the morning.

The principal residents at St. Paulo were the priest, a white from Pará, who spent his days and most of his nights in gambling and rum-drinking, corrupting the young fellows and setting the vilest example to the Indians ; the sub-delegado, an upright, open-hearted, and loyal negro, whom I have before mentioned, Senhor José Patricio ; the Juiz de Paz, a half-caste named Geraldo, and lastly, Senhor Antonio Ribeiro, who was Director of the Indians. Geraldo and Ribeiro were my near neighbours, but they took offence at me after the first few days, because I would not join them in their drinking bouts, which took place about every third day. They used to begin early in the morning with Cashaça mixed with grated ginger, a powerful drink which used to excite them almost to madness. Neighbour Geraldo, after these morning potations, used to station himself opposite my house and rave about foreigners, gesticulating in a threatening manner towards me, by the hour. After becoming sober in the evening, he usually came to offer me the humblest apologies, driven to it, I believe, by his wife, he himself being quite unconscious of this breach of good manners. The wives of the St. Paulo worthies, however, were generally as bad as their husbands ; nearly all the women being hard drinkers, and corrupt to the last degree. Wife-beating naturally flourished under such a state of things. I found it always best to lock myself in-doors after sunset, and take no notice of the thumps and screams which used to rouse the village in different quarters throughout the night, especially at festival times.

The only companionable man I found in the place, except José Patricio, who was absent most part of the time, was the negro tailor of the village, a tall, thin, grave young man, named Mestre Chico (Master Frank), whose acquaintance I had made at Pará several years previously. He was a free negro by birth, but had had the advantage of kind treatment in his younger days, having been brought up by a humane and sensible man, one Captain Basilio, of Pernambuco, his padrinho, or godfather. He neither drank, smoked, nor gambled, and was thoroughly disgusted at the depravity of all classes in this wretched little settlement, which he intended to quit as soon as possible. When he visited me at night, he used to knock at my shutters in a manner we had agreed on, it being necessary to guard against admitting drunken neighbours, and we then spent the long evenings most pleasantly, working and conversing. His manners were courteous, and his talk well worth listening to, for the shrewdness and good sense of his remarks. I first met Mestre Chico at the house of an old negress of Pará, Tia Rufina (Aunt Rufina), who used to take charge of my goods when I was absent on a voyage, and this affords me an opportunity of giving a few further instances of the excellent qualities of free negroes in a country where they are not wholly condemned to a degrading position by the pride or hatred of the white race. This old woman was born a slave, but like many others in the large towns of Brazil, she had been allowed to trade on her own account, as market-woman, paying a fixed sum daily to her owner, and keeping for herself all her surplus gains. In a few years she had saved sufficient money to purchase her freedom, and that of her grown-up son. This done, the old lady continued to strive until she had earned enough to buy the house in which she lived, a considerable property situated in one of the principal streets. When I returned from the interior, after seven years' absence from Pará, I found she was still advancing in prosperity, entirely through her own exertions (being a widow) and those of her son, who continued, with the most regular industry, his trade as blacksmith, and was now building a number of small houses on a piece of unoccupied land attached to her property. I found these and many other free negroes most trustworthy people, and admired the constancy of their friendships and the gentleness and cheerfulness of their manners towards each other. They showed great disinterestedness in their dealings with me, doing me many a piece of service without a hint at remuneration; but this may have been partly due to the name of Englishman, the knowledge of our national generosity towards the African race being spread far and wide amongst the Brazilian negroes.

I remained at St. Paulo five months; five years would not have been sufficient to exhaust the treasures of its neighbourhood in Zoology and Botany. Although now a forest-rambler of ten years' experience, the beautiful forest which surrounds this settlement gave me as much enjoyment as if I had only just landed for the first time in a tropical country. The Zoology revealed plainly the nearer proximity of the locality to the eastern slopes of the Andes than any I had yet visited, by the first appearance of many of the peculiar and richly-coloured forms (especially of insects), which are known only as inhabitants of the warm and moist valleys of New Granada and Peru. The plateau on which the village is built extends on one side nearly a mile into the forest, but on the other side the descent into the lowland begins close to the streets; the hill sloping abruptly towards a boggy meadow surrounded by woods, through which a narrow winding path continues the slope down to a cool shady glen, with a brook of icy-cold water flowing at the bottom. At midday the vertical sun penetrates into the gloomy depths of this romantic spot, lighting up the leafy banks of the rivulet and its clean sandy margins, where numbers of scarlet, green, and black tanagers and brightly-coloured butterflies sport about in the stray beams. Sparkling brooks, large and small, traverse the glorious forest in almost every direction, and one is constantly meeting, whilst rambling through the thickets, with trickling rills and bubbling springs, so well-provided is the country with moisture. Some of the rivulets flow over a sandy and pebbly bed, and the banks of all are clothed with the most magnificent vegetation conceivable. I had the almost daily habit, in my solitary walks, of resting on the clean banks of these swift-flowing streams, and bathing for an hour at a time in their bracing waters; hours which now remain amongst my most pleasant memories. The broad forest roads continue, as I was told, a distance of several days' journey into the interior, which is peopled by Tucúnas and other Indians, living in scattered houses and villages nearly in their primitive state, the nearest village lying about six miles from St. Paulo. The banks of all the streams are dotted with palm-thatched dwellings of Tucúnas, all half-buried in the leafy wilderness, the scattered families having chosen the coolest and shadiest nooks for their abodes.

I frequently heard in the neighbourhood of these huts, the "realejo" or organ bird (Cyphorhinus cantans), the most remarkable songster, by far, of the Amazonian forests. When its singular notes strike the ear for the first time, the impression cannot be resisted that they are produced by a human voice. Some musical boy must be gathering fruit in the thickets, and is singing a few notes to cheer himself. The tones become more fluty and plaintive; they are now those of a flageolet, and notwithstanding the utter impossibility of the thing, one is for the moment convinced that somebody is playing that instrument. No bird is to be seen, however closely the surrounding trees and bushes may be scanned, and yet the voice seems to come from the thicket close to one's ears. The ending of the song is rather disappointing. It begins with a few very slow and mellow notes, following each other like the commencement of an air; one listens expecting to hear a complete strain, but an abrupt pause occurs, and then the song breaks down, finishing with a number of clicking unmusical sounds like a piping barrel-organ out of wind and tune. I never heard the bird on the Lower Amazons, and very rarely heard it even at Ega; it is the only songster which makes an impression on the natives, who sometimes rest their paddles whilst travelling in their small canoes along the shady by-streams, as if struck by the mysterious sounds.

The Tucúna Indians are a tribe resembling much the Shumánas, Passés, Jurís, and Mauhés in their physical appearance and customs. They lead like those tribes a settled agricultural life, each horde obeying a chief of more or less influence, according to his energy and ambition, and possessing its pajé or medicine-man, who fosters its superstitions; but they are much more idle and debauched than other Indians belonging to the superior tribes. They are not so warlike and loyal as the Mundurucús, although resembling them in many respects, nor have they the slender figures, dignified mien, and gentle disposition of the Passés; there are, however, no trenchant points of difference to distinguish them from these highest of all the tribes. Both men and women are tattooed, the pattern being sometimes a scroll on each cheek, but generally rows of short straight lines on the face. Most of the older people wear bracelets, anklets and garters of tapir-hide or tough bark; in their homes they wear no other dress except on festival days, when they ornament themselves with feathers or masked cloaks made of the inner bark of a tree. They were very shy when I made my first visits to their habitations in the forest, all scampering off to the thicket when I approached, but on subsequent days they became more familiar, and I found them a harmless, good-natured people.

A great part of the horde living at the first Maloca or village dwell in a common habitation, a large oblong hut built and arranged inside with such a disregard of all symmetry, that it appeared as though constructed by a number of hands each working independently, stretching a rafter or fitting in a piece of thatch, without reference to what his fellow-labourers were doing. The walls as well as the roof are covered with thatch of palm-leaves; each piece consisting of leaflets plaited and attached in a row to a lath many feet in length. Strong upright posts support the roof, hammocks being slung between them, leaving a free space for passage and for fires in the middle, and on one side is an elevated stage (*girao*) overhead, formed of split palm stems. The Tucúnas excel most of the other tribes in the manufacture of pottery. They make broad-mouthed jars for Tucupí sauce, caysúma or mandioca beer, capable of holding twenty or more gallons, ornamenting them outside with crossed diagonal streaks of various colours. These jars, with cooking-pots, smaller jars for holding water, blow-pipes, quivers, matirí bags* full of small articles, baskets, skins of animals, and so forth, form the principal part of the furniture of their huts both large and small. The dead bodies of their chiefs are interred, the knees doubled up, in large jars under the floors of their huts.

The semi-religious dances and drinking bouts usual amongst the settled tribes of Amazonian Indians are indulged in to greater excess by the Tucúnas than they are by most other tribes. The Juruparí or Demon is the only superior being they have any conception of, and his name is mixed up with all their ceremonies, but it is difficult to ascertain what they consider to be his attributes. He seems to be believed in simply as a mischievous imp, who is at the bottom of all those mishaps of their daily life, the causes of which are not very immediate or obvious to their dull under-

* These bags are formed of remarkably neat twine made of Bromelia fibres elaborately knitted, all in one piece, with sticks ; a belt of the same material, but more closely woven, being attached to the top to suspend them by. They afford good examples of the mechanical ability of these Indians. The Tucúnas also possess the art of skinning and stuffing birds, the handsome kinds of which they sell in great numbers to passing travellers.

standings. It is vain to try to get information out of a Tucúna on this subject; they affect great mystery when the name is mentioned, and give very confused answers to questions: it was clear, however, that the idea of a spirit as a beneficent God or Creator had not entered the minds of these Indians. There is great similarity in all their ceremonies and mummeries, whether the object is a wedding, the celebration of the feast of fruits, the plucking of the hair from the heads of their children, or a holiday got up simply out of a love of dissipation. Some of the tribe on these occasions deck themselves with the bright-coloured feathers of parrots and macaws. The chief wears a head-dress or cap made by fixing the breast-feathers of the Toucan on a web of Bromelia twine, with erect tail plumes of macaws rising from the crown. The cinctures of the arms and legs are also then ornamented with bunches of feathers. Others wear masked dresses: these are long cloaks reaching below the knee and made of the thick whitish-coloured inner bark of a tree, the fibres of which are interlaced in so regular a manner, that the material looks like artificial cloth. The cloak covers the head; two holes are cut out for the eyes, a large round piece of the cloth stretched on a rim of flexible wood is stitched on each side to represent ears, and the features are painted in exaggerated style with yellow, red, and black streaks. The dresses are sewn into the proper shapes with thread made of the inner bark of the Uaissíma tree. Sometimes grotesque head-dresses, representing monkeys' busts or heads of other animals, made by stretching cloth or skin over a basket-work frame, are worn at these holidays. The biggest and ugliest mask represents the Juruparí. In these festival habiliments the Tucúnas go through their monotonous see-saw and stamping dances accompanied by singing and drumming, and keep up the sport often for three or four days and nights in succession, drinking enormous quantities of caysúma, smoking tobacco, and snuffing paricá powder.

I could not learn that there was any deep symbolical meaning in these masked dances, or that they commemorated any past event in the history of the tribe. Some of them seem vaguely intended as a propitiation of the Juruparí, but the masker who represents the demon sometimes gets drunk along with the rest, and is not treated with any reverence. From all I could make out, these Indians preserve no memory of events going beyond the times of their fathers or grandfathers. Almost every joyful event is made the occasion of a festival: weddings amongst the rest. A young man who wishes to wed a Tucúna girl has to demand her hand of her parents, who arrange the rest of the affair, and fix a day for the marriage ceremony. A wedding which took place in the Christmas week whilst I was at St. Paulo, was kept up with great spirit for three or four days; flagging during the heats of mid-day, but renewing itself with increased vigour every evening. During the whole time the bride, decked out with feather ornaments, was under the charge of the older squaws, whose business seemed to be, sedulously to keep the bridegroom at a safe distance until the end of the dreary period of dancing and boosing. The Tucúnas have the singular custom, in common with the Collínas and Mauhés, of treating their young girls, on their showing the first signs of womanhood, as if they had committed some crime. They are sent up to the girao under the smoky and filthy roof, and kept there on very meagre diet, sometimes for a whole month. I heard of one poor girl dying under this treatment.

The original territory of the Tucúna tribe embraced the banks of most of the by-streams, from forty miles below St. Paulo to beyond Loreto in Peru, a distance of about 200 miles; the tribe, however, is not well-demarcated from that of the Collínas, who appear to be a section of Tucúnas, and whose home extends 200 miles further to the east. The only other tribe of this neighbourhood concerning which I obtained any information were the Majerónas, whose territory embraces several hundred miles of the western bank of the river Jauarí, an affluent of the Solimoens, 120 miles beyond St. Paulo. These are a fierce, indomitable, and hostile people, like the Aráras of the river Madeira; they are also cannibals. The navigation of the Jauarí is rendered impossible on account of the Majerónas lying in wait on its banks to intercept and murder all travellers, especially whites.

Four months before my arrival at St. Paulo, two young half-castes (nearly white) of the village went to trade on the Jauarí; the Majerónas having shown signs of abating their hostility for a year or two previously. They had not been long gone, when their canoe returned with the news that the two young fellows had been shot with arrows, roasted and eaten by the savages. José Patricio, with his usual activity in the cause of law and order, despatched a party of armed men of the National Guard to the place to make inquiries, and, if the murder should appear to be unprovoked, to retaliate. When they reached the settlement of the horde who had eaten the two men, it was found evacuated, with the exception of one girl, who had been in the woods when the rest of her people had taken flight, and whom the guards brought with them to St. Paulo. It was gathered from her, and from other Indians on the Jauarí, that the young men had brought their fate on themselves through improper conduct towards the Majeróna women. The girl, on arriving at St. Paulo, was taken care of by Senhor José Patricio, baptised under the name of Maria, and taught Portuguese. I saw a good deal of her, for my friend sent her daily to my house to fill the water-jars, make the fire, and so forth. I also gained her good-will by extracting the grub of an Œstrus fly * from her back, and thus cured her of a painful tumour. She was decidedly the best-humoured and, to all appearance, the kindest-hearted specimen of

* A species of Œstrus or gadfly, on the upper Amazons, fixes on the flesh of man as breeding place for its grub. I extracted five at different times from my own flesh. The first was fixed in the calf of my leg, causing there a suppurating tumour, which, being unaware of the existence of this Œstrus, I thought at first was a common boil. The tumour grew and the pain increased until I became quite lame, and then, on carefully examining the supposed boil, I saw the head of a grub moving in a small hole at its apex. The extraction of the animal was a difficult operation, it being an inch in length and of increasing breadth from head to tail, besides being secured to the flesh of the inside of the tumour by two horny hooks. An old Indian of Ega showed me the most effective way of proceeding, which was to stupefy the grub with strong tobacco juice, causing it to relax its grip in the interior, and then pull it out of the narrow orifice of the tumour by main force.

her race I had yet seen. She was tall, and very stout; in colour much lighter than the ordinary Indian tint, and her ways altogether were more like those of a careless, laughing country wench, such as might be met with any day amongst the labouring class in villages in our own country, than a cannibal. I heard this artless maiden relate, in the coolest manner possible, how she ate a portion of the bodies of the young men whom her tribe had roasted. But what increased greatly the incongruity of this business, the young widow of one of the victims, a neighbour of mine, happened to be present during the narrative, and showed her interest in it by laughing at the broken Portuguese in which the girl related the horrible story.

In the fourth month of my sojourn at St. Paulo I had a serious illness, an attack of the "sizoens," or ague of the country, which, as it left me with shattered health and damped enthusiasm, led to my abandoning the plan I had formed of proceeding to the Peruvian towns of Pebas and Moyobamba, 250 and 600 miles further west, and so completing the examination of the Natural History of the Amazonian plains up to the foot of the Andes. I made a very large collection at St. Paulo, and employed a collector at Tabatinga and on the banks of the Jauarí for several months, so that I acquired a very fair knowledge altogether of the productions of the country bordering the Amazons to the end of the Brazilian territory, a distance of 1900 miles from the Atlantic at the mouth of the Pará; but beyond the Peruvian boundary I found now I should be unable to go. My ague seemed to be the culmination of a gradual deterioration of health, which had been going on for several years. I had exposed myself too much in the sun, working to the utmost of my strength six days a week, and had suffered much, besides, from bad and insufficient food. The ague did not exist at St. Paulo; but the foul and humid state of the village was, perhaps, sufficient to produce ague in a person much weakened from other causes. The country bordering the shores of the Solimoens is healthy throughout; some endemic diseases certainly exist, but these are not of a fatal nature, and the epidemics which desolated the Lower Amazons from Pará to the Rio Negro, between the years 1850 and 1856, had never reached this favoured land. Ague is known only on the banks of those tributary streams which have dark-coloured water.

I always carried a stock of medicines with me, and a small phial of quinine, which I had bought at Pará in 1851, but never yet had use for, now came in very useful. I took for each dose as much as would lie on the tip of a penknife-blade, mixing it with warm camomile tea. The first few days after my first attack I could not stir, and was delirious during the paroxysms of fever; but the worst being over, I made an effort to rouse myself, knowing that incurable disorders of the liver and spleen follow ague in this country if the feeling of lassitude is too much indulged. So every morning I shouldered my gun or insect-net, and went my usual walk in the forest. The fit of shivering very often seized me before I got home, and I then used to stand still and brave it out. When the steamer ascended in January, 1858, Lieutenant Nunes was shocked to see me so much shattered, and recommended me strongly to return at once to Ega. I took his advice, and embarked with him, when he touched at St. Paulo on his downward voyage, on the 2nd of February. I still hoped to be able to turn my face westward again, to gather the yet unseen treasures of the marvellous countries lying between Tabatinga and the slopes of the Andes; but although, after a short rest in Ega, the ague left me, my general health remained in a state too weak to justify the undertaking of further journeys. At length I left Ega, on the 3rd of February, 1859, *en route* for England.

I arrived at Pará on the 17th of March, after an absence in the interior of seven years and a half. My old friends, English, American, and Brazilian, scarcely knew me again, but all gave me a very warm welcome, especially Mr. G. R. Brocklehurst (of the firm of R. Singlehurst and Co., the chief foreign merchants, who had been my correspondents), who received me into his house, and treated me with the utmost kindness. I was rather surprised at the warm appreciation shown by many of the principal people of my labours; but, in fact, the interior of the country is still the "sertaõ" (wilderness),—a terra incognita to most residents of the seaport,—and a man who had spent seven and a half years in exploring it solely with scientific aims was somewhat of a curiosity. I found Pará greatly changed and improved. It was no longer the weedy, ruinous, village-looking place that it appeared when I first knew it in 1848. The population had been increased (to 20,000) by an influx of Portuguese, Madeiran, and German immigrants, and for many years past the provincial government had spent their considerable surplus revenue in beautifying the city.* The streets, formerly unpaved or strewn with loose stones and sand, were now laid with concrete in a most complete manner; all the projecting masonry of the irregularly-built houses had been cleared away, and the buildings made more uniform. Most of the dilapidated houses were replaced by handsome new edifices, having long and elegant balconies fronting the first floors, at an elevation of several feet above the roadway. The large, swampy squares had been drained, weeded, and planted with rows of almond and casuarina trees, so that they were now a great ornament to the city, instead of an eyesore as they

* The revenue of the province of Pará, derived almost wholly from high custom-house duties, had averaged for some years past about £1000,000 sterling. The import duties vary from 18 to 80 per cent. ad valorem; export duties from 5 to 10 per cent., the most productive article being india-rubber.

The total value of exports for 1858 was £355,905 4*s.* 0*d.*, employing 104 vessels of 29,493 total tonnage. More than half the foreign trade was done with Great Britain; the principal nations in order of amount of import trade ranking as follows:—

1. Great Britain.
2. United States.
3. France.
4. Portugal.
5. Hanse Towns.

As most of the articles of consumption are imported and most of those produced exported, the foreign trade of Pará is larger, compared with the internal trade, than it is in most countries. The insignificance of the trade of a country of such vast extent and resources becomes very apparent from the totals here quoted.

formerly were. My old favourite road, the Monguba avenue, had been renovated and joined to many other magnificent rides lined with trees, which in a very few years had grown to a height sufficient to afford agreeable shade; one of these, the Estrada de São José, had been planted with coco-nut palms. Sixty public vehicles, light cabriolets (some of them built in Pará), now plied in the streets, increasing much the animation of the beautified squares, streets, and avenues.

I found also the habits of the people considerably changed. Many of the old religious holidays had declined in importance and given way to secular amusements; social parties, balls, music, billiards, and so forth. There was quite as much pleasure-seeking as formerly, but it was turned in a more rational direction, and the Paraenses seemed now to copy rather the customs of the northern nations of Europe, than those of the mother-country, Portugal. I was glad to see several new booksellers' shops, and also a fine edifice devoted to a reading-room supplied with periodicals, globes, and maps, and a circulating library. There were now many printing-offices, and four daily newspapers. The health of the place had greatly improved since 1850, the year of the yellow fever, and Pará was now considered no longer dangerous to new comers.

So much for the improvements visible in the place, and now for the dark side of the picture. The expenses of living had increased about fourfold, a natural consequence of the demand for labour and for native products of all kinds having augmented in greater ratio than the supply, through large arrivals of non-productive residents, and considerable importations of money on account of the steamboat company and foreign merchants. Pará, in 1848, was one of the cheapest places of residence on the American continent; it was now one of the dearest. Imported articles of food, clothing, and furniture were mostly cheaper, although charged with duties varying from 18 to 80 per cent., besides high freights and large profits, than those produced in the neighbourhood. Salt codfish was twopence per pound cheaper than the vile salt pirarucú of the country. Oranges, which could formerly be had almost gratis, were now sold in the streets at the rate of three for a penny; large bananas were a penny each fruit; tomatos were from two to three pence each, and all other fruits in this fruit-producing country had advanced in like proportion. Mandioca-meal, the bread of the country, had become so scarce and dear and bad that the poorer classes of natives suffered famine, and all who could afford it were obliged to eat wheaten bread at fourpence to fivepence per pound, made from American flour, 1200 barrels of which were consumed monthly; this was now, therefore, a very serious item of daily expense to all but the most wealthy. House-rent was most exorbitant; a miserable little place of two rooms, without fixtures or conveniences of any kind, having simply blank walls, cost at the rate of 18*l.* sterling a year. Lastly, the hire of servants was beyond the means of all persons in moderate circumstances; a lazy cook or porter could not be had for less than three or four shillings a day, besides his board and what he could steal. It cost me half-a-crown for the hire of a small boat and one man to disembark from the steamer, a distance of 100 yards.

In rambling over my old ground in the forests of the neighbourhood, I found great changes had taken place—to me, changes for the worse. The mantle of shrubs, bushes, and creeping plants which formerly, when the suburbs were undisturbed by axe or spade, had been left free to arrange itself in rich, full and smooth sheets and masses over the forest borders, had been nearly all cut away, and troops of labourers were still employed cutting ugly muddy roads for carts and cattle, through the once clean and lonely woods. Houses and mills had been erected on the borders of these new roads. The noble forest-trees had been cut down, and their naked, half-burnt stems remained in the midst of ashes, muddy puddles, and heaps of broken branches. I was obliged to hire a negro boy to show me the way to my favourite path near Una, which I have described in the second chapter of this narrative; the new clearings having quite obliterated the old forest roads. Only a few acres of the glorious forest near Una now remained in their natural state. On the other side of the city near the old road to the rice mills, several scores of woodsmen were employed under Government, in cutting a broad carriage-road through the forest to Maranham, the capital of the neighbouring province, distant 250 miles from Pará, and this had entirely destroyed the solitude of the grand old forest path. In the course of a few years, however, a new growth of creepers will cover the naked tree-trunks on the borders of this new road, and luxuriant shrubs form a green fringe to the path: it will then become as beautiful a woodland road as the old one was. A naturalist will have, henceforward, to go farther from the city to find the glorious forest scenery which lay so near in 1848, and work much more laboriously than was formerly needed, to make the large collections which Mr. Wallace and I succeeded in doing in the neighbourhood of Pará.

June 2, 1859.—At length, on the second of June, I left Pará, probably for ever; embarking in a North American trading-vessel, the "Frederick Demming," for New York, the United States' route being the quickest as well as the pleasantest way of reaching England. My extensive private collections were divided into three portions and sent by three separate ships, to lessen the risk of loss of the whole. On the evening of the third of June, I took a last view of the glorious forest for which I had so much love, and to explore which I had devoted so many years. The saddest hours I ever recollect to have spent were those of the succeeding night when, the mameluco pilot having left us free of the shoals and out of sight of land though within the mouth of the river at anchor waiting for the wind, I felt that the last link which connected me with the land of so many pleasing recollections was broken. The Paraenses, who are fully aware of the attractiveness of their country, have an alliterative proverb, "Quem vai para (o) Pará para," "He who goes to Pará stops there," and I had often thought I should myself have been added to the list of examples.

The desire, however, of seeing again my parents and enjoying once more the rich pleasures of intellectual society, had succeeded in overcoming the attractions of a region which may be fittingly called a Naturalist's Paradise. During this last night on the Pará river, a crowd of unusual thoughts occupied my mind. Recollections of English climate, scenery, and modes of life came to me with a vividness I had never before experienced, during the eleven years of my absence. Pictures of startling clearness rose up of the gloomy winters, the long grey twilights, murky atmosphere, elongated shadows, chilly springs, and sloppy summers; of factory chimneys and crowds of grimy operatives, rung to work in early morning by factory bells; of union workhouses, confined rooms, artificial cares and slavish conventionalities. To live again amidst these dull scenes I was quitting a country of perpetual summer, where my life had been spent like that of three-fourths of the people in gipsy fashion, on the endless streams or in the boundless forests. I was leaving the equator, where the well-balanced forces of Nature maintained a land-surface and climate that seemed to be typical of mundane order and beauty, to sail towards the North Pole, where lay my home under crepuscular skies somewhere about fifty-two degrees of latitude. It was natural to feel a little dismayed at the prospect of so great a change, but now, after three years of renewed experience of England, I find how incomparably superior is civilised life, where feelings, tastes, and intellect find abundant nourishment, to the spiritual sterility of half-savage existence, even if it were passed in the garden of Eden.

What has struck me powerfully is the immeasurably greater diversity and interest of human character and social conditions in a single civilised nation, than in equatorial South America where three distinct races of man live together. The superiority of the bleak north to tropical regions however is only in their social aspect, for I hold to the opinion that although humanity can reach an advanced state of culture only by battling with the inclemencies of nature in high latitudes, it is under the equator alone that the perfect race of the future will attain to complete fruition of man's beautiful heritage, the earth.

The following day, having no wind, we drifted out of the mouth of the Pará with the current of fresh water that is poured from the mouth of the river, and in twenty-four hours advanced in this way seventy miles on our road. On the 6th of June, when in 7° 55′ N. lat. and 52° 30′ W. long., and therefore about 400 miles from the mouth of the main Amazons, we passed numerous patches of floating grass mingled with tree-trunks and withered foliage. Amongst these masses I espied many fruits of that peculiarly Amazonian tree the Ubussú palm; and this was the last I saw of the Great River.

INDEX.

THE END.

BRADBURY AND EVANS, PRINTERS, WHITEFRIARS.